Special and General Relativity

Springer Nature More Media App

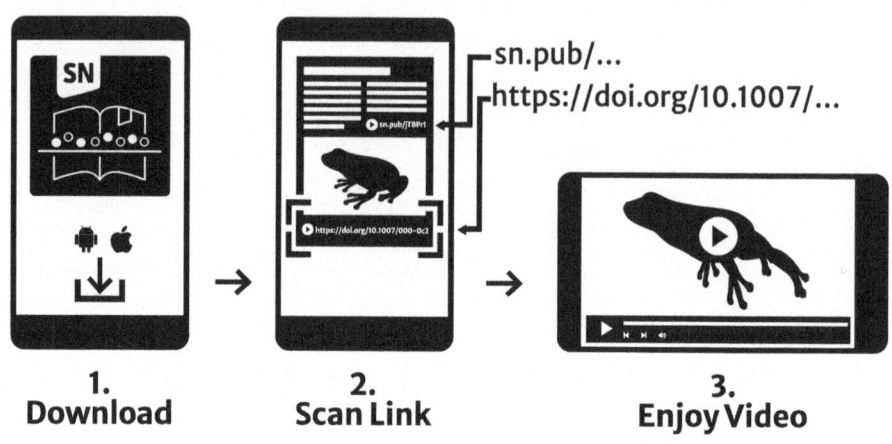

sn.pub/...
https://doi.org/10.1007/...

1.
Download

2.
Scan Link

3.
Enjoy Video

Support: customerservice@springernature.com

Sebastian Boblest • Thomas Müller •
Günter Wunner

Special and General Relativity

Fundamentals, Applications in Astrophysics and Cosmology and Relativistic Visualization

Sebastian Boblest
Dürnau, Baden-Württemberg, Germany

Günter Wunner
Institut für Theoret. Physik I
Universität Stuttgart
Stuttgart, Baden-Württemberg, Germany

Thomas Müller
Haus der Astronomie
Max Planck Institute for Astronomy
Heidelberg, Baden-Württemberg, Germany

This work contains media enhancements, which are displayed with a "play" icon. Material in the print book can be viewed on a mobile device by downloading the Springer Nature "More Media" app available in the major app stores. The media enhancements in the online version of the work can be accessed directly by authorized users.

ISBN 978-3-662-71331-0   ISBN 978-3-662-71332-7   (eBook)
https://doi.org/10.1007/978-3-662-71332-7

Translation from the German language edition: "Spezielle und allgemeine Relativitätstheorie" by Sebastian Boblest et al., © Springer-Verlag GmbH Deutschland, ein Teil von Springer Nature 2022. Published by Springer Berlin Heidelberg. All Rights Reserved.

Cover illustration: Sebastian Boblest, Thomas Müller, Günter Wunner

Planning/Editing: Gabriele Ruckelshausen
This Springer imprint is published by the registered company Springer-Verlag GmbH, DE, part of Springer Nature.
The registered company address is: Heidelberger Platz 3, 14197 Berlin, Germany

If disposing of this product, please recycle the paper.

We dedicate this edition of our textbook to the memory of our coauthor Prof. Günter Wunner, who passed away shortly before we received the proof. This English edition exists because it was his wish to make our textbook accessible to a broader readership and he translated and updated the vast majority of the original German text.

We hope that this book will make you appreciate how much mankind has learned about the universe and also how much there is still to discover. We think this would make Günter happy.

Preface

The extremely positive response to the first two editions of this textbook in German has prompted us to embark on a translation into English to make the book accessible to a worldwide readership. The first edition of this book was published in 2015. In the following 6 years, huge progress was made in the applications of general relativity in astronomy and astrophysics, which absolutely had to be incorporated into a second German edition in 2021, and now also into the English translation.

We highlight only a few of these groundbreaking discoveries. In 2015, the Laser Interferometer Gravitational-Wave Observatory (LIGO) provided the first direct evidence of gravitational waves, predicted by Einstein in 1917, with frequencies in the range of a hundred Hertz (Nobel Prize in Physics for this discovery in 2017). Moreover, in 2023 a candidate for a 5 nanohertz gravitational wave was detected by pulsar timing observations carried out over more than two decades. In 2017, the Event Horizon Telescope captured the first image of a black hole, in the center of the galaxy Messier 87. In the meantime, the Event Horizon Collaboration has been able to produce images also of the ring around the black hole in the center of the Milky Way.

The GRAVITY instrument operated by ESO succeeded in detecting the gravitational redshift and the Schwarzschild precession of the highly elliptical orbit of a star orbiting the black hole at the center of the Milky Way. This also provided definitive proof of the existence of the galactic black hole (Nobel Prize in Physics 2020).

The well-established value of the Hubble constant obtained from analyzing the anisotropy of the cosmic microwave background radiation from the early universe is in striking conflict with the significantly larger value derived from measurements in the local universe ("The Hubble controversy").

The book also contains examples of visualizations in special and general relativity. Videos for further visualization are available for some figures. Using the Springer Nature More Media app, you can access these videos quickly and conveniently. In the app, you can choose between scanning the figures, or scanning the corresponding links. You can also find videos and software for the relativistic simulations at https://www.haus-der-astronomie.de/publications/tmueller/boblest-mueller-wunner.

We attach an abridged version of the preface to the first edition, which outlines the purpose, the objectives, and the concept of this book. There we also gratefully acknowledge valuable advice that we received from colleagues and students. Finally, we would like to thank Sonam Yadav, Anand Ventakachalam, Anja Groth, and Gabriele Ruckelshausen from Springer-Verlag for their excellent cooperation.

Dürnau, Germany   Sebastian Boblest
Heidelberg, Germany   Thomas Müller
Stuttgart, Germany   Günter Wunner
June 2024

Preface to the First German Edition

This book is the outcome of lectures on special and general relativity and on astrophysics that Günter Wunner has given to physics students at the University of Stuttgart over the past two decades. In no other branch of science the general theory of relativity is more important than in astrophysics. It therefore seemed appropriate to combine relativity and astrophysics in one book, and to highlight the interconnections. However, astrophysics is such a wide field of research that a complete presentation is not possible in the framework of an introductory book. Therefore, our primary focus within astrophysics will be on the formation and evolution of stars, all the way to their end products—white dwarfs, neutron stars, and black holes—, as well as on cosmology.

In astrophysics in particular, pioneering new discoveries were made in recent decades. Today well over 1000 neutron stars are known, and their properties are studied extensively both by observation and by theory. Even within the subject areas covered, only a small selection of the phenomena that astrophysics deals with today can be discussed. Of course, there are many textbooks available today on the topics covered in this book, and in some cases we have based our discussion on other texts. We would like to particularly emphasize the books "Astronomy: Principles and practice" by David Clarke and Archie E. Roy, "Introduction to Cosmology" by Barbara Ryden, "Stellar Evolution and Nucleosynthesis" by Sean Ryan and Andrew Norton, and "Geometry, Topology and Physics" by Mikio Nakahara.

In this book we try to give a first introduction to some of the fascinating areas of physics, but without claiming to cover them exhaustively in the style of specialized textbooks. In particular, the book is not a basic course in astronomy, but rather focuses on the physical and above all, theoretical physics principles that are important in traditional and modern astrophysics for understanding the processes in the cosmos.

The theory of relativity deals with situations far beyond our everyday experience, and many of its predictions are difficult to reconcile with our human imagination. Visualizing relativistic effects on the computer can help here. Professor Hanns Ruder initiated this branch of research at the University of Tübingen, and today it is an active field of research at the Institute for Visualization and Interactive Systems at the University of Stuttgart. Thanks to the rapid development of computer technology, it is now possible to generate very detailed and complex visualizations of relativity. These methods make relativistic effects tangible in a way that would otherwise hardly be possible. Therefore, we dedicate two separate chapters to relativistic visualization.

The mathematical background is introduced to the extent necessary for further discussion; but the focus of the text is clearly on the treatment of physical effects and the comparison with observations and experiments.

With this compilation of topics, the book is aimed at science students at the graduate level, but also at the interested layperson. As we are unable to address interesting details in many places, references to further reading can be found throughout the text. In particular, we have also included references to original publications in specialized journals. The reader can use these references to gain a more comprehensive picture of topics that are of particular interest to him or her. We have added exercises for some topics. Of course, exercises help you to work through a topic even better. However, we have made sure that the reader can follow the text without having to solve the exercises. This means we deal with important calculations in detail directly in the text, and restrict ourselves to additional aspects in the exercises. The solutions to the exercises can also be found on the website mentioned above.

During the writing of this text, we benefited greatly from discussions with colleagues. In particular, we would like to thank Holger Cartarius and Jörg Main. We would also like to thank Michael Bussler, Dennis Dast, Daniel Haag, Rüdiger Fortanier, Robin Gutöhrlein, Markus Huber, Andrej Junginger, Andreas Löhle, Dirk Meyer, Katrin Scharnowski, and Christoph Schimeczek. In addition, many students also gave us valuable feedback. Parts of the text are also based on earlier elaborations of the lectures by Swantje Bebenburg, Dominique Dudowski, Alexander Herzog and Michael Klas.

Explanation of the Picture on the Front Cover

The cover image shows an accretion disk around a static black hole; hot areas appear bluish and cooler areas reddish. In addition to the seeming geometric distortion caused by the curved spacetime, the Doppler effect can be clearly recognized. Material moving toward the observer appears hotter than material moving away from the observer. A description of the geometric distortion is given in Sect. 16.2.4.

Remarks on the Notation

Ordinary three-component vectors $v = (v_1, v_2, v_3)^T$ are set in boldface, and their components $v_i, i = \{1, 2, 3\}$, are designated by a Latin subscript. The notation $(\ldots)^T$ means the transpose of a vector. For better distinction, four-vectors $\underline{x} = x^\mu \partial_\mu$ are underlined, and their components $x^\mu, \mu = \{0, 1, 2, 3\}$, are characterized by Greek indices. If we only refer to the spatial components of a four-vector, we use the usual Latin indices. We give a detailed description of four-vectors in Chap. 5.

Both in special and general relativity we choose the metric tensor with signature +2, for example $\eta_{\mu\nu} = \mathrm{diag}(-1, 1, 1, 1)$ (see Chap. 5). This choice of sign has an effect on the form of various equations in this text. Readers who, when comparing with other literature, encounter different signs, should bear this in mind. An overview of the numerous possible other metric conventions can be found in the book *Gravitation* by Misner, Thorne, and Wheeler (W.H. Freeman and Company, New York, 1973).

Declarations

Competing Interests The authors have no competing interests to declare that are relevant to the content of this manuscript.

Contents

About the Authors

Sebastian Boblest studied physics at the University of Stuttgart and earned his PhD in Theoretical Physics in 2013. He then worked as a postdoctoral researcher at the Institute for Visualization and Interactive Systems at the University of Stuttgart. His research interests are visualization in special and general relativity, as well as the visualization of flow simulations on supercomputers. He now works at Robert Bosch GmbH, Reutlingen, in the field of embedded artificial intelligence.

Thomas Müller is a research associate at the House of Astronomy and at the Max Planck Institute for Astronomy in Heidelberg. He studied physics at the Eberhard Karl University of Tübingen, where he also earned his PhD. He then worked as a postdoctoral researcher at the Institute for Visualization and Interactive Systems at the University of Stuttgart. His research interests are visualization in special and general relativity, astronomy, and astrophysics, the visualization of high-resolution LIDAR data for geomorphological analyses, and the development of educational software.

Günter Wunner (1949–2025) studied physics at the University of Erlangen-Nuremberg, where he earned his PhD in 1976 and his habilitation degree in 1982. From 1984 to 1990, he worked as an Associate Professor at the Institute for Theoretical Astrophysics at the University of Tübingen. From 1990 to 1997 he held a chair for Theoretical Plasma and Atomic Physics at the University of the Ruhr at Bochum. From 1997 to 2018 he was head of the Institute for Theoretical Physics I at the University of Stuttgart. His research areas included nonlinear dynamics with reference to quantum physics, non-Hermitian quantum mechanics, as well as atomic and astrophysics.

List of Abbreviations

BB	Big Bang
BC	Big Crunch
CDM	Cold Dark Matter
CMB	Cosmic Microwave Background
COBE	Cosmic Background Explorer
CPU	Central Processing Unit
EHT	Event Horizon Telescope
FLRW Metric	Friedmann-Lemaître-Robertson-Walker Metric
FV	False Vacuum
GPU	Graphics Processing Unit
GR	General Relativity
GUT	Grand Unified Theory
HDM	Hot Dark Matter
HRD	Hertzsprung-Russell Diagram
LIGO	Laser Interferometer Gravitational-Wave Observatory
LISA	Laser Interferometer Space Antenna
PBF	Photon-Baryon Fluid
PK	Post-Keplerian Parameter
PSR	Pulsar
SDSS	Sloan Digital Sky Survey
SN	Supernova
SR	Special Relativity
TOV Equation	Tolman-Oppenheimer-Volkoff Equation
WMAP	Wilkinson Microwave Anisotropy Probe

List of Figures

List of Tables

Introduction

1

Contents

In this chapter we will give a first impression of the type of questions that we shall deal with in this textbook. We shall also put together a compendium of concepts and relationships which we can refer to in later chapters. Some of the topics in this introduction may not seem central to the subject of this textbook, for example the structure of the solar system, or the origin of the seasons. But these topics are so basic in astronomy that we will summarize a few facts about them.

1.1   Summary

This textbook is divided into four parts.

- The first part deals with Einstein's theory of special relativity (SR). By reviewing a few important experiments, we demonstrate that the speed of light, or more precisely, the velocity of light in vacuum, has the same constant value c in all frames of reference, regardless of the magnitude of their relative velocities. This finding is difficult to understand, and it can be explained consistently only in a theory which no longer considers space and time as distinct independent quantities, but combines them in a four-dimensional *spacetime*. We shall discuss the physical consequences of this theory.
- In Part II we shall see that we have to extend special relativity to a theory of general relativity (GR) in order to describe gravitation. General relativity predicts novel physical phenomena, in particular for very massive bodies, which we shall consider in detail.
  The phenomena predicted by SR and GR differ completely from our everyday experience, and are hard to imagine. However, using computer simulations, it is possible to visualize what an observer would actually see in these relativistic situations. Such visualizations help to gain a deeper understanding of relativity. Therefore, we dedicate a chapter of its own to visualization in both, SR and GR, and also discuss some details of visualization techniques.
- Part III deals with the evolution of stars from their very beginning to their final stages. This part of the book may seem unrelated to the first two parts. We shall see, however, that in the description of white dwarf stars in Chap. 20, and in particular of neutron stars in Chap. 21, both the effects of special relativity and general relativity are fundamental. Moreover, binary systems of neutron stars offer the opportunity to test the predictions of general relativity. The corresponding observations are discussed in Sect. 21.5.2.
- The last part is devoted to cosmology, i.e., the science of the universe as a whole. Gravitation is the dominant interaction on cosmic scales, therefore we need general relativity for a correct description.

1.2   Light and the Speed of Light

Sight is one of the most important human senses since through it we receive the optical information about our environment. The overwhelming part of information on celestial bodies, and ultimately also about the universe, is also based on the observation of electromagnetic waves, over the entire spectrum, from radio waves, via optical radiation, to gamma radiation.

No experimental setups are needed to realize that the velocity of rays of light must be much bigger than, e.g., the velocity of sound in air: $v_S \approx 343 \, \mathrm{m\,s^{-1}}$. This is evident already from the time difference between watching a flash of lightning in a thunderstorm, and hearing the thunder. Whether light has a finite velocity at all, or is infinitely fast, was a controversy among scholars over centuries. *Galilei*[1] tried to answer the question by an experiment in which signals of light were sent to and fro between hills with known distances. Using this setup, he could only show that the speed of light must be at least several kilometers per second.

The first proof of the finiteness of the speed of light gave *Rømer*[2] [14]. He measured the time that it takes the Jupiter moon Io to make a transit through the shadow of the planet and noticed annual periodic changes of the transit time. He observed longer transit times when the Earth moved away from Jupiter on its orbit around the Sun, and shorter transit times when it moved towards Jupiter, see Fig. 1.1. The reason of the effect is the finite speed of light. When the Earth moves away from Jupiter the last ray of light from Io before its entry into Jupiter's shadow travels a shorter distance to Earth than the ray after the exit from the shadow. The apparent time for the transit through the shadow therefore becomes longer by this difference of the time of flight of light. The reverse is the case when the Earth moves towards Jupiter: The transit time is shortened by this difference in time. The speed of light cannot be determined by the measurement alone, because additionally the diameter of Earth's orbit must be known. *Huygens*[3] used the relatively inaccurate measurements of Rømer together with his equally inaccurate

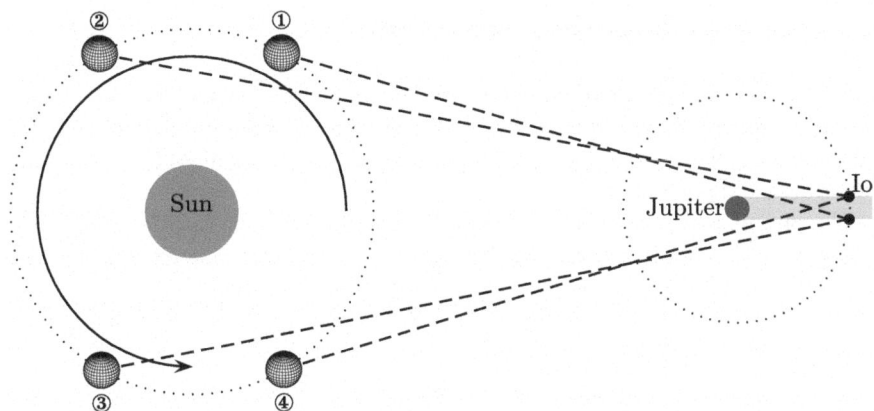

Fig. 1.1 Measurement of the speed of light by Rømer and Huygens through observations of the Jupiter moon Io. When the Earth moves away from Jupiter during a transit of Io through the shadow of Jupiter (situations ① and ②), the time for the transit is longer. On the other hand, when the Earth moves towards Jupiter (situations ③ and ④), the transit time is shorter. From this time difference Rømer could approximately determine the time that it takes light to travel through Earth's orbit

---

[1] Galileo Galilei, 1564–1642, Italian mathematician, physicist, and astronomer.

[2] Ole Rømer, 1644–1710, Danish astronomer.

[3] Christiaan Huygens, 1629–1695, Dutch astronomer, mathematician, and physicist.

estimate of the diameter of the orbit to find an approximate value of the speed of light of $220{,}000$ km s^{-1} [4].

In 1983 the speed of light in vacuum has been assigned the fixed value [8]

$$c = 299{,}792{,}458 \text{ m s}^{-1}. \tag{1.1}$$

Of course this does not mean that the speed of light can be measured without any uncertainties. Rather, on account of its central importance in physics, c was used to define the physical unit of length, meter. Therefore, more precise measurements of the speed of light lead to a change of the unit meter, instead of a change of the value of c.

Of course, it is important to know the speed of light as precisely as possible to interpret correctly, e.g., observations of stars, or experiments on Earth. But in physics the importance of the speed of light goes far beyond its mere numerical value. This is because its value does not depend on whether an observer measuring the speed of light is moving towards or away from a source of light: She always measures exactly the value c. This *observer independence* of the speed of light is the starting point of special relativity. In Chap. 2 we shall consider this aspect in more detail, and present an in-depth discussion of SR in the subsequent chapters.

1.3   Brief Outline of Electrodynamics

Our current understanding of nature is that there exist four elementary interactions: the strong interaction, responsible for the binding of protons and neutrons in atomic nuclei, the weak interaction, important in beta decay, the electromagnetic interaction, and gravitation.

The strong and the weak interaction are short-range interactions, they only act on length scales in the range of the diameters of atomic nuclei. By contrast, electromagnetic forces and gravitation are long-range interactions.

The electric field is defined via the force $\boldsymbol{F}_{\mathrm{el}}$ exerted on a small test charge q (at rest):

$$\boldsymbol{F}_{\mathrm{el}}(\boldsymbol{r}) = q\,\boldsymbol{E}(\boldsymbol{r}). \tag{1.2}$$

The magnetic induction is defined via the force $\boldsymbol{F}_{\mathrm{m}}$ exerted on a test charge q, moving at a velocity $\boldsymbol{v}$:

$$\boldsymbol{F}_{\mathrm{m}} = q\,[\boldsymbol{v} \times \boldsymbol{B}(\boldsymbol{r})]. \tag{1.3}$$

Combining the two definitions leads to the *Lorentz force*, the force acting on a test charge q in an electromagnetic field,

$$\boldsymbol{F} = q[\boldsymbol{E}(\boldsymbol{r}) + \boldsymbol{v} \times \boldsymbol{B}(\boldsymbol{r})]. \tag{1.4}$$

The electric field E and the magnetic flux density B fulfill *Maxwell's equations*.[4] The *homogeneous Maxwell equations*, which include only the two fields, are

$$\nabla \times E + \dot{B} = 0 \tag{1.5a}$$

and

$$\nabla \cdot B = 0. \tag{1.5b}$$

According to (1.5a), time-dependent magnetic fields generate electric vortex fields, while the vanishing of the divergence of the magnetic flux density in (1.5b) implies the non-existence of magnetic monopoles. The two *inhomogeneous Maxwell equations* are

$$\nabla \times B = \mu_0(j + \varepsilon_0 \dot{E}), \tag{1.5c}$$

and

$$\nabla \cdot E = \frac{\rho_{el}}{\varepsilon_0}. \tag{1.5d}$$

In (1.5c) and (1.5d) the quantity [8]

$$\varepsilon_0 = 8.854\,187\,8188(14) \cdot 10^{-12} \, \mathrm{F\,m^{-1}} \tag{1.6}$$

is the *electric field constant*, which is related to the speed of light and the *magnetic field constant*[5]

$$\mu_0 = 1.256\,637\,061\,27(20) \cdot 10^{-6} \, \mathrm{NA^{-2}}$$
$$= 0.999\,999\,999\,87(16) \cdot 4\pi \cdot 10^{-7} \, \mathrm{N\,A^{-2}} \tag{1.7}$$

by the equation

$$c = \frac{1}{\sqrt{\varepsilon_0 \mu_0}}. \tag{1.8}$$

According to (1.5c), electric currents j and time-dependent electric fields generate magnetic vortex fields. Finally, Eq. (1.5d) identifies electric charge densities ρ_{el} as

---

[4] James Clerk Maxwell, 1831–1879, Scottish physicist. Besides Maxwell's equations, he is known for his contributions to kinetic gas theory. The Maxwell-Boltzmann distribution is named after him and Ludwig Boltzmann.

[5] The unusual form of μ_0 as a constant times 4π stems from the old definition of the physical unit *ampere*. Until 2019, both ε_0 and μ_0, as c, were defined as exact with $\mu_0 = 4\pi \cdot 10^{-7} \, \mathrm{N\,A^{-2}}$.

the sources of electric fields. In Chap. 7 we shall see how these relations can be formulated and understood in a very elegant and compact way in the framework of special relativity.

In electrostatics, the electric field is a conservative force field, therefore it can be expressed as the negative gradient of the electrostatic potential ϕ_{el},

$$E_{el} = -\nabla \phi_{el}. \tag{1.9}$$

Inserting (1.9) into the inhomogeneous Maxwell equation (1.5d) leads to *Poisson's Equation*[6] of electrostatics

$$\Delta \phi_{el} = -\frac{\rho_{el}}{\varepsilon_0}. \tag{1.10}$$

A fundamental law of electrostatics is the force acting between two point charges. When we place a point charge with electric charge Q in the origin of the coordinate system, and assume a test charge q (at rest) in a distance of r, then the test charge q feels the *Coulomb force*

$$F_{el}(r) = \frac{1}{4\pi \varepsilon_0} \frac{qQ}{r^2} e_r. \tag{1.11}$$

Here e_r designates the unit vector from the charge Q to the charge q. Equation (1.11) is an immediate consequence of (1.5d). To see this we first integrate (1.5d) over some arbitrary volume V. Then, we use the *Gauss integral theorem*,[7] to transform the volume integral over the divergence of the electric field strength into a surface integral over the flux of the field strength through the surface ∂V of the volume:

$$\int_V \nabla \cdot E \, dV = \oint_{\partial V} E \cdot df = \frac{1}{\varepsilon_0} \int_V \rho_{el} \, dV = Q(V)/\varepsilon_0. \tag{1.12}$$

The volume integral over the charge density yields the total charge enclosed by the volume, and $df = n \, dS$ is an oriented surface element, i.e., a vector normal to the surface of the volume V, see Fig. 1.2.

We can apply this general result to our special case. For a point charge Q in the origin we have

$$\rho_{el}(r) = Q\delta(r). \tag{1.13}$$

For the volume, we choose a sphere S around the origin with radius r, and find

$$\int_K \nabla \cdot E \, dV = \oint_{\partial K} E \cdot df = 4\pi r^2 E(r) = Q/\varepsilon_0. \tag{1.14}$$

---

[6] Siméon Denis Poisson, 1781–1840, French physicist and mathematician.

[7] Carl Friedrich Gauß, 1777–1855, German mathematician, astronomer, and physicist.

Fig. 1.2 Illustration of the
Gauss integral theorem

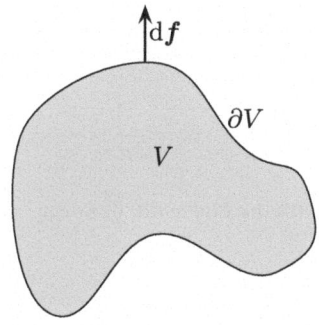

The surface of the sphere is $A = 4\pi r^2$. Remembering the definition $F(r) = qE(r)$
and solving with respect to $E(r)$ we obtain the Coulomb law (1.11).

1.4   Brief Outline of Newton's Theory of Gravitation

In Newtonian mechanics,[8] the gravitational force is defined in complete analogy
with the Coulomb force. The force between a mass M in the origin and a test mass
m at a distance r is given by

$$\boldsymbol{F}_{\mathrm{m}}(\boldsymbol{r}) = -G\frac{mM}{r^2}\mathbf{e}_{\mathrm{r}} \tag{1.15}$$

with the *gravitational constant* [8]

$$G = 6.674\,30(15) \cdot 10^{-11}\ \mathrm{m^3\,kg^{-1}\,s^{-2}}. \tag{1.16}$$

The Coulomb law (1.11) turns into the law of gravitation (1.15) when we exchange
electric charges with masses and make the replacement

$$\frac{1}{4\pi\varepsilon_0} \iff -G. \tag{1.17}$$

Thus, the sources of the gravitational field strength are masses, just like electric
charges are the sources of the electric field strength. Therefore, the property of a
body to exert a gravitational force is called its gravitational charge, or *gravitational
mass*, in contradistinction to its *inertial mass*, which we will come to in Sect. 1.4.1.

---

[8] Isaac Newton, 1642–1727, English physicist and mathematician. His opus magnum "Philosophiæ
Naturalis Principia Mathematica" was the corner stone of classical mechanics.

We can push this analogy even one step further. The coupling constant of the electromagnetic interaction is *Sommerfeld's[9] fine structure constant*

$$\alpha = \frac{1}{4\pi\varepsilon_0}\frac{e^2}{\hbar c} = 7.297\,352\,5643(11)\cdot 10^{-3} \approx \frac{1}{137} \tag{1.18}$$

with the *elementary charge*

$$e = 1.602\,176\,634 \cdot 10^{-19}\,\text{C} \tag{1.19}$$

and the *reduced Planck quantum of action*[10]

$$\hbar = \frac{h}{2\pi} = 1.054\,571\,817\ldots \cdot 10^{-34}\,\text{J s}, \tag{1.20}$$

which is defined using the original *Planck quantum of action*

$$h = 6.626\,070\,15 \cdot 10^{-34}\,\text{J s}. \tag{1.21}$$

Using the replacement rule (1.17) we can also define (apart from the minus sign) a fine structure constant for gravitation. In the place of the elementary charge we need an elementary mass. It is reasonable to choose for this purpose the mass of one of the most stable elementary building blocks of matter, the mass of the proton m_p. We then obtain

$$\alpha_G = G\frac{m_p^2}{\hbar c} = 5.90574 \cdot 10^{-39}. \tag{1.22}$$

When we shall discuss the final stages of stars in Chap. 20, we shall come across this constant again.

The Coulomb force is much stronger than the gravitational force. However, since in nature both positive and negative charges exist, on macroscopic scales the Coulomb interaction is less important: When a body aggregates a great amount of, e.g., negative charges, then according to (1.11) it will attract positive charges in its environment, and thus reduce, or even neutralize, its negative charge. By contrast, no negative masses exist, therefore the "gravitational charge" cannot be neutralized. This is the reason why gravitation is the dominant interaction in the universe. Therefore, it is also crucial for the evolution of the universe as a whole, as we shall see later in Chap. 23 on cosmology. It should also be pointed out that

---

[9] Arnold Johannes Wilhelm Sommerfeld, 1868–1951, German mathematician and theoretical physicist.

[10] Max Planck, 1858–1947, German theoretical physicist, considered as a co-founder of quantum mechanics. Nobel Prize 1918.

because of the minus sign in (1.15), masses always attract, whereas electric charges with equal sign always repel.

When we introduce a gravitational field strength—in the same way as we did in electrostatics—as the force per gravitational test charge, i.e., $E_m = F_m/m$, then the replacement rule (1.17) yields an analogous inhomogeneous equation for E_m:

$$\nabla \cdot E_m = -4\pi G \rho_m. \tag{1.23}$$

According to the replacement rules, ρ_m must denote the *mass density*. It follows that a conservative gravitational field strength possesses a gravitational potential ϕ_m with

$$E_m = -\nabla \phi_m. \tag{1.24}$$

When we insert (1.24) in (1.23), we arrive at Poisson's equation of Newtonian gravitation theory

$$\Delta \phi_m = 4\pi G \rho_m. \tag{1.25}$$

This equation allows calculating the gravitational potential, and from it with (1.24) the gravitational field strength, in the whole space for an arbitrary given mass density distribution $\rho_m(r)$.

As in electrostatics, Eq. (1.23) can be brought in integral form:

$$-4\pi G \int_V \rho_m dV = \oint_{\partial V} E_m \cdot df = -4\pi G M(V). \tag{1.26}$$

Instead of the charge enclosed by the volume, we have the mass $M(V)$ enclosed by the volume.

For the Coulomb force, we had evaluated this expression for a point charge. Another important and more general case is that of a spherically symmetric mass distribution with radius R,

$$\rho_m(r) = \rho_m(r), \tag{1.27}$$

Because of the spherical symmetry, the gravitational field can only depend on the distance from the center of the mass distribution, and can only have a radial component:

$$E_m = E_m(r)e_r. \tag{1.28}$$

We first consider the case $r \leq R$, inside the mass distribution (cf. Fig. 1.3). We integrate (1.26) over a sphere of radius r

$$\oint_{\partial K} E_m \cdot df = 4\pi r^2 E_m(r) = -4\pi G M(r), \tag{1.29}$$

Fig. 1.3 Illustration for the
calculation of E_m for a
spherically symmetric mass
distribution, at the radial
coordinate r with $r < R$.
Only the mass inside the
sphere with radius r
contributes to the
gravitational field strength

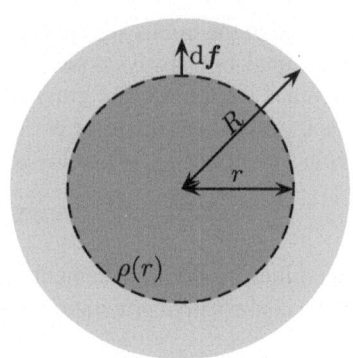

where $M(r)$ is the mass enclosed by the sphere with radius r,

$$M(r) = \int_0^r 4\pi r'^2 \rho_m(r')\mathrm{d}r'. \tag{1.30}$$

Therefore, the gravitational field strength is

$$E_m = -G\frac{M(r)}{r^2}\mathbf{e}_r. \tag{1.31}$$

Note that only the layers below the radius r have a gravitational effect on E_m, the forces exerted by the layers above cancel each other, see Exercise 1.2.

Outside the mass distribution, for $r > R$, the enclosed mass becomes the total mass, and we find the well known result that a spherically symmetric mass distribution acts as if the total mass were assembled at the center,

$$E_m = -G\frac{M}{r^2}\mathbf{e}_r, \tag{1.32}$$

the same result as for a point mass. The *gravitational potential* belonging to this field strength is

$$\phi_m = -G\frac{M}{r}. \tag{1.33}$$

1.4.1 Equivalence Principle

The mass appearing in all formulas of the previous section is the gravitational charge, or *gravitational mass*. There exists yet another type of mass. It is an ingredient of Newton's second law

$$F = m\ddot{r} \tag{1.34}$$

which relates the force F acting on a body with the acceleration $\ddot{r}$; it is called the *inertial mass*. This name can be understood, when we write (1.34) in the form $\ddot{r} = F/m$. The bigger the inertial mass, the smaller the acceleration of a body for a given force acting on the body. Gravitational mass and inertial mass are originally completely different physical quantities. However, already Galilei could show by his falling bodies experiments that the ratio between gravitational and inertial mass is identical for all bodies, and, in particular, does not depend on their chemical composition. This *equivalence principle* is one of the corner stones of general relativity. A comprehensive discussion of this principle will follow in Chap. 10.

1.4.2 Schwarzschild Radius

We see from (1.32) that the gravitational field strength on the surface of a spherically symmetric body with given mass M is the larger the smaller its radius R, or because of $M = (4/3)\pi R^3 \rho_m$, the higher its density.

In this section we will derive a characteristic length scale for the gravitational interaction, which tells us whether the gravitational effect of a body is strong or weak. To do so we have to go a little ahead of our present knowledge. In Chap. 6 we shall come across the most famous formula of special relativity,

$$E = mc^2, \tag{1.35}$$

which links the rest mass m of a body with its rest energy.

The potential energy of a test mass m in the potential (1.33) of a spherically symmetric mass distribution is

$$V_m(r) = m\phi(r) = -\frac{GMm}{r}. \tag{1.36}$$

We can measure this potential also in units of the rest energy:

$$V_m = -mc^2 \frac{GMm}{mc^2 r} = -\frac{1}{2}mc^2 \frac{r_s}{r}. \tag{1.37}$$

Here we have introduced the *Schwarzschild radius*[11]

$$r_s = 2\frac{GM}{c^2}. \tag{1.38}$$

The additional factor of two presently can be seen as a convention, but will find a natural explanation in Chap. 13, when we discuss the spherically symmetric mass distribution in the framework of general relativity.

---

[11] Karl Schwarzschild, 1873–1916, German physicist and astronomer.

Table 1.1 Numerical values of the radius, the mass, the escape velocities, which we shall discuss in Sect. 1.4.4, and the Schwarzschild radius, for different cosmic objects. Betelgeuse is a Red Giant star in the constellation Orion. The numerical values for Betelgeuse are taken from Joyce et al. [5] or derived from values therein

Object	Radius	Mass M	v_{c_2}	r_s
Small planetoid	3 km	$3.4 \cdot 10^{14}$ kg	$3.9 \, \text{ms}^{-1}$	0.5 pm
Average planetoid	20 km	$1.0 \cdot 10^{17}$ kg	$26 \, \text{m s}^{-1}$	0.15 nm
Big planetoid	350 km	$5.4 \cdot 10^{20}$ kg	$450 \, \text{m s}^{-1}$	0.8 μm
Moon	1740 km	$7.53 \cdot 10^{22}$ kg	$2.4 \, \text{km s}^{-1}$	0.11 mm
Earth	6378 km	$5.973 \cdot 10^{24}$ kg	$11.2 \, \text{km s}^{-1}$	8.9 mm
Jupiter	$7.1 \cdot 10^4$ km	$1.9 \cdot 10^{27}$ kg	$60 \, \text{km s}^{-1}$	2.8 m
Sun	$6.96 \cdot 10^5$ km	$1.99 \cdot 10^{30}$ kg	$620 \, \text{km s}^{-1}$	2950 m
Betelgeuse	$5.3^{+0.8}_{-0.4} \cdot 10^8$ km	$(3.3\text{-}3.8) \cdot 10^{31}$ kg	$(87\text{-}106) \, \text{km s}^{-1}$	$(49\text{-}56)$ km

Apart from the factor $1/2$, the potential energy can be expressed as the rest energy multiplied by the ratio of the Schwarzschild radius and the actual radius of the object. The gravitational potential ϕ_m, which is independent of the test mass, can also be written as

$$\phi_m = -\frac{GM}{r} = -\frac{1}{2}c^2\frac{r_s}{r}. \tag{1.39}$$

Thus the Schwarzschild radius is the characteristic length scale for the strength of the gravitational effect of a given mass. With the values of the gravitational constant from (1.16) and the speed of light (1.1), the Schwarzschild radius can be calculated for any given mass M. Table 1.1 shows examples for different cosmic objects. The Schwarzschild radius of the Sun is roughly three kilometers, and of the Earth slightly below one centimeter.

The fact that the Schwarzschild radius of the Earth is so small compared with its radius

$$R_{\oplus} = 6378 \text{ km} \tag{1.40}$$

(at the equator) demonstrates why gravitation is called a very weak interaction.

This holds true also for other objects in the table. The ratio of the Schwarzschild radii of two objects can be different from the ratio of their actual radii. For example, the ratios of the radii of the Earth and the Sun, and the Schwarzschild radii are

$$\frac{r_s^{\oplus}}{r_s^{\odot}} = \frac{8.9 \text{ mm}}{2950 \text{ m}} \approx 3 \cdot 10^{-6} \quad \text{and} \quad \frac{R_{\oplus}}{R_{\odot}} = \frac{6378 \text{ km}}{696,000 \text{ km}} \approx 9 \cdot 10^{-3}. \tag{1.41}$$

The reason is that the Schwarzschild radii depend linearly on the mass, but the actual radii only on the cubic root of the mass (assuming a homogeneous mass density), because the volume of the object is $\sim r^3$.

1.4.3   Gravitational Binding Energy

For an accumulation of mass in the cosmos, e.g., a galaxy, a cloud of gas, a star, or a planetoid, the gravitational binding energy is a measure of the energy content which these objects possess due to the mutual gravitational attraction of their individual mass elements. It corresponds to the energy necessary to separate the individual mass elements, and to transport them to infinity. The potential energy of a point mass m in the gravitational field of another point mass M at a distance r is $E_m = -GMm/r$. Correspondingly, the binding energy of an arbitrary mass distribution with total mass M should be of the form

$$E_m \sim \frac{\text{gravitational constant} \cdot \text{mass}^2}{\text{characteristic length}}. \tag{1.42}$$

It can be shown that quite generally the relation

$$E_m = \frac{1}{2} \int \phi_m(\boldsymbol{r}) \rho_m(\boldsymbol{r}) \, dV, \tag{1.43}$$

holds for the total binding energy of the mass distribution. Here $\phi_m(\boldsymbol{r})$ is the gravitational potential, and the integral is taken over the volume enclosed by the mass distribution. For arbitrary mass distributions this can be done only numerically.

A simple, but important example is that of a homogeneous sphere, i.e., with mass M, radius R, and the constant density

$$\rho_m = \frac{M}{\frac{4}{3}\pi R^3}.$$

In this case the separation of the individual mass elements and their transport to infinity can be visualized as always removing the outermost shell of the sphere with mass $dM = (4/3)\pi\rho_m r^2 \, dr$ from the potential $\phi_m(r) = -GM(r)/r = -G(4/3)\pi\rho_m r^2$. Equation (1.43) then simplifies to

$$E_m = -\int_0^R G \frac{16\pi^2}{3} \rho_m^2 r^4 \, dr = -\frac{R^5}{5} G \frac{16\pi^2}{3} \rho_m^2 = -\frac{3}{5} G \frac{M^2}{R}. \tag{1.44}$$

This result agrees with the form expected from (1.42). We can compare the gravitational binding energy of the sphere with its relativistic rest energy, Mc^2,

$$\frac{E_m}{Mc^2} = -\frac{3}{5} \frac{GM^2}{RMc^2} = -\frac{3}{10} \frac{r_s}{R}. \tag{1.45}$$

Again the Schwarzschild radius r_s from (1.38) appears as a characteristic length scale. The gravitational binding energy is smaller than the rest energy by the ratio of the Schwarzschild radius and the actual radius of the sphere.

For the Sun the ratio is

$$\frac{E_m}{Mc^2} = -\frac{3}{10}\frac{r_s^\odot}{R_\odot} = -\frac{3}{10}\frac{2950\,\text{m}}{6.9599 \cdot 10^8\,\text{m}} \approx -1.3 \cdot 10^{-6}. \tag{1.46}$$

Thus, the binding energy is smaller than the rest energy by roughly a factor of one over a million. The binding energy can also be regarded as a "mass defect" of the rest mass.

The binding energy of a mass distribution diminishes its rest energy by the factor (1.46). If, e.g., a cloud of gas contracts to eventually form stars, this energy must be released. We shall see in Chap. 17 that this happens in the form of heat and radiation.

1.4.4   Cosmic Velocities

In astrophysics, there exist two important velocities which are related to the gravitational interaction between masses.

We consider again a body with a spherically symmetric mass distribution, total mass M, and radius R, and a test mass m. We now ask: What is the velocity that the test mass must have, starting from a distance r, to overcome the gravitational field of M, i.e., to fly to infinity, without falling back again. Then, at launch, its kinetic energy must be at least equal to the (modulus) of its potential energy $m\phi_m$, with ϕ_m from (1.33), at the position r. This means, its total energy must at least be zero, which is the condition for unbound motion. This leads to the equation

$$\frac{1}{2}mv^2 = \frac{mMG}{r}. \tag{1.47}$$

This yields the velocity necessary at launch

$$v_0 = \sqrt{\frac{2GM}{r}}, \tag{1.48}$$

or, expressed in units of the speed of light,

$$v_0 = c\sqrt{\frac{2MG}{rc^2}} = c\sqrt{\frac{r_s}{r}}. \tag{1.49}$$

Again the Schwarzschild radius is the decisive length scale. When the test mass starts from the surface of the body with radius R, then v_0 is the *escape velocity* of the mass M

$$v_{c_2} = c\sqrt{\frac{r_s}{R}}. \tag{1.50}$$

Fig. 1.4 A test mass which orbits a spherically symmetric body with mass M near its surface needs the first cosmic velocity $v_{c_1} = c\sqrt{r_s/2r}$. To escape from the gravitational field of the body it must have the velocity $v_{c_2} = c\sqrt{r_s/r}$

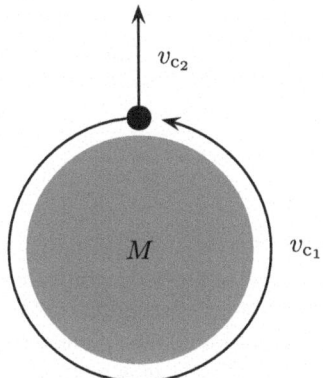

We use the index c_2, since the escape velocity is also called the second cosmic velocity. Numerical values for the escape velocities of different cosmic objects are listed in Table 1.1. We see, in the case of the Earth, that a space probe which is launched to another planet, must have an initial velocity of approximately 11.2 km s^{-1}. At the same time, the escape velocity is also the velocity of a body, originally at rest at infinity, which the body has when it impacts the surface in free fall.

The first cosmic velocity v_{c_1} is the velocity a test mass needs to circle the spherical object at a distance not far above its surface. The condition is that the centripetal force is provided by the gravitational force,

$$\frac{mv_{c_1}^2}{R} = \frac{mMG}{R^2}, \tag{1.51}$$

from which follows

$$v_{c_1} = c\sqrt{\frac{MG}{Rc^2}} = c\sqrt{\frac{r_s}{2R}} = \frac{1}{\sqrt{2}}v_{c_2}. \tag{1.52}$$

Here we have used the relation $v = \omega r$ between angular velocity and velocity. Note that in writing down Eq. (1.51) for the balance of forces, we have to insert the *inertial mass* in the centripetal force, but the *gravitational mass* in the gravitation. In this simple example we already see the great importance of the equivalence principle from Sect. 1.4.1. The velocity on a circular orbit is smaller than the escape velocity by a factor of $1/\sqrt{2} \approx 0.707$, see also Fig. 1.4.

Relativistic Escape Velocities

For the cosmic objects considered in Table 1.1 the escape velocities are always well below the speed of light. The reason is that their Schwarzschild radii are much smaller than their actual radii. The situation is different for neutron stars, which are a possible final stage of stellar evolution. These objects have masses on the order of magnitude of the mass of the Sun, and therefore also Schwarzschild radii in the range of $r_s \approx 3$ km. The radii of these compact objects, which we shall treat in Chap. 21, have values of only $r \sim 10$–20 km. One obtains escape velocities

$$v_{c_2} = c\sqrt{\frac{r_s}{R}} \lesssim c\sqrt{\frac{3 \text{ km}}{10 \text{ km}}} \approx \frac{1}{2}c. \qquad (1.53)$$

In a binary system consisting of a neutron star and a normal star, the neutron star can pull matter from the companion star because of its strong gravitation. The material then impacts the surface of the neutron star with roughly half the speed of light. The enormous amounts of energy released in the impact can actually be observed in the form of X-rays in accreting X-ray pulsars, as we shall discuss in Chap. 21.

Strictly speaking, at such high velocities it would be necessary to work with the relativistic expression (6.39) for the kinetic energy in (1.47). But our results yield the correct order of magnitude.

Black Holes

The radii of neutron stars are bigger than their Schwarzschild radii only by a relatively small factor. When we imagine an even more extreme object with $r = r_s$, then the escape velocity (1.50) becomes $v = c$! For still smaller radii, according to our simple consideration, superluminal velocities would be necessary to escape from the gravitational field of the object. Even light could not escape from such an object, the object necessarily would appear dark.

The designation *black hole* for these objects was first introduced by John Archibald Wheeler in 1967.[12] But back in 1784, *John Michell*[13] had already conjectured that bodies must exist that are so heavy that not even light can escape. By far more well known is the work of *Laplace*.[14] He noted in 1796, on the basis of Newton's theory of gravitation, that for a sufficiently large mass, concentrated in a very small region of space, the escape velocity becomes bigger than the speed of light [7]. He concluded that supermassive stars must exist from which no light can escape, and which therefore must be dark. A star with the average mass density of the Earth would appear dark, if it had a radius roughly 250 times bigger than the radius of the Sun, see Exercise 1.4.

Because of their extreme gravitation, black holes cannot be described quantitatively in Newton's gravitational theory. The correct description will be given in Chaps. 13 and 14 in the framework of general relativity. It is misleading, however, to speak of the "surface" of such objects. As we shall see, if matter crosses the Schwarzschild radius, it will inevitably fall down to the origin $r = 0$.

---

[12] Reference: https://www.worldwidewords.org/topicalwords/tw-bla1.htm.

[13] John Michell, 1724–1793, English natural philosopher.

[14] Pierre Simon Laplace, 1749–1827, French mathematician, physicist, and astronomer.

1.5   Astrophysical Basics

The scales occurring in astrophysics, for example for masses and distances, hugely differ from the orders of magnitude that we are used to in everyday life. In this section we shall introduce appropriate units for astronomical scales, and discuss other important astronomical topics, such as suitable astronomical coordinate systems.

1.5.1   A Brief Overview of the Solar System

The size, the mass, and further properties of the Sun and its planets, will serve as useful points of reference when we later discuss other astrophysical objects.

The Sun
The Sun is the star closest to us, and evidently of paramount importance to us. Without the Sun, no life would be possible on Earth, at least not in the form we know. Because of its much smaller distance in comparison with other stars, the Sun can be studied in great detail. The mass of the Sun is [10]

$$M_\odot = 1.9885 \cdot 10^{30} \text{ kg,} \tag{1.54}$$

the approximate value $M_\odot \sim 2 \cdot 10^{30}$ kg is easy to memorize. The radius of the Sun is [10]

$$R_\odot = 6.96 \cdot 10^8 \text{ m,} \tag{1.55}$$

therefore $R_\odot \sim 700{,}000$ km, again a value that is easy to remember. We can gain an impression of the size of the Sun by comparing with the radius $R_\oplus = 6378$ km of the Earth from (1.40) and its mass

$$M_\oplus = 5.9726 \cdot 10^{24} \text{ kg.} \tag{1.56}$$

The radius of the Sun corresponds to about 109 Earth radii, and the mass of the Sun is $3.3 \cdot 10^5$ times bigger than the mass of the Earth.

The average distance of the Earth from the Sun is approximately $149.6 \cdot 10^6$ km. This distance defines the *astronomical unit* (AU). Since August 2012 it has been assigned the exact value [2]

$$1 \text{ AU} = 1.495978707 \cdot 10^{11} \text{ m.} \tag{1.57}$$

Another parameter of the Sun, important for us on the Earth, is its *luminosity*

$$L_\odot = 3.86 \cdot 10^{26} \text{ W,} \tag{1.58}$$

Fig. 1.5 The power arriving
at the Earth on an area of one
square meter can be
determined from the total
power radiated by the Sun
through the surface of a
sphere with radius $r = 1$ AU

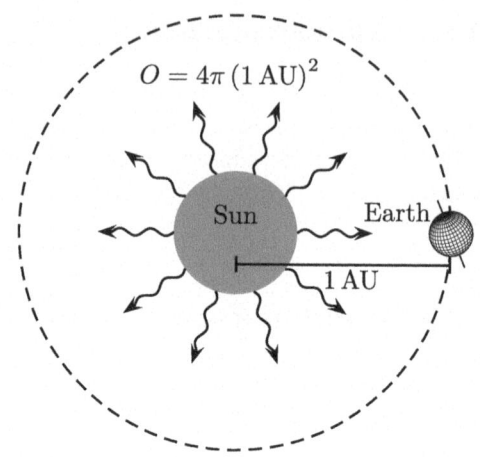

which is the total energy radiated by the Sun per second.

From $L_\odot$ and the value of the astronomical unit we can calculate the intensity of
the radiation which arrives at the Earth from the Sun, that is, the flux of radiation
per second and unit area averaged over all wavelengths. This quantity is called the
solar constant $S_\odot$. To calculate $S_\odot$, we note that the total energy radiated by the Sun
must flow through the surface of a sphere with radius 1 AU. We find

$$S_\odot = \frac{L_\odot}{4\pi(1 \text{ AU [m]})^2} = 1.372 \text{ kW m}^{-2}, \tag{1.59}$$

see Fig. 1.5. The value of $S_\odot$ shows a slight variability, since $L_\odot$, too, is not exactly
constant. The value $S_\odot = 1.3608 \pm 0.0005$ kW m^{-2} results from measurements
around the year 2008, when the luminosity of the Sun was in a minimum [6],
and deviates slightly from the result of our calculation. The solar constant is the
fundamental quantity for all calculations regarding the use of solar energy on the
Earth. Note that a considerable amount of the radiation, roughly 50%, is absorbed
by the atmosphere or reflected.

Using the value of the astronomical unit, we can determine the angular diameter
of the Sun as seen from the Earth. The result is

$$\alpha_\odot = \frac{2R_\odot}{1 \text{ AU}} \approx 0.0093 \frac{180°}{\pi} \approx 0.53° = 32', \tag{1.60}$$

thus approximately half a degree, and almost equal to the angular diameter of the
Moon, with $R_{\text{☾}} = 1737$ km and an average distance $d_{\text{☾}} = 384,400$ km.

The sign $'$ denotes arcminutes. Correspondingly, 60 arcminutes yield one degree,
and 60 arcseconds ($60''$) are equal to one arcminute ($1'$).

Table 1.2 Important properties of the Sun [6, 10]. As the Sun is the closest and most important star to us, we shall later compare the properties of other astronomical objects with parameters of the Sun. Apparent and absolute magnitude will be discussed in Sect. 1.5.3

Size	Symbol	Value
Radius	$R_\odot$	696,000 km
Mass	$M_\odot$	$1.9885 \cdot 10^{30}$ kg
Angular diameter at 1 AU	–	$32'$
Average distance	1 AU	$149.6 \cdot 10^6$ km
Apparent magnitude	$m_\odot$	-26.87^m
Absolute magnitude	$M_\odot$	4.7^m
Luminosity	$L_\odot$	$3.86 \cdot 10^{26}$ W
Solar constant	$S_\odot$	1.3608 ± 0.0005 kW m^{-2}

In radians, one arcsecond is given by

$$1'' = \pi/(180 \cdot 60 \cdot 60) \approx 1/206{,}265 \approx 4.85 \cdot 10^{-6}. \tag{1.61}$$

Therefore one arcsecond corresponds to the angle that an object of size 1 m subtends at a distance of 206,265 m.

The average distance of Mars from the Sun is 1.52 AU, from there the apparent angular diameter of the Sun is about $20'$, on Jupiter at a distance of 5.2 AU, it is still $6'$, on Saturn (9.576 AU) about $3'$, and from distant Pluto (30.14 AU) only $1'$. In Table 1.2 the most important characteristic numerical values of the Sun are summarized.

The Planets of the Solar System

A planet is defined as an almost spherical object which has "cleaned up" its orbit from other objects. The Sun has 8 planets: On the basis of this definition, the most distant Pluto is no longer regarded as a planet since 2006, but rather the eponymous giving member of the group of *plutoids*, the dwarf planets that circle the Sun outside the orbit of Neptune.

Table 1.3 lists important physical parameters of the planets, of Pluto, and of the Moon. Jupiter is by far the most massive and biggest planet in the solar system, its mass being three times bigger than the mass of all other planets together. Still, its mass is only roughly 0.1% the mass of the Sun. Figure 1.6 gives an impression of the proportions of the planets in the solar system.

Kepler's Laws

Apart from small corrections, the properties of the orbits of the planets around the Sun are described by *Kepler's laws*.[15]

According to Kepler's first law, the orbits of the planets are ellipses, with the Sun in one of the foci. The deviation from a circular orbit is characterized by the numerical *eccentricity*

$$\varepsilon = \frac{\sqrt{a^2 - b^2}}{a} \in [0, 1), \tag{1.62}$$

[15] Johannes Kepler, 1571–1630, German mathematician, astronomer, and theologian.

Table 1.3 The planets of our solar system. The second column shows the astronomical symbol of the planet, the remaining columns give the mass M in units of the mass of the Earth, $M_{\oplus} = 5.9726 \cdot 10^{24}$ kg, the size d of the major axis of the orbit in AU, the eccentricity ε of the orbit, the orbital period P, and the radius in kilometers. According to a new definition, Pluto is classified as a dwarf planet, and, strictly speaking, would not belong to this list. For the Moon its distance from the Earth [9] is given

Planet	Symbol	$M\,[M_{\oplus}]$	d [AU]	ε	P [y]	R [km]
Mercury	☿	0.0553	0.387	0.2056	0.24	2440
Venus	♀	0.815	0.723	0.0067	0.62	6052
Earth	♁	1	1	0.0167	1	6378
Moon	☾	0.0123	0.00257	0.0549	0.075	1738
Mars	♂	0.107	1.524	0.0935	1.88	3396
Jupiter	♃	317.83	5.204	0.0489	11.9	71,492
Saturn	♄	95.159	9.582	0.0565	29.4	60,268
Uranus	♅	14.536	19.201	0.0457	83.8	25,559
Neptune	♆	17.147	30.047	0.677	163.7	24,764
Pluto	♇	0.0022	39.482	0.2488	248.1	1195

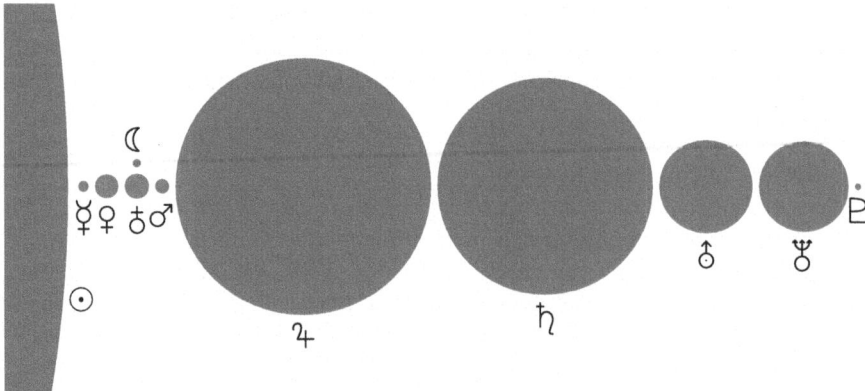

Fig. 1.6 The size of the planets, of the dwarf planet Pluto, and of the Moon in comparison with the Sun

where a and b designate the major and minor axes of the ellipse. The smaller ε, the better the resemblance to a circle. Because of the elliptic form of the orbit, the distance of a planet from the Sun changes during one orbital cycle. The point on the orbit closest to the Sun is called the *perihelion*, the point farthest away the *aphelion*. In the perihelion, the distance of the Earth from the Sun is $147.09 \cdot 10^6$ km, while in the aphelion it is $152.10 \cdot 10^6$ km, with an eccentricity of $\varepsilon = 0.0167$ [9].

Kepler's second law states that the line segment joining a planet and the Sun sweeps out equal areas A during equal intervals of time Δt, see Fig. 1.7.

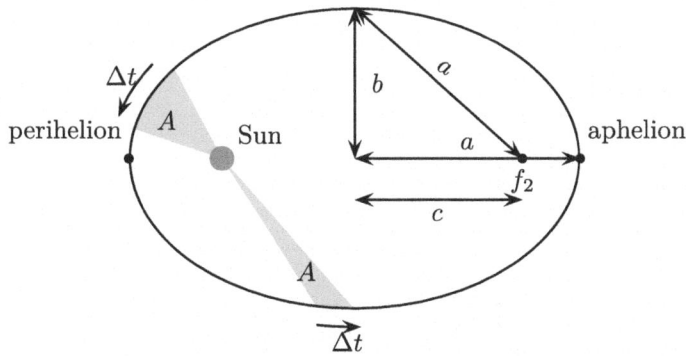

Fig. 1.7 The orbits of the planets are ellipses with a major axis a and a minor axis b, and a distance c of the foci from the center. The Sun is in one of the two foci. During the motion, the line segment joining a planet and the Sun sweeps out equal areas A during equal intervals of time Δt

Finally, Kepler's third law links the orbital periods with the sizes of the major axes. For every planet, the relation

$$GM = \omega^2 a^3 = \left(\frac{2\pi}{P}\right)^2 a^3 \tag{1.63}$$

holds. Inserting in (1.63), e.g., the orbital period of the Earth $P = 3.15 \cdot 10^7$ s and the average distance from the Sun, $a = 1$ AU, leads to a solar mass of $M_\odot \approx 2 \cdot 10^{30}$ kg.

Kepler's third law can be derived for circular orbits in an elementary way. We consider the circular orbit of a satellite with mass m circling a central object with mass M at the distance r. As in the derivation of the first cosmic velocity, the centripetal force acting on m must be equal to the gravitational attraction,

$$m\omega^2 r = \frac{GmM}{r^2}. \tag{1.64}$$

Cancelling m on both sides and multiplying by r^2 yields

$$GM = \omega^2 r^3. \tag{1.65}$$

In this form, Kepler's third law is also called the "123 law", because of the powers of M^1, ω^2 and r^3.

1.5.2 Units of Length in Astronomy and Astrophysics

Because of the huge distances in outer space and in the universe, the meter is no longer a suitable unit of length. On length scales in the solar system, it is mostly the

Fig. 1.8 At a distance of
1 parsec, the distance 1 AU of
the Earth from the Sun
subtends an angle of one
arcsecond (1″)

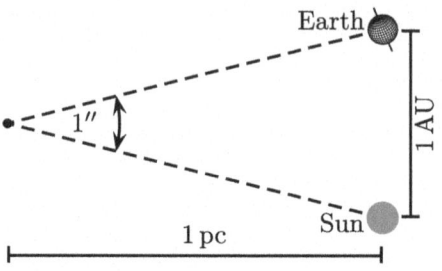

astronomical unit from Eq. (1.57) that can be used as a measure of length. But this
is not a suitable unit for the much bigger distances to stars and galaxies. There, the
light year (1 ly) provides a good reference. One light year is the distance covered
by light in vacuum during one terrestrial year. For this definition, the International
Astronomical Union has fixed the length of the Julian year to 365.25 days. The
advantage of this specification is that one light year corresponds to an integer
number in meters:

$$1 \text{ ly} = c \times 365.25 \text{ d} = 9460{,}730{,}472{,}580{,}800 \text{ m} \approx 9.46 \cdot 10^{15} \text{ m}, \tag{1.66}$$

roughly 10 trillion kilometers.

The unit *parsec* (1 pc) is used even more frequently in astrophysics. Parsec is the
abbreviation for "parallax second", it is defined as the distance at which the average
distance of Sun and Earth, i.e., one astronomical unit from Eq. (1.57), subtends
an angle of one arcsecond, with its definition from (1.61), see Fig. 1.8. Since an
arcsecond is a very tiny angle, the triangle shown in Fig. 1.8 can be approximated
by a rectangular triangle. In radians, the angle α which the length a extends at a large
distance d is given by $\alpha \approx a/d$, because of $\tan(\alpha) \approx \sin(\alpha) \approx \alpha$ for $\alpha \ll 1$. For
these small angles this approximation is completely legitimate. Thus, for $a = 1$ AU
and $\alpha = 1''$ we obtain

$$1 \text{ pc}_{geo} \approx 3.086 \cdot 10^{16} \text{ m} \approx 206{,}265 \text{ AU} \approx 3.2616 \text{ ly}. \tag{1.67}$$

With a slight deviation from this geometric definition, the International Astronomical Union fixed the value to exactly [15]

$$1 \text{ pc} = 3.0857 \cdot 10^{16} \text{ m}. \tag{1.68}$$

In Table 1.4, the conversion factors between the most important units of length
in astronomy are summarized.

Table 1.4 Conversion factors of the most important units of length in astronomy

	m	AU	ly	pc
m	–	$6.6846 \cdot 10^{-12}$	$1.0570 \cdot 10^{-16}$	$3.2408 \cdot 10^{-17}$
AU	$1.4960 \cdot 10^{11}$	–	$1.5813 \cdot 10^{-5}$	$4.8481 \cdot 10^{-6}$
ly	$9.4607 \cdot 10^{15}$	$6.3241 \cdot 10^{4}$	–	0.3066
pc	$3.0857 \cdot 10^{16}$	$2.0627 \cdot 10^{5}$	3.2616	–

1.5.3 The Brightness of Stars

The brightness of a star, more precisely, its total radiation power, is an important parameter. How brightly we see a star does not solely depend on its radiation power, but also on its spectrum, and, above all, on its distance from us. Therefore, the brightness of a star in the sky tells us nothing about its actual luminosity. For this reason two measures of brightness are used in astrophysics. The *apparent magnitude* is a measure of how bright a star shines in the sky. The *absolute magnitude*, on the other hand, is a measure of its intrinsic luminosity. Instead of directly specifying the physical radiation power that reaches us, the apparent brightness of a star is given in magnitude scales.

Let us consider a star S_1 with the apparent magnitude m_1 and radiation flux I_1 (energy per unit time and unit area) arriving at the Earth, and let us consider a second star S_2 with apparent magnitude m_2 and radiation flux I_2, then the difference of the apparent magnitudes is defined by

$$m_2 = m_1 - 2.5 \, \log_{10} \frac{I_2}{I_1}. \tag{1.69}$$

To illustrate: If the radiation flux of star 2 is 100 times stronger than that of star 1, i.e. $I_2/I_1 = 100$, then $\log_{10}(I_2/I_1) = \log_{10}(100) = 2$, and $m_2 = m_1 - 5$. Thus, brighter stars have a smaller magnitude than darker stars. The definition (1.69) takes into account a fundamental psychophysical law: The physiologically perceived strength of a stimulus, here the physical intensity of the electromagnetic radiation, is proportional to the logarithm of the stimulus. This is true, e.g., also in acoustics. The factor 2.5 appearing in front of the logarithm in (1.69) makes sure that the apparent magnitudes which Arabian and Babylonian astronomers had defined on the basis of their physiological perception, and which are still being used in star maps, are correctly reproduced.

Table 1.5 shows a few examples of apparent magnitudes of well-known astronomical objects. The faintest stars visible to the naked eye in a very dark night have apparent magnitudes of about 6^m, the Hubble Space Telescope could detect distant objects with an apparent magnitude of 31.5^m, and modern Earth-bound telescopes can identify even fainter objects.

Table 1.5 Apparent magnitude of a few objects. The data for the stars are from the *Hipparcos star catalogue* [11]. Sirius is the brightest star in the northern sky. Polaris is the North Star or Pole Star

Object	Apparent magnitude
Vega	0.03^m
Polaris	1.97^m
Sirius	-1.44^m
Full Moon	-12.5^m
Sun	-26.87^m

The range of magnitudes of astronomical objects visible to the naked eye spans over 32 magnitudes, from the Sun to the faintest recognizable stars, corresponding to 12 powers of ten in the radiation flux $((10^2)^{32/5} \approx 10^{12})$.

We can reformulate (1.69) as

$$I_2 = I_1 \cdot 10^{0.4(m_1 - m_2)} = I_1 \cdot 2.512^{(m_1 - m_2)}. \tag{1.70}$$

Thus, the decrease, or increase, of the apparent magnitude by 1 means an increase, or decrease, of the radiation intensity by a factor of 2.512. Note that $2.512 \approx \sqrt[5]{100}$. In fact, the apparent magnitude depends on the range of wavelengths. It must be added that in astronomy, apart from the *visual magnitude* m_{visual} of a star discussed so far, apparent magnitudes m_λ in well-defined wavelength windows are considered.

To obtain a distance-independent parameter for the luminosity of a star, the apparent magnitude at a fixed standard distance of 10 parsec is calculated and defined as the *absolute magnitude* M of the object. The standard distance of $10\,\text{pc} = 32.6\,\text{ly}$ used for the specification of the absolute magnitude is chosen to be typical for visible stars in the neighbourhood of the Sun. Sirius, for example, the nearest fixed star in the northern sky, is at a distance of approximately 8.6 ly [13], and the nearest star in the southern sky, Proxima Centauri, at about 4.22 ly [12]. With the definition (1.69) we can calculate the *absolute magnitude of the Sun*

$$M_\odot = m_\odot - 2.5 \log_{10} \frac{I_{10\,\text{pc}}}{I_{1\,\text{AU}}} = m_\odot - 2.5 \log_{10} \frac{(1\,\text{AU})^2}{(10\,\text{pc})^2} = +4.7^m, \tag{1.71}$$

where we have used the relation $I \sim r^{-2}$ as in (1.59). The Sun would be a faint star, just visible to the naked eye.

The difference of the apparent magnitude and the absolute magnitude $m - M$ provides a measure for the distance d of a celestial object. Let I_0 be the intensity emitted at the position of the object. When we write $m_1 = m$ in (1.69) for the apparent magnitude, and for a distance 10 pc, $m_2 = M$, then we have

$$M = m - 2.5 \log_{10} \frac{I_0}{(10\,\text{pc})^2} \frac{d^2}{I_0} = m - 5 \log_{10} \frac{d}{10\,\text{pc}} = m - 5 \log_{10} \frac{d}{1\,\text{pc}} + 5 \tag{1.72}$$

For a galaxy, e.g., at a distance of 1 Mpc $= 10^6$ pc, we find

$$M = m - 5 \log_{10} 10^6 + 5 = m - 30 + 5 = m - 25, \tag{1.73}$$

i.e., m − M = 25. Thus, the absolute magnitude of the galaxy is bigger by 25 magnitudes than when observed from the Earth, corresponding to a 10^{10} times stronger radiation flux. The difference m − M is called the *distance modulus*.

1.5.4 The Starry Sky in the Course of a Year

Even though our focus lies not on the actual observation of stars, the most basic facts of the motion of the Earth in the solar system belong to the elementary knowledge of astronomy, and we therefore discuss them. It is an everyday experience that the apparent course of the Sun over the sky changes according to the seasons. Also, we have stars that are visible only in winter, or in summer. Both phenomena have the same cause.

We have already learned that the Earth orbits the Sun at an average distance of 1 AU. At the same time the Earth revolves around its axis during one day. Therefore, we have two planes: the equatorial plane perpendicular to the rotation axis, and the orbital plane. These two planes are not parallel but are inclined to each other by an angle of $\xi = 23.44°$.

The Origin of the Seasons
The orientation of the rotational axis of the Earth relative to the orbital plane changes within one year and leads to the emergence of the seasons. When the northern hemisphere of the Earth is facing the Sun, then the Sun rises high above the horizon on the northern hemisphere during the day. At the beginning of summer, the Sun is in the tropic of Cancer, at a northern latitude of 23.44°, and stands in the *zenith*, perpendicular above the observer, and the days are longer than the nights. At the North Pole the Sun does not set at all. On this day, at the Antarctic Circle, at a southern latitude of 66.34°, the Sun does not rise at all, and at the South Pole it does not rise for half a year around that date. This is the beginning of winter on the Southern Hemisphere, and at the South Pole the Sun is 23.44° below the horizon during the whole day. At the beginning of winter on the Northern Hemisphere, these statements hold vice versa, by exchanging the Northern and Southern Hemispheres. In Fig. 1.9, these relations are outlined.

As the Earth does not have a perfect spherical shape, Sun and Moon exert a torque on its rotational axis. This leads to a precession, and the axis of the Earth revolves once in about 25,800 years on a cone around the vertical to the orbital plane, i.e., around the polar axis of the ecliptic, which we shall define in the next section. This also leads to a gradual shift of the position of stars in the sky. Because of its long period, the precession has no noticeable effect on the formation of the seasons.

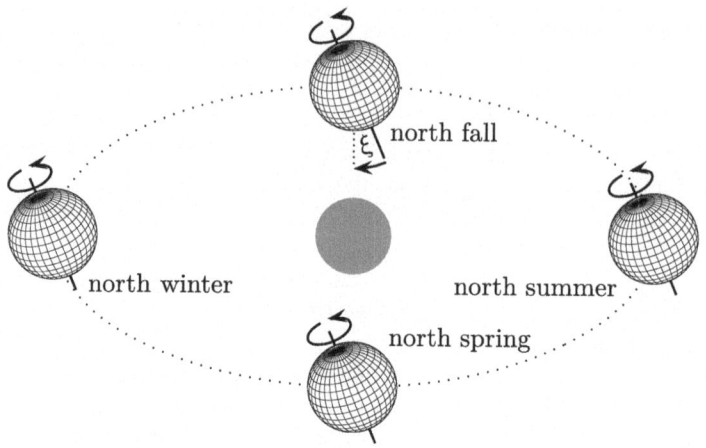

Fig. 1.9 Emergence of the seasons because of the inclination of the equatorial plane relative to the orbital plane by $\xi = 23.44°$. The orientation of the Earth relative to the orbital plane is shown for the respective seasons on the Northern Hemisphere. In summer, the Northern Hemisphere is facing the Sun. During the day, the Sun rises high above the horizon, and the days are longer than the nights. The opposite holds true in winter

Apparent Course of the Sun in the Sky

In the preceding section, we have considered the Earth-Sun system from an outside perspective which helped us to explain the emergence of the seasons. Now we go back to the perspective of an observer on Earth. As the Earth turns around its axis, all stars and also the Sun seem to move across the sky within one day. When we imagine the stars to be attached to the inner surface of an infinitely distant sphere, then the axis of the Earth extended to infinity intersects this celestial sphere at the *celestial north pole* or *celestial south pole*. To an observer on the Earth, all the stars and the Sun visible in the sky seem to revolve around these celestial poles. Because of the inclination of the orbital plane relative to the equator plane the Sun seems to move along a curve on the celestial sphere during one year. This curve is called the *ecliptic*, see Fig. 1.10. The ecliptic intersects the celestial equator at the *vernal equinox* and the *fall equinox*. During winter, the Sun moves below, in summer above the celestial equator. This explains why during the entire north winter, precisely speaking from the beginning of fall until the beginning of spring, the Sun does not rise at the North Pole, and does not set there during all of summer, precisely from the beginning of spring to the beginning of fall. From the perspective of an observer at the North Pole, the Sun moves on a circle at the angular height $\xi = 23.44°$ above the horizon at the beginning of summer, and sinks deeper and deeper towards the horizon within the next half year. From the vernal equinox, angles are measured along the celestial equator, and the ecliptic, to specify the position of a star in the sky. The angle along the celestial equator is called *right ascension*, the angle along the ecliptic *ecliptic longitude*. The corresponding perpendicular angles are

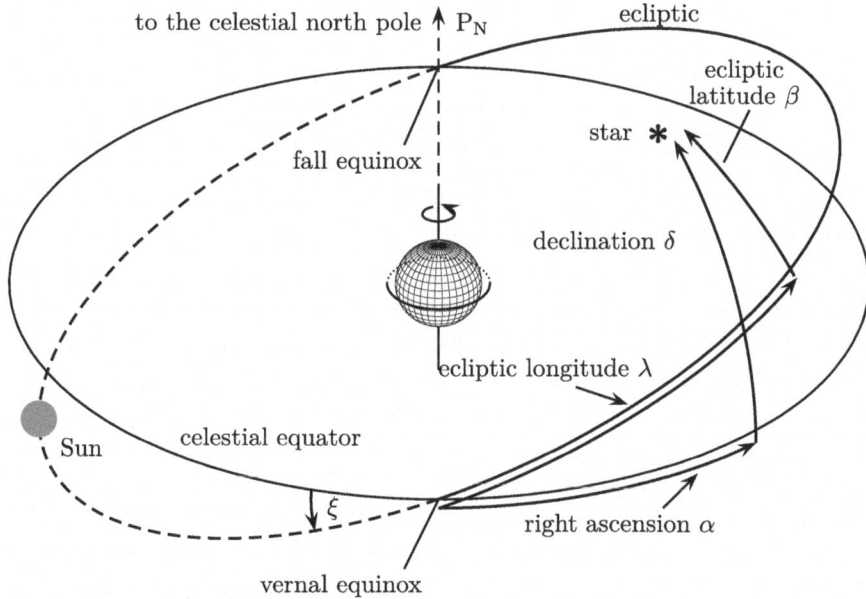

Fig. 1.10 The ecliptic is the apparent path of the Sun in the sky during one year. The ecliptic plane is inclined relative to the equatorial plane by an angle $\xi = 23.44°$, therefore the diurnal course of the Sun in the day sky changes within one year. Different coordinate systems are defined with respect to the equator plane and the ecliptic plane for determining the position of a star, see Sect. 1.5.7

called *declination* and *ecliptic latitude*. In Sect. 1.5.7 we shall discuss the different coordinate systems in more detail.

Visibility of Stars

During one year, due to the changing position of the Earth with respect to the Sun, different stars are visible in the sky in changing seasons. This is illustrated in Fig. 1.11. The star S in the figure rises together with the Sun in the north winter, and therefore is not visible. However, during north summer, the star rises in the night, and can be observed. This distinction is not valid for *circumpolar stars*. These are stars in the sky so close to the celestial north or south pole that they never sink below the horizon. Which stars are circumpolar depends on the position of the observer on Earth [3].

1.5.5 Aberration of Starlight

If a nearby star is observed over weeks or months, it turns out that its position in the sky changes slightly and periodically in the course of one year. This was first

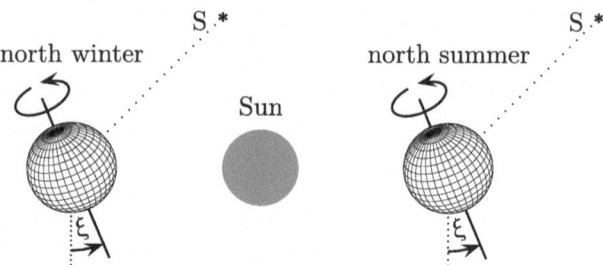

Fig. 1.11 The star S cannot be observed in the north winter as it rises jointly with the Sun. In summer, however, it rises in the night and is visible

Fig. 1.12 Aberration of starlight. Because of the motion of the Earth, stars seem to be shifted by a small angle $\vartheta = \arctan(v/c) \approx v/c$ with respect to their actual position

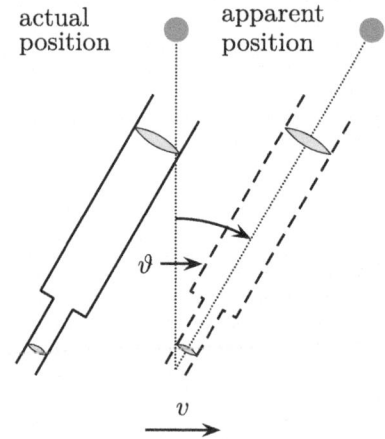

discovered in 1728 by the British astronomer *James Bradley*[16] [1]. The change in the position of the star is proportional to the ratio of the orbital velocity of the Earth and the speed of light. The reason for this seeming change in position is the finiteness of the speed of light. To demonstrate this we consider, for the sake of simplicity, a star standing in the zenith, see Fig. 1.12. If an observer were at rest relative to the star it would be necessary for him to position his telescope at an angle of 90° relative to the ground to observe the star. However, if the observer moves at a velocity v, a photon entering the middle of the telescope, would not impinge in the middle of the objective, since during the time $t = l/c$, which it took the photon to cross the length l of the telescope, the telescope has traveled a length $s = tv = lv/c$. To compensate for this change in the position, the telescope must be turned by an angle $\vartheta = \arctan(v/c) \approx v_{\oplus}/c$ relative to the direction of motion.

As the direction of motion of the Earth changes within a year, stars seem to move on small ellipses in the sky because of the aberration. The exact form of the ellipse depends on the position of the star in the sky, see Fig. 1.13. A star exactly in the

---

[16] James Bradley, 1693–1762, English astronomer and theologian.

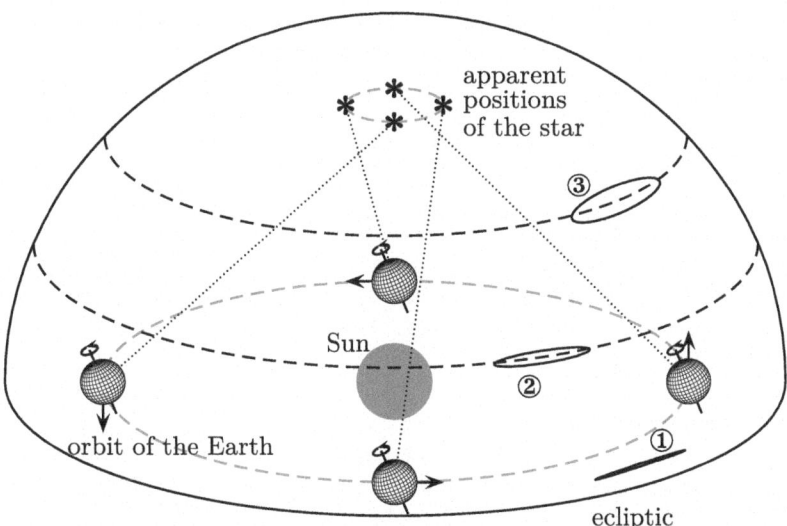

Fig. 1.13 Aberration of starlight. As the Earth moves at a velocity of $v_{\oplus}$, stars appear to be shifted by a small angle $\vartheta = \arctan(v_{\oplus}/c) \approx v_{\oplus}/c$ relative to their actual position. Therefore, within one year, stars move on ellipses in the sky. For stars close or at the ecliptic pole, the ellipses gradually turn into circles (situations ①–③)

orbital plane of the Sun, the *ecliptic*, changes its position only inside the orbital plane. A motion of the Earth toward the star or away from it does not induce a change in position. For stars outside the ecliptic plane, the change in position is an ellipse, the eccentricity of which is the smaller the closer the star is to the *ecliptic pole*, i.e., vertically above the ecliptic. Stars at this position seem to move on a circle. The angular diameter 2ϑ of the circle is given by the ratio $\tan(\vartheta) = v_{\oplus}/c$.

With Earth's velocity

$$v_{\oplus} \approx \frac{2\pi \, 1 \, \text{AU}}{1 \, \text{y}} \approx 30 \, \text{km s}^{-1} \tag{1.74}$$

one finds

$$\vartheta_{\text{aberration}} \approx 20.5''. \tag{1.75}$$

1.5.6 Measuring Distances on Astronomical Scales

Measuring precisely the distance to a star, a planet, or any other object in the solar system, is of central importance to many investigations. If one wishes to determine the absolute magnitude of a star, one needs, besides its apparent magnitude, a very precise knowledge of its distance. Depending on the order of magnitude of the distance to be determined, astronomers use different methods.

Radar Measurements in the Solar System

For objects in the solar system it is possible to send radar signals to, e.g., a planet. There the signals are reflected, and the reflected signal is detected on Earth. The distance is obtained from the time of flight Δt

$$d = c \frac{\Delta t}{2}. \tag{1.76}$$

Using this method, distances can be measured very precisely. In this way, *Irwin I. Shapiro*[17] could determine the distance to Venus, and measure deviations from the value expected from classical physics. These deviations could be explained by general relativity, see Sect. 13.4.4.

The Parallax Technique

Stars are so much farther away from the Earth than even the most distant objects in the solar system, that the radar method cannot be applied. The reflected signals would have to travel many years, and would be too faint to detect.

The *parallax technique* can be used to measure the distance of not too distant stars. When a star is observed in the course of one year, the angle ϑ, under which it is observed, seems to change a little, the reason being that the position of the Earth on its orbit changes, see Fig. 1.14.

The value of the astronomical unit could be determined very precisely using the radar method, and now is defined exactly (cf. (1.57)), therefore the distance to the star can be determined very precisely via the angle difference $\Delta\vartheta$, as long as $\Delta\vartheta$ is not too small. By convention, the angle difference is chosen with respect to the distance of the Earth from the Sun, therefore it is the semiannual angle difference. With elementary trigonometry, the distance is found to be

$$d = 1 \, \mathrm{pc} \, (\Delta\vartheta \, [\mathrm{arcsec}])^{-1}. \tag{1.77}$$

Fig. 1.14 In the parallax technique, a star is observed twice within an interval of six months. Since the position of the Earth on its orbit changes during this period, the star appears under two slightly different angles

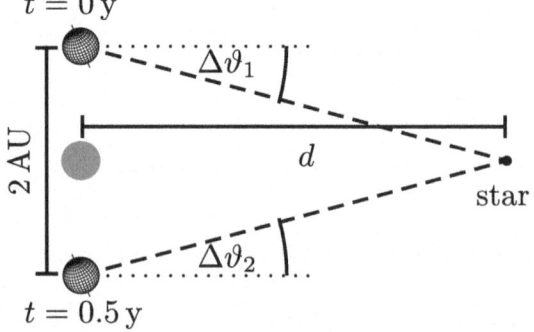

[17] Irwin Ira Shapiro, ⋆1929, US-American astrophysicist.

The expedience of the definition of the unit parsec in Sect. 1.5.2 can be seen in this result. The distance of a star in parsec is the inverse of its parallax in arcseconds. The nearest star Proxima Centauri has a parallax of approximately 0.772″ [12], therefore its distance is

$$d = \frac{1}{0.772} \, \text{pc} \approx 1.3 \, \text{pc}. \tag{1.78}$$

All other stars are farther away from the Earth, and therefore have smaller parallaxes. In star catalogs, such as the Hipparcos catalog, mostly the parallax in arcseconds is tabulated, instead of the distance.

The change in the angle on account of the aberration is superimposed on the parallax. According to (1.75) aberration causes much bigger effects than the parallax. In fact, Bradley, the discoverer of aberration [1] originally wanted to measure the parallax of a star.

Distance Measurements Using Brightness
For objects farther away than roughly 100 pc, the change in angle $\Delta\vartheta$ becomes immeasurably small. A possibility to measure the distance to such objects is provided by the radiation flux that arrives at the Earth. If we know the absolute magnitude of an object, we can obtain its distance with the relation

$$d = \left(\frac{L}{4\pi S} \right)^{1/2}, \tag{1.79}$$

where L is the luminosity of the object, in analogy with the relation between the luminosity of the Sun and the solar constant in (1.59).

The obvious problem of this method is of course that in general the absolute magnitude of an object is not known. To use this method, one needs *standard candles*, i.e., objects of known brightness. These are classes of objects whose absolute magnitude is always the same, or can be derived from other measurable properties.

One possible type of standard candles are *cepheids*, variable stars whose brightness oscillates with a definite period, see Sect. 23.3. There exists a well-known relation between the period and the luminosity, i.e., the absolute magnitude, for these stars. And yet at least one such star must be found whose distance can be determined using the parallax technique to gauge the period-luminosity relation. The search for such stars is impeded by the fact that they are relatively rare.

A further type of standard candles is provided by supernovae of type Ia, which are important in particular in cosmology, when very large distances must be measured. In these very bright events, a white dwarf star explodes, see Sect. 20.4.

Fig. 1.15 The celestial
meridian is the imaginary
projection of the terrestrial
meridian of an observer B
onto the celestial sphere. It
intersects the north and south
point at the horizon of the
fundamental plane of the
observer. It also runs through
the zenith and nadir, and the
celestial north and south
poles, P_N and P_S. In the
figure, the angle φ
corresponds to $\pi/2 - \phi$,
where ϕ is the northern
latitude of the observer

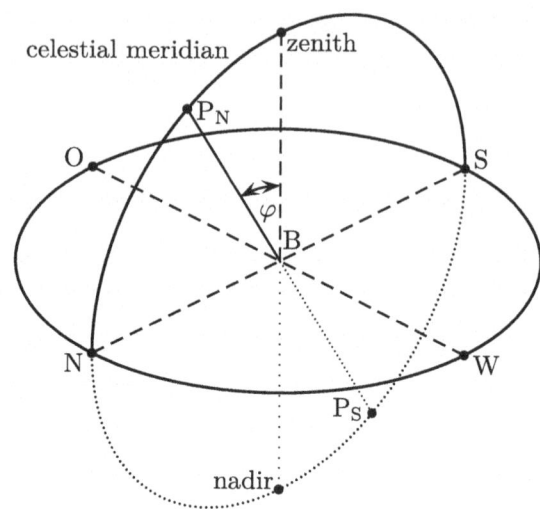

1.5.7   Coordinate Systems

To specify the position of stars, various coordinate systems can be used, which all
have advantages and disadvantages of their own. In this section we shall briefly
present the most important of these coordinate systems. A more detailed account
can be found, e.g., in Clarke [3].

We already mentioned in Sect. 1.5.4 that angles can be defined along the celestial
equator and the ecliptic, which can be used for identifying the position of stars on
the celestial sphere or *hemisphere*.

Horizontal System

The horizontal coordinate system is a celestial coordinate system that uses the
observer's local horizon as the fundamental plane to define two angles: altitude
and azimuth, see below. One introduces the *celestial meridian* perpendicular to the
horizontal plane. It is a great circle on the celestial sphere, which passes through
the north point of the horizon, the celestial north pole, the zenith, the south point
of the horizon, the celestial south pole, and the *nadir*, the point on the celestial
sphere directly opposite to the zenith, see Fig. 1.15. Thus, the celestial meridians
are a projection of the terrestrial meridians onto the celestial sphere. Observers at
the same longitude share the same celestial meridian, only the angles φ between the
zenith and the celestial north pole differ. In the horizontal system, the position of
the star is specified by its *altitude* above the horizon, $h \in [0, \pi/2]$, and its *azimuth*
$A \in [0, 2\pi)$. The azimuth is the angle between the line from the north point in the
horizontal plane and the vertical circle through the star. It is measured from the north
point in eastern direction. Figure 1.16 illustrates these definitions in the horizontal
system.

Fig. 1.16 Specifying the position of a star in the horizontal system. The coordinates A and h depend on the location of the observer on the Earth, and, because of the rotation of the Earth, also on the time of the observation

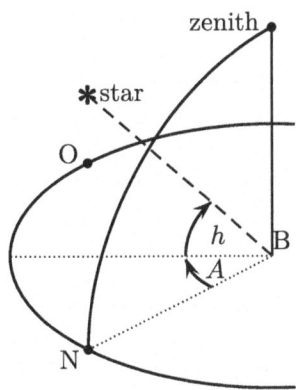

The horizontal system is used in alt-azimuth mounts of telescopes, which is a simple two-axis mount. In this case, tracking a star in the sky requires to rotate the instrument about two perpendicular axes, one vertical and the other horizontal. These mounts are used for example in amateur telescopes, but also in large radio telescopes and large optical telescopes. In these telescopes the tracking of a star is controlled automatically by microprocessors. The horizontal system has two disadvantages. On the one hand, the coordinates change because of the rotation of the Earth. On the other hand, the coordinates differ from one observer on the Earth to another. Therefore, these coordinates are not suitable for a unified description of stellar positions. To catalog objects in the sky, other coordinates are necessary.

Equatorial System

The equatorial system is defined by the rotational axis of the Earth and the celestial equator. In principle, it is the projection of the geographic grid of the Earth onto the celestial sphere. The hour circles correspond to the longitudes, and the circles parallel to the celestial equator correspond to the latitudes. There is a distinction between the "fixed" and the "moving" equatorial system. Both systems are best understood when we go back to Fig. 1.10.

In the *fixed equatorial system* the position of an object is specified by its *declination δ* and its *hour angle t*. The hour angle is measured along the celestial equator. Its zero is the point of intersection of the celestial equator and the celestial meridian. It therefore depends on the geographical longitude of the observer, and cannot be displayed in Fig. 1.10. As the Earth turns on its axis within 24 hours, i.e., by 360°, the hour angle of a star changes during one hour by 15°, and therefore is time-dependent. In addition, it changes in the course of one year. The zero of the system, because of its definition via the celestial meridian, is coupled to the position of the observer, which is the reason for the designation "fixed equatorial system".

Most telescopes have an "equatorial", or "parallactic", mount. One axis, the hour axis, is oriented parallel to the axis of the Earth. In this case a rotation of only one axis is needed to compensate for the rotation of the Earth, i.e., for tracking. However, due to its time and position dependence, the fixed equatorial coordinate system also is not suitable for a unified, observer-independent description of the starry sky.

In the *moving equatorial system* the zero for the hour angle is chosen such that the coordinates of an object become time-independent. This means the system moves in accordance with the apparent rotation of the celestial sphere, therefore the name "moving equatorial system". The position of the vernal equinox is chosen as the zero of the hour angle, therefore the hour angle is the right ascension measured in units of time, $\alpha \in [0\,h,\ 24\,h]$.

In most catalogs of stars and galaxies, this equatorial system is used because of its independence of the location of the observer and of the time.

Ecliptic System

The quantities needed in this system also have been introduced in Fig. 1.10. In the ecliptic system the fundamental plane is the ecliptic, i.e., the orbital plane of the Sun. In this system, the ecliptic latitude β corresponds to the declination in the equatorial system. The ecliptic longitude λ is measured along the ecliptic. As in the equatorial system, the vernal equinox is chosen as the zero of the ecliptic longitude. It is the analogue of the right ascension in the equatorial plane. Because of its relation to the Sun, the ecliptic coordinate system is important for the description of the motion of objects in the solar system, such as planets, asteroids, or comets. In particular, the orbital path of these objects is measured relative to the ecliptic, as discussed already for the Earth.

Galactic System

The fundamental plane of the galactic system is the plane of the Milky Way, the *galactic equatorial plane*. The origin of the coordinate system is the center of the Milky Way. The *galactic latitude b* is the angle between the object and the plane of the Milky Way, in analogy with the ecliptic latitude and the declination. The *galactic longitude l* is the angle between the line connecting the Sun and the Earth and the intersection of the meridian of the object with the galactic equatorial plane; it corresponds to the ecliptic longitude or the right ascension. One can visualize the galactic system if one imagines the Milky Way to be in the center of Fig. 1.10 instead of the Earth, and the position of the Sun as a fixed point instead of the vernal equinox. Galactic coordinates are mainly used when the spatial distribution of objects in our galaxy is important.

Perturbations of the Coordinates

The motion of the Earth is subject to effects that produce long-term variations. Therefore, it is not sufficient just to specify the coordinates of celestial objects. In star catalogs also the *equinox* is given, i.e., the point in time, or the epoch, to which the measurement of the coordinates is referring to. One of the most important sources of perturbations is the *precession* caused by the gravitational forces of the Moon acting on the Earth. We have mentioned this precession already in the context of the emergence of seasons. As in the precession of a heavy top in classical mechanics, an additional *nutation* of the axis of the Earth is superimposed on the

Fig. 1.17 From the source of light, the ray of light goes to the rotating mirror and from there to a spherical mirror at a distance r, from where it returns to the rotating mirror

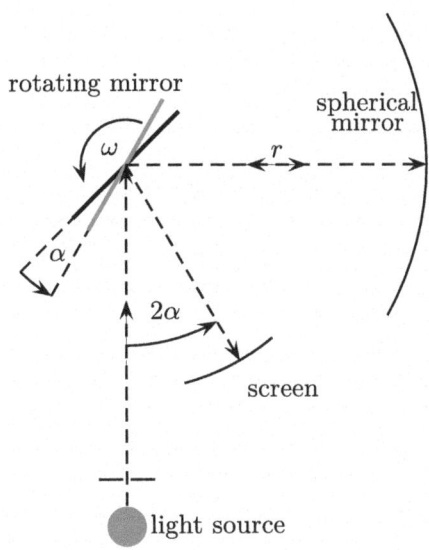

precession, with a period of 18.6 years. It is also mainly caused by the Moon. But there are even more effects which cause perturbations, even though not as drastic as those caused by the Moon, which change right ascension and declination over long times. The position of the vernal equinox and Polaris also change for the same reasons.

1.6   Exercises

1.1. Foucault's rotating mirror Rømer's measurement of the speed of light was subject to large uncertainties. A much more precise measurement dates back to *Foucault*.[18] A possible setup is sketched in Fig. 1.17. A ray of light is emitted from a source of light, impinges on a rotating mirror, is reflected by a spherical mirror, and again hits the rotating mirror. Since the mirror has kept rotating, the ray of light is not exactly reflected back to the source of light, but arrives with a slight angular offset.

Calculate the distance r at which the spherical mirror has to be placed, when your rotating mirror reaches 20,000 revolutions per second, and you wish to attain an angular displacement of $2\alpha = 1°$.

1.2. Gravitational potential of a hollow sphere Show that the gravitational potential inside a hollow sphere is constant.

---

[18] Léon Foucault, 1819–1868, French physicist.

1.3. Gravitation and the Coulomb force in the hydrogen atom In this chapter we discussed that the Coulomb force is much stronger than the gravitation. As an example, calculate the ratio of gravitational and Coulomb force of proton and electron in the ground state of the hydrogen atom.

1.4. Black hole of water The radius of a spherical mass with uniform mass density increases as $r \sim M^{1/3}$, but its Schwarzschild radius as $r_s \sim M$. Therefore, it is possible, for any value of the uniform mass density, to find a sphere with radius $r < r_s$, if it is only large enough.

1. What must be the minimum size of a sphere of water which fulfills this condition?
2. Verify the statement in Sect. 1.4.4 that a sphere with the mass density of the Earth and $r \approx 250\, r_\odot$ fulfills the condition $r = r_s$.
3. We shall see in Chap. 26 that the average mass density in the Universe is approximately 2 protons per cubic meter. What radius is obtained in this case?

References

1. Bradley, J.: A new apparent motion discovered in the fixed stars; its cause assigned; the velocity and equable motion of light deduced. Philos. Trans. R. Soc. Lond. **8**, 308–321 (1728)
2. Brumfiel, G.: The astronomical unit gets fixed. Nature News (2012)
3. Clarke, D., Roy, A.E.: Astronomy: Principles and Practice, 4th edn. Institute of Physics Publications, Bristol (2003)
4. Huygens, C.: Traitée de la lumière. Chez Pierre van der Aa, Leiden (1690)
5. Joyce, M., Leung, S.-C., Molnár, L., Ireland, M., Kobayashi, Ch., Nomoto, K.: Standing on the shoulders of giants: new mass and distance estimates for Betelgeuse through combined evolutionary, asteroseismic, and hydrodynamic simulations with MESA. Astrophys. J. **902**(1), 63 (2020)
6. Kopp, G., Lean, J.L.: A new, lower value of total solar irradiance: evidence and climate significance. Geophys. Res. Lett. **38**(1), L01706 (2011)
7. Laplace, P.S.: Exposition du Système du Monde – Livre quatrième: De la théorie de la pesanteur universelle. Imprimerie du Cercle Social (1796)
8. Mohr, P.J., Tiesinga, E., Newell, D.B., Taylor, B.N.: Codata Internationally Recommended [sic!] 2022 Values of the Fundamental Physical Constants. https://physics.nist.gov/constants
9. Nasa Planetary Fact Sheet: https://nssdc.gsfc.nasa.gov/planetary/factsheet
10. Nasa Sun Fact Sheet: https://nssdc.gsfc.nasa.gov/planetary/factsheet/sunfact.html
11. Perryman, M.A.C., et al.: The Hipparcos Catalogue. Astron. Astrophys. **323**, L49–L52 (1997)
12. Perryman, M.A.C., et al.: Data for "Proxima Centauri" from the Hipparcos catalog on August 12, 2013, Identifier 70890. https://simbad.u-strasbg.fr/simbad/sim-id?Ident=Hip+70890
13. Perryman, M.A.C., et al.: Data for "Sirius" from the Hipparcos catalog on August 9, 2013, Identifier 32349. https://simbad.u-strasbg.fr/simbad/sim-id?Ident=Hip+32349
14. Rømer, O.: A demonstration concerning the motion of light, communicated from Paris, in the Journal des Scavans, and here made English. Philos. Trans. **12**, 893–894 (1677)
15. Wilkins, G.A.: The IAU style manual (1989) – the preparation of astronomical papers and reports. IAU (1989). https://www.iau.org/static/publications/stylemanual1989.pdf

Part I

Special Relativity

The Road to Special Relativity

2

Contents

Einstein's[1] publication of the theory of special relativity (SR) in 1905 [2] came at a time when physics was undergoing dramatic changes through revolutionary developments within only a few decades. Besides special relativity, these include the theory of general relativity, and the final formulation of quantum mechanics by Heisenberg,[2] Schrödinger,[3] and Dirac.[4]

The theory of special relativity revolutionized the concept of space and time. Before, space and time had been treated as strictly separate entities; now special relativity combined them in a four-dimensional spacetime. The theory of general relativity, which Einstein worked out in the following 10 years, extended the concept of spacetime even further.

It is not within the scope of this chapter, and also not its objective, to put Einstein's work into the historical context of the work of other scientists at the time, even though such a review would be very informative. Readers interested in the topic are referred to, e.g., an article by Darrigol [1].

---

[1] Albert Einstein, 1879–1955, German-American physicist. Nobel Prize in 1921 for the explanation of the photoelectric effect.

[2] Werner Heisenberg, 1901–1976, German theoretical physicist. Nobel Prize in 1933.

[3] Erwin Schrödinger, 1887–1961, Austrian theoretical physicist. Nobel Prize in 1933.

[4] Paul Adrien Maurice Dirac, 1902–1984, British theoretical physicist, Nobel Prize in 1933.

© The Editor(s) (if applicable) and The Author(s), under exclusive license to
Springer-Verlag GmbH, DE, part of Springer Nature 2026
S. Boblest et al., *Special and General Relativity*,
https://doi.org/10.1007/978-3-662-71332-7_2

Instead, we shall highlight—not necessarily in historically correct order—a few essential experiments whose results were in blatant contradiction to the current theory at the time. These results motivated the formulation of SR, which after all could resolve the seeming contradictions.

2.1 Models of Light Propagation in the Nineteenth Century

Sound is a wave phenomenon. Perturbations of the otherwise uniform density and pressure propagate through air, water, and other media as waves. One could also say that sound is produced by the oscillations of its carrier medium. In the nineteenth century almost every scientist was convinced that light also is a wave phenomenon, proved, e.g., by the interference experiments of *Hertz*.[5] But the question was: What is the carrier medium in which the oscillations of light take place? An answer to this question was made even more difficult since it was known that the light from the Sun or the stars propagates through empty space, vacuum, before it reaches the Earth.

To solve the problem, physicists postulated the existence of a carrier medium of light, the *world ether*, or, as Einstein called it, the *light medium*. The existence of a light medium of this kind was almost taken for granted among scientists. Several hypotheses were put forward as to the specific properties of this light medium. For example, Maxwell proposed that ether and matter form one entity, and that moving matter, even if very dilute, should completely "drag" the ether. This is the *ether drag hypothesis*. However, this concept contradicted the observed *aberration* of starlight, which we have discussed in Sect. 1.5.5. This contradiction gave rise to the alternative idea that the Earth is moving through an ether *at rest*. Yet the "ether wind" that should blow on the surface of the Earth because of this motion, could never be observed in any optical experiments. To solve this problem, *Fresnel*[6] proposed the hypothesis of a partial ether drag.

One of the prominent objectives of experimental physics in the late nineteenth century was to decide experimentally, which of these hypotheses was the correct one. *Michelson*[7] in particular stood out in these experiments.

2.2 Michelson-Morley Experiment

The experiment carried out by Michelson in 1881 [3], and in an improved version together with Morley in 1887 [5], is certainly the most famous experiment that

---

[5] Heinrich Rudolf Hertz, 1857–1894, German physicist.

[6] Augustin J. Fresnel, 1788–1827, French physicist and engineer.

[7] Albert A. Michelson, 1852–1931, US-American physicist of German origin. The experiment described in Michelson [3] was actually carried out in Potsdam, Germany. Nobel Prize in physics 1907.

caused big difficulties to the hypothesis of the existence of a light medium. The original objective of the experiment, though, was to determine the velocity of the Earth relative to the assumed light medium. The basic idea behind the experiment was the assumption that the time it takes for a ray of light to cover a certain distance on the Earth should depend on the direction of propagation of the ray of light. Let us consider a ray of light that flies from a point A to another point B, a distance l away, where it is reflected back. We also assume that the path of the light is parallel to the (yet unknown) direction of the motion of the Earth through the light medium. If the Earth moves with a velocity v relative to the light medium, the velocity of light is effectively reduced by v, and therefore the light front approaches the point B with a velocity of $c - v$. The time of flight is

$$t_{AB} = \frac{l}{c - v}. \tag{2.1}$$

In the opposite direction, the point B runs towards the light front with the velocity v, therefore the time of flight becomes

$$t_{BA} = \frac{l}{c + v}. \tag{2.2}$$

The total round-trip time is

$$t_{ABA} = 2l \frac{c}{c^2 - v^2}. \tag{2.3}$$

On the other hand, if the Earth were at rest relative to the light medium, the time of flight in either direction would be

$$t_0 = \frac{l}{c}. \tag{2.4}$$

If one can determine either the difference $t_{ABA} - 2t_0$ or the difference $t_{AB} - t_{BA}$ in an experiment, then one can determine the velocity v since

$$t_{AB} - t_{BA} = 2l \frac{v}{c^2 - v^2} \approx 2\frac{vl}{c^2} \left[1 + \mathcal{O}\left(\frac{v^2}{c^2} \right) \right] \approx 2t_0 \frac{v}{c}, \tag{2.5}$$

where c and t_0 are known. Michelson realized that by measuring this time difference it should be possible to determine the velocity v.

His experiment consisted of an interferometer with two perpendicular arms. Figure 2.1 shows a very simplified sketch of his experimental setup. A ray of light is guided from an optical source to a beam splitter. Part of the light is reflected at the beam splitter, flies upward to the mirror M_1 on the path l, is reflected and returns to the beam splitter. There it passes through the beam splitter and enters a telescope. The part of light that is reflected by the beam splitter in this case is unimportant for

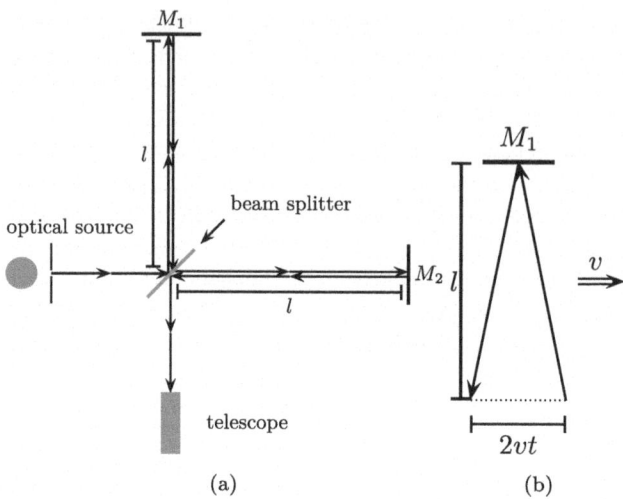

Fig. 2.1 Sketch for the Michelson-Morley experiment. (**a**) Simply speaking, the setup consists of an optical source from which a ray of light is divided into two rays by a beam splitter. The two rays are guided to the mirrors M_1 and M_2. Then they interfere in a telescope where the interference fringes can be observed. (**b**) Since the interferometer moves though the light medium, a longer optical path results in the arm of the interferometer standing perpendicular to the direction of motion

the experiment. At the same time, at the first contact with the beam splitter, part of the light is transmitted and flies to the mirror M_2 on a path with equal length l, there it is reflected and flies back to the beam splitter. Part of the ray of light is reflected down by the beam splitter into the telescope where it interferes with the first ray of light. The total setup can be rotated around a perpendicular axis. During a rotation, the interference fringes should change. To understand this we begin with the case that the axis from the optical source to the mirror M_2 is aligned exactly parallel to the direction of motion of the Earth. For the ray of light which flies from the beam splitter to the mirror M_2 and back, the time of flight is, as calculated above, $t_{ABA} = 2lc/(c^2 - v^2)$. The distance covered relative to the light medium during this time then is

$$d_\parallel = 2l\frac{c^2}{c^2 - v^2} \approx 2l\left(1 + \frac{v^2}{c^2}\right) + \mathcal{O}\left(\frac{v^4}{c^4}\right). \qquad (2.6)$$

For the arm of the interferometer perpendicular to the direction of motion, the distance covered also changes. This can be easily seen from Fig. 2.1b. Since the experimental setup moves along with the Earth relative to the light medium, the path of the ray of light to the mirror M_1 is not straight but slightly inclined. When the ray arrives at the mirror after a time t, the experimental setup has moved forward by the distance vt. Thus, for the distance covered by the ray of light we have two relations, which we can equate to eliminate t:

$$d = ct = \sqrt{l^2 + v^2 t^2}. \tag{2.7}$$

From this we easily calculate

$$t = \frac{l}{c} \frac{1}{\sqrt{1 - v^2/c^2}}. \tag{2.8}$$

The total distance, covered by the ray of light to the mirror and back, therefore is

$$d_\perp = 2l \frac{1}{\sqrt{1 - v^2/c^2}} \approx 2l \left(1 + \frac{1}{2} \frac{v^2}{c^2}\right) + \mathcal{O}\left(\frac{v^4}{c^4}\right). \tag{2.9}$$

We see that in the perpendicular case the effect of the moving Earth on the distance is not zero, but only smaller by one half compared to the parallel case. This renders the detection of the difference even more difficult. It now becomes

$$\Delta d = |d_\perp - d_\parallel| = l \frac{v^2}{c^2} + \mathcal{O}\left(\frac{v^4}{c^4}\right). \tag{2.10}$$

If we estimate the velocity of the Earth relative to the light medium by its orbital velocity $v_\oplus \approx 30\,\mathrm{km\,s^{-1}}$ from (1.74), this leads to a difference of $\Delta d \approx 10^{-8}\,l$. Assuming an arm length of $l = 1.2\,\mathrm{m}$ and using light with a wavelength of $\lambda \sim 600\,\mathrm{nm}$, we find $l \approx 2 \cdot 10^6 \lambda$ and thus $\Delta d \approx 0.02\,\lambda$.

After a rotation by 90° the situation is exactly reversed. Now the path of the light in the other interferometer arm is longer. After a quarter rotation we obtain a total shift of

$$\Delta d \approx 0.04\lambda. \tag{2.11}$$

The correction of the time of flight in perpendicular direction was overlooked by Michelson when evaluating the results of his first experiment in 1881. Because of that, he assumed twice the difference. The setup of this experiment was very similar to the sketch shown in Fig. 2.1. When the error was brought to Michelson's attention, it became evident that in a repetition of the experiment a more sophisticated setup would be necessary to detect the corresponding effects. Figure 2.2 shows the setup used by Michelson and Morley in 1887. They employed several mirrors in order to increase the length of the path of the rays of light, which increased the path difference, in comparison with (2.11), and thus the measuring accuracy. Experiences from the first experiment had shown that the crucial point of the experiment was to rotate the setup as smoothly as possible. To avoid any disturbances, the new setup was mounted swimming on mercury. The whole setup was rotated continuously at a velocity of 1 rotation in 6 minutes. It was expected that during the rotation of the

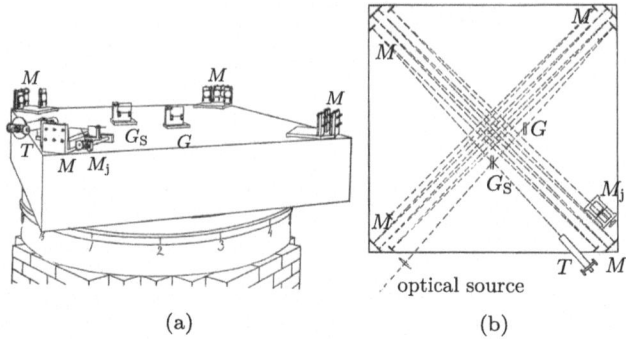

Fig. 2.2 Setup of the Michelson-Morley experiment. (**a**) Perspective view, (**b**) top view of the optical path. The interferometer is mounted on a square stone with lateral length $a = 1.5\,\text{m}$. The stone is supported by a wooden ring swimming on mercury. This makes it possible to rotate the experiment as smoothly as possible. To maximize the optical path, the actual interferometer consists of several mirrors M placed at the corners of the stone. In the center of the stone are two glasses G. G_S serves as beam splitter. From Michelson and Morley [5], with kind permission

device the observed interference fringes would shift periodically. In spite of all these efforts, Michelson and Morley could never see any shift of the interference fringes. The hypothesis of a light medium, through which the Earth is moving, could no longer be maintained. Michelson received the Nobel Prize in 1907 for his precise spectroscopic investigations.

2.3   The Fizeau Experiment on Ether Drag

This experiment was first carried out in 1850 by *Fizeau*,[8] which is the reason why it is named after him. An improved version of this experiment was repeated in 1886 by Michelson and Morley [4]. The objective of the experiment was to test the ether drag in optically transparent media. Figure 2.3 shows a sketch of the experimental setup. Water is guided through a pipe with four 90°-bends. In this way, two parallel sections are present in which water flows in opposite directions. A ray of light is divided into two partial rays by a beam splitter, which pass through the two parallel sections of the water flow in opposite directions. Therefore, one of the rays propagates in the flow direction, and the other opposite to it. Subsequently, part of the two rays of light is deflected by the beam splitter to the observer, where the rays interfere. The velocity of light in a medium with refractive index n is

$$c_{\text{med}} = \frac{c}{n}. \tag{2.12}$$

---

[8] Armand Hippolyte Louis Fizeau, 1819–1896, French physicist.

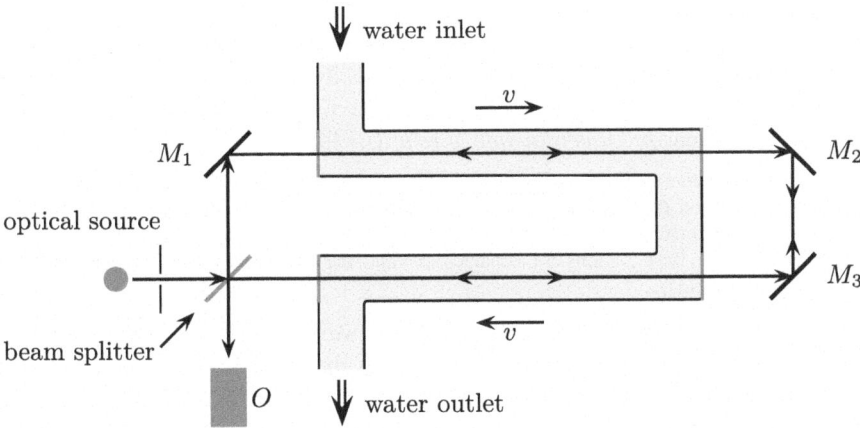

Fig. 2.3 Sketch of the Fizeau experiment. A beam splitter divides the ray of light from an optical source into two rays. Both rays pass through a setup of mirrors M_i and water pipes. In these pipes water flows with velocity v. One of the rays traverses the water in flow direction, the other opposite to it. Subsequently, the rays interfere at the observer O

When we assume that the light is "dragged" by the water, we expect velocities in the flow direction and opposite to it

$$c_{\text{med}\pm} = \frac{c}{n} \pm v. \tag{2.13}$$

Instead, Fizeau, and also Michelson and Morley, found

$$c_{\text{Med}\pm} = \frac{c}{n} \pm v \left(1 - \frac{1}{n^2}\right). \tag{2.14}$$

This result was in agreement with Fresnel's assumption of a partial ether drag in the medium, but was in contradiction to the results of the Michelson experiment. We shall see that the result of the Fizeau experiment can be explained very simply in special relativity, see Exercise 3.2.

2.4   The Principle of Relativity, and Inertial Systems

The repeated failures to prove the existence of a world ether in experiments, led Einstein to the conclusion that there is no such thing as a world ether, and the speed of light has the same value for any observer, irrespective of his state of motion relative to the source of light.

   If a world ether existed, motion could be defined in absolute terms. The velocity of an observer would be zero if he found himself at rest relative to the world ether. Therefore, if we assume, following Einstein, that no world ether exists, we must also relinquish our concept of absolute motion. Let us consider, e.g., a space traveler in a region of the universe where there are no stars close enough that he could observe them. The engines of his spacecraft are cut off, therefore he does not experience any acceleration. How is he to decide what is the velocity he is traveling at, and relative to what? It is only motion relative to another body which can be noticed.

Einstein summarized this *principle of relativity* like this [2]:

> [...] the failed attempts to state a motion of the Earth relative to the "light medium" lead to the assumption that not only in mechanics but also in electrodynamics no properties of phenomena correspond to the concept of absolute rest, but that rather in all coordinate systems in which the equations of mechanics are valid, the same electrodynamic and optical laws are also valid. [...]

In Chap. 29 we shall learn about the *cosmic microwave background radiation, CMB.* The CMB is clear evidence that, in an early phase, the universe was in a state of very high temperature and density, from which it cools down and expands ever since. The CMB is very isotropic, but is red or blue shifted in parts of the celestial sphere on account of the Doppler effect caused by the orbital motion of the Earth, see the discussion in Sect. 29.1. We shall treat the Doppler effect in Sect. 8.1. Therefore, we could use the CMB to define a state of absolute rest. To test the issue at hand, we could ask our space traveler to decide by some physical experiment in his closed-up spacecraft whether he is at rest, or moving at constant velocity, relative to the CMB. The principle of relativity states that he will fail in this task.

The *inertial system* is one of the most important concepts of Newtonian mechanics. They are defined in Newton's first law:

> An object will remain at rest or in uniform motion in a straight line unless acted upon by an external force.

This statement does not hold for objects moving on the surface of the Earth since fictitious forces appear such as the *Coriolis force.*

The principle of relativity now states that we cannot distinguish any particular inertial system, namely one that is at absolute rest. Therefore, all inertial systems are equivalent, and equally suited to formulate the laws of physics, and in all inertial systems they assume the same form.

With our present knowledge of the principle of relativity, and the observer-independence of the speed of light, we can turn to a detailed discussion of the foundations of special relativity in the following chapters.

References

1. Darrigol, O.: The Genesis of the theory of relativity. Sémin. Poincaré **1**, 1–22 (2005)
2. Einstein, A.: Zur Elektrodynamik bewegter Körper (On the electrodynamics of moving bodies). Ann. Phys. **17**(10), 891–921 (1905)
3. Michelson, A.A.: The relative motion of the Earth and the luminiferous ether. Am. J. Sc. **22**(128), 120–129 (1881)
4. Michelson, A.A., Morley, E.W.: Influence of the motion of the medium on the velocity of light. Am. J. Sc. **31**(185), 377–386 (1886)
5. Michelson, A.A., Morley, E.W.: On the relative motion of the Earth and the luminiferous ether. Am. J. Sc. **34**(203), 333–345 (1887)

Lorentz Transformations

3

Contents

In this chapter, we explore the consequences of the postulate of a constant, observer-independent speed of light. To do this, we will first briefly recapitulate the essential ingredients of Newtonian mechanics and electrodynamics, and then motivate the introduction of the Lorentz transformation.

3.1   Transformation Between Different Frames of Reference

In Newtonian mechanics, every point in space is characterized by its coordinate vector $r = (x, y, z)^{\mathrm{T}} \in \mathbb{R}^3$ in some coordinate system. The time t can be counted from an arbitrary point in time.

The motion of a point particle is governed by Newton's second law, which states that the rate of change of a particle's momentum $p = m\dot{r}$ is given by the force

F acting on the particle, $\dot{p} = F$. If the mass m is constant[1] the law assumes the familiar form

$$F = m\ddot{r} \quad \text{or} \quad \ddot{r}(t) = \frac{1}{m}F(r,t). \tag{3.1}$$

Thus, the acceleration $a = \ddot{r}$ of the point particle depends on the external force F and its inertial mass m. Solving this system of differential equations yields the trajectory $r(t)$ of the particle. In the classical understanding of space and time, the laws of Newtonian mechanics are valid in any *inertial system*, see Sect. 2.4.

The transition from one inertial system to another inertial system, moving with relative velocity v, is accomplished by the general *Galilei transformation*

$$\begin{aligned} r' &= D \cdot r - vt - r_0 \\ t' &= t - t_0. \end{aligned} \tag{3.2}$$

Here D denotes a rotation of the inertial system, v is the relative velocity of the inertial systems, and r_0 and t_0 are the relative displacements of the origins of space and time.

The Galilei transformation (3.2) contains a total of 10 free parameters: Three angles of rotation in D, three components of the velocity in v, three components of translation in r_0, and the time difference t_0. The set of all Galilei transformations is a 10-parameter *group* since the composition of two Galilei transformations yields another Galilei transformation. The neutral element of the group is the identity operation $D = 1$, $v = 0$, $r_0 = 0$, $t_0 = 0$. The inverse element is given by the inverse transformation. It is important to note that classical mechanics is invariant under all transformations of this group: When we insert the Galileo transformation in the equations of motion (3.1), their form remains unchanged.

How do the basic equations of electrodynamics behave under Galilei transformations?

Maxwell's equations (1.5)

$$\nabla \times E + \dot{B} = 0,$$

$$\nabla \cdot B = 0,$$

$$\frac{1}{\mu_0}\nabla \times B - \varepsilon_0\dot{E} = j, \tag{3.3}$$

$$\nabla \cdot E = \frac{\rho_{\text{el}}}{\varepsilon_0}$$

---

[1] Problems in which the mass is not constant are, e.g., the launch of a rocket, which ejects fuel, or the gradual slipping of a chain from a table.

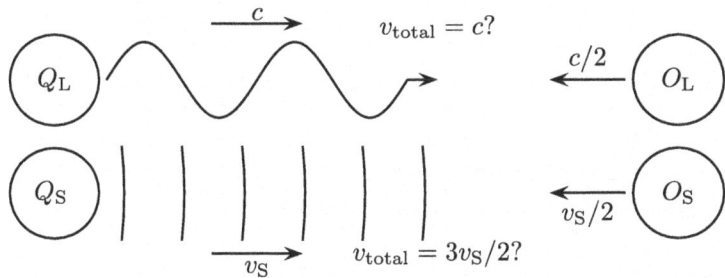

Fig. 3.1 If an observer O_S moves towards a source of sound with half the velocity of sound, he observes, after a Galilei transformation, a velocity of sound of $v_{\text{total}} = 3v_S/2$, in comparison with an observer at rest. Since the speed of light is a constant, the same cannot be true for a source of light and an observer O_L moving towards it with half the speed of light. Consequently, the Galilei transformation can only be an approximation for small velocities of the correct transformation

are a system of differential equations which allow for the calculation of the electric and magnetic fields $\boldsymbol{E}(\boldsymbol{r}, t)$ and $\boldsymbol{B}(\boldsymbol{r}, t)$ for a given distribution of electric charges $\rho_{\text{el}}(\boldsymbol{r}, t)$ (space charge density) and currents $\boldsymbol{j}(\boldsymbol{r}, t)$. The solutions can be static fields, treated in electrostatics and magnetostatics, or electromagnetic waves, which in vacuum propagate with the speed of light $c = 1/\sqrt{\mu_0 \varepsilon_0}$. We know already from (1.8) that the speed of light is related to the electric field constant ε_0 and the magnetic field constant μ_0.

It is easy to see that Maxwell's equations are *not* invariant under Galilei transformations. Let us consider a plane electromagnetic wave propagating in the x-direction. In an inertial system moving with $\boldsymbol{r}' = \boldsymbol{r} - v\,\boldsymbol{e}_x t$, the wave would propagate with a velocity $c' = c - v \neq c$. Therefore, the wave with velocity c' cannot be a solution of Maxwell's equations.

The statement that the speed of light has the same value c in all inertial systems was the inescapable conclusion of the experiments discussed in Chap. 2, and of many others. Maxwell's equations are also valid in any inertial system. Then there can be only one consequence: In electrodynamics the transition from one inertial system to another is not described by Galilei transformations!

Here we have arrived at a problematic point: If in classical mechanics an observer moves towards a source of sound with half the velocity of sound, then he observes, after a Galilei transformation, a velocity of sound of $v_{\text{total}} = 3v_S/2$, in comparison with an observer at rest. Since the speed of light is constant, the same cannot be true for a source of light and an observer moving towards it with half the speed of light, see Fig. 3.1. It feels hard to accept that the transformation behavior in classical mechanics should be different from that in electrodynamics.

We can resolve this contradiction if we assume that the Galilei transformation is only the approximation of that transformation which transforms the equations of electrodynamics. This transformation is called the *Lorentz transformation*.[2]

---

[2] Hendrik Antoon Lorentz, 1853–1928, Dutch mathematician and physicist. Besides the Lorentz transformation, the Lorentz force of electrodynamics is also named after him.

It is mainly Einstein's merit to realize that the Lorentz transformation is not restricted to electrodynamics, but describes a general property of space and time. The familiar properties of space and time related to the Galilei transformation, e.g., the existence of an absolute time, are no longer present in the theory of relativity. Obviously, one feature of the Lorentz transformation must be that velocities cannot add up to a total velocity exceeding the speed of light. This will be possible only if length and time scales can differ from one inertial system to another.

3.2   Motivating the Lorentz Transformation

The invariance of Maxwell's equations under Lorentz transformations was already known prior to Einstein's paper "Zur Elektrodynamik bewegter Körper"[3] [1].

We will now motivate a derivation of the Lorentz transformation. Our starting point is a specific Galilei transformation from an inertial system S to a system S' with $D = 1$, $v = v e_x$, $r_0 = 0$, $t_0 = 0$, viz.

$$x' = x - vt, \quad y' = y, \quad z' = z, \quad t' = t. \tag{3.4}$$

We consider a ray of light starting at the spacetime origin, $r = 0, t = 0$. It moves at the speed of light, and after a time t reaches the point (x, y, z) with $\sqrt{x^2 + y^2 + z^2} = ct$. Since the speed of light is supposed to be constant in all systems, we must also have $\sqrt{x'^2 + y'^2 + z'^2} = ct'$, where (x', y', z') and t' describe the same point in the spacetime, but in other coordinates. Squaring the expressions and rearranging terms yields the two relations

$$x^2 + y^2 + z^2 - c^2 t^2 = 0 \quad \text{and} \quad x'^2 + y'^2 + z'^2 - c^2 t'^2 = 0, \tag{3.5}$$

where the first equation is trivial, while the second is a requirement on the transformation. Inserting the Galilei transformation (3.4) in this equation leads to

$$\begin{aligned}
x'^2 + y'^2 + z'^2 - c^2 t'^2 &= (x - vt)^2 + y^2 + z^2 - c^2 t^2 \\
&= x^2 + y^2 + z^2 - c^2 t^2 - 2xvt + v^2 t^2 \\
&= -2xvt + v^2 t^2 \neq 0.
\end{aligned} \tag{3.6}$$

The Galilei transformation does not fulfill the requirement of a constant speed of light. To eliminate the residual term $-2xvt + v^2 t^2$ by a transformation of time, we will test two approaches.

---

[3] "On the electrodynamics of moving bodies".

In the first approach we allow a simple displacement α of time, i.e., instead of (3.4) we have

$$x' = x - vt, \quad y' = y, \quad z' = z, \quad t' = t - \alpha. \tag{3.7}$$

We insert this and obtain instead of (3.6)

$$x'^2 + y'^2 + z'^2 - c^2 t'^2 = x^2 + y^2 + z^2 - c^2 t^2 - 2xvt + v^2 t^2 + 2\alpha t c^2 - c^2 \alpha^2$$
$$= \left(1 - \frac{v^2}{c^2}\right) x^2 + y^2 + z^2 - \left(1 - \frac{v^2}{c^2}\right) c^2 t^2. \tag{3.8}$$

In the second line we have set $\alpha = xv/c^2$. Now factors $\left(1 - v^2/c^2\right)$ have appeared which we also have to get rid of. By the above choice for α we have achieved that x^2 and $c^2 t^2$ have the same prefactors. This helps us to find an improved approach.

This approach is

$$x' = \frac{1}{\sqrt{1 - \frac{v^2}{c^2}}} (x - vt), \quad y' = y, \quad z' = z, \quad t' = \frac{1}{\sqrt{1 - \frac{v^2}{c^2}}} \left(t - \frac{xv}{c^2}\right), \tag{3.9}$$

which leads to the desired result,

$$x'^2 + y'^2 + z'^2 - c^2 t'^2 = x^2 + y^2 + z^2 - c^2 t^2. \tag{3.10}$$

We see that Lorentz transformations leave the quantity

$$s^2 = -c^2 t^2 + x^2 + y^2 + z^2 \tag{3.11}$$

invariant, i.e., $s^2 = s'^2$. The choice of the sign of s^2 is purely arbitrary, and it could be chosen differently. We follow the sign convention of Misner,[4] Thorne[5] and Wheeler[6] [3].

If there were only plus signs in (3.11) we would have

$$\|\boldsymbol{x}'\|^2 = \|\boldsymbol{x}\|^2, \tag{3.12}$$

---

[4] Charles William Misner, 1932–2023, US-American theoretical physicist. Student of John Archibald Wheeler.

[5] Kip Stephen Thorne, ⋆1940, US-American theoretical physicist. Student of John Archibald Wheeler. Nobel Prize in Physics 2017 for his contributions to the detection of gravitational waves.

[6] John Archibald Wheeler, 1911–2008, US-American theoretical physicist.

i.e., the Lorentz transformation could be interpreted as a rotation matrix which conserves the norm of a vector, just like in *Euclidean geometry*.[7] In fact, in a certain sense, we can consider Lorentz transformations as rotations, but with an imaginary angle for the time coordinate, see Sect. 5.1.2.

In (3.9) we have introduced the specific Lorentz transformation for velocities $v = v \, \mathbf{e}_x$ in the x-direction. Here, specific means that the transformation is into a system which moves parallel to a coordinate axis and is not rotated relative to the original system at rest.

Why do we not feel the effects of the Lorentz transformation in our everyday life? If we are very generous with regard to velocities, we can choose as the maximum velocity which we are confronted with on Earth the second cosmic velocity $v \approx 11 \, \text{km} \, \text{s}^{-1}$, which a rocket needs to escape the gravitational field of the Earth, see Table 1.1. For our daily routine this is an extremely high velocity, yet the ratio is $v/c \approx 4 \cdot 10^{-5}$ and thus $1/\sqrt{1 - v^2/c^2} \approx 1 + 7 \cdot 10^{-10}$! The velocities in our everyday experience are way too low to recognize the difference between the Lorentz and the Galilei transformation. On the other hand, we also see that

$$\lim_{v \to c} \frac{1}{\sqrt{1 - v^2/c^2}} = \infty. \tag{3.13}$$

Thus, there exists no Lorentz transformation to an inertial system moving at the speed of light. The ultimate conclusion is that a photon, which does move at the speed of light, possesses no rest frame; as we know, in any system light moves at exactly the speed of light.

The invariance of Maxwell's equations is not evident from the notation (3.3). But the equations can be brought into a mathematically very elegant form, which clearly reveals their Lorentz invariance. We will see this in Chap. 7.

3.3   Matrix Representation

In special relativity the speed of light serves as an important point of reference for velocities, no velocity can exceed it. It is therefore expedient to measure velocities in units of the speed of light. For this reason we introduce the velocity parameter

$$\beta \equiv \frac{v}{c} \quad \text{with} \quad 0 \le |\beta| < 1. \tag{3.14}$$

The notation of the Lorentz transformation (3.9) becomes even more compact when we introduce a symbol of its own for the square root expression:

$$\gamma \equiv \frac{1}{\sqrt{1 - v^2/c^2}} = \frac{1}{\sqrt{1 - \beta^2}}. \tag{3.15}$$

---

[7] Euclid of Alexandria lived around 300 BC. His opus magnum "Elements" is the main source of ancient geometry, and was used as a textbook even in later times.

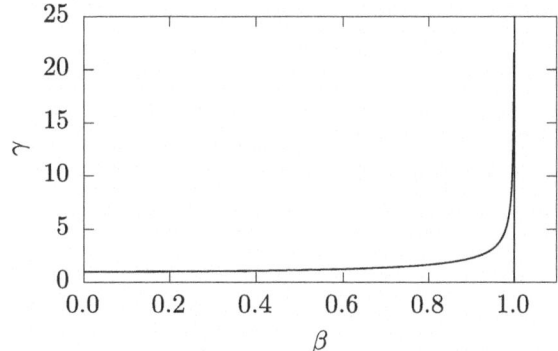

Fig. 3.2 For velocities $\beta \ll 1$, the γ factor has values close to one, but it diverges for $\beta \to 1$

It is called the *Lorentz factor*. For $\beta \to 1$ it diverges as $\gamma \to \infty$, see Fig. 3.2. Moreover, it is no longer reasonable to use separate notations for the spatial coordinates and the time coordinate, since they are connected by the Lorentz transformation. Therefore, we combine position and time in 4-dimensional *spacetime* and write the coordinates as

$$x^\mu = (x^0, x^1, x^2, x^3) \equiv (ct, x, y, z) \in \mathbb{R}^{1,3}. \qquad (3.16)$$

The quantity x^μ is called a *four-vector*, with $\mu \in \{0, 1, 2, 3\}$. Note that the time coordinate has the index 0. We have multiplied the time by the speed of light, therefore all coordinates have the same dimension length. From now on Greek indices always denote four-vectors, or components of four-vectors, depending on the context. When we shall refer only to the spatial coordinates of a four-vector, we shall use Latin indices, x^i with $i \in \{1, 2, 3\}$. Detailed information on why we write μ as an upper index, and why the set of all spacetime points is in $\mathbb{R}^{1,3}$, and not simply in $\mathbb{R}^4$, will be given in Chap. 5.

In the new notation, the specific Lorentz transformation (3.9) reads

$$x'^0 = \gamma(x^0 - \beta x^1), \quad x'^1 = \gamma(x^1 - \beta x^0), \quad x'^2 = x^2, \quad x'^3 = x^3. \qquad (3.17)$$

As the Lorentz transformation is a linear mapping of the spacetime coordinates, we can write (3.17) in the form of a simple matrix-vector multiplication

$$\begin{pmatrix} x'^0 \\ x'^1 \\ x'^2 \\ x'^3 \end{pmatrix} = \begin{pmatrix} \gamma & -\beta\gamma & 0 & 0 \\ -\beta\gamma & \gamma & 0 & 0 \\ 0 & 0 & 1 & 0 \\ 0 & 0 & 0 & 1 \end{pmatrix} \cdot \begin{pmatrix} x^0 \\ x^1 \\ x^2 \\ x^3 \end{pmatrix}, \qquad (3.18)$$

or, in shorthand notation,

$$\underline{x}' = \Lambda \cdot \underline{x}. \qquad (3.19)$$

We shall use the capital Greek letter Λ (Lambda) for the matrices of Lorentz transformations. The four-vector $\underline{x}$ has the components x^μ from (3.16).

3.3.1 Special Lorentz Transformations

Equation (3.9) is not the only possible Lorentz transformation. It connects two
inertial systems one of which moves with velocity β in x-direction relative to the
other system. Transformations into a system which moves with a certain velocity
relative to the current inertial system are called *Lorentz boosts*. Other important
Lorentz transformations are the boosts Λ_x, Λ_y, and Λ_z in x-, y-, and z-direction,
plus pure rotations Λ_{D_i}:

$$\Lambda_x = \begin{pmatrix} \gamma & -\beta\gamma & 0 & 0 \\ -\beta\gamma & \gamma & 0 & 0 \\ 0 & 0 & 1 & 0 \\ 0 & 0 & 0 & 1 \end{pmatrix}, \quad \Lambda_y = \begin{pmatrix} \gamma & 0 & -\beta\gamma & 0 \\ 0 & 1 & 0 & 0 \\ -\beta\gamma & 0 & \gamma & 0 \\ 0 & 0 & 0 & 1 \end{pmatrix},$$

$$\Lambda_z = \begin{pmatrix} \gamma & 0 & 0 & -\beta\gamma \\ 0 & 1 & 0 & 0 \\ 0 & 0 & 1 & 0 \\ -\beta\gamma & 0 & 0 & \gamma \end{pmatrix}, \quad \Lambda_{D_i} = \begin{pmatrix} 1 & 0 & 0 & 0 \\ 0 & & & \\ 0 & & D_i & \\ 0 & & & \end{pmatrix},$$

$$(3.20)$$

with a rotation matrix D_i, which fulfills $D_i^T D_i = D_i D_i^T = 1$. The standard (3×3)-
matrices of rotations about the x-, y-, and z-axis read

$$D_x(\alpha) = \begin{pmatrix} 1 & 0 & 0 \\ 0 & \cos(\alpha) & -\sin(\alpha) \\ 0 & \sin(\alpha) & \cos(\alpha) \end{pmatrix}, \quad D_y(\alpha) = \begin{pmatrix} \cos(\alpha) & 0 & \sin(\alpha) \\ 0 & 1 & 0 \\ -\sin(\alpha) & 0 & \cos(\alpha) \end{pmatrix},$$

$$D_z(\alpha) = \begin{pmatrix} \cos(\alpha) & -\sin(\alpha) & 0 \\ \sin(\alpha) & \cos(\alpha) & 0 \\ 0 & 0 & 1 \end{pmatrix}.$$

$$(3.21)$$

The determinants of the individual boosts and of pure rotations all have the value
one. The inverse matrices are obtained by interchanging the signs of the velocity,
$\beta \mapsto -\beta$, or of the angles $\alpha \mapsto -\alpha$, respectively.

3.3.2 Comparison with the Galilei Transformation

The form (3.2) of the Galilei transformation is not well suited to recognize its similarity to Lorentz transformations. But Galilei transformations without displacement
of the origin can also be written in matrix form. The Galilei transformation

$$G_x = \begin{pmatrix} 1 & 0 & 0 & 0 \\ -\beta & 1 & 0 & 0 \\ 0 & 0 & 1 & 0 \\ 0 & 0 & 0 & 1 \end{pmatrix}.$$

$$(3.22)$$

corresponds to the boost in x-direction in (3.20). In fact, in the limit of small velocities, $\beta \ll 1$, we have $\beta\gamma \approx \beta$. However, for the Galilei transformation the "boost" is not a symmetric matrix, since the coupling between time and space components is neglected. In that sense, the Galilei transformation is *not* the rigorous mathematical limit of the Lorentz transformation.

3.3.3   Rapidity

It is also possible to express the Lorentz transformation using the *rapidity* θ, which is defined by the relation

$$\tanh(\theta) = \beta \quad \text{or} \quad \theta = \operatorname{artanh}(\beta), \tag{3.23}$$

see Fig. 3.3. The point of this definition becomes apparent when we consider that these relations follow from (3.23):

$$\cosh(\theta) = \frac{1}{\sqrt{1-\beta^2}} = \gamma \quad \text{and} \quad \sinh(\theta) = \frac{\beta}{\sqrt{1-\beta^2}} = \gamma\beta. \tag{3.24}$$

Using the rapidity we can write the Lorentz boost Λ_x in the form

$$\Lambda_x = \begin{pmatrix} \cosh(\theta) & -\sinh(\theta) & 0 & 0 \\ -\sinh(\theta) & \cosh(\theta) & 0 & 0 \\ 0 & 0 & 1 & 0 \\ 0 & 0 & 0 & 1 \end{pmatrix}. \tag{3.25}$$

For comparison: In three-dimensional Euclidean space, the elementary matrix of a rotation about the z-axis is given by (3.21) (where we have chosen the anti-clockwise direction of rotation). The inverse of the rotation matrix follows directly from rotating back by the same angle, $D_z^{-1}(\varphi) = D_z(-\varphi)$. One recognizes the

Fig. 3.3 In contrast to the velocity, for which $\beta \in (-1, 1)$, the rapidity is an unbounded function

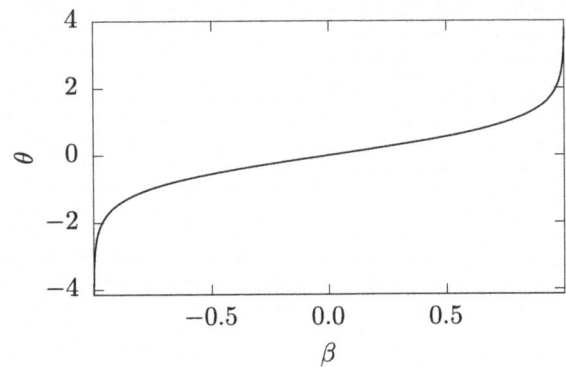

great similarity between the rotation matrix and the boost (3.25). The similarity becomes even more apparent when we remember how the trigonometric functions "sin" and "cos" are related to the hyperbolic functions "sinh" and "cosh"; for example $\cosh(\mathrm{i}x) = \cos(x)$. Thus, Lorentz boosts are rotations by a *hyperbolic angle* in $\mathbb{R}^{1,3}$ or *Minkowski space*, see Sect. 5.1 for the introduction of this term.

3.3.4   Composition of Lorentz Transformations

Of course Lorentz transformations can be performed one after the other. As an example we consider the derivation of the Lorentz boost into a system S', moving with an arbitrary velocity $\boldsymbol{v} = (v_x, v_y, v_z)^{\mathrm{T}}$, or $\boldsymbol{\beta} = (\beta_x, \beta_y, \beta_z)^{\mathrm{T}}$, relative to a system S. The axes of the two systems are assumed to remain parallel. The basic idea of the calculation is to first rotate the system S in such a way that the rotated x-axis points into the direction of the velocity vector $\boldsymbol{\beta}$. Then a boost is performed into the x-direction, and finally the system is rotated back again. The first step is a rotation by an angle φ_1 around the z-axis, followed by a rotation around the resulting y-axis by an angle φ_2, see Fig. 3.4. With the definition

$$\beta_\perp = \sqrt{\beta_x^2 + \beta_y^2} \tag{3.26}$$

we find the relations between the angles of rotation and the velocity components

$$\sin(\varphi_1) = \frac{\beta_y}{\beta_\perp}, \quad \cos(\varphi_1) = \frac{\beta_x}{\beta_\perp}, \tag{3.27a}$$

and

$$\sin(\varphi_2) = \frac{\beta_z}{\beta}, \quad \cos(\varphi_2) = \frac{\beta_\perp}{\beta}. \tag{3.27b}$$

Fig. 3.4 Definition of the velocity components β_x, β_y, and β_z and their relation to the angles of rotation φ_1 and φ_2

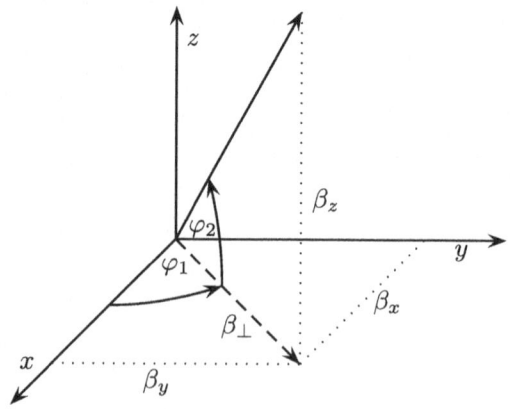

From these relations, we immediately obtain the rotation matrices

$$
\boldsymbol{\Lambda}_{D_z^{-1}} = \begin{pmatrix} 1 & 0 & 0 & 0 \\ 0 & \frac{\beta_x}{\beta_\perp} & \frac{\beta_y}{\beta_\perp} & 0 \\ 0 & -\frac{\beta_y}{\beta_\perp} & \frac{\beta_x}{\beta_\perp} & 0 \\ 0 & 0 & 0 & 1 \end{pmatrix} \quad \text{and} \quad \boldsymbol{\Lambda}_{D_y^{-1}} = \begin{pmatrix} 1 & 0 & 0 & 0 \\ 0 & \frac{\beta_\perp}{\beta} & 0 & \frac{\beta_z}{\beta} \\ 0 & 0 & 1 & 0 \\ 0 & -\frac{\beta_z}{\beta} & 0 & \frac{\beta_\perp}{\beta} \end{pmatrix}.
\tag{3.28}
$$

Note that here the mathematically anti-clockwise direction of rotation D_z must be reversed, since the velocity vector $\boldsymbol{\beta}$ has to be rotated back to the x-axis.

Then the general Lorentz boost is given by

$$
\boldsymbol{\Lambda}_B = \boldsymbol{\Lambda}_R^T \cdot \boldsymbol{\Lambda}_x \cdot \boldsymbol{\Lambda}_R,
\tag{3.29}
$$

with the resulting total rotation matrix

$$
\boldsymbol{\Lambda}_R = \boldsymbol{\Lambda}_{D_y^{-1}} \cdot \boldsymbol{\Lambda}_{D_z^{-1}} = \begin{pmatrix} 1 & 0 & 0 & 0 \\ 0 & \frac{\beta_x}{\beta} & \frac{\beta_y}{\beta} & \frac{\beta_z}{\beta} \\ 0 & -\frac{\beta_y}{\beta_\perp} & \frac{\beta_x}{\beta_\perp} & 0 \\ 0 & -\frac{\beta_x\beta_z}{\beta\beta_\perp} & -\frac{\beta_y\beta_z}{\beta\beta_\perp} & \frac{\beta_\perp}{\beta} \end{pmatrix}.
\tag{3.30}
$$

Finally, we find the explicit form for this Lorentz transformation:

$$
\boldsymbol{\Lambda}_B = \begin{pmatrix} \gamma & -\gamma\beta_x & -\gamma\beta_y & -\gamma\beta_z \\ -\gamma\beta_x \\ -\gamma\beta_y & \frac{\beta_i\beta_j(\gamma-1)}{\beta^2} + \delta_{ij} \\ -\gamma\beta_z \end{pmatrix} = \begin{pmatrix} \gamma & -\gamma\beta_x & -\gamma\beta_y & -\gamma\beta_z \\ -\gamma\beta_x \\ -\gamma\beta_y & \frac{\gamma^2}{\gamma+1}\beta_i\beta_j + \delta_{ij} \\ -\gamma\beta_z \end{pmatrix}.
\tag{3.31}
$$

The matrix belonging to the general Lorentz boost (3.31) is symmetric, accordingly every boost is represented by a symmetric matrix. This is not true for Lorentz transformation that also contain rotations, since rotations are not described by symmetric matrices.

3.4 Composition Law of Velocities

Let us consider three coordinate systems S, S' and S''. S' moves relative to S with velocity β_1 parallel to the x-axis, and S'' moves relative zu S' with velocity β_2 parallel to the x'-axis. What is the velocity of S'' relative to S?

For Galilei transformations we just would have to add the velocities, viz. $\beta_3 = \beta_1 + \beta_2$. In special relativity, however, we must work with Lorentz transformations, and we have $\Lambda_3 = \Lambda_2 \cdot \Lambda_1$. From the form of the boosts (3.20) in x-direction follows, for the relevant $(ct\text{-}x)$-components.

$$
\begin{aligned}
\Lambda_3 &= \begin{pmatrix} \gamma_2 & -\beta_2\gamma_2 \\ -\beta_2\gamma_2 & \gamma_2 \end{pmatrix} \cdot \begin{pmatrix} \gamma_1 & -\beta_1\gamma_1 \\ -\beta_1\gamma_1 & \gamma_1 \end{pmatrix} \\
&= \begin{pmatrix} \gamma_1\gamma_2(1+\beta_1\beta_2) & -\gamma_1\gamma_2(\beta_1+\beta_2) \\ -\gamma_1\gamma_2(\beta_1+\beta_2) & \gamma_1\gamma_2(1+\beta_1\beta_2) \end{pmatrix}.
\end{aligned}
\tag{3.32}
$$

The fact that Λ_3 is a symmetric matrix suggests that the composition of the two boosts yields another boost. From (3.32) we obtain the equations

$$
\gamma_3 = \gamma_1\gamma_2(1+\beta_1\beta_2),
\tag{3.33a}
$$

$$
\beta_3\gamma_3 = \gamma_1\gamma_2(\beta_1+\beta_2).
\tag{3.33b}
$$

With the definition of γ in (3.15), Eq. (3.33a) yields

$$
\gamma_3 = \frac{1}{\sqrt{1-\beta_3^2}} = \frac{1+\beta_1\beta_2}{\sqrt{(1-\beta_1^2)(1-\beta_2^2)}}.
\tag{3.34}
$$

We rearrange the right-hand side in accordance with the left-hand side via

$$
\frac{1+\beta_1\beta_2}{\sqrt{(1-\beta_1^2)(1-\beta_2^2)}} = \frac{1}{\sqrt{\frac{(1-\beta_1^2)(1-\beta_2^2)}{(1+\beta_1\beta_2)^2}}} = \frac{1}{\sqrt{1-\frac{(\beta_1+\beta_2)^2}{(1+\beta_1\beta_2)^2}}},
\tag{3.35}
$$

therefore

$$
\gamma_3 = \frac{1}{\sqrt{1-\left(\frac{\beta_1+\beta_2}{1+\beta_1\beta_2}\right)^2}}.
\tag{3.36}
$$

Thus, the velocity of S'' relative to S is

$$
\beta_3 = \frac{\beta_1+\beta_2}{1+\beta_1\beta_2}, \quad \text{or} \quad v_3 = \frac{v_1+v_2}{1+(v_1 v_2)/c^2}.
\tag{3.37}
$$

By inserting into (3.33b) we can easily check that the value found for β_3 also fulfills that equation. Equation (3.37) is called the *composition law of velocities* of special relativity for the case of parallel velocities. The law is also called the *velocity-addition formula*.

Setting, e.g., $v_1 < c$ and $v_2 = c$, we obtain

$$
v_3 = \frac{v_1+c}{1+\frac{v_1 c}{c^2}} = \frac{v_1+c}{\frac{v_1+c}{c}} = c.
\tag{3.38}
$$

Fig. 3.5 Addition of two equal velocities β_1 and β_2, in special relativity, and according to the Galilei transformation

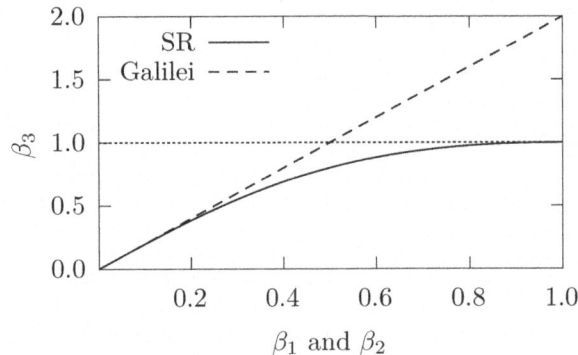

Thus, the maximum value of the velocity v_3 is equal to the speed of light, as it must be. Figure 3.5 shows the result of the addition of two equal velocities β_1 and β_2 according to the Galilei transformation, and the Lorentz transformation. Of course, for small velocities both transformations yield the same result.

3.4.1 Addition of Velocities and Rapidity

In (3.23) we introduced the rapidity $\theta = \mathrm{artanh}(\beta)$ which allows formulating Lorentz boosts in a way analogous to rotation matrices, with the hyperbolic angle θ. For the composition of two rotations, e.g., around the z-axis with angles of rotation φ_1 and φ_2 we have

$$\boldsymbol{D}_z(\varphi_2) \cdot \boldsymbol{D}_z(\varphi_1) = \boldsymbol{D}_z(\varphi_1 + \varphi_2). \tag{3.39}$$

The angles of rotation are simply added. Similarly, we have for the rapidity

$$\boldsymbol{\Lambda}(\theta_1) \cdot \boldsymbol{\Lambda}(\theta_2) = \boldsymbol{\Lambda}(\theta_1 + \theta_2), \tag{3.40}$$

as is to be shown in Exercise 3.3. Thus, in the composition of two parallel boosts the rapidity of the total boost is just the sum of the rapidities of the two individual boosts. This clearly demonstrates the analogy with rotations.

3.4.2 Addition of Non-parallel Velocities

In Galilei transformations, the addition of non-parallel velocities is simply done by adding the velocity vectors, $v_3 = v_1 + v_2$.

We now study the addition of non-parallel velocities in special relativity. We shall restrict ourselves to the special case of a combination of a boost in the x-direction, with a subsequent boost in the y-direction. We closely follow the discussion of Ferraro and Thibeault [2].

The Lorentz matrices of the two boosts Λ_x and Λ_y are

$$\Lambda_x = \begin{pmatrix} \gamma_x & -\gamma_x\beta_x & 0 & 0 \\ -\gamma_x\beta_x & \gamma_x & 0 & 0 \\ 0 & 0 & 1 & 0 \\ 0 & 0 & 0 & 1 \end{pmatrix} \quad \text{and} \quad \Lambda_y = \begin{pmatrix} \gamma_y & 0 & -\gamma_y\beta_y & 0 \\ 0 & 1 & 0 & 0 \\ -\gamma_y\beta_y & 0 & \gamma_y & 0 \\ 0 & 0 & 0 & 1 \end{pmatrix}, \tag{3.41}$$

where $\gamma_x = 1/\sqrt{1-\beta_x^2}$ and $\gamma_y = 1/\sqrt{1-\beta_y^2}$. The composition of the two boosts yields

$$\Lambda_y \cdot \Lambda_x = \begin{pmatrix} \gamma_y\gamma_x & -\gamma_y\gamma_x\beta_x & -\gamma_y\beta_y & 0 \\ -\gamma_x\beta_x & \gamma_x & 0 & 0 \\ -\gamma_x\gamma_y\beta_y & \gamma_x\beta_x\gamma_y\beta_y & \gamma_y & 0 \\ 0 & 0 & 0 & 1 \end{pmatrix}. \tag{3.42}$$

This is not a symmetric matrix, and therefore cannot be written as a proper boost, since, as we know, according to (3.31) every boost can be represented by a symmetric matrix. The trick is to introduce an additional rotation to recover a proper boost transformation. To find the corresponding angle of rotation ϑ, we try the approach

$$\Lambda_{\text{Boost}} = \Lambda_{D_z} \cdot \Lambda_y \cdot \Lambda_x, \tag{3.43}$$

with the rotation

$$\Lambda_{D_z} = \begin{pmatrix} 1 & 0 & 0 & 0 \\ 0 & \cos(\vartheta) & -\sin(\vartheta) & 0 \\ 0 & \sin(\vartheta) & \cos(\vartheta) & 0 \\ 0 & 0 & 0 & 1 \end{pmatrix}, \tag{3.44}$$

In this general matrix we have to determine the angle ϑ in such a way that Λ_{Boost} becomes symmetric. The calculation yields

$$\Lambda_{\text{Boost}} =$$

$$\begin{pmatrix} \gamma_y\gamma_x & -\gamma_y\gamma_x\beta_x & -\gamma_y\beta_y & 0 \\ -\gamma_x\beta_x\cos(\vartheta) + \gamma_y\gamma_x\beta_y\sin(\vartheta) & \gamma_x\cos(\vartheta) - \gamma_y\gamma_x\beta_y\beta_x\sin(\vartheta) & -\gamma_y\sin(\vartheta) & 0 \\ -\gamma_x\beta_x\sin(\vartheta) - \gamma_x\gamma_y\beta_y\cos(\vartheta) & \gamma_x\sin(\vartheta) + \gamma_x\gamma_y\beta_x\beta_y\cos(\vartheta) & \gamma_y\cos(\vartheta) & 0 \\ 0 & 0 & 0 & 1 \end{pmatrix}. \tag{3.45}$$

The condition that this be a symmetric matrix leads to the three equations:

$$\cos(\vartheta) = \frac{\gamma_y (\beta_y \sin(\vartheta) + \beta_x)}{\beta_x},$$ (3.46a)

from the comparison of the 01- and the 10-component,

$$\cos(\vartheta) = \frac{\gamma_y \beta_y - \gamma_x \beta_x \sin(\vartheta)}{\gamma_y \gamma_x \beta_y}$$ (3.46b)

from the comparison of the 02- and the 20-component, and

$$\tan(\vartheta) = -\frac{\gamma_x \beta_x \gamma_y \beta_y}{\gamma_x + \gamma_y}$$ (3.46c)

from the comparison of the 12- and the 21-component.

The conditions (3.46a) and (3.46b) lead to

$$\sin(\vartheta) = \frac{(1 - \gamma_x \gamma_y)\gamma_y \beta_x \beta_y}{\gamma_x (\beta_x^2 + \gamma_y^2 \beta_y^2)} = -\frac{\gamma_x \gamma_y \beta_x \beta_y}{\gamma_x \gamma_y + 1}.$$ (3.47)

In the second step we have expanded the fraction by γ_x and used the relation $\beta^2 \gamma^2 = \gamma^2 - 1$, and the third binomial formula. With $\cos(\vartheta) = \sqrt{1 - \sin^2(\vartheta)}$ for $\vartheta \in [0, \pi/2]$ we obtain

$$\cos(\vartheta) = \frac{\gamma_x + \gamma_y}{\gamma_x \gamma_y + 1}.$$ (3.48)

Using the identity $\sin(x) = \tan(x)/\sqrt{1 + \tan^2(x)}$, it is possible to show that (3.46c) also leads to (3.48).

Inserting (3.47) and (3.48) in (3.45) finally leads to

$$\Lambda_{\text{Boost}} = \begin{pmatrix} \gamma_y \gamma_x & -\gamma_y \gamma_x \beta_x & -\gamma_y \beta_y & 0 \\ -\gamma_y \gamma_x \beta_x \left(1 + \frac{\gamma_x^2 \beta_x^2 \gamma_y^2}{\gamma_x \gamma_y + 1}\right) & \frac{\gamma_x \beta_x \gamma_y^2 \beta_y}{\gamma_x \gamma_y + 1} & 0 \\ -\gamma_y \beta_y & \frac{\gamma_x \beta_x \gamma_y^2 \beta_y}{\gamma_x \gamma_y + 1} & \frac{\gamma_y (\gamma_x + \gamma_y)}{\gamma_x \gamma_y + 1} & 0 \\ 0 & 0 & 0 & 1 \end{pmatrix}.$$ (3.49)

This matrix is symmetric, as required, and therefore represents a proper boost. By comparing with the expression (3.31) for the general boost we can easily find the direction and the modulus of the corresponding velocity vector. We find the relations

$$\Lambda_{00} = \gamma, \quad \Lambda_{01} = -\gamma \beta_x, \quad \Lambda_{02} = -\gamma \beta_y, \quad \Lambda_{03} = -\gamma \beta_z,$$ (3.50)

where of course $\beta_z = 0$. The boost velocity is found from the comparison of the 00-components, viz.

$$\gamma = \gamma_x \gamma_y, \quad \text{or} \quad \beta = \sqrt{\beta_x^2 + \beta_y^2 - \beta_x^2 \beta_y^2}. \tag{3.51}$$

The two remaining equations yield

$$\beta = \begin{pmatrix} \beta_x \\ \beta_y / \gamma_x \\ 0 \end{pmatrix}. \tag{3.52}$$

The factor $1/\gamma_x$ in the y-component of the velocity distinguishes the result from the non-relativistic case.

To illustrate these results we consider the special case $\beta_x = \beta_y$, as we did in the addition of parallel velocities. The Galilei transformation would lead to the modulus of the velocity $\beta = \sqrt{2}\beta_x$. Instead, we find $\beta = \beta_x \sqrt{2 - \beta_x^2}$. Of course, again the total velocity cannot exceed the speed of light, $\beta = 1$. Figure 3.6 compares this result with that of parallel addition in special relativity, and with that of the Galilei transformation. It is not only the modulus of the velocity but also its direction which is different from the non-relativistic result. In the Galilei transformation the angle of the velocity vector relative to the x-axis is always $\varphi_{\text{Gal}} = \arctan(\beta_y/\beta_x) = \pi/4$. By contrast, in special relativity for large velocities β_x the limiting angle tends to

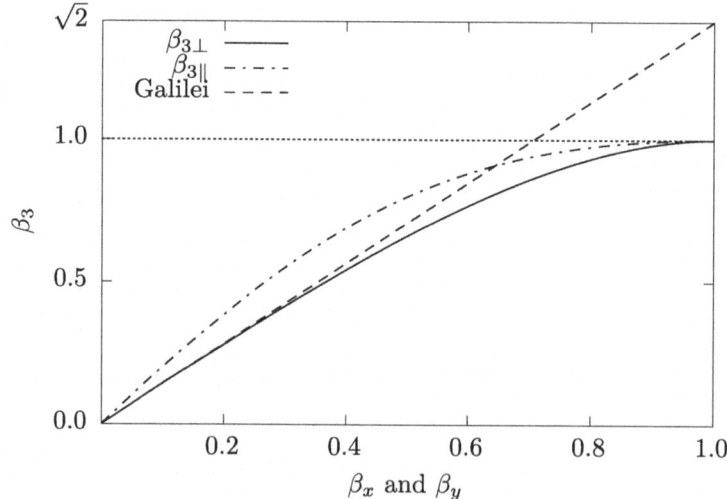

Fig. 3.6 Modulus of the total velocity resulting from the addition of two equal velocities β_x and β_y perpendicular to each other according to special relativity, and according to the Galilei transformation. The result for parallel addition is also shown (dash-dotted curve)

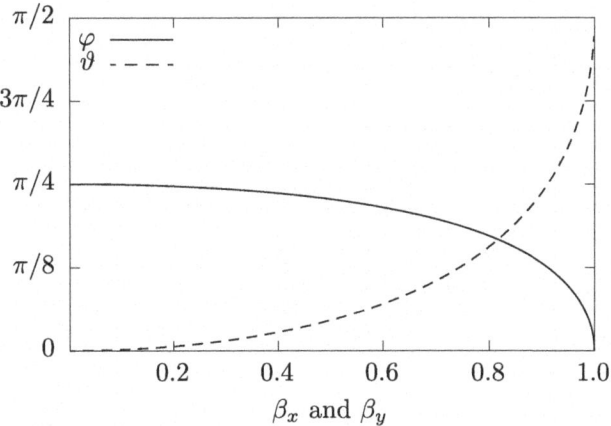

Fig. 3.7 Angle of rotation ϑ and direction of the velocity φ, when adding two equal velocities β_x and β_y, perpendicular to each other

$\varphi \to 0$, because of $1/\gamma_x \to 0$. The resulting velocity almost points in the direction of the x-axis.

At the same time, from (3.48) we find the angle of rotation

$$\vartheta = \arccos\left(\frac{2\gamma_x}{1 + \gamma_x^2}\right). \tag{3.53}$$

For large velocities the argument of the arccos tends to zero, i.e., in this limit the angle becomes $\vartheta \to \pi/2$: The resulting coordinate system is rotated by an angle of $90°$, see Fig. 3.7.

3.5 Minkowski Diagram

Usually the graphical representation of one-dimensional motions can be given by means of a space-time diagram, with time on the vertical axis, and space on the horizontal axis. A point in the space-time diagram designates an "event" which happens at a definite time and a definite place. The corresponding axis intercepts are called the *coordinates* of the event.

With regard to the theory of relativity, we must bear in mind that an event occurs of course independently of any coordinates. But in order to carry out calculations, we need to agree on a coordinate system in order to assign coordinates to the event.

The exact position of the origin of the coordinate system can be chosen arbitrarily, and any uniform movement of the coordinate system can be transformed away by a Galilei transformation, at least as long as we remain in Newtonian mechanics. However, in special relativity we have to pay attention also to the way an observer moves in spacetime.

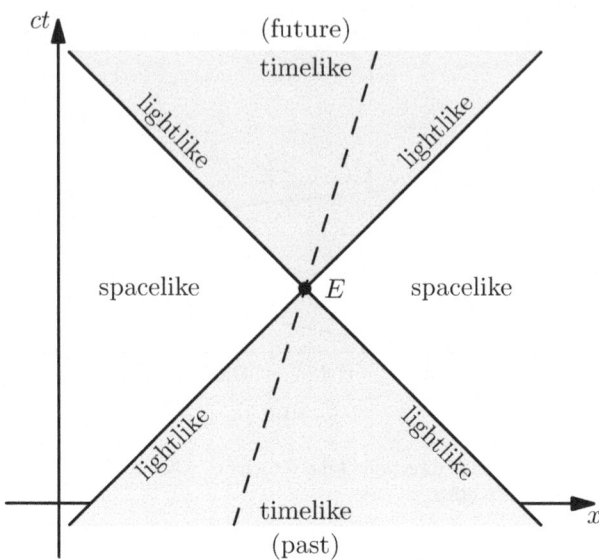

Fig. 3.8 Minkowski diagram with an event E and the corresponding light cones, which separate spacelike and timelike events. The *dashed line* shows the world line of an observer moving with constant velocity in the positive x-direction and passes through the event E

The *Minkowski diagram*[8] is an extension of the usual space-time diagram which allows for the description of how events are seen in frames of reference moving with a certain velocity relative to each other.

The transformation of the coordinates of the two systems is provided by the Lorentz transformation, which we now will depict in the Minkowski diagram. We will restrict ourselves to the $1 + 1D$-Minkowski diagram, which consists of one space axis and one time axis. The Lorentz transformation for a motion in the x-direction was

$$ct' = \gamma \, (ct - \beta x), \quad x' = \gamma \, (x - \beta ct), \quad y' = y, \quad z' = z. \qquad (3.54)$$

We multiply the time axis by the speed of light c and use the same length scale on the x- and the ct-axis. In this diagram, rays of light move along straight lines with slope $\pm 45°$. When we plot the straight lines corresponding to rays of light propagating from a certain event E into the positive and negative x-direction, the Minkowski diagram is divided into three different sectors which reflect the causal structure of spacetime, see Fig. 3.8.

The rays of light separating the different sectors are also called the *future* or *past light cones*, respectively. All events lying within the future light cone can in principle

[8] Hermann Minkowski, 1864–1909, German mathematician and physicist of Russian origin.

be affected causally by the event E. This can be realized either by a massive particle moving on a *timelike* trajectory with a velocity $v < c$, or by a ray of light moving on a *lightlike* trajectory. Correspondingly, all events in the past light cone can have a causal effect on the event E. All other events are causally decoupled from the event E, and are called *spacelike* with regard to E.

We now will find out how two inertial systems S and S', connected by a Lorentz transformation, can be plotted into the diagram. Let S be an inertial system at rest with the coordinate axes ct and x, and S' be the system in relative motion with respect to S. For an observer sitting in the origin of S' we trivially have $x' = \gamma(x - \beta ct) = 0$. Therefore, his timelike trajectory, also called his *world line*, is described by the equation $ct = x/\beta$. All events happening simultaneously in S' at time $t' = 0$ are defined by the relation $ct' = \gamma(ct - \beta x) = 0$, which leads us to the equation for the straight line $ct = \beta x$. Both straight lines provide the coordinate axes of the system S', where the angle

$$\psi = \arctan(\beta) \tag{3.55}$$

between the ct-axis and the ct'-axis is equal to the angle between the x-axis and the x'-axis, see Fig. 3.9.

The ticks along the coordinate axes can also be derived directly from the Lorentz transformation. For the ct'-axis we still have $x' = 0$, i.e., $ct = x/\beta$, and furthermore $ct' = \gamma(ct - \beta x) = n$ with $n \in \mathbb{N}$ (number of the tick). From this we can obtain the coordinates $(ct = n\gamma, x = n\beta\gamma)$ of the ticks relative to S. An analogous calculation yields the ticks along the x'-axis, $(ct = n\beta\gamma, x = n\gamma)$. The ticks also could have been obtained by intersecting the hyperbolas $s^2 = |ct^2 - x^2| = |ct'^2 - x'^2| = s'^2 = n^2$ with the coordinate axes.

Now we can determine the coordinates of a given event E both with respect to the system S and the system S'. The orthogonal projection of E onto the axes x and ct yields the coordinates in S. We obtain the axis intercepts referred to S' by a parallel shift of the x'- and ct'-axis through E. This construction is necessary since events at the same time and the same place must lie parallel to the corresponding axes.

3.6   Exercises

3.1. Galilei invariance of Newton's equations of motion Show that Newton's equations of motion (3.1) are invariant under Galilei transformations.

3.2. Velocity-addition formula and the Fizeau experiment In Sect. 2.3 we have learned about the experiment carried out by Fizeau, which disproved the ether-drag hypothesis. Show that the velocity-addition formula (3.37) reproduces Eq. (2.14).

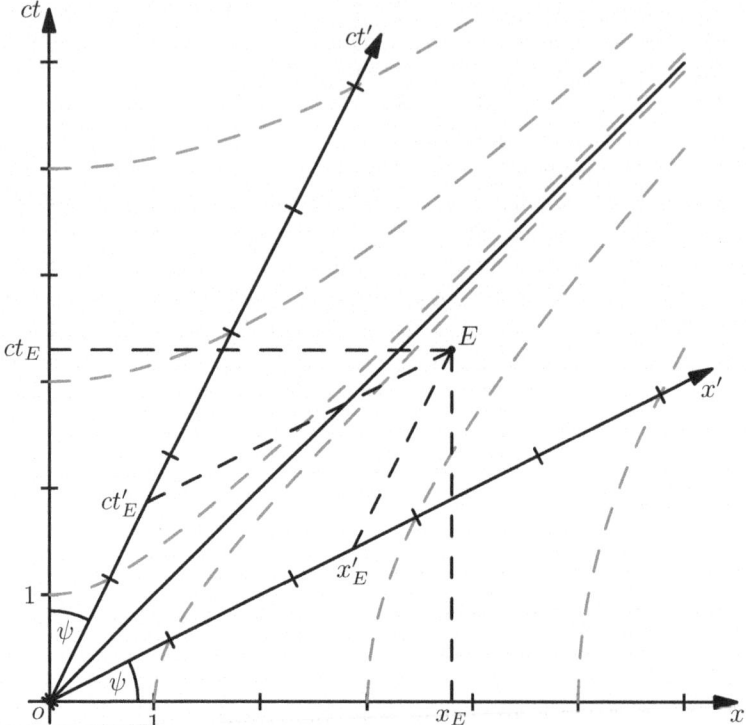

Fig. 3.9 Minkowski diagram for an inertial system S' moving with velocity $\beta = 0.5$ relative to S. The dashed hyperbolas represent the condition of invariance $s^2 = |ct^2 - x^2| = |ct'^2 - x'^2| = s'^2$. The angle bisector corresponds to a ray of light which is emitted from the observer in the origin at time $t = t' = 0$ into the positive x-direction

3.3. Composition law of velocities in terms of rapidity Prove the relation (3.40), which states that in the composition of two boosts in the same direction the rapidities of the boosts add up to the rapidity of the total boost.

Make use of the addition theorems for hyperbolic functions

$$\cosh(\theta_1)\cosh(\theta_2) + \sinh(\theta_1)\sinh(\theta_2) = \cosh(\theta_1 + \theta_2) \tag{3.56a}$$

$$\cosh(\theta_1)\sinh(\theta_2) + \sinh(\theta_1)\cosh(\theta_2) = \sinh(\theta_1 + \theta_2), \tag{3.56b}$$

and the definitions

$$\text{artanh}(x) = \frac{1}{2}\ln\left(\frac{1+x}{1-x}\right) \quad \text{and} \quad \tanh(x) = \frac{e^{2x} - 1}{e^{2x} + 1}. \tag{3.57}$$

References

1. Einstein, A.: Zur Elektrodynamik bewegter Körper (On the electrodynamics of moving bodies). Ann. Phys. **17**(10), 891–921 (1905); English translation for example: https://www.fourmilab.ch/etexts/einstein/specrel/www/
2. Ferraro, R., Thibeault, M.: Generic composition of boosts: an elementary derivation of the Wigner rotation. Eur. J. Phys. **20**(3), 143 (1999)
3. Misner, C.W., Thorne, K.S., Wheeler, J.A.: Gravitation. W.H. Freeman, New York (1973)

Physical Consequences of Lorentz Invariance

4

Contents

The Lorentz invariance of the physical laws leads to a variety of phenomena absent in non-relativistic physics, which is invariant only under Galileo transformations. Almost all the consequences of special relativity contradict our everyday experience, since they manifest themselves only at very high velocities. The contradiction to our everyday experience becomes obvious, in particular, in the formulation of apparent paradoxes, which we analyze and resolve in detail.

4.1 Loss of Simultaneity

All of us have a very clear and elementary understanding of what it means when we say two events happen "simultaneously". Today, practically every person possesses at least one accurate watch, and we only need to compare the clock times on the watches of two observers seeing one and the same event from different places in order to decide whether they saw the event simultaneously (at least within the measuring accuracy).

Fig. 4.1 Scheme of
synchronizing clocks. The
time on two clocks, at a
distance d, is set to $t = 0$,
when the two rays of light,
emitted in the middle at one
and the same moment, arrive
at the clocks

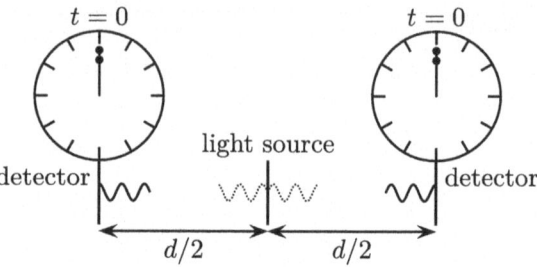

But how can we tell that two clocks are synchronous? Obviously this question is irrelevant to the accuracies needed in everyday life; all we have to do is to bring the clocks to one place and compare their clock times. However, on a more fundamental level this question is by no means trivial since to bring the two clocks to one place, at least one of them has to be moved out of its rest frame. Then its coordinates, including the time coordinate, transform according to a Lorentz transformation, and thus the transport may spoil the synchronicity.

One possibility for synchronizing two clocks at different places is shown in Fig. 4.1. The two clocks are at rest, with a fixed distance d. A device in the middle emits light rays simultaneously to the clock on the left and on the right. The time on the clocks is set to $t = 0$, when the light rays arrive. Note that radio-controlled clocks operate on this principle: They regularly compare their clock time with that of a central time signal transmitter. For receivers at different locations the inaccuracy caused by the different distances to the transmitter is usually negligible, but, in principle, could be corrected for. We study how the simultaneity of events is experienced in different frames of reference. We consider two systems S and S', and assume that S' moves along the x-axis of S with velocity $\beta > 0$. Two events E_1 ("clock 1 shows $t'_1 = 3\,\mathrm{s}$") and E_2 ("clock 2 shows $t'_2 = 3\,\mathrm{s}$") are supposed to happen simultaneously in S', but at different places. The spacetime coordinates of the events are

$$E_1 : \left(ct'_1, x'_1\right) \quad \text{and} \quad E_2 : \left(ct'_2, x'_2\right), \quad \text{with} \quad x'_1 \neq x'_2. \tag{4.1}$$

We are interested in the time coordinates of the two events with respect to the system S. We can calculate them using the inverse Lorentz transformation

$$ct = \gamma \left(ct' + \beta x'\right), \quad x = \gamma \left(x' + \beta ct'\right). \tag{4.2}$$

From the difference of the time coordinates in S

$$ct_2 - ct_1 = \gamma \left(ct'_2 + \beta x'_2 - ct'_1 - \beta x'_1\right) = \gamma \beta \left(x'_2 - x'_1\right) \tag{4.3}$$

follows $t_2 \neq t_1$, since $t'_1 = t'_2$ and $x'_1 \neq x'_2$. The two events do not happen simultaneously in S, see Fig. 4.2.

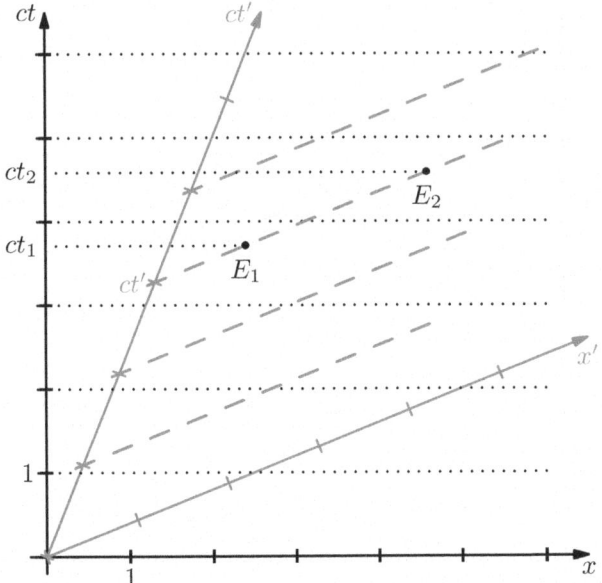

Fig. 4.2 The system S' moves with velocity $\beta = 0.5$ relative to S. Two events E_1 and E_2 which happen at the same time in S', but at different places, do *not* happen simultaneously in S. The *grey dashed lines* are the places of equal time in S', while the *black dotted lines* are places with equal time in S

If the notion of simultaneity of events depends on the choice of the inertial system, the chronological order of events could also differ from one inertial system to another. At first glance this may seem unphysical since an arbitrary chronological order of events appears to violate the causality principle of cause and effect. This seeming contradiction is resolved by realizing that the points are always separated by space-like intervals. Two events occurring at the same time but at different places in one and the same inertial system are not causally connected. However, the *chronological order* of time-like, possibly causally connected events, is the same in every inertial system, even though different time intervals between the events are possible.

4.2   Moving Rods: Lorentz Contraction

The effect of *Lorentz contraction* is closely related to the loss of simultaneity discussed in the previous section. Measuring the length of a rod means that we determine, *at a fixed time*, the coordinates of the endpoints of the rod, be it at rest or in motion, and calculate from this data the distance of the endpoints.

Let us consider a system S' moving along the x-axis of S with velocity $\beta > 0$. A rod of length $l' = |x'_B - x'_A|$ rests in S', and therefore moves with the velocity

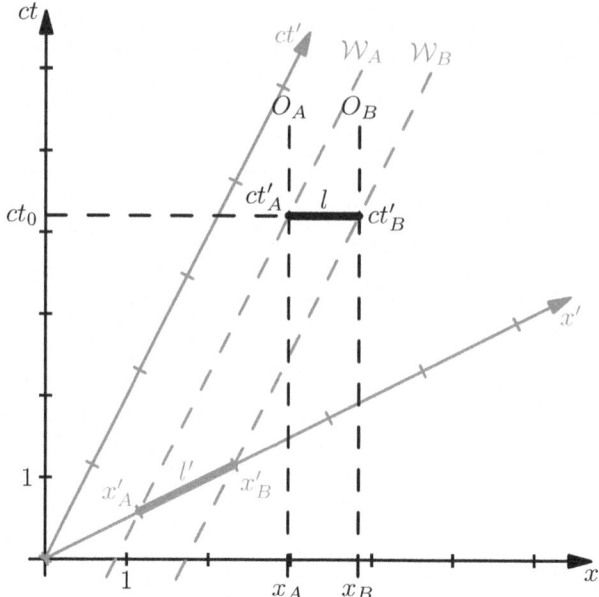

Fig. 4.3 Length contraction of a rod with length l, at rest in S'. Two observers O_A and O_B, at rest in S, measure the apparent length l of the rod at time ct_0. In the figure, the velocity of S' relative to S is $\beta = 0.5$

$\beta > 0$ relative to S. In S', its endpoints are described by the world lines $\mathcal{W}_A$ and $\mathcal{W}_B$, see Fig. 4.3. Two observers in the moving system S' measure the positions of the two endpoints at the fixed time $ct' = 0$. Referred to S, however, these two measuring events do not happen simultaneously.

For the measurement of the length in S we need two observers O_A and O_B and their coordinate time ct_0 at the moment the endpoints of the rod are passing by them. The spatial distance $l = |x_B - x_A|$ in S then obviously is what we would call the length of the rod in S.

To establish the relation between the rest length l' of the rod and the length l, we need the world lines $\mathcal{W}_A$ and $\mathcal{W}_B$ of the endpoints. We parametrize these world lines using the coordinate times t'_A and t'_B,

$$\mathcal{W}_A\colon ct_A = \gamma\left(ct'_A + \beta x'_A\right), \quad x_A = \gamma\left(x'_A + \beta ct'_A\right), \tag{4.4}$$

$$\mathcal{W}_B\colon ct_B = \gamma\left(ct'_B + \beta x'_B\right), \quad x_B = \gamma\left(x'_B + \beta ct'_B\right). \tag{4.5}$$

The length is measured at a fixed time $t_A = t_B = t_0$ in S. We now can determine the times t'_A and t'_B and insert these into the equations for x_A and x_B. The coordinate difference is

$$l = |x_B - x_A| = \gamma |x'_B(1 - \beta^2) - x'_A(1 - \beta^2)| = \frac{|x'_B - x'_A|}{\gamma} = \frac{l'}{\gamma} \qquad (4.6)$$

Since $\gamma > 1$ for $\beta > 0$, the length l measured in S is always shorter than the rest length l'. Of course, the physical length of the rod has not changed, it is only the measuring method that produces a shorter result.

We shall later visualize in detail what Lorentz contraction implies for the *visual appearance* of fast moving bodies. A remarkable result will be that the Lorentz contraction cannot be observed directly because of the different times of flight for light rays starting from different points of the bodies.

4.3 Moving Clocks: Time Dilation

The loss of simultaneity disagrees with our everyday experience of the passing of time. The situation becomes even more mind-boggling when we measure time differences from different frames of reference. Two identical clocks rest in the origin of S and S', respectively. The system S' moves with velocity $\beta > 0$ relative to S. We consider the clock which rests in S'; obviously its world line corresponds to the ct'-axis, see Fig. 4.4. Referred to the system S, its world line has the coordinates $(ct = \gamma ct', x = \gamma \beta ct')$. We can convert the time difference $c\Delta t' = c(t'_2 - t'_1)$ into the time difference in S,

$$c\Delta t = ct_2 - ct_1 = \gamma ct'_2 - \gamma ct'_1 = \gamma c\Delta t'. \qquad (4.7)$$

We measure a shorter time difference in the moving system S' than in the system S.

Without further remarks we would run into a paradox. We could also consider the situation where S' is at rest and S is in motion. This would be completely legitimate on the basis of the principle of relativity. Then the time difference in the system S would be shorter than in S', which obviously is a contradiction.

The crucial point is that we cannot assess the time difference in S only with a single clock in the origin. As already seen in the context of the Lorentz contraction, at least two synchronized clocks, at rest at different positions, are needed in S, from which we read the clock times ct'_1 and ct'_2 of the moving clock at their individual positions $x_1 = \gamma \beta ct'_1$ and $x_2 = \gamma \beta ct'_2$.

This effect of *time dilation* could be proved in experiments with short-lived elementary particles. When cosmic radiation hits the Earth's atmosphere, high-energy elementary particles are created, for example muons. These have velocities on the order of the speed of light and possess average lifetimes of $\Delta t \approx 2 \cdot 10^{-6}$ s.

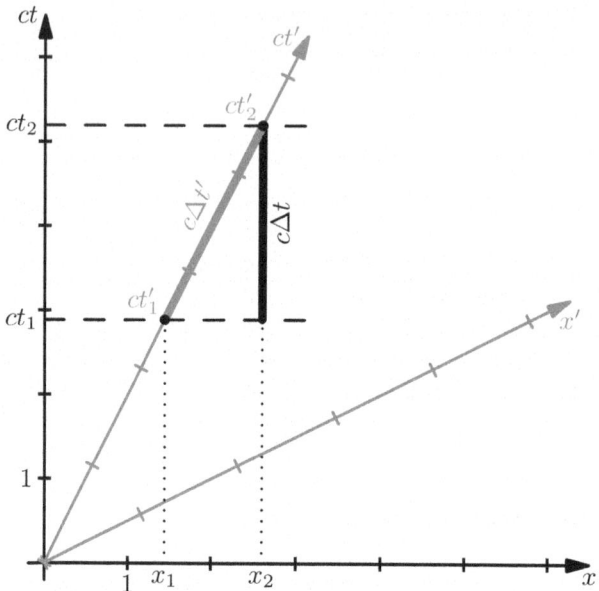

Fig. 4.4 The clock which rests at the origin of the system S' shows a time difference of $c\Delta t'$. In the system S this corresponds to a time difference $c\Delta t$

In 1941, Rossi and Hall [5] used these muons to measure the relativistic time dilation for the first time, see Fig. 4.5. Without time dilation, one would calculate an average mean-free path of the muons from the average lifetime

$$L < c\Delta t \approx 600\,\text{m}. \tag{4.8}$$

Since muons are produced at a height of $9 - 12$ km, none of these particles would reach the ground. However, seen from the rest frame of the Earth, the muons live much longer, because of time dilation, and their average mean-free path increases to

$$L = \beta\gamma c\Delta t. \tag{4.9}$$

For a velocity of $\beta = 0.995$ this yields roughly 6000 m, sufficient for an appreciable portion of muons to reach the ground.

Rossi and Hall exploited the dependence of the lifetime on the velocity to prove time dilation. For this purpose they measured the amount of muons arriving at two points with different heights above sea level. On the one hand in Denver at a height of 1616 m, on the other at the height of 3240 m at the nearby Echo Lake. Because of the larger height, there more muons were detected than in Denver.

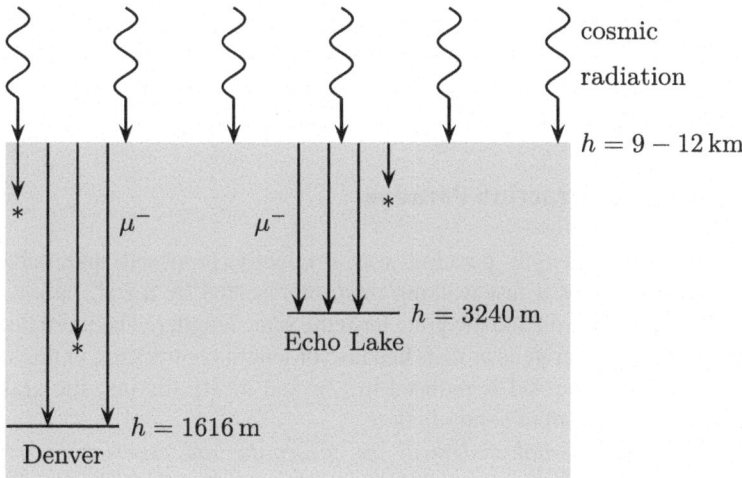

Fig. 4.5 Sketch of the experiment of Rossi and Hall [5]. When cosmic radiation hits the Earth's atmosphere short-lived high-energy muons (μ^-) are produced, with a lifetime of roughly $2 \cdot 10^{-6}$ s. Depending on the height of the observation points above sea level, different amounts of particles reach the ground

To gain additional information, in some observation runs iron plates were placed in front of the detectors, which only muons with a definite sufficiently high velocity could pass. By comparing the decrease of the numbers of muons with different velocities, they could check and confirm the predictions of special relativity. Similar experiments were carried out later with higher precision, and excellently confirmed the predictions of special relativity, see, e.g., Frisch and Smith [2]. A very compact apparatus for measuring the lifetimes of muons is described in Coan et al. [1].

A beautiful experiment demonstrating time dilation was performed with real clocks in October 1971 by Hafele and Keating [3, 4]. They flew 4 cesium atomic clocks aboard commercial planes twice around the world, one in the direction of Earth's rotation (east) and one against the direction of Earth's rotation (west). At the end, they compared with a clock that had remained on the ground. The time dilations of the moving clocks were as expected from special relativity. In particular, the clock flying eastward, i.e., faster, showed a greater loss of time than the one flying westward. In this experiment, due to Earth's gravitation, effects of general relativity were also important, see Sect. 13.4.6.

4.4   Paradoxes of Special Relativity

The properties of the Lorentz invariance very easily lead to apparent contradictions. But contradictory predictions would render special relativity absurd, making it useless as a physical theory.

Special relativity is probably the one theory in which the greatest number of paradoxes can be discussed. In this section we shall deal with two of these seeming contradictions. With the information of the previous sections, we are well prepared to resolve the contradictions.

4.4.1   Length Contraction Paradox

This version of the length paradox was originally proposed and solved by Rindler[1] [6]: It involves a fast walking man, represented by a rod, falling into a grate. Both the moving rod and the grate have the same length l. The inertial system is the one in which the grate is at rest. Because of length contraction, in this inertial system the length of the rod is reduced to l/γ and nicely fits into the grate. We immediately see an apparent contradiction:

"*Relative to the inertial system of the grate, the rod experiences a length contraction and fits into the grate. Relative to the rest frame of the rod, however, it is the grate which experiences a length contraction. Therefore, the rod does not fit into the grate.*"

To resolve the paradox we have to specify precisely what we mean when we say "fits into the grate". What we mean is that the two endpoints of the rod are *simultaneously* inside the grate. But we have seen that simultaneity is a property that strongly depends on the specific inertial system. This is the key to resolving the paradox.

We assume that the rod moves in positive x-direction with velocity β with respect to the inertial system S of the grate. Accordingly, in the rest frame of the rod, the grate moves in negative x'-direction, with velocity $-\beta$. In the rest frame S, its two edges are parametrized by the world lines

$$\mathcal{R}_l : \left(ct_{gl}, x_{gl}(t_{gl}) = r_l = \text{const}\right) \quad \text{and} \quad \mathcal{R}_r : \left(ct_{gr}, x_{gr}(t_{gr}) = r_r = \text{const}\right),$$
$$(4.10)$$

where the first index of the coordinates means "grate" and the second index designates "left" or "right". We obtain the corresponding parametrization with respect to the rest frame of the rod by a Lorentz transformation

$$\mathcal{R}_l : \left(ct'_{gl}, x'_{gl}(t'_{gl}) = \frac{r_l}{\gamma} - \beta ct'_{gl}\right) \quad \text{and} \quad \mathcal{R}_r : \left(ct'_{gr}, x'_{gr}(t'_{gr}) = \frac{r_r}{\gamma} - \beta ct'_{gr}\right)$$
$$(4.11)$$

with $ct'_{gl} = \gamma(ct_{gl} - \beta r_l)$ and $ct'_{gr} = \gamma(ct_{gr} - \beta r_r)$. In an analogous way we can obtain the world lines of the endpoints of the rod in its rest frame S'

$$\mathcal{S}_l : \left(ct'_{rl}, x'_{rl}(t'_{rl}) = s'_l = \text{const}\right) \quad \text{and} \quad \mathcal{S}_r : \left(ct'_{rr}, x'_{rr}(t'_{rr}) = s'_r = \text{const}\right).$$
$$(4.12)$$

---

[1] Wolfgang Rindler, 1924–2019, US-American theoretical physicist of Austrian origin.

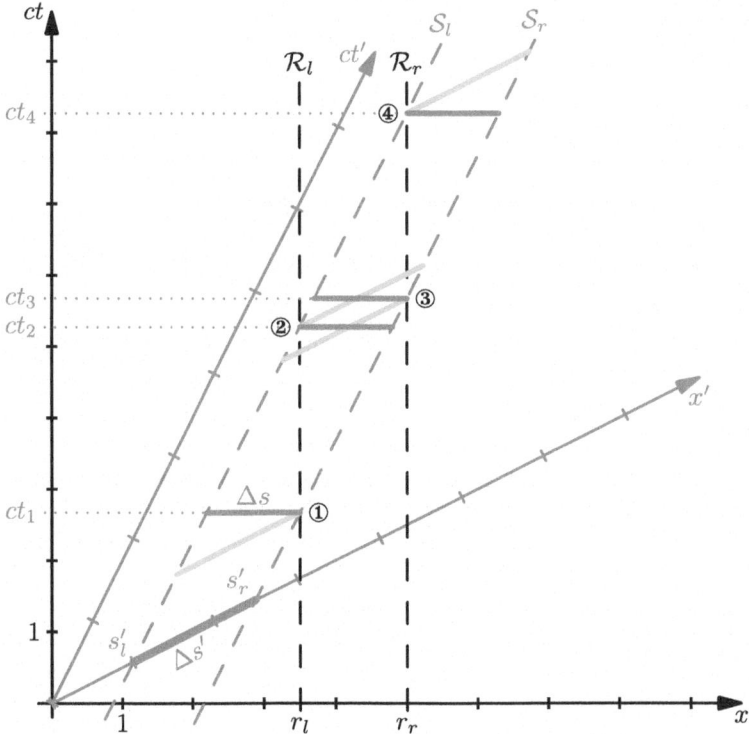

Fig. 4.6 Length contraction paradox as seen from the rest frame S of the grate. The world lines $\mathcal{R}_l$ and $\mathcal{R}_r$ denote the edges of the grate, while the world lines $\mathcal{S}_l$ and $\mathcal{S}_r$ designate the endpoints of the rod

The inverse transformation from S' to S yields

$$\mathcal{S}_l : \left(ct_{rl}, x_{rl}(t_{rl}) = \frac{s_l'}{\gamma} + \beta ct_{rl} \right) \quad \text{and} \quad \mathcal{S}_r : \left(ct_{rr}, x_{rr}(t_{rr}) = \frac{s_r'}{\gamma} + \beta ct_{rr} \right)$$

(4.13)

with $ct_{rl} = \gamma(ct_{rl}' + \beta s_l')$ and $ct_{rr} = \gamma(ct_{rr}' + \beta s_r')$.

Figure 4.6 shows the Minkowski diagram of the situation as seen from the rest frame S of the grate, with the world lines $\mathcal{R}_l$ and $\mathcal{R}_r$ of its edges. The world lines of the left and right endpoints of the rod are described by $\mathcal{S}_l$ and $\mathcal{S}_r$. The length $\Delta s' = |s_r' - s_l'|$ of the rod measured in S' is equal to the distance $\Delta r = |r_r - r_l|$ of the edges of the grate measured in S; therefore we set $\Delta r = \Delta s' =: \ell$. As we know from Sect. 4.2, *measuring* means that we have to determine the coordinates of the endpoints of the rod at one and the same moment. Consequently, in the rest frame of the grate the rod has a measured length of $\Delta s = \Delta s'/\gamma = \ell/\gamma$, and therefore fits into the grate.

The events ①–④ in Fig. 4.6 characterize the four essential phases of the motion. In ① the right-hand side of the rod (dark gray bar) hits the left-hand edge of the grate. Between the events ② and ③ the rod "resides" inside the grate. In ④, the rod has left the grate completely. The times of these events can be determined by intersecting the corresponding world lines of the rod with those of the grate. We obtain

$$
ct_1 = \frac{r_l}{\beta} - \frac{s'_r}{\beta\gamma}, \quad ct_2 = \frac{r_l}{\beta} - \frac{s'_l}{\beta\gamma}, \quad ct_3 = \frac{r_r}{\beta} - \frac{s'_r}{\beta\gamma}, \quad ct_4 = \frac{r_r}{\beta} - \frac{s'_l}{\beta\gamma}.
$$
(4.14)

For every event, the bars in light gray show that position of the rod, at which it is presently located in its rest frame S'. We see that, when in event ③ the right endpoint of the rod touches the right edge of the grate, its left endpoint has not yet entered the grate. Again, the reason for this seeming contradiction is the dependence of *simultaneity* on the inertial system. From the point of view of the grate system, at fixed time ct_3 the rod is completely inside the grate. However, with respect to the rest frame of the rod, its endpoints are not measured at the same time, but at different times.

Figure 4.7 shows the situation seen from the rest frame of the rod at the identical events ①–④. Their coordinates can be obtained by a Lorentz transformation from the system S to the system S',

$$
ct'_1 = \frac{r_l}{\beta\gamma} - \frac{s'_r}{\beta}, \quad ct'_2 = \frac{r_l}{\beta\gamma} - \frac{s'_l}{\beta}, \quad ct'_3 = \frac{r_r}{\beta\gamma} - \frac{s'_r}{\beta}, \quad ct'_4 = \frac{r_r}{\beta\gamma} - \frac{s'_l}{\beta}.
$$
(4.15)

The event ① designates the moment at which the right endpoint of the rod hits the left edge of the grate. The horizontal bars in light gray indicate the position of the rod in its rest frame, whereas the oblique bars in dark gray describe the position of the rod measured in the rest frame of the grate. When we look at the events ② and ③, we have the surprising result that they are chronologically interchanged: the event ③ happens before ②. We can immediately derive their time differences from the relations (4.14) and (4.15),

$$
c\Delta t_{3,2} = ct_3 - ct_2 = \frac{\ell}{\beta}\left(1 - \frac{1}{\gamma}\right) > 0
$$
(4.16)

and

$$
c\Delta t'_{3,2} = ct'_3 - ct'_2 = \frac{\ell}{\beta}\left(\frac{1}{\gamma} - 1\right) < 0.
$$
(4.17)

In other words, the right endpoint of the rod first hits the right edge of the grate in the event ③, before the left endpoint of the rod touches the left edge of the grate

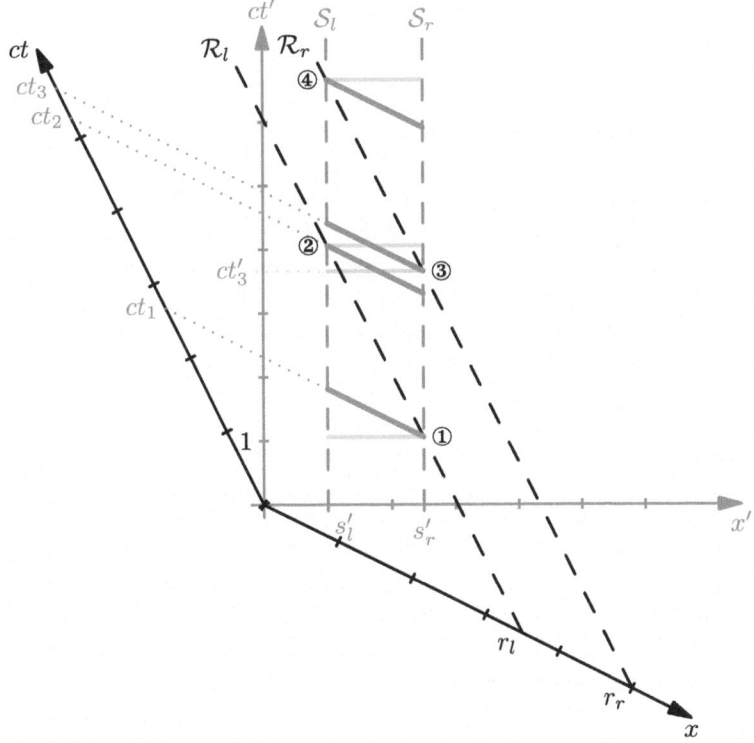

Fig. 4.7 Length contraction paradox as seen from the rest fame S'

in ②. We could already observe this sequence of events in Fig. 4.6 from the light gray bars.

The seeming paradox is that the rod is longer than the distance measured for the edges of the grate, e.g., at time ct'_3, and therefore should not fit into the grate. To resolve this paradox, we must relinquish the notion of a rigid rod and a solid grate, and consider the respective endpoints of the rod and the edges of the grate as point events. This is necessary since we are considering only a one-dimensional motion in space, and therefore a solid grate would stop the rod at the first contact.

We can resolve the original problem posed by the paradox, namely, whether or not the rod "fits" into the grate as follows. From the point of view of the grate system S the rod spatially "fits" into the grate since the endpoints of the rod pass through the grate during the events ② and ③. From the point of view of the rod system S' the rod chronologically "fits" into the grate since the endpoints of the rod can pass through the grate during the events ② and ③.

We note that in the literature the length contraction paradox is also known as the "ladder-barn", or "pole-barn" paradox. Depending on the frame of reference, a moving ladder, or pole fits into the barn, or not. In these versions of the paradox, the ladder, or pole, replaces our rod, and the barn the grate. Of course the resolution of the paradox is the same as discussed in this section.

4.4.2 The Clock Paradox

We consider two identical clocks. The first clock rests in the system S, and the second in the system S'. We assume that the system S' moves along the x-axis of the system S with velocity β. The popular statement now would be that, because of time dilation, the second clock runs more slowly, since, unlike the first clock, it is in motion. A seeming contradiction arises when we consider the situation from the rest frame of the second clock. Referred to this system, the first clock is in motion and experiences a time dilation. Which clock runs more slowly, the first one or the second?

First we must realize that we cannot tell whether the second clock runs more slowly using just the first clock. It is only at the one moment when the second clock passes through the position of the first clock that we can read its clock time. To measure a *time interval* we need to compare with another clock which is placed at a different location in the system S. Therefore, we place a second clock in the system S. To assess the situation in both systems on an equal footing, a second clock is also placed in the system S' at a different position. In both systems the individual clocks (at rest) are assumed to be synchronized among each other and placed at the same distance. With these provisions, we can use the results of the previous section, where $|s_r' - s_l'| = |r_r - r_l| =: \ell$. We imagine that the clocks are attached to both edges of the grate, and both endpoints of the rod, respectively.

Figure 4.8 shows the Minkowski diagram for the clock paradox from the point of view of the rest system S, with the two world lines S_l and S_r of the clocks, at rest in S', but moving in the positive x-direction relative to S.

To determine the time interval that has passed for the moving clock on the world line S_r, we calculate the time difference between the events ① and ③, first from the point of view of the moving clocks, and then the time difference on the clocks at rest in S at the coordinates $x = r_l$ and $x = r_r$. Since in the present discussion the events coincide with those of the previous section, we can directly transfer the relations (4.14) and (4.15). The time differences are

$$c\Delta t_{1,3} = ct_3 - ct_1 = \frac{r_r - r_l}{\beta} = \frac{\ell}{\beta} \quad \text{and} \quad c\Delta t_{1,3}' = ct_3' - ct_1' = \frac{r_r - r_l}{\beta\gamma} = \frac{\ell}{\beta\gamma}.$$
$$(4.18)$$

Even though the distance between ① and ③ appears longer than the vertical distance in the Minkowski diagram, we still have to take into account the scaling of the time axis. Obviously $c\Delta t_{1,3}' < c\Delta t_{1,3}$, and therefore time passes more slowly in S' than in S. We obtain the same result for the clock on the world line S_l.

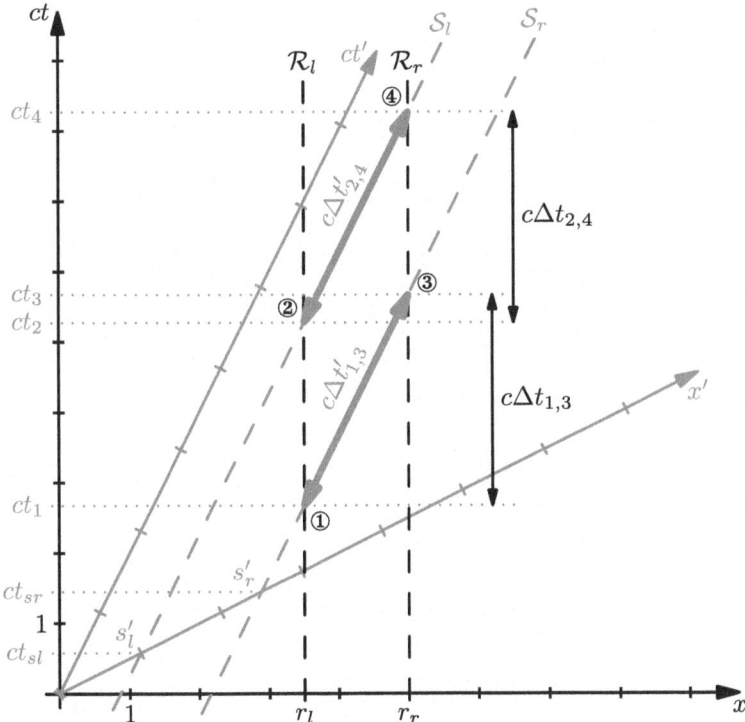

Fig. 4.8 Clock paradox from the point of view of the rest system S. The clocks at rest in S follow the world lines $\mathcal{R}_l$ and $\mathcal{R}_r$. The world lines of the clocks moving relative to S are given by $\mathcal{S}_l$ and $\mathcal{S}_r$

From the point of view of the system S', see Fig. 4.9, we now have to calculate the time difference of the moving clock along the world line $\mathcal{R}_l$ by inspecting the events ① and ②. Using the relations (4.14) and (4.15) we find

$$c\Delta t_{1,2} = ct_2 - ct_1 = \frac{s'_r - s'_l}{\beta \gamma} = \frac{\ell}{\beta \gamma} \quad \text{and} \quad c\Delta t'_{1,2} = ct'_2 - ct'_1 = \frac{s'_r - s'_l}{\beta} = \frac{\ell}{\beta}.$$
(4.19)

Since $c\Delta t_{1,2} < c\Delta t'_{1,2}$, time in the system S passes more slowly than in the system S'. This is in sharp contrast to our previous statement that time passes more slowly in S' than in S. The cause of this discrepancy lies in the definition of simultaneity in the two systems. The two clocks in the system S, at the positions $x = r_l$ and $x = r_r$, are synchronized with respect to $ct_{sl} = ct_{sr} = 0$. Using a Lorentz transformation, we calculate the corresponding times in the system S'. We have

$$ct'_{rl}(ct_{rl} = 0) = -\gamma \beta r_l \quad \text{and} \quad ct'_{rr}(ct_{rr} = 0) = -\gamma \beta r_r.$$
(4.20)

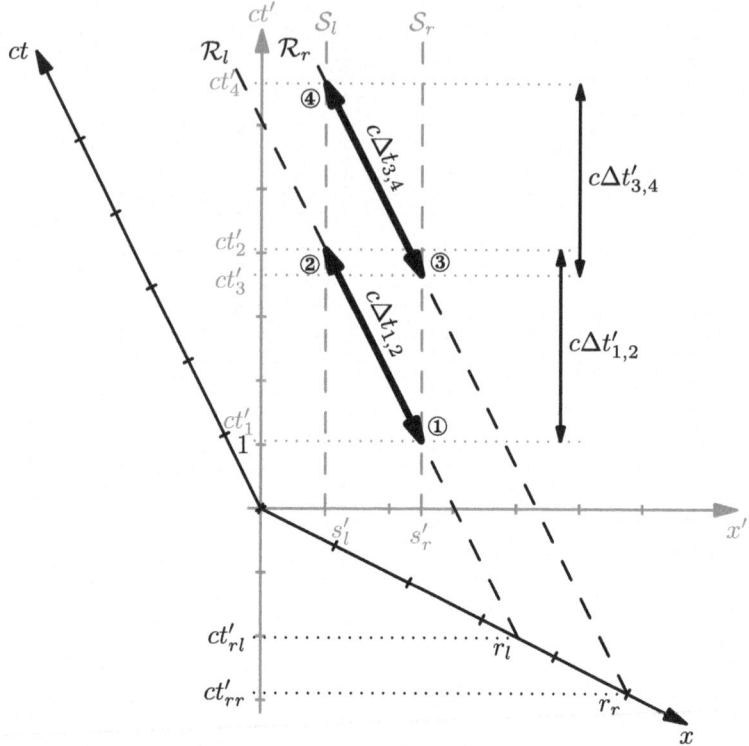

Fig. 4.9 Clock paradox from the point of view of the rest frame S'. Here the two clocks of the system S move along the world lines $\mathcal{R}_l$ and $\mathcal{R}_r$

From $r_r > r_l$ follows $ct'_{rr} < ct'_{rl} < 0$, see also Fig. 4.9. As seen from the system S', the two clocks are not running in synchrony.

On the other hand, the two clocks in the system S placed at the positions $x' = s'_l$ and $x' = s'_r$ are synchronized with respect to $ct'_{sl} = ct'_{sr} = 0$. As seen from the system S, this corresponds to times

$$ct_{sl}(ct'_{sl} = 0) = \gamma\beta s'_l \quad \text{and} \quad ct_{sr}(ct'_{sr} = 0) = \gamma\beta s'_r, \tag{4.21}$$

i.e., the moment of synchronization in S' happens at different times in S. From $s'_r > s'_l$ follows $ct_{sr} > ct_{sl} > 0$, see also Fig. 4.8.

Now we see the reason underlying the clock paradox: The two systems S and S' can never agree, as a matter of principle, on a joint synchronization of times. Which of the two clocks runs more slowly therefore depends on which frame of reference is used to determine time differences.

4.4.3 The Twin Paradox

This is probably the most prominent paradox of special relativity. A pair of twins is considered. One of the twins remains back on Earth, the other travels at a high velocity into the vast expanse of the Universe, turns around and returns to Earth. Because of the time dilation, on Earth a much longer time has passed than in the spacecraft.

The paradox arises when one assesses the situation from the point of view of the traveling twin. To him, the twin staying back on Earth moves at a high velocity, thus the time dilation should occur for the twin on Earth.

However, in this case the rest frame of the twin on Earth and that of the traveling twin are not equivalent. During the journey, the traveling twin is not always in the same inertial system, since, in order to return, he must brake, turn around, and then accelerate again. Due to the forces acting during these maneuvers, it is completely unambiguous, which of the two twins was in motion, and which of them was at rest. We shall consider the twin paradox quantitatively in the framework of relativistic mechanics at the end of Chap. 6.

4.5 Exercises

4.1. The "muon" paradox. Returning to the muon experiments, we point out an apparent problem. In the rest frame of the muons their lifetime is $\tau \approx 2 \cdot 10^{-6}$ s, because in their rest frame the clocks do not run more slowly. Why is it possible for them still to reach the ground?

References

1. Coan, T., Liu, T., Ye, J.: A compact apparatus for muon lifetime measurement and time dilation demonstration in the undergraduate laboratory. Am. J. Phys. **74**(2), 161–164 (2006)
2. Frisch, D.H., Smith, J.H.: Measurement of the relativistic time dilation using μ-mesons. Am. J. Phys. **31**(5), 342–355 (1963)
3. Hafele, J.C., Keating, R.E.: Around-the-world atomic clocks: Observed relativistic time gains. Science **177**(4044), 168–170 (1972)
4. Hafele, J.C., Keating, R.E.: Around-the-world atomic clocks: predicted relativistic time gains. Science **177**(4044), 166–168 (1972)
5. Rossi, B., Hall, D.B.: Variation of the rate of decay of mesotrons with momentum. Phys. Rev. **59**, 223–228 (1941)
6. Rindler, W.: Length contraction paradox. Am. J. Phys. **29**(6), 365–366 (1961)

Mathematical Formalism of Special Relativity

5

Contents

So far, our mathematical notation was sufficient to understand the basic features of special relativity and their consequences. To the reader, many of the topics covered in this chapter may seem an "over-mathematization", or obsessive introduction of mathematical notation. For later applications, however, this mathematical notation is indispensable. With a view to relativistic mechanics, but in particular to the covariant formulation of electrodynamics, and later of general relativity, we must avail ourselves of a more compact and general notation. We have already written four-vectors with a Greek index in the form $x^\mu = (x^0, x^1, x^2, x^3) = (ct, x, y, z)$, and introduced the concept that, depending on the context, x^μ may denote a vector $\underline{x}$, or the component of a vector. The transformation to another frame of reference then is simply represented by the multiplication of a matrix with a vector.

From now on, we shall also write the matrices of Lorentz transformations with indices. If a vector has one index, then obviously a matrix must have two indices: one denoting the row, and the other the column of the matrix. Therefore, Λ now becomes $\Lambda^\mu{}_\nu$.

For brevity of notation, we shall also introduce Einstein summation. The essential rule of Einstein summation notation is that in mathematical expressions repeated indices are implicitly summed over, with the restriction that each index can appear at most twice in any term. With this convention, a Lorentz transformation from one inertial system to another is written as

$$x'^\mu = \Lambda^\mu{}_\nu x^\nu \equiv \sum_{\nu=0}^{3} \Lambda^\mu{}_\nu x^\nu. \tag{5.1}$$

5.1   Minkowski Space

When introducing Lorentz transformations we had already postulated that the quantity

$$s^2 = -c^2 t^2 + x^2 + y^2 + z^2 \tag{5.2}$$

remains invariant under Lorentz transformations. Evidently there is a similarity with the distance squared in Euclidean space,

$$l^2 = x^2 + y^2 + z^2. \tag{5.3}$$

Therefore, (5.2) can be considered as the distance squared of two events in four-dimensional spacetime, the difference being that the sign of the time coordinate is negative. The distance l between two points in Euclidean space is left invariant under arbitrary rotational matrices D. In that sense we can say that Lorentz transformations are characterized by the property that they leave the "distance" (5.2) invariant. In mathematical terms, we say that this space has a *metric* different from the Euclidean one. The metric pertinent to special relativity is the *Minkowski metric*, and accordingly the corresponding space is called the *Minkowski space*.

5.1.1   Definition of the Minkowski Space

The Minkowski space is a four-dimensional real vector space equipped with the following *scalar product*: Let $\underline{a}$ and $\underline{b}$ be two four-vectors with components a^μ and b^μ. The scalar product $\langle \underline{a}, \underline{b} \rangle$ is defined by

$$\begin{aligned}
\langle \underline{a}, \underline{b} \rangle = a_\mu b^\mu &= a_0 b^0 + a_1 b^1 + a_2 b^2 + a_3 b^3 \\
&= -a^0 b^0 + a^1 b^1 + a^2 b^2 + a^3 b^3 \\
&= \eta_{\mu\nu} a^\mu b^\nu,
\end{aligned} \tag{5.4}$$

with the *Minkowski metric*

$$\eta_{\mu\nu} = \begin{pmatrix} -1 & 0 & 0 & 0 \\ 0 & 1 & 0 & 0 \\ 0 & 0 & 1 & 0 \\ 0 & 0 & 0 & 1 \end{pmatrix}. \tag{5.5}$$

The quantity b^μ with an upper index is called a *contravariant vector*, the quantity a_μ with a lower index is called a *covariant vector*. We shall give a precise definition of these notations in Sect. 5.2. The matrix $\eta_{\mu\nu} = \eta^{\mu\nu}$ allows raising and lowering indices in Minkowski space. Thus, the quantities a^μ and a_μ are related by

$$a_\mu = \eta_{\mu\nu}a^\nu = \begin{pmatrix} -a^0 \\ a^1 \\ a^2 \\ a^3 \end{pmatrix} \quad \text{with} \quad a^\nu = \begin{pmatrix} a^0 \\ a^1 \\ a^2 \\ a^3 \end{pmatrix}. \tag{5.6}$$

Using the Minkowski metric we can write the square of the distance s^2 as

$$s^2 = \eta_{\mu\nu}x^\mu x^\nu. \tag{5.7}$$

In Euclidean space we have

$$\langle x, x \rangle > 0 \quad \text{for all} \quad x \neq 0 \quad \text{and} \quad \langle x, x \rangle = 0 \Leftrightarrow x = 0. \tag{5.8}$$

The scalar product of a vector with itself is always greater than zero (unless it is the null vector, for which the scalar product is zero), which means that the scalar product is positive definite. By contrast, the scalar product in Minkowski space is not positive definite, as can be directly seen from its definition. As mentioned before, vectors with $\langle \underline{x}, \underline{x} \rangle < 0$ are called *timelike*, vectors with $\langle \underline{x}, \underline{x} \rangle > 0$ *spacelike*, and those with $\langle \underline{x}, \underline{x} \rangle = 0$ *lightlike*.

5.1.2 Definition of the Lorentz Transformation

Using the Minkowski metric, we can derive the condition necessary for a matrix to be a Lorentz transformation, i.e., a matrix that leaves the distance s invariant. We consider two inertial frames S and S' and a four-vector with coordinates x^ν in S and x'^μ in S'. Furthermore, we consider the Lorentz transformation which leads from S to S', i.e., $x'^\mu = \Lambda^\mu{}_\nu x^\nu$.

The requirement of the invariance of $s^2 = x^\mu x_\mu$ under Lorentz transformations then implies

$$\eta_{\alpha\beta}x^\alpha x^\beta = \eta_{\mu\nu}x'^\mu x'^\nu. \tag{5.9}$$

For x'^μ we insert the expression defined by the Lorentz transformation, and obtain

$$\eta_{\mu\nu}x'^\mu x'^\nu = \eta_{\mu\nu}\left(\Lambda^\mu{}_\alpha x^\alpha\right)\left(\Lambda^\nu{}_\beta x^\beta\right). \tag{5.10}$$

We use this result in (5.9), and shift all terms to the left-hand side of the equation. Then

$$\xi_{\alpha\beta}x^\alpha x^\beta = 0 \quad \text{with} \quad \xi_{\alpha\beta} = \eta_{\mu\nu}\Lambda^\mu{}_\alpha\Lambda^\nu{}_\beta - \eta_{\alpha\beta}. \tag{5.11}$$

Since $\xi_{\alpha\beta}x^\alpha x^\beta = 0$ must apply for arbitrary x^α and x^β, we have the relation

$$\Lambda^\mu{}_\alpha\eta_{\mu\nu}\Lambda^\nu{}_\beta = \eta_{\alpha\beta}. \tag{5.12}$$

Note that we have changed the order of the factors on the left-hand side. The first multiplication in this equation is not a matrix multiplication, as μ is a column index in both the Lorentz transformation and the Minkowski metric. Mathematical operations cannot always, or directly, be represented as matrix equations. If we wish to represent Eq. (5.12) as a matrix multiplication, the first Lorentz matrix must be transposed. We then obtain the condition characterizing Lorentz transformations

$$\boldsymbol{\Lambda}^{\mathrm{T}}\boldsymbol{\eta}\boldsymbol{\Lambda} = \boldsymbol{\eta}. \tag{5.13}$$

This equation is analogous to the definition $\boldsymbol{D}^{\mathrm{T}}\mathbf{1}\boldsymbol{D} = \mathbf{1}$ of orthogonal rotational matrices in Euclidean space. There, the identity matrix $\mathbf{1}$ appears instead of the Minkowski metric $\boldsymbol{\eta}$.

We calculate the determinant of (5.13):

$$\det(\boldsymbol{\Lambda}^{\mathrm{T}}\boldsymbol{\eta}\boldsymbol{\Lambda}) = \det(\boldsymbol{\Lambda}^{\mathrm{T}})\det(\boldsymbol{\eta})\det(\boldsymbol{\Lambda}) \stackrel{!}{=} \det(\boldsymbol{\eta}). \tag{5.14}$$

With $\det(\boldsymbol{\eta}) = -1$ and $\det(\boldsymbol{\Lambda}^{\mathrm{T}}) = \det(\boldsymbol{\Lambda})$ we obtain

$$\det \boldsymbol{\Lambda} = \pm 1. \tag{5.15}$$

When we let the Minkowski metric operate once more on (5.12), we are led to

$$\eta^{\kappa\alpha}\Lambda^\mu{}_\alpha\eta_{\mu\nu}\Lambda^\nu{}_\beta = \eta^{\kappa\alpha}\eta_{\alpha\beta}. \tag{5.16}$$

Therefore, with $\eta^{\kappa\alpha}\eta_{\alpha\beta} = \delta^\kappa_\beta$ and $\eta^{\kappa\alpha}\Lambda^\mu{}_\alpha\eta_{\mu\nu} = \Lambda_\nu{}^\kappa$,

$$\Lambda_\nu{}^\kappa\Lambda^\nu{}_\beta = \delta^\kappa_\beta = \begin{cases} 1, & \text{if } \kappa = \beta, \\ 0, & \text{otherwise,} \end{cases} \tag{5.17}$$

where δ^κ_β is the *Kronecker symbol*.[1] Obviously the quantity

$$\Lambda_\nu{}^\kappa = \eta^{\kappa\alpha}\Lambda^\mu{}_\alpha\eta_{\mu\nu} \tag{5.18}$$

---

[1] Leopold Kronecker, 1823–1891, German mathematician.

is the inverse of the Lorentz transformation $\Lambda^\mu{}_\alpha$. It is important to notice the position of the indices.

We evaluate this equation for a Lorentz boost in the x-direction, see Eq. (3.20). Raising the second index can be done via

$$\Lambda^{\alpha\mu} = \eta^{\beta\mu} \Lambda^\alpha{}_\beta = \Lambda^\alpha{}_\beta \eta^{\beta\mu}. \tag{5.19}$$

In matrix notation, this equation reads

$$\Lambda^{\alpha\mu} = \begin{pmatrix} \gamma & -\beta\gamma & 0 & 0 \\ -\beta\gamma & \gamma & 0 & 0 \\ 0 & 0 & 1 & 0 \\ 0 & 0 & 0 & 1 \end{pmatrix} \cdot \begin{pmatrix} -1 & 0 & 0 & 0 \\ 0 & 1 & 0 & 0 \\ 0 & 0 & 1 & 0 \\ 0 & 0 & 0 & 1 \end{pmatrix} = \begin{pmatrix} -\gamma & -\beta\gamma & 0 & 0 \\ \beta\gamma & \gamma & 0 & 0 \\ 0 & 0 & 1 & 0 \\ 0 & 0 & 0 & 1 \end{pmatrix}. \tag{5.20}$$

Lowering the first index is accomplished by

$$\Lambda_\lambda{}^\mu = \eta_{\lambda\alpha} \Lambda^{\alpha\mu}. \tag{5.21}$$

This equation can also be written in matrix notation

$$\Lambda_\lambda{}^\mu = \begin{pmatrix} -1 & 0 & 0 & 0 \\ 0 & 1 & 0 & 0 \\ 0 & 0 & 1 & 0 \\ 0 & 0 & 0 & 1 \end{pmatrix} \cdot \begin{pmatrix} -\gamma & -\beta\gamma & 0 & 0 \\ \beta\gamma & \gamma & 0 & 0 \\ 0 & 0 & 1 & 0 \\ 0 & 0 & 0 & 1 \end{pmatrix} = \begin{pmatrix} \gamma & \beta\gamma & 0 & 0 \\ \beta\gamma & \gamma & 0 & 0 \\ 0 & 0 & 1 & 0 \\ 0 & 0 & 0 & 1 \end{pmatrix}. \tag{5.22}$$

We can now evaluate Eq. (5.17) for the Lorentz boost in matrix notation. As required, we find

$$\Lambda_\lambda{}^\mu \Lambda^\lambda{}_\nu = \begin{pmatrix} \gamma & \beta\gamma & 0 & 0 \\ \beta\gamma & \gamma & 0 & 0 \\ 0 & 0 & 1 & 0 \\ 0 & 0 & 0 & 1 \end{pmatrix} \cdot \begin{pmatrix} \gamma & -\beta\gamma & 0 & 0 \\ -\beta\gamma & \gamma & 0 & 0 \\ 0 & 0 & 1 & 0 \\ 0 & 0 & 0 & 1 \end{pmatrix}$$

$$= \begin{pmatrix} \gamma^2(1-\beta^2) & 0 & 0 & 0 \\ 0 & \gamma^2(1-\beta^2) & 0 & 0 \\ 0 & 0 & 1 & 0 \\ 0 & 0 & 0 & 1 \end{pmatrix} = \begin{pmatrix} 1 & 0 & 0 & 0 \\ 0 & 1 & 0 & 0 \\ 0 & 0 & 1 & 0 \\ 0 & 0 & 0 & 1 \end{pmatrix} = \delta^\mu_\nu. \tag{5.23}$$

Therefore, as expected, the inverse Lorentz transformation of a given boost is a boost with the negative velocity $-\beta$. For a simple spatial rotation $\boldsymbol{\Lambda}_R$ from (3.20) we also have, with $\boldsymbol{D}\boldsymbol{D}^\mathsf{T} = \boldsymbol{D}^\mathsf{T}\boldsymbol{D} = 1$

$$\Lambda_\lambda{}^\mu \Lambda^\lambda{}_\nu = \begin{pmatrix} 1 & 0 \\ 0 & \boldsymbol{D}^\mathsf{T} \end{pmatrix} \cdot \begin{pmatrix} 1 & 0 \\ 0 & \boldsymbol{D} \end{pmatrix} = \delta^\mu_\nu. \tag{5.24}$$

The set of all rotational matrices form the rotational group $SO(3)$ (*special orthogonal group*), the set of all Lorentz matrices form the *Lorentz group* $O(3, 1)$. In these groups, pure translations of the origin $x^\mu \mapsto x^\mu + a^\mu$ are not yet included. When they are taken into account, one obtains the *Poincaré group*,[2] which, like the group of Galilei transformations, possesses 10 free parameters: Three rotations, three boosts, three spatial translations, and one time translation. The transformations of this group are called *Poincaré transformations*.

5.2   Contravariant and Covariant Vectors

We have to specify the relations between quantities with upper and lower indices. Let $a^\mu \in V$ be a contravariant vector in the vector space V. Then $a_\mu = \eta_{\mu\nu} a^\nu \in V^*$ is a covariant vector, and an element of the *dual space* V^* of 1-forms, i.e.,

$$\varphi_a : V \to \mathbb{R} \quad \text{is a linear mapping with} \quad a_\mu : b^\mu \mapsto \langle \underline{a}, \underline{b} \rangle = a_\mu b^\mu \in \mathbb{R}. \tag{5.25}$$

Any four-component quantity a^μ which transforms according to the Lorentz matrix

$$a'^\mu = \Lambda^\mu{}_\nu a^\nu \tag{5.26}$$

is called a *contravariant tensor of first rank*, or *of rank 1*. Now let a^μ be a contravariant vector with $a'^\mu = \Lambda^\mu{}_\nu a^\nu$, then

$$a'_\mu = \eta_{\mu\alpha} a'^\alpha = \eta_{\mu\alpha} \Lambda^\alpha{}_\nu a^\nu = \eta_{\mu\alpha} \eta^{\nu\beta} \Lambda^\alpha{}_\nu a_\beta = \Lambda_\mu{}^\beta a_\beta, \tag{5.27}$$

with the inverse Lorentz transformation $\Lambda_\mu{}^\beta$. In the second step we have inserted $a^\nu = \eta^{\nu\beta} a_\beta$. In this way we have derived the transformation behavior of the quantity a_β. Any four-component quantity which transforms according to the inverse Lorentz matrix

$$a'_\mu = \Lambda_\mu{}^\nu a_\nu \tag{5.28}$$

is called a *covariant tensor of first rank*, or *of rank 1*.

---

[2] Henri Poincaré, 1854–1912, French mathematician, theoretical physicist, and philosopher.

5.2.1 Transformation Behavior of Differentials and Coordinate Derivatives

Let x^μ be a contravariant vector. Then

$$x'^\mu = \Lambda^\mu{}_\nu x^\nu, \quad \text{and also} \quad dx'^\mu = \Lambda^\mu{}_\nu dx^\nu. \tag{5.29}$$

We see that differentials dx^μ transform like contravariant vectors.

Let $f = f(x^\mu)$ be a scalar function. Then the differential is

$$df = \frac{\partial f}{\partial x^\mu} dx^\mu. \tag{5.30}$$

With $x'^\mu = \Lambda^\mu{}_\nu x^\nu$ or $x^\nu = \Lambda_\mu{}^\nu x'^\mu$ follows directly

$$df = \frac{\partial f}{\partial x'^\mu} dx'^\mu = \frac{\partial f}{\partial x^\nu} \frac{\partial x^\nu}{\partial x'^\mu} dx'^\mu = \left[\Lambda_\mu{}^\nu \frac{\partial}{\partial x^\nu} \right] f dx'^\mu. \tag{5.31}$$

From now on, we shall abbreviate derivatives with respect to coordinates by

$$\frac{\partial}{\partial x^\mu} \equiv \partial_\mu = \left(\frac{1}{c} \partial_t, \nabla \right)^{\mathrm{T}} \tag{5.32a}$$

and

$$\partial^\mu = \eta^{\mu\nu} \partial_\nu = \left(-\frac{1}{c} \partial_t, \nabla \right)^{\mathrm{T}}, \tag{5.32b}$$

with the *nabla operator* $\nabla = (\partial_x, \partial_y, \partial_z)^{\mathrm{T}}$. The comparison of the different expressions in (5.31) shows that coordinate derivatives ∂_μ transform like covariant vectors,

$$\partial'_\mu = \Lambda_\mu{}^\nu \partial_\nu \tag{5.33a}$$

and accordingly ∂^μ like a contravariant vector,

$$\partial'^\mu = \Lambda^\mu{}_\nu \partial^\nu. \tag{5.33b}$$

The fact that we write coordinate derivatives with a lower index accounts for this property.

5.2.2   Tensor Algebra

To describe all phenomena in special relativity, we can not get along just with covariant and contravariant vectors, i.e., tensors of rank 1. We have to generalize our previous definitions of covariant and contravariant tensors.

A *tensor of type* (r, s) is a multilinear mapping

$$
T : \underbrace{V^* \times V^* \times \ldots \times V^*}_{r \text{ times}} \times \underbrace{V \times V \times \ldots \times V}_{s \text{ times}} \to \mathbb{R}
$$

$$
\underbrace{(\varphi, \chi, \ldots, \omega}_{r \text{ times}}; \underbrace{u, v, \ldots, w)}_{s \text{ times}} \mapsto T(\varphi, \chi, \ldots, \omega; u, v, \ldots, w) \in \mathbb{R}.
$$

(5.34)

Here, Greek letters denote elements of the dual space V^*, that is, covariant vectors, and Latin letters elements of V, i.e., contravariant vectors. The quantity T is called an r-times contravariant and an s-times covariant tensor. Equation (5.34) is the generalization of the scalar product in (5.25).

Multilinearity designates the property of a tensor that it is linear in every argument (while keeping all other arguments fixed). Therefore, a tensor χ can be decomposed in

$$
\chi = a\chi_1 + b\chi_2 + \ldots, \tag{5.35}
$$

with the coefficients, in our case, a and $b \in \mathbb{R}$. The set of all tensors of type (r, s) constitutes a *vector space* V_s^r. In index notation, this mapping can be represented in the form

$$
\overset{r\text{-times}}{\overbrace{T^{\alpha_1 \ldots \alpha_r}}} \underbrace{{}_{\beta_1 \ldots \beta_s}}_{s\text{-times}} \varphi_{\alpha_1} \chi_{\alpha_2} \ldots \omega_{\alpha_r} u^{\beta_1} v^{\beta_2} \ldots w^{\beta_s} = c \in \mathbb{R}. \tag{5.36}
$$

Here, $\varphi_{\alpha_1}, \chi_{\alpha_2}, \ldots, \omega_{\alpha_r}$ are covariant vectors, and $u^{\beta_1}, v^{\beta_2}, \ldots, w^{\beta_s}$ contravariant vectors.

We can combine tensors in a way that another tensor of different type arises. If $T \in V_s^r$ and $S \in V_{s'}^{r'}$, then we can define a new tensor

$$
T \otimes S \in V_s^r \times V_{s'}^{r'} = V_{s+s'}^{r+r'}. \tag{5.37}
$$

Thus, a tensor of type $(r + r', s + s')$ is formed from a tensor of type (r, s) and a tensor (r', s'). We can represent this relation also in index notation:

$$
T^{\alpha_1 \ldots \alpha_r}{}_{\beta_1 \ldots \beta_s} S^{\alpha_1' \ldots \alpha_{r'}'}{}_{\beta_1' \ldots \beta_{s'}'} = U^{\alpha_1 \ldots \alpha_r}{}_{\beta_1 \ldots \beta_s}{}^{\alpha_1' \ldots \alpha_{r'}'}{}_{\beta_1' \ldots \beta_{s'}'}. \tag{5.38}
$$

The operation "$\otimes$" is called the *tensor product*, or *direct product*.

Let $T \in V_s^r$ again be a tensor of type (r, s). We denote the k-th covariant and the j-th contravariant index by the same symbol, and sum over these two indices. We obtain a tensor in V_{s-1}^{r-1}:

$$T^{\alpha_1 \ldots \alpha_j \ldots \alpha_r}{}_{\beta_1 \ldots \beta_j \ldots \beta_s} \in V_s^r \quad \text{and} \quad T^{\alpha_1 \ldots \beta_j \ldots \alpha_r}{}_{\beta_1 \ldots \beta_j \ldots \beta_s} = \text{Tr}_j^k T \in V_{s-1}^{r-1}.$$

(5.39)

This operation is called *tensor contraction*, or just *contraction*.

Let us look at a few examples for these definitions. Let a^μ and $b^\mu \in V_0^1$ be contravariant four-vectors. Then

- $\eta_{\mu\nu} a^\nu = a_\mu \in V_1^0$ is a covariant vector,
- $a^\mu b^\nu \in V_0^2$ is a direct product and contravariant tensor of rank 2,
- $c^\mu{}_\nu = a^\mu b_\nu \in V_1^1$ is a direct product, and a first-rank contravariant and a first-rank covariant tensor,
- $c^\mu{}_\mu = a^\mu b_\mu \in V_0^0 = \mathbb{R}$ is a contraction and a tensor of rank 0, i.e., a scalar, see Sect. 5.2.4.

More information on tensor calculus will be presented in Sect. 11.2 within the framework of general relativity.

5.2.3 Tensor Property of the Differential Operator

We have not yet discussed the tensor property of the *differential operator*:

$$\partial_\mu = \frac{\partial}{\partial x^\mu} = \left(\frac{\partial}{\partial(ct)}, \frac{\partial}{\partial x}, \frac{\partial}{\partial y}, \frac{\partial}{\partial z} \right),$$

(5.40)

which is also called the *four-gradient*. If we let the four-gradient operate on a Lorentz scalar, we obtain a covariant vector:

$$\partial_\mu \varphi \equiv \varphi_{,\mu}.$$

(5.41)

Note that we have introduced the notation $X_{,\mu}$ for differentiation with respect to coordinates. We shall use this notation in parallel with the notation $\partial_\mu X$. Letting the four-gradient act on a four-vector a^μ yields a Lorentz scalar:

$$\partial_\mu a^\mu.$$

(5.42)

This expression is also denoted the *four-divergence*. Furthermore,

$$\partial_\mu a_\nu - \partial_\nu a_\mu$$

(5.43)

is an antisymmetric covariant tensor of rank 2, it is called the *four-curl*.

5.2.4   Proper Time

If we consider only infinitesimal distances in Minkowski space, the distance squared is

$$ds^2 = -c^2 dt^2 + dx^2 + dy^2 + dz^2 = -c^2 dt^2 + d\boldsymbol{x}^2, \tag{5.44}$$

with the spacelike distances dx, dy, dz, and the timelike distance dt, see also (5.7). The expression $ds^2 = dx_\mu dx^\mu = \eta_{\mu\nu} dx^\mu dx^\nu$, with $x^\mu = (ct, x, y, z)$, is also called the infinitesimal line element. Evidently it is invariant under Lorentz transformations.

For an observer moving along a world line with velocity $\boldsymbol{v}(t)$, the infinitesimal spatial distance is $d\boldsymbol{x} = \boldsymbol{v}(t)dt$. Therefore, we can write the line element (5.44) in the form

$$ds^2 = -c^2 \left(1 - \frac{\boldsymbol{v}(t)^2}{c^2}\right) dt^2 = -c^2 \left(1 - \beta(t)^2\right) dt^2. \tag{5.45}$$

If the velocity is $\beta = 0$, then $ds^2 = -c^2 dt^2$. Since it is always possible to find an instantaneous rest frame for the observer, we have quite generally $ds^2 = -c^2 d\tau^2$, with the *proper time* τ, which elapses in the rest frame of the observer. Obviously, it is a Lorentz scalar.

Let τ denote the proper time of the observer moving at velocity β relative to a system of synchronized clocks at rest. If we compare with the time of these clocks, we can deduce from (5.45)

$$-c^2 \left(1 - \beta^2\right) dt^2 = -c^2 d\tau^2 \quad \text{or} \quad d\tau = \sqrt{1 - \beta^2}\, dt = \frac{1}{\gamma} dt. \tag{5.46}$$

Since $\gamma \geq 1$, the time for the moving observer passes more slowly than it passes in the system of synchronized clocks at rest.

Relativistic Mechanics

6

Contents

We have already seen that Newtonian mechanics is not covariant under Lorentz transformations. For example, a constant acceleration a leads to a velocity $v(t) = at > c$ for $t > c/a$.

The challenge then, is to find a formulation of a relativistic, Lorentz covariant mechanics, which in the limit of small velocities turns again into Newton's mechanics. For this purpose we consider point particles in the four-dimensional spacetime. The *world line* of a particle is given by

$$x^\mu = x^\mu(t) = \begin{pmatrix} ct \\ \boldsymbol{r}(t) \end{pmatrix}. \tag{6.1}$$

The trajectory is defined in the same way as in Newtonian mechanics, and corresponds to the set of all events lying on the trajectory of the particle. Because of time dilation, time in the rest frame of the particle passes differently from the time of an external observer. Therefore, a parametrization of the trajectory in terms of the proper time is necessary,

$$x^{\mu}[t(\tau)] = \begin{pmatrix} ct(\tau) \\ r[t(\tau)] \end{pmatrix}. \tag{6.2}$$

Our first goal is to formulate the quantities of classical mechanics: velocity, acceleration, momentum, and energy in a Lorentz covariant way. Then we will try to understand the consequences of this formulation by studying specific examples.

A really comprehensive representation of relativistic mechanics is out of the scope of this book. For example, we shall omit the generalization of angular momentum. The interested reader will find more details, e.g., in Goldstein [3].

6.1   Four-Velocity

In trying to define a four-velocity, we face the problem that the coordinate time t is not a Lorentz scalar. Therefore, dx^{μ}/dt, too, is not Lorentz covariant. But we know that the proper time τ is a Lorentz scalar. Therefore,

$$u^{\mu} = \frac{dx^{\mu}}{d\tau} \tag{6.3}$$

is a contravariant four-vector, the *four-velocity*. The differential of the proper time follows from the expression (5.46)

$$u^{\mu} = \gamma(t) \frac{dx^{\mu}}{dt}, \tag{6.4}$$

and using the abbreviations $dx^{i}/dt = \dot{x}^{i}$ and $\dot{r} = \left(\dot{x}^1, \dot{x}^2, \dot{x}^3\right)^{\mathsf{T}}$, we can write

$$u^{\mu} = \gamma(t) \begin{pmatrix} c \\ \dot{r} \end{pmatrix} \quad \text{and} \quad u_{\mu} = \eta_{\mu\nu} u^{\nu} = \gamma(t) \begin{pmatrix} -c \\ \dot{r} \end{pmatrix}. \tag{6.5}$$

The contraction of the four-velocity yields a Lorentz scalar:

$$u_{\mu} u^{\mu} = -\gamma^2 c^2 + \gamma^2 \dot{r}^2 = \gamma^2 c^2 (-1 + \beta^2) = -c^2 < 0. \tag{6.6}$$

Thus, in any case $u_{\mu} u^{\mu} < 0$, and therefore u^{μ} is a timelike vector.

6.2   Four-Acceleration

We arrive at the *four-acceleration* in a completely analogous way:

$$b^\mu = \frac{\mathrm{d}u^\mu}{\mathrm{d}\tau} = \frac{\mathrm{d}^2 x^\mu}{\mathrm{d}\tau^2}. \tag{6.7}$$

When we evaluate this expression, we find

$$b^\mu = \gamma \frac{\mathrm{d}u^\mu}{\mathrm{d}t} = \gamma \frac{\mathrm{d}}{\mathrm{d}t}\left[\gamma \begin{pmatrix} c \\ \dot{r} \end{pmatrix}\right] = \gamma\dot{\gamma}\begin{pmatrix} c \\ \dot{r} \end{pmatrix} + \gamma^2 \begin{pmatrix} 0 \\ \ddot{r} \end{pmatrix}. \tag{6.8}$$

The time derivative of γ is given by

$$\dot{\gamma} = \frac{\mathrm{d}}{\mathrm{d}t}\frac{1}{\sqrt{1-\beta^2}} = \gamma^3 \boldsymbol{\beta} \cdot \dot{\boldsymbol{\beta}}. \tag{6.9}$$

Therefore, the explicit form of the four-acceleration is

$$b^\mu = \gamma^4 \boldsymbol{\beta} \cdot \dot{\boldsymbol{\beta}} c \begin{pmatrix} 1 \\ \boldsymbol{\beta} \end{pmatrix} + \gamma^2 c \begin{pmatrix} 0 \\ \dot{\boldsymbol{\beta}} \end{pmatrix} = c\gamma^4 \begin{pmatrix} \boldsymbol{\beta} \cdot \dot{\boldsymbol{\beta}} \\ \dot{\boldsymbol{\beta}}/\gamma^2 + (\boldsymbol{\beta} \cdot \dot{\boldsymbol{\beta}})\boldsymbol{\beta} \end{pmatrix}. \tag{6.10}$$

We recognize that in the limit $\beta \to 0$ all terms containing β vanish, and the four-acceleration reduces to

$$b^\mu \to \begin{pmatrix} 0 \\ \ddot{r} \end{pmatrix}, \quad \text{for} \quad \beta \ll 1. \tag{6.11}$$

Thus, in the limit of small velocities we obtain the correct non-relativistic result.

6.3   Four-Momentum

The rest mass m_0 of a particle is a Lorentz scalar. Therefore, we can introduce the *four-momentum* directly as

$$p^\mu = m_0 u^\mu = m_0 \gamma \begin{pmatrix} c \\ \dot{r} \end{pmatrix}. \tag{6.12}$$

A remark is in place: In the literature one often finds the statement that the mass of a particle is velocity-dependent via $m(\gamma) = m_0 \gamma$. In many cases one can work with this definition. Strictly speaking, however, the factor γ in (6.12) belongs to the four-velocity and not to the mass. By definition, the rest mass of a particle does not dependent on the velocity, and therefore is a Lorentz scalar.

6.4   Four-Force

The basic equation of Newton's mechanics is

$$F^N = \dot{p}, \tag{6.13}$$

with the Newtonian force F^N. An obvious relativistic generalization could be

$$F^\mu = \frac{dp^\mu}{d\tau} = \gamma \frac{dp^\mu}{dt} = m_0 b^\mu \tag{6.14}$$

This equation, however, cannot be proven rigorously, and can only be verified experimentally. We insert (6.10) and obtain

$$F^\mu = m_0 c \gamma^4 \begin{pmatrix} \boldsymbol{\beta} \cdot \dot{\boldsymbol{\beta}} \\ \dot{\boldsymbol{\beta}}/\gamma^2 + (\boldsymbol{\beta} \cdot \dot{\boldsymbol{\beta}})\boldsymbol{\beta} \end{pmatrix} = \begin{pmatrix} m_0 c \gamma^4 \boldsymbol{\beta} \cdot \dot{\boldsymbol{\beta}} \\ \gamma\, F^N \end{pmatrix}. \tag{6.15}$$

In the second step, we have used Newton's relation $F^N = dp/dt$ in the spatial components, together with (5.46), which of course still is valid.

For the spatial components of (6.15) we can recognize the relation

$$F^N = m_0 c \gamma (\dot{\boldsymbol{\beta}} + \gamma^2 (\boldsymbol{\beta} \cdot \dot{\boldsymbol{\beta}})\boldsymbol{\beta}). \tag{6.16}$$

We use the relation to obtain an expression for the 0-component:

$$\begin{aligned}
\gamma \boldsymbol{\beta} \cdot F^N &= m_0 c \gamma^2 \boldsymbol{\beta} \cdot \dot{\boldsymbol{\beta}} + m_0 c \gamma^4 (\boldsymbol{\beta} \cdot \dot{\boldsymbol{\beta}})\beta^2 \\
&= m_0 c \gamma^2 \boldsymbol{\beta} \cdot \dot{\boldsymbol{\beta}}(1 + \gamma^2 \beta^2) = m_0 c \gamma^4 \boldsymbol{\beta} \cdot \dot{\boldsymbol{\beta}} = F^0.
\end{aligned} \tag{6.17}$$

Therefore,

$$F^\mu = \gamma \begin{pmatrix} \boldsymbol{\beta} \cdot F^N \\ F^N \end{pmatrix} = m_0 b^\mu. \tag{6.18}$$

The relativistic equations of motion can be formulated using this four-force.

6.5   Force-Free Motion

The force-free motion can be described as the shortest trajectory between two spacetime events A and B. The calculation is done by the variation of the trajectory, see Fig. 6.1, i.e.,

$$\delta \int_A^B |ds| = 0. \tag{6.19}$$

Fig. 6.1 Variation of the trajectory. We consider small variations $\delta r(t)$ of the trajectory $r(t)$ from event A to event B, under the constraint $\delta r(t_A) = \delta r(t_B) = 0$

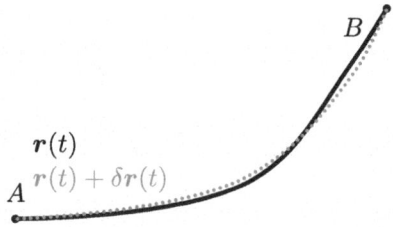

We have to use the modulus of $\mathrm{d}s$ since for timelike intervals one has $\mathrm{d}s^2 < 0$. Then, we parametrize the trajectory by the time t and insert the definition of the line element,

$$\delta \int_A^B |\mathrm{d}s| = \delta \int_A^B \sqrt{c^2\mathrm{d}t^2 - \mathrm{d}x^2 - \mathrm{d}y^2 - \mathrm{d}z^2} = \delta \int_A^B \sqrt{c^2 - \dot{r}^2}\,\mathrm{d}t$$

$$= -\int_A^B \frac{\dot{r}\delta\dot{r}}{\sqrt{c^2 - \dot{r}^2}}\,\mathrm{d}t. \tag{6.20}$$

In the second step we have exploited the usual rules for derivatives. To evaluate the integral, we apply product integration. We set

$$p = \frac{\dot{r}}{\sqrt{c^2 - \dot{r}^2}} \quad \text{and} \quad \mathrm{d}q = \delta\dot{r} \cdot \mathrm{d}t \tag{6.21}$$

and obtain, after differentiation and integration,

$$\mathrm{d}p = \left[\frac{\mathrm{d}}{\mathrm{d}t} \frac{\dot{r}}{\sqrt{c^2 - \dot{r}^2}} \right] \mathrm{d}t \quad \text{and} \quad q = \delta r. \tag{6.22}$$

We have used the fact that differentiation and variation commute, i.e.,

$$\delta\dot{r} = \delta\frac{\mathrm{d}r}{\mathrm{d}t} = \frac{\mathrm{d}}{\mathrm{d}t}\delta r. \tag{6.23}$$

Inserting these results we find

$$-\frac{\dot{r}}{\sqrt{c^2 - \dot{r}^2}} \cdot \delta r \Big|_A^B + \int_A^B \delta r(t) \left[\frac{\mathrm{d}}{\mathrm{d}t} \frac{\dot{r}}{\sqrt{c^2 - \dot{r}^2}} \right] \mathrm{d}t = 0. \tag{6.24}$$

Because of $\delta r(A) = \delta r(B) = 0$, the first term vanishes. Now we have

$$\int_A^B \delta r(t) \left[\frac{d}{dt} \frac{\dot{r}}{\sqrt{c^2 - \dot{r}^2}} \right] dt = 0 \quad \text{for arbitrary} \quad \delta r(t). \tag{6.25}$$

This equation can be fulfilled for arbitrary variations $\delta r(t)$ only if the time derivative in the square brackets vanishes. Therefore, the condition is

$$\frac{d}{dt} \frac{\dot{r}}{\sqrt{c^2 - \dot{r}^2}} = \frac{\ddot{r}}{\sqrt{c^2 - \dot{r}^2}} + \frac{\ddot{r}\dot{r}^2}{(c^2 - \dot{r}^2)^{3/2}} = \ddot{r} \cdot \frac{c^2}{(c^2 - \dot{r}^2)^{3/2}} \overset{!}{=} \mathbf{0}. \tag{6.26}$$

From this follows $\ddot{r} = 0$, or $\dot{r} = \dot{r}_0 = $ const. When we multiply (6.26) by the rest mass m_0 we obtain

$$\frac{d}{dt} \frac{m_0 \cdot \dot{r}}{\sqrt{c^2 - v^2}} = \frac{d}{dt} m_0 \gamma \boldsymbol{\beta} = \mathbf{0}. \tag{6.27}$$

Note that we used $\dot{r}/c = \boldsymbol{\beta}$. This is the equation for the relativistic momentum for force-free motion. Evidently, it is a conserved quantity.

6.6   Relativistic Energy

Let us have a look at the four-force $F^\mu = \gamma \left(v \cdot F^N / c, \, F^N \right)^T$ from (6.18). For the 0-component we have:

$$F^0 = m_0 \frac{du^0}{d\tau} = m_0 \gamma \frac{d}{dt} (\gamma c) = \gamma \frac{v \cdot F^N}{c}. \tag{6.28}$$

We can recognize the relation

$$\frac{d}{dt} (m_0 \gamma c^2) = v \cdot F^N = \frac{F^N \cdot dx}{dt} = \frac{dW}{dt}, \tag{6.29}$$

with the work W. Noting this relation, we can write for the *relativistic energy*

$$W = mc^2 = \gamma m_0 c^2 = E. \tag{6.30}$$

When integrating (6.29) we have chosen an integration constant in such a way that in (6.30) the energy is that of a particle at rest with rest mass m_0:

$$E = m_0 c^2. \tag{6.31}$$

6.6.1 Equivalence of Mass and Energy

It is fair to say that Eq. (6.31) is probably the most famous formula of physics. By this relation, an energy content can be attributed to every mass. For 1 kg of matter, e.g., the equivalent energy is

$$E_{1\,kg} \approx 8.988 \cdot 10^{16}\,J. \tag{6.32}$$

For comparison: In Germany the total primary energy consumption was roughly $1.24 \cdot 10^{19}$ J in 2022 [2]. If one could convert matter in large amounts into energy, then roughly 138 kg would be sufficient to supply the total energy demand.

It is by no means trivial to derive the relation (6.31) in a theoretically rigorous way. Einstein himself devoted a number of papers to this problem, without being completely successful. A review of his attempts can be found in an article by Hecht [4].

In experiments, however, the relation has been confirmed convincingly. For example, Rainville et al. [7] used nuclear reactions in which a nuclide captures a neutron, and then emits a γ-quantum. The mass difference Δm between the nucleus plus the neutron before the reaction, and the mass of the resulting nucleus after capture should exactly be the energy of the γ-quantum. The group found the result

$$1 - \frac{\Delta m c^2}{E} = (-1.4 \pm 4.4) \cdot 10^{-7}. \tag{6.33}$$

When we compare the general result for the relativistic energy with the four-momentum in (6.12), we recognize the relation

$$E = \gamma m_0 c^2 = c p^0 \quad \text{or} \quad p^0 = \frac{E}{c}. \tag{6.34}$$

Thus, we can write the four-momentum as

$$p^\mu = \begin{pmatrix} E/c \\ m_0 \gamma \boldsymbol{v} \end{pmatrix} = \begin{pmatrix} E/c \\ \boldsymbol{p} \end{pmatrix}. \tag{6.35}$$

In this form we call it the *energy-momentum vector*. Contracting with its contravariant counterpart leads to

$$p_\mu p^\mu = -\frac{E^2}{c^2} + \boldsymbol{p}^2. \tag{6.36}$$

Since $p_\mu p^\mu$ is a Lorentz scalar, it has the same value in every inertial system. In the rest frame of the particle, $\boldsymbol{p} = \boldsymbol{0}$, and the energy is the rest energy

$$-\frac{E^2}{c^2} + \boldsymbol{p}^2 = -m_0^2 c^2. \tag{6.37}$$

In this way we find the general relation between relativistic energy and relativistic momentum

$$E^2 = m_0^2 c^4 + c^2 \boldsymbol{p}^2 \quad \text{or} \quad E = \sqrt{m_0^2 c^4 + c^2 \boldsymbol{p}^2}. \tag{6.38}$$

Equation (6.38) is the *relativistic law of conservation of energy*. For small values of the momentum the expression for E in (6.38) can be expanded in a Taylor series:

$$E = m_0 c^2 \sqrt{1 + \frac{\boldsymbol{p}^2}{m_0^2 c^2}} \approx m_0 c^2 \left(1 + \frac{\boldsymbol{p}^2}{2 m_0^2 c^2} + \dots \right) = \frac{\boldsymbol{p}^2}{2 m_0} + m_0 c^2 + \mathcal{O}(\boldsymbol{p}^4),$$
$$\tag{6.39}$$

with the rest energy $m_0 c^2$ and the kinetic energy $\boldsymbol{p}^2/(2 m_0) + \mathcal{O}(\boldsymbol{p}^4)$. We see that, in general, the relativistic energy is the sum of the kinetic energy and the rest energy. This must be taken into account when comparing with the non-relativistic kinetic energy.

What happens at very large velocities? In the case $m_0 \neq 0$ we have

$$E \to \infty \tag{6.40}$$

for $|\boldsymbol{v}| \to c$, because of $\gamma \to \infty$. The relativistic energy diverges when the velocity approaches the speed of light. As a consequence, particles with non-vanishing rest mass always have to move slower than light. To accelerate them to the speed of light, an infinite amount of energy would be necessary.

Here is an illustration: The rest energy of the proton is [10]

$$E_\mathrm{p} = m_\mathrm{p} c^2 = 938.27208943(29) \,\text{MeV}. \tag{6.41}$$

If the proton moves, e.g., with $\beta = 0.9999999725$, its total energy is $E = 4$ TeV, and thus more than 4000 times as big as its rest energy. In the *Large Hadron Collider*, presently the most powerful particle accelerator, protons hit each other with this energy. For two protons this corresponds to a collision energy of 8 TeV. In Fig. 6.2 the relativistic energy is shown for a proton, in comparison with its Newtonian kinetic energy.

6.6.2 Photons and the Compton Effect

Photons have no rest mass. Therefore,

$$p_\mu p^\mu = \frac{E^2}{c^2} - \boldsymbol{p}^2 = m_0^2 c^2 = 0. \tag{6.42}$$

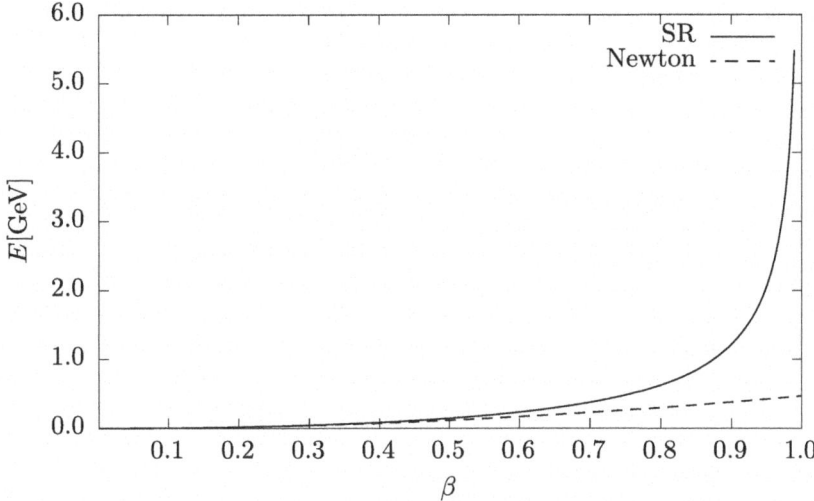

Fig. 6.2 Comparison of relativistic and Newtonian kinetic energy for a proton. In the relativistic expression, the rest energy is subtracted. For $\beta \to 1$ the relativistic energy diverges

Then

$$E = c|\boldsymbol{p}| = \hbar\omega = \frac{hc}{\lambda}. \tag{6.43}$$

We therefore have, with the relations $\boldsymbol{p} = \hbar\boldsymbol{k}$ and $E = h\nu = \hbar\omega$, for the momentum and the wavevector, or energy and frequency,

$$p^\mu = \begin{pmatrix} \hbar\omega/c \\ \hbar\boldsymbol{k} \end{pmatrix} = \hbar \begin{pmatrix} \omega/c \\ \boldsymbol{k} \end{pmatrix} = \hbar k^\mu, \tag{6.44}$$

with the *four-wavevector* k^μ. From (6.43) we can see that

$$|\boldsymbol{k}| = \frac{\omega}{c}, \tag{6.45}$$

i.e., k^μ is a lightlike vector. We now consider some particle with mass m_1, which emits photons. This is, e.g., the situation in excited atoms. To simplify the calculation we assume that the particle is at rest and emits two photons of equal energy in opposite directions. Then the particle experiences no recoil, i.e., no force, and the components of its four-momentum do not change, see Fig. 6.3. The conservation of the total four-momentum leads to

$$\begin{pmatrix} E_1/c \\ 0 \end{pmatrix} = \begin{pmatrix} |\boldsymbol{p}| \\ -\boldsymbol{p} \end{pmatrix} + \begin{pmatrix} E_2/c \\ 0 \end{pmatrix} + \begin{pmatrix} |\boldsymbol{p}| \\ \boldsymbol{p} \end{pmatrix}. \tag{6.46}$$

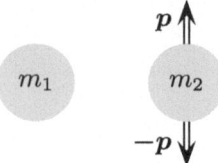

Fig. 6.3 Illustration of the equivalence of mass and energy: A particle which emits two photons of equal energy in opposite directions does not change its momentum, and its kinetic energy. Therefore, its rest mass must decrease

From this follows

$$E_2 = E_1 - 2c|\mathbf{p}| \quad \text{or} \quad m_2c^2 = m_1c^2 - 2c|\mathbf{p}|. \tag{6.47}$$

We can conclude that

$$m_2 = m_1 - 2\frac{|\mathbf{p}|}{c} = m_1 - 2\frac{E_{\text{photon}}}{c^2}. \tag{6.48}$$

The emission of energy, in this case in the form of photons or electromagnetic radiation, decreases the rest mass of the particle.

A very important physical process is the scattering of photons off particles, in particular electrons. Let us assume that the photon moves along the x-axis, and the electron is initially at rest. The four-momenta of the photon and the electron then are

$$p^\mu = \begin{pmatrix} \hbar\omega/c \\ \hbar\omega/c \\ 0 \\ 0 \end{pmatrix} \quad \text{and} \quad e^\mu = \begin{pmatrix} m_e c \\ 0 \\ 0 \\ 0 \end{pmatrix}. \tag{6.49}$$

Therefore, the total four-momentum of the system photon plus electron is

$$g^\mu = p^\mu + e^\mu. \tag{6.50}$$

The total momentum is conserved in the collision, i.e.,

$$p^\mu + e^\mu = p'^\mu + e'^\mu. \tag{6.51}$$

We are not primarily interested in the four-momentum e'^μ of the electron after the collision. Therefore, we eliminate it using the relation (6.51). To do so, we write down the scalar product with p'_μ and exploit the relation $p'_\mu p'^\mu = 0$. Then

$$p'_\mu p^\mu + p'_\mu e^\mu = p'_\mu e'^\mu. \tag{6.52}$$

Furthermore, because of $p_\mu e^\mu = p^\mu e_\mu$, we have

$$g_\mu g^\mu = m_e^2 c^2 + 2 p_\mu e^\mu + 0 = m_e^2 c^2 + 2 p'_\mu e'^\mu + 0 = g'_\mu g'^\mu, \tag{6.53}$$

and thus

$$p_\mu e^\mu = p'_\mu e'^\mu. \tag{6.54}$$

We insert this in (6.52), and find

$$p'_\mu p^\mu + p'_\mu e^\mu = p_\mu e^\mu. \tag{6.55}$$

After the collision the photon can have changed its energy (i.e., its frequency or wavelength) but also its direction of flight. Therefore, we may write

$$p'^\mu = \begin{pmatrix} \hbar \omega'/c \\ (\hbar \omega'/c)\cos(\theta) \\ (\hbar \omega'/c)\sin(\theta) \\ 0 \end{pmatrix}. \tag{6.56}$$

Without loss of generality we may choose the coordinate system in such a way that after the collision the photon moves in the xy-plane. We insert (6.49) and (6.56) in (6.55) and form the scalar products:

$$\frac{\hbar^2}{c^2} \omega \omega' (\cos(\theta) - 1) - m \hbar \omega' = -m \hbar \omega. \tag{6.57}$$

We are interested in the ratio of the wavelengths of the photon prior to and after the collision. Therefore, we set $\omega = 2\pi c/\lambda$, and accordingly $\omega' = 2\pi c/\lambda'$, and find

$$\lambda' - \lambda = \frac{h}{m_e c} (1 - \cos(\theta)). \tag{6.58}$$

In this equation,

$$\lambda_e := \frac{h}{m_e c} = 2.42631023538(76) \cdot 10^{-12} \text{ m} \tag{6.59}$$

is the *Compton wavelength*[1] of the electron [10], and the effect itself is called the Compton effect.

---

[1] Arthur Holly Compton, 1892–1962, US-American physicist, Nobel Prize in physics 1927.

Note that (6.59) is a special case of the *de Broglie relation*[2] $p \cdot \lambda = h$ for relativistic momenta $p \simeq m_e c$.

The intuitive meaning of the Compton wavelength is the increase of the wavelength of a photon which scatters perpendicularly off an electron ($\theta = \pi/2$ in (6.58)). Note that the Compton wavelength does not depend on the energy of the incident photon. Therefore, high-energy photons, with short wavelengths, lose much energy in a collision. On the other hand, low-energy photons can be upscattered to higher energies in collisions with high-energy electrons (inverse Compton effect).

6.6.3   More Examples

The equivalence of mass and energy is manifest in many other physical processes.

Excited States of Atoms and Molecules

Atoms and molecules in excited states are heavier than their counterparts in the ground state. Let us consider, e.g., the hydrogen atom: When a proton and an electron combine to form the ground state of the atom, its binding energy of 13.6 eV is released. The mass of the hydrogen atom is therefore smaller than the sum of the masses of the free proton and the free electron. This is called the *mass defect*; it has its cause in the negative binding energy of proton and electron. For the hydrogen atom the effect is very small since the rest energy of the proton is roughly 70 million times bigger than the binding energy of the hydrogen atom.

Atomic Nuclei

As already mentioned, atomic nuclei also exhibit a mass defect. The total mass of an atomic nucleus is smaller than the sum of the individual masses of the protons and neutrons that constitute the nucleus. The mass defect results from the binding energy E_B/c^2 caused by the strong interaction between the nucleons. Therefore, the mass of the atomic nucleus is

$$m(A, Z) = Zm_p + (A - Z)m_n + \frac{E_B}{c^2}, \tag{6.60}$$

where $E_B < 0$. As an example, let us consider the nuclide ^{12}C. The mass number is $A = 12$, and the nuclear charge $Z = 6$. The *atomic mass unit* is [10]

$$m_u = \frac{1}{12}m(^{12}C) = 1.66053906892(52) \cdot 10^{-27} \text{kg}. \tag{6.61}$$

---

[2] Louis-Victor Pierre Raymond de Broglie, 1892–1987, French physicist. Nobel Prize in physics 1929.

The total mass of the individual constituents of the carbon atom is given by

$$\frac{\frac{1}{12}(6m_p + 6m_n + 6m_e)}{m_u} \approx 1.008, \tag{6.62}$$

with the proton mass

$$m_p = 1.67262192595(52) \cdot 10^{-27} \text{ kg}, \tag{6.63}$$

the neutron mass

$$m_n = 1.67492750056(85) \cdot 10^{-27} \text{ kg}, \tag{6.64}$$

and the electron mass

$$m_e = 9.1093837139(28) \cdot 10^{-31} \text{ kg}. \tag{6.65}$$

We see that in the formation of the carbon atom roughly 0.8% of the total mass of the protons and the neutrons goes into the binding energy.

Nuclear Fusion and Fission Reactions
In both nuclear fission and fusion reactions, a large amount of energy can be released. In Chap. 19 we shall see that the luminosity of stars is due to fusion processes.

Particle-Antiparticle Annihilation
Particle and antiparticle pairs can be produced or annihilated, for example in the reaction

$$e^+ + e^- \longleftrightarrow 2\gamma. \tag{6.66}$$

When an electron and a positron annihilate, they create two photons. The electron rest energy is roughly 511 keV. Therefore,

$$E_\gamma \geq 511 \text{ keV}. \tag{6.67}$$

In the annihilation, 100% of the mass is transformed into energy. The rest energy of the electron is a lower bound on the energy released, since electrons still can have kinetic energy.

6.7 Traveling with Constant Acceleration

As an application of the relativistic mechanics derived in this chapter, we shall consider an astronaut, whose spacecraft is accelerated in his rest frame with constant acceleration $a = g = 9.81 \text{ m s}^{-2}$.

6.7.1   Equations of Motion

We shall formulate the equations of motion in an inertial system S which is at rest on the Earth. The rest frame of the rocket will be denoted S'. It is assumed to move along the x-axis of S. Then the four-acceleration in S' is $b'^{\mu} = (0, g, 0, 0)^{\mathrm{T}}$. The transformation back into the Earth system can be obtained by the inverse Lorentz transformation (5.22)

$$b^{\mu} = \Lambda_{\alpha}{}^{\mu} b'^{\alpha} = \begin{pmatrix} \gamma & \beta\gamma & 0 & 0 \\ \beta\gamma & \gamma & 0 & 0 \\ 0 & 0 & 1 & 0 \\ 0 & 0 & 0 & 1 \end{pmatrix} \cdot \begin{pmatrix} 0 \\ g \\ 0 \\ 0 \end{pmatrix} = \begin{pmatrix} \beta\gamma g \\ \gamma g \\ 0 \\ 0 \end{pmatrix}. \tag{6.68}$$

Let us evaluate this expression. We first note that in the Earth system $\boldsymbol{\beta} = \beta\, \mathbf{e}_x$. Then

$$b^x = \gamma g = \frac{\mathrm{d}}{\mathrm{d}\tau} u^x = \frac{\mathrm{d}}{\mathrm{d}\tau}(\gamma v) = \gamma \frac{\mathrm{d}}{\mathrm{d}t}(\gamma v) = c\gamma \frac{\mathrm{d}}{\mathrm{d}t}(\gamma \beta), \tag{6.69}$$

and thus

$$g = \frac{\mathrm{d}}{\mathrm{d}t}(\gamma v). \tag{6.70}$$

With the initial conditions $x(0) = 0$ and $v(0) = 0$ we have the result

$$\gamma v = gt = \gamma \frac{\mathrm{d}x}{\mathrm{d}t}, \quad \text{or} \quad \frac{\mathrm{d}x}{\mathrm{d}t} = v = \frac{gt}{\gamma} = \sqrt{1 - \beta^2}\, gt. \tag{6.71}$$

Equation (6.71) yields $c^2\beta^2 = (1 - \beta^2)g^2t^2$, or $(c^2 + g^2t^2)\beta^2 = g^2t^2$, and thus finally

$$\beta(t) = \frac{gt}{\sqrt{c^2 + g^2t^2}} = \frac{1}{\sqrt{c^2/(g^2t^2) + 1}} < 1. \tag{6.72}$$

We see that at all times the velocity remains below the speed of light, see Fig. 6.4. We obtain the trajectory $x(t)$ by integrating over the velocity $v(t) = c\beta(t)$. The result is

$$x(t) = \int_0^t v(t')\,\mathrm{d}t' = c\sqrt{t'^2 + \frac{c^2}{g^2}}\Bigg|_0^t = \frac{c^2}{g}\left[\sqrt{1 + \left(\frac{gt}{c}\right)^2} - 1\right]. \tag{6.73}$$

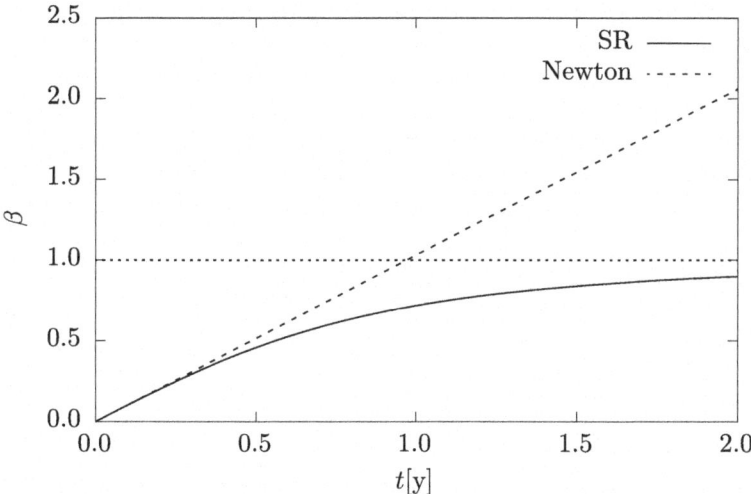

Fig. 6.4 Temporal evolution of the relativistic velocity with constant acceleration. While in Newtonian mechanics the velocity increases beyond all bounds *(dashed line)*, in special relativity the speed of light $\beta = 1$ constitutes an upper bound *(solid line)*

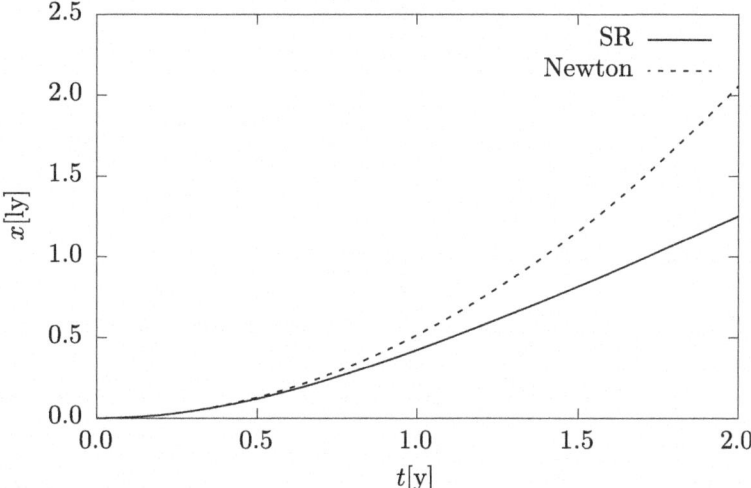

Fig. 6.5 Relativistic equations of motion of the rocket: In Newtonian mechanics, the distance covered increases for all times t as $gt^2/2$ *(dashed line)*. In special relativity, the distance is proportional to ct in the limit of large t, independent of the acceleration g

An expansion of (6.73) for small times t yields the result of Newtonian mechanics, $x(t) \approx \frac{g}{2}t^2$. By contrast, for very long times, $x(t) \approx ct$, independent of the value of the acceleration, see Fig. 6.5.

Traveling with Constant Acceleration from the Point of View of the Astronaut

So far we have studied the motion with constant acceleration from the rest frame of an observer that remains back on Earth. But how does the astronaut himself experience the motion? We can find this out by expressing both the velocity and the distance covered as functions of the proper time. According to (5.46), the relation between the time t and the proper time τ is

$$\tau = \int_0^t \frac{1}{\gamma(t')}\, dt' = \int_0^t \sqrt{1 - \beta^2(t')}\, dt'. \tag{6.74}$$

We insert $\beta(t)$ from (6.72) and obtain

$$\tau = \int_0^t \sqrt{1 - \frac{g^2 t'^2}{c^2 + g^2 t'^2}}\, dt' = \frac{c}{g} \ln\left[\frac{gt}{c} + \sqrt{1 + \left(\frac{gt}{c}\right)^2}\right] = \frac{c}{g} \operatorname{arsinh}\left(\frac{gt}{c}\right). \tag{6.75}$$

Solving with respect to t, we find

$$t(\tau) = \frac{c}{g} \sinh\left(\frac{g}{c}\tau\right). \tag{6.76}$$

Using this result, we can express the world line $x(\tau)$ of the rocket also referred to its own proper time. Inserting (6.76) into (6.73) yields

$$x(\tau) = \frac{c^2}{g}\left[\cosh\left(\frac{g}{c}\tau\right) - 1\right]. \tag{6.77}$$

We recognize that both the time t elapsed in the Earth system and the distance covered depend exponentially on the duration τ of the acceleration. In the literature this is called *hyperbolic motion*, since $t(\tau)$ and $x(\tau)$ are defined in terms of hyperbolic functions.

Because of the exponential relation between the proper time and the distance covered, an astronaut, traveling only with Earth's constant gravitational acceleration g, could reach very far away destinations in the Universe in relatively short times. However, the main obstacle to the actual realization of such a journey is, apart from technical problems, the immense supply of energy necessary to permanently maintain the acceleration.

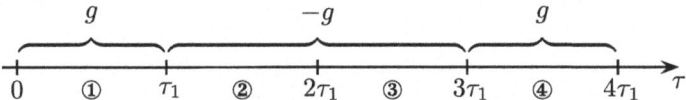

Fig. 6.6 Parts of a round trip in the twin paradox. In part ① the astronaut accelerates in the direction to his destination. In part ② he breaks on his way to the destination and finally comes to a halt at the destination. In part ③ he accelerates again towards the Earth, and in part ④ he breaks, to stop at the Earth

6.7.2   Application to the Twin Paradox

In Sect. 4.4 we had already briefly addressed the twin paradox. Using Eqs. (6.76) and (6.77) for time and position, we can now discuss the twin paradox quantitatively. Our experimental setup is as follows: One of the two twins travels, in his rest frame, for a period τ_1 with constant acceleration. Then he brakes for the same period τ_1 with acceleration $-g$ to arrive at his destination point $\tau = 2\tau_1$ with velocity zero. He starts his way back by accelerating in the direction of the Earth with constant g, and begins to break after another period τ_1 with constant g in order to arrive at the Earth with zero velocity. Thus, in his rest frame he returns to the Earth after a proper time of $4\tau_1$.

Since the sign of the acceleration changes during the journey, the functions $t_{tw}(\tau)$ and $x_{tw}(\tau)$ are defined on different sections. The periods ② and ③ of the journey depicted in Fig. 6.6 can be treated together. Solving the corresponding equations for the different branches yields

$$t_{tw}(\tau) = \frac{c}{g} \begin{cases} \sinh\left(\frac{g}{c}\tau\right) & \text{for } \tau \le \tau_1, \\ \sinh\left[\frac{g}{c}(\tau - 2\tau_1)\right] + 2\sinh\left(\frac{g}{c}\tau_1\right) & \text{for } \tau_1 < \tau \le 3\tau_1, \\ \sinh\left[\frac{g}{c}(\tau - 4\tau_1)\right] + 4\sinh\left(\frac{g}{c}\tau_1\right) & \text{for } 3\tau_1 < \tau \le 4\tau_1 \end{cases} \qquad (6.78)$$

and

$$x_{tw}(\tau) = \frac{c^2}{g} \begin{cases} \cosh\left(\frac{g}{c}\tau\right) - 1 & \text{for } \tau \le \tau_1, \\ 2\cosh\left(\frac{g}{c}\tau_1\right) - \cosh\left[\frac{g}{c}(\tau - 2\tau_1)\right] - 1, & \text{for } \tau_1 < \tau \le 3\tau_1, \\ \cosh\left[\frac{g}{c}(\tau - 4\tau_1)\right] - 1 & \text{for } 3\tau_1 < \tau \le 4\tau_1. \end{cases}$$
$$(6.79)$$

The integration constants in the different branches result from the constraint that these are continuous functions.

We use the relation (5.46) to determine the velocity $\beta(\tau)$ during the journey

$$\gamma(\tau) = \frac{dt}{d\tau} = \frac{1}{\sqrt{1 - \beta(\tau)^2}}, \quad \text{or} \quad \beta(\tau) = \pm\frac{\sqrt{\gamma^2 - 1}}{\gamma}, \qquad (6.80)$$

with

$$\gamma(\tau) = \cosh\left(\frac{g}{c}\tau\right), \tag{6.81}$$

which we find using (6.76). Note that in the second branch we have to choose the negative sign. Using the relation $\cosh^2(x) - \sinh^2(x) = 1$ of hyperbolic functions, we can write

$$\beta_{tw}(\tau) = \begin{cases} \tanh\left(\frac{g}{c}\tau\right) & \text{for} \quad \tau \leq \tau_1, \\ -\tanh\left[\frac{g}{c}(\tau - 2\tau_1)\right], & \text{for} \quad \tau_1 < \tau \leq 3\tau_1, \\ \tanh\left[\frac{g}{c}(\tau - 4\tau_1)\right] & \text{for} \quad 3\tau_1 < \tau \leq 4\tau_1. \end{cases} \tag{6.82}$$

Since $\tanh(y) \in (-1, 1)$, we recognize again that the velocity always remains below the speed of light.

As an example, we consider a journey with an acceleration period τ_1 of 2 years. Then, for the traveling twin the journey lasts 8 years. From (6.78) we can calculate the duration of the journey for the twin remaining on Earth, it is[3]

$$t(4\tau_1) = 15.03 \, \text{y}. \tag{6.83}$$

It is only after this time that the traveling twin returns to Earth. Of course the maximum distance from the Earth is reached after half the journey. From (6.79) follows

$$x(2\tau_1) = 5.82 \, \text{ly}. \tag{6.84}$$

The maximum velocity on the journey according to (6.82) is

$$\beta_{max} = \beta(\tau_1) = -\beta(3\tau_1) = 0.968. \tag{6.85}$$

This corresponds to a value of $\gamma(\tau_1) = 4.00$.

Figure 6.7 visualizes the journey of the twin. The top panel shows the relation between the coordinate time and the distance from the Earth. The points of maximum velocity on the way to the destination and on the way back are marked by black dots, as is the point of return. The panel in the middle of Fig. 6.7 shows the relation between the coordinate time t and the proper time τ. The black dots correspond to the positions in the top panel. The slope of this curve is given by $\gamma(\tau)$. At the point of return, $\beta = 0$, and consequently $\gamma = 1$. It is only at this point that coordinate time and proper time coincide. The exact opposite is the case at the

---

[3] A note which helps to simplify the calculations: The results given can be found in a very convenient way by measuring time in years and distances using light years. In these units, $c = 1 \, \text{ly} \, \text{y}^{-1}$, and the acceleration is simply given by $g = 9.81 \, \text{m} \, \text{s}^{-2} = 1.03 \, \text{ly} \, \text{y}^{-2}$.

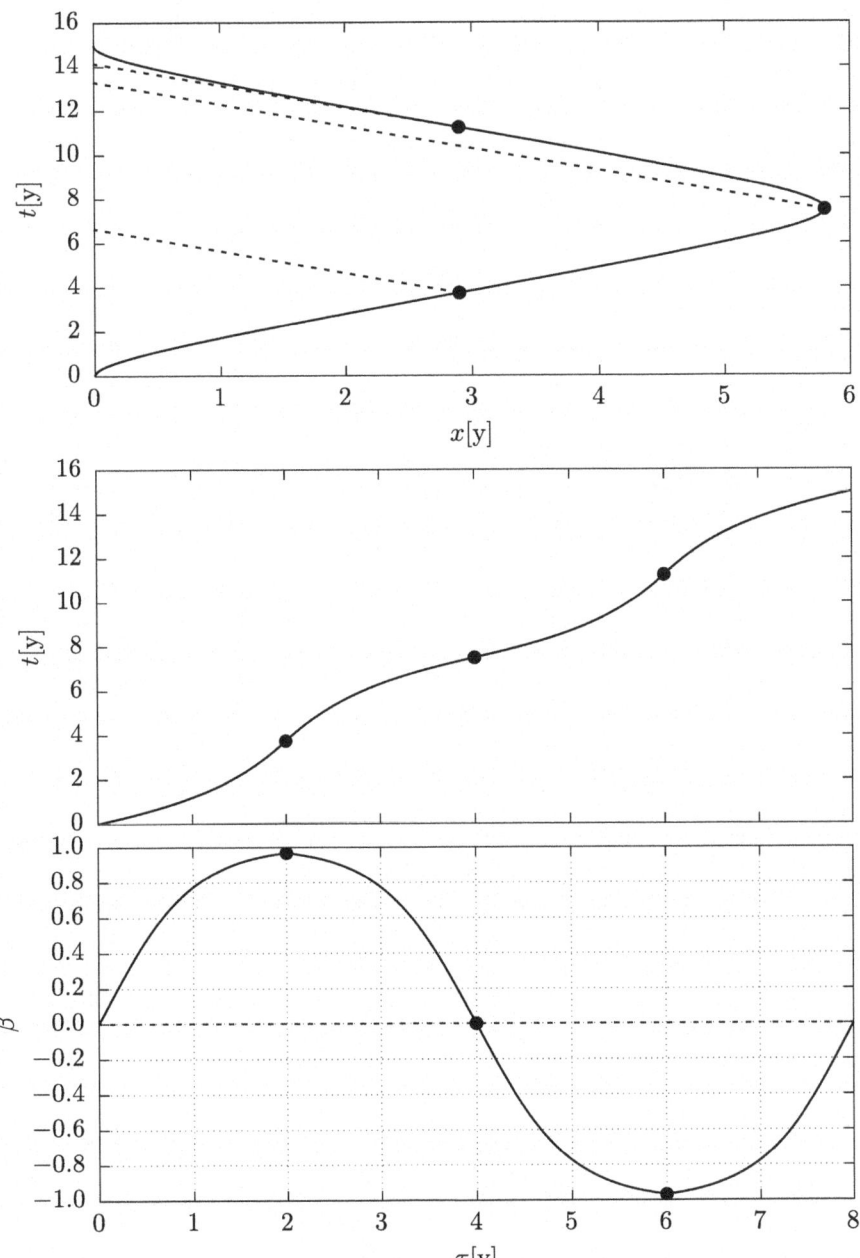

Fig. 6.7 World line $x(t)$ (*top*), coordinate time $t(\tau)$ (*middle*), and velocity $\beta(\tau)$ (*bottom*) of the traveling twin. In the top panel, light signals sent from the traveling twin to the Earth at times $t = \tau_1$, $2\tau_1$, and $3\tau_1$ are plotted as dashed lines

two points of maximum velocity. There $\gamma = 4.00$, and thus the difference between proper time and coordinate time is maximal. Finally, the bottom panel in Fig. 6.7 shows the velocity during the journey as a function of the proper time.

Because of the time dilation, it would be possible, at least in principle, to travel to the edge of the observable Universe during a lifetime. The total duration τ_f of a round trip to an object at the distance s from the Earth can be determined easily from (6.79). To see this, we solve the condition

$$x(2\tau_1) = 2x(\tau_1) = s \tag{6.86}$$

with respect to τ_1, and find the total duration of the journey

$$\tau_f = 4\tau_1(s) = 4\frac{c}{g}\operatorname{arcosh}\left(\frac{1}{2}\frac{gs}{c^2} + 1\right). \tag{6.87}$$

Table 6.1 lists the total travel times, the elapsed coordinate time, the maximum velocity, and the maximal Lorentz factor γ for typical distances to nearby stars, to the Andromeda Galaxy, and to the edge of the observable Universe. The latter distance we roughly estimate by multiplying the age of the Universe [6] by the speed of light, which essentially corresponds to the *Hubble distance* (cf. Sect. 23.3).

As mentioned, because of the extreme time dilation, an astronaut could cover even most extreme distances in his lifetime, even in the range of billions of light years. However, in our later discussion of cosmology, we shall see that our Universe is not a static Minkowski spacetime, but must be described as a space expanding forever. In a spacetime like this, additional aspects arise in the twin paradox, since during an extended, long journey, spacetime expands. In this situation one needs a concept of constant acceleration in a curved spacetime, as it is discussed in Rindler [8]. However, we shall not elaborate on this point, since it could be understood only within the framework of general relativity, and using the resulting equations of cosmology. The reader interested in this topic will find the relevant discussion in Boblest et al. [1].

Another serious problem is the incredible amount of energy necessary for such a journey. The relativistic energy in (6.30) is given by the Lorentz factor γ times the

Table 6.1 Comparison of round trips to different destinations in the classical twin paradox. The table lists the proper time τ_f of the astronaut and the coordinate time $t(\tau_f)$ at the end of the journey, the maximum velocity β_{max}, and the maximum Lorentz factor γ. The results are shown for typical distances to nearby stars, for a journey to the Andromeda Galaxy, roughly two million light years away, and to the edge of the visible Universe; see also Müller et al. [5]

Distance [ly]	τ_f[y]	$t(\tau_f)$ [y]	β_{max}	γ_{max}
5	7.55	13.32	0.9602	3.582
100	18.03	203.84	0.9998	52.64
$2.00 \cdot 10^6$	56.31	$4.00 \cdot 10^6$	$1 - 4.69 \cdot 10^{-13}$	$1.033 \cdot 10^6$
$1.3784 \cdot 10^{10}$	90.54	$2.76 \cdot 10^{10}$	$1 - 9.87 \cdot 10^{-21}$	$7.12 \cdot 10^9$

rest energy. Thus, if a Lorentz factor in the range of 10^6 or higher occurs during such a journey, one would need enough fuel to produce more than a million times the rest energy of the spacecraft. Even if one were able to convert matter and antimatter into radiation, the ratio of the amounts of payload and of fuel would be way above one into a million, in particular, since the fuel for later phases of the space travel also has to be accelerated from the start.

6.8   Relativistic Circular Orbit

As a short supplement, we will discuss the motion of a particle with constant velocity on a circular orbit. In this case, the modulus of the acceleration is constant, but its acceleration vector is always perpendicular to the direction of motion.

We assume that the motion is in the xy-plane, i.e., $z = 0$. At time $\tau = t = 0$, the particle is located at the coordinates $x = R$, $y = z = 0$, where R is the radius of the circular orbit. As in the previous section, we choose the modulus of the four-acceleration $a = g$.

The world line of the particle has the form

$$x^\mu(\tau) = \begin{pmatrix} c\gamma\tau \\ R\cos(\omega\tau) \\ R\sin(\omega\tau) \end{pmatrix}. \tag{6.88}$$

Note that γ is time-independent since the velocity is constant. The same is true for the angular velocity ω. The four-velocity then is

$$u^\mu(\tau) = \frac{dx^\mu(\tau)}{d\tau} = \begin{pmatrix} c\gamma \\ -R\omega\sin(\omega\tau) \\ R\omega\cos(\omega\tau) \end{pmatrix}. \tag{6.89}$$

To determine the angular velocity ω we use the constraint $u^\mu u_\mu = -c^2$, which leads to

$$-c^2 = -c^2\gamma^2 + R^2\omega^2. \tag{6.90}$$

This implies

$$\omega = \frac{c\gamma\beta}{R}. \tag{6.91}$$

We have used the relation $\gamma^2 - 1 = \gamma^2\beta^2$. Therefore, we obtain the four-acceleration

$$b^\mu(\tau) = \frac{du^\mu(\tau)}{d\tau} = \begin{pmatrix} 0 \\ -R\omega^2\cos(\omega\tau) \\ -R\omega^2\sin(\omega\tau) \end{pmatrix}. \tag{6.92}$$

As in (6.68), the four-acceleration is related to the acceleration in the rest frame of the observer by a Lorentz transformation, with the subtle difference, that the acceleration is always perpendicular to the direction of motion. When we insert the acceleration $b'^\mu = (0, -g, 0, 0)$ at $\tau = t = 0$ in (6.68), we immediately recognize that in this case $b^\mu = (0, -g, 0, 0)$, without an additional factor γ. For all other times the situation is analogous, and can be obtained by an additional rotation of the frame of reference.

As a consequence we can simply set $|\underline{b}| = \sqrt{b_\mu b^\mu} = g$. Together with the relations (6.91) and (6.92) we directly obtain

$$|\underline{b}| = \frac{c^2 \gamma^2 \beta^2}{R} \tag{6.93}$$

or

$$R(\beta) = \frac{c^2 \gamma^2 \beta^2}{g}. \tag{6.94}$$

In the last step we have found the radius of the circular orbit. In classical mechanics we can calculate the radius of the corresponding circular orbit by setting the acceleration $a = g$ equal to the centripetal acceleration $a_{\mathrm{cp}} = v^2 / R$, which leads to

$$R = \frac{v^2}{g} = \frac{c^2 \beta^2}{g}. \tag{6.95}$$

We see that in the relativistic case an additional factor γ^2 appears. For $v \ll c$ the factor $\gamma \to 1$, and the classical result turns up as the non-relativistic limit, see Fig. 6.8.

It must be noted that in this situation another effect would have to be considered, which we will not treat in the present context. This effect is the *Thomas precession*[4] of the frame of reference of the astronaut. This precession has the effect that after a complete cycle the frame of reference is no longer parallel to the original one, unless the astronaut corrects this by additional navigation. The Thomas precession [9] is also important in atomic physics as a relativistic correction to spin-orbit coupling.

6.9   Exercises

6.1. Round trip on a circular orbit (for motivated math freaks) In this exercise, the results of Sects. 6.7 and 6.8 are combined in a round trip, which begins and ends with a phase of linear constant acceleration, but contains an interim phase in

---

[4] Llewellyn Hilleth Thomas, 1903–1992, British theoretical physicist and mathematician.

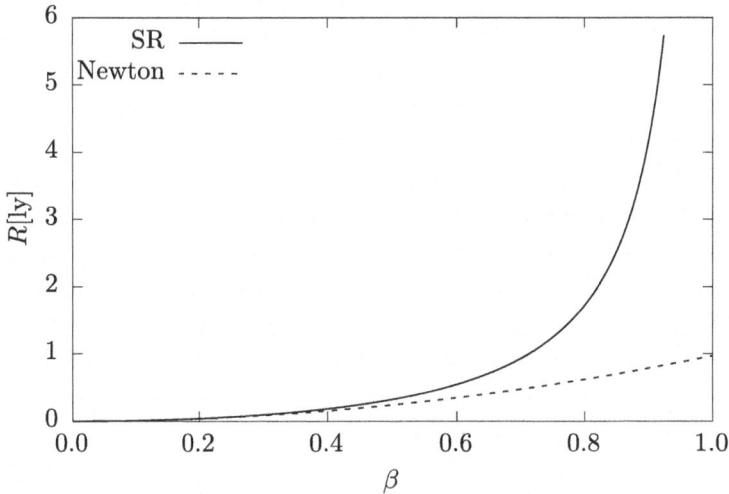

Fig. 6.8 Radius $R(\beta)$ of a circular orbit at relativistic velocities and for constant acceleration $a = g$. In the relativistic calculation (*solid line*) the radius diverges at $\beta = 1$. The radius according to Newtonian mechanics (*dashed line*) increases quadratically with the velocity. For small velocities $v \ll c$ the relativistic result and the classical result agree

which the astronaut turns around by flying on a circular orbit with constant velocity, see Fig. 6.9.

Instead of starting along the x-axis, the astronaut must start under an angle δ with respect to the x-axis. This means that before swinging into the turning cycle the world line has the form

$$\begin{pmatrix} x(\tau) \\ y(\tau) \end{pmatrix} = d(\tau) \begin{pmatrix} \cos[\delta(\tau_1)] \\ \sin[\delta(\tau_1)] \end{pmatrix}. \tag{6.96}$$

Here we have introduced the dependence of the starting angle on τ_1 to demonstrate that this angle depends on the length of the acceleration phase τ_1 via the radius of the turning cycle, which also depends on τ_1.

(a) Determine the mathematical representation of the total world line of the astronaut.
(b) Calculate the distances the astronaut can reach for a given length of the acceleration phase τ_1.
(c) For a given distance, determine the necessary length of the acceleration phase. Calculate the total proper time that it takes the astronaut to return to his origin, and the time measured by an observer remaining on Earth. What are the differences compared to a strictly linear journey?

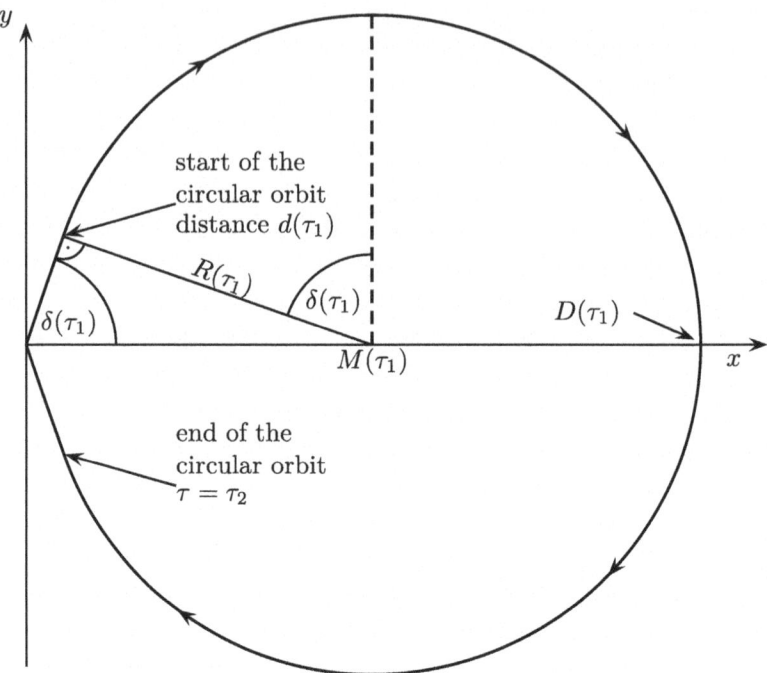

Fig. 6.9 Sketch for an astronaut on a turning cycle

6.2. Relativistic acceleration Calculate the two scalar products $u_\mu b^\mu$ and $b_\mu b^\mu$ and discuss the results.

References

1. Boblest, S., Müller, T., Wunner, G.: Twin paradox in de Sitter spacetime. Eur. J. Phys. **32**(5), 1117–1142 (2011)
2. Bundesministerium für Wirtschaft und Klimaschutz: Energieeffizienz in Zahlen: Entwicklungen und Trends in Deutschland (Federal Ministry for Economic Affairs and Climate Protection: Energy efficiency in figures: Developments and trends in Germany) 2022
3. Goldstein, H.: Classical Mechanics. Addison-Wesley, Reading (2001)
4. Hecht, E.: How Einstein confirmed $E_0 = mc^2$. Am. J. Phys. **79**(6), 591–600 (2011)
5. Müller, T., King, A., Adis, D.: A trip to the end of the universe and the twin "paradox". Am. J. Phys. **76**, 360–373 (2008)
6. Planck Collaboration: Planck 2013 results. I. Overview of products and scientific result. Astron. Astrophys. **571**, A1 (2014)
7. Rainville, S., et al.: World year of physics: a direct test of $E = mc^2$. Nature **438**, 1096–1097 (2005)
8. Rindler, W.: Hyperbolic motion in curved space time. Phys. Rev. **119**, 2082–2089 (1960)
9. Thomas, L.H.: The motion of the spinning electron. Nature **117**, 514 (1926)
10. Mohr, P.J., Tiesinga, E., Newell, D.B., Taylor, B.N.: Codata Internationally Reconmmended [sic!] 2022 Values of the Fundamental Physical Constants. https://physics.nist.gov/constants

Covariant Formulation of Electrodynamics

7

Contents

Newtonian mechanics is only Galilei invariant. Therefore, it was necessary to find a covariant formulation of relativistic mechanics. This led to a modification of the classical equations of motion.

By contrast, electrodynamics, i.e., Maxwell's equations, are already inherently Lorentz covariant. This, however, is not evident from the formulation of the equations in terms of the electric field E, magnetic flux density B, electric currents j, and charge densities ρ_{el}. In particular, neither E, B nor j are four-vectors, and ρ_{el} is not a Lorentz scalar. In this chapter we will present Maxwell's equations in a covariant form. This will allow us to study the transformation behavior of electric and magnetic fields, as well as of charges and currents.

© The Editor(s) (if applicable) and The Author(s), under exclusive license to
Springer-Verlag GmbH, DE, part of Springer Nature 2026
S. Boblest et al., *Special and General Relativity*,
https://doi.org/10.1007/978-3-662-71332-7_7

7.1   Potentials in Classical Electrodynamics

In Sect. 1.3 we had already briefly discussed electrodynamics. In this section, we shall extend that discussion. We already know Maxwell's equations. These can be split into the *homogeneous Maxwell equations*

$$\nabla \times \boldsymbol{E} + \dot{\boldsymbol{B}} = \boldsymbol{0}, \tag{7.1}$$

$$\nabla \cdot \boldsymbol{B} = 0, \tag{7.2}$$

and the *inhomogeneous Maxwell equations*

$$\nabla \times \boldsymbol{B} = \mu_0(\boldsymbol{j} + \varepsilon_0 \dot{\boldsymbol{E}}), \tag{7.3}$$

$$\nabla \cdot \boldsymbol{E} = \frac{\rho_{\text{el}}}{\varepsilon_0}. \tag{7.4}$$

Integrating (7.1) over some surface F and using Stokes' theorem yields Faraday's law of induction

$$\int_F (\nabla \times \boldsymbol{E})\, \mathrm{d}f = \int_{\partial F} \boldsymbol{E} \cdot \mathrm{d}\boldsymbol{s} = U_{\text{ind}} = -\int_F \dot{\boldsymbol{B}} \cdot \mathrm{d}f = -\dot{\varPhi}, \tag{7.5}$$

with the *magnetic flux* through the surface F

$$\varPhi = \int_F \boldsymbol{B} \cdot \mathrm{d}f. \tag{7.6}$$

Equation (7.2) implies that no magnetic monopoles exist. Vector analysis tells us that, because of (7.2), $\boldsymbol{B}$ can be represented as the curl of some *vector potential*,

$$\boldsymbol{B} = \nabla \times \boldsymbol{A}. \tag{7.7}$$

From (7.1) then follows

$$\nabla \times \boldsymbol{E} + \nabla \times \dot{\boldsymbol{A}} = \nabla \times (\boldsymbol{E} + \dot{\boldsymbol{A}}) = \boldsymbol{0}. \tag{7.8}$$

Thus, the curl of the vector field $\boldsymbol{E} + \dot{\boldsymbol{A}}$ vanishes. Vector analysis also tells us that such a vector field can be represented, on a simply connected manifold, as the gradient of some scalar field. Therefore, we can also write

$$\boldsymbol{E} + \dot{\boldsymbol{A}} = -\nabla \varphi, \tag{7.9}$$

with the *electric potential* φ. Solved with respect to the electric field this leads to

$$E = -\nabla \varphi - \dot{A}. \tag{7.10}$$

Neither the scalar potential nor the vector potential are unique. We therefore have a *gauge freedom*. If we set $A' = A + \nabla \chi$, with some arbitrary scalar function $\chi(r, t)$, we obtain

$$\nabla \times A' = \nabla \times A + \underbrace{\nabla \times (\nabla \chi)}_{0} = \nabla \times A = B. \tag{7.11}$$

In this case, the electric field is

$$E = -\nabla \varphi - \dot{A}' + \nabla \dot{\chi} = -\nabla \varphi' - \dot{A}' \tag{7.12}$$

with

$$\varphi' = \varphi - \dot{\chi}. \tag{7.13}$$

The quantity $\chi(r, t)$ is called the *gauge function*, and the mapping $(A, \varphi) \mapsto (A', \varphi')$ is called a *gauge transformation*.

7.2   Formulation Using the Four-Potential

When we insert the potentials into (7.3) and (7.4) we find with $\mu_0 \varepsilon_0 = 1/c^2$ from (1.8),

$$\nabla \times (\nabla \times A) + \frac{1}{c^2}(\nabla \dot{\varphi} + \ddot{A}) = \mu_0 j \tag{7.14a}$$

and

$$\nabla \cdot \left(-\nabla \varphi - \dot{A} \right) = \frac{\rho_{el}}{\varepsilon_0}. \tag{7.14b}$$

Using the relation $a \times (b \times c) = b(a \cdot c) - c(a \cdot b)$, we can write (7.14a) as

$$\nabla (\nabla \cdot A) - \Delta A + \frac{1}{c^2} \left(\nabla \dot{\varphi} + \ddot{A} \right) = \mu_0 j. \tag{7.15}$$

We now exploit the gauge freedom, and choose the potentials in such a way that

$$\frac{\dot{\varphi}}{c^2} + \nabla \cdot A = 0. \tag{7.16}$$

This means that the first term in (7.15) and the first term in the bracket in (7.15) should cancel each other. Then just remains

$$- \Delta A + \frac{\ddot{A}}{c^2} = \Box A = \mu_0 j. \tag{7.17}$$

In this equation, we have introduced the *d'Alembert operator*[1]

$$\partial_\mu \partial^\mu = -\frac{1}{c^2} \frac{\partial^2}{\partial t^2} + \Delta = -\Box, \tag{7.18}$$

with the differential operators ∂_μ and ∂^μ from (5.32).

The gauge used above is called the *Lorenz gauge condition*.[2] In this case, the gauge function $\chi(r, t)$ is the solution of the differential equation

$$\frac{1}{c^2} \ddot{\chi} - \Delta \chi = \frac{1}{c^2} \dot{\varphi} + \nabla \cdot A. \tag{7.19}$$

Let us reformulate the condition (7.16) once again:

$$\frac{1}{c} \frac{\partial}{\partial t} \frac{\varphi}{c} + \nabla \cdot A = 0. \tag{7.20}$$

We recognize that this equation corresponds to the requirement that the covariant derivative ∂_μ of the four-vector

$$A^\mu = \begin{pmatrix} \varphi/c \\ A \end{pmatrix} \tag{7.21}$$

should vanish. The vector field A^μ is called the *four-potential*. The Lorenz gauge condition can then be written as

$$\partial_\mu A^\mu = 0. \tag{7.22}$$

The Lorenz gauge condition can be postulated in this form in *any* frame of reference. Therefore, $\partial_\mu A^\mu$ is a Lorentz scalar. Accordingly, A^μ is a contravariant Lorentz tensor of first rank. Consequently, under a Lorentz transformation it transforms as the coordinates.

---

[1] Jean-Baptiste le Rond d'Alembert, 1717–1783, French mathematician and philosopher. D'Alembert's principle of classical mechanics is named after him.

[2] Ludvig Lorenz, 1829–1891, Danish physicist.

From (7.14b) follows

$$-\nabla \cdot \dot{A} - \Delta\varphi = \frac{\rho_{el}}{\varepsilon_0}. \tag{7.23}$$

Taking the time derivative of $\partial_\mu A^\mu = 0$, we find

$$\frac{1}{c^2}\ddot{\varphi} = -\nabla \cdot \dot{A}. \tag{7.24}$$

Using Eqs. (7.23) and (7.24), we finally end up with

$$\frac{1}{c^2}\frac{\partial^2}{\partial t^2}\frac{\varphi}{c} - \Delta\frac{\varphi}{c} = -\partial_\mu\partial^\mu\frac{\varphi}{c} = \Box\frac{\varphi}{c} = \frac{1}{c}\frac{\rho_{el}}{\varepsilon_0} = \sqrt{\frac{\mu_0}{\varepsilon_0}}\rho_{el}. \tag{7.25}$$

In the final step we have again used $c^2 = 1/\mu_0\varepsilon_0$ from (1.8). Combining (7.17) with (7.25), we have the result

$$\Box A^\mu = \begin{pmatrix} \sqrt{\frac{\mu_0}{\varepsilon_0}}\rho_{el} \\ \mu_0 \cdot j \end{pmatrix} = \mu_0\begin{pmatrix} c\rho_{el} \\ j \end{pmatrix} = \mu_0 j^\mu, \tag{7.26}$$

with the *four-current* j^μ. Evidently it is a contravariant Lorentz tensor of first rank. Therefore,

$$\partial_\mu j^\mu = \dot{\rho}_{el} + \nabla \cdot j \tag{7.27}$$

is a Lorentz scalar.

We can still extract some more information from this expression. Knowing that the divergence of a vector field with vanishing curl is zero, we can write

$$\nabla \cdot (\nabla \times B) = 0 = \mu_0(\nabla \cdot j + \varepsilon_0 \nabla \cdot \dot{E}) = \mu_0(\nabla \cdot j + \dot{\rho}_{el}). \tag{7.28}$$

Since the right-most part of (7.28) is a Lorentz scalar, the equation must apply in all frames of reference. Therefore, this *continuity equation*

$$\partial_\mu j^\mu = \dot{\rho}_{el} + \nabla \cdot j = 0 \tag{7.29}$$

is valid in all frames of reference.

7.2.1   Wave Equation

We consider (7.26) in vacuum. Then $j^\mu = 0$, and we obtain the *wave equation*

$$\Box A^\nu = 0. \tag{7.30}$$

The most general solution of (7.30) is a superposition of plane waves,

$$f(x^\mu) = e^{-ik_\mu x^\mu} = e^{i(k \cdot x - \omega t)}. \tag{7.31}$$

Here the *four-wavevector* from (6.44) appears in the covariant form

$$k_\mu = \begin{pmatrix} -\omega/c \\ k \end{pmatrix}. \tag{7.32}$$

Inserting (7.31) into (7.30), we see that f has to fulfill the equation:

$$\Box f(x^\mu) = \left(\frac{1}{c^2} \frac{\partial^2}{\partial t^2} - \Delta \right) f(x^\mu) = \left(k^2 - \frac{\omega^2}{c^2} \right) f(x^\mu) \overset{!}{=} 0. \tag{7.33}$$

Therefore,

$$\omega = c|k|. \tag{7.34}$$

This is already known from (6.45). It is also known that k_μ is a lightlike vector. Here we see that the lightlike character of k^μ is a necessary condition for photons to fulfill their own wave equation. The most general solution of the wave equation of electrodynamics in vacuum can then be written in the form

$$A^\nu(x^\mu) = \int_{\mathbb{R}^{1,3}} \tilde{A}^\nu(k^\mu)\delta[(k^0)^2 - k^2]e^{-ik_\mu x^\mu} d^4k, \tag{7.35}$$

where the amplitudes $\tilde{A}^\nu(k^\mu)$ can be arbitrary.

7.3   Formulation Using the Field Strength Tensor

As already mentioned, neither the electric nor the magnetic field are Lorentz covariant. Yet, with (7.26) we have found a relation equivalent to Maxwell's equations which is manifestly Lorentz covariant. The Lorentz covariance, however, is formulated for the four-potential, and not for the electric and magnetic field. If one wished to better understand how these fields transform in special relativity, a formulation would be desirable in which these fields show up explicitly. We now work out this kind of formulation.

7.3.1   Field Strength Tensor

The central object in the covariant formulation of electrodynamics is the *field strength tensor*. Let us calculate the *four-dimensional curl* of the four-potential:

$$F_{\mu\nu} = \partial_\mu A_\nu - \partial_\nu A_\mu. \tag{7.36}$$

Evidently $F_{\mu\nu}$ is an antisymmetric covariant Lorentz tensor of second rank. Note that in (7.36)

$$A_\nu = \eta_{\mu\nu} A^\mu = \begin{pmatrix} -\frac{\varphi}{c} \\ A \end{pmatrix}. \tag{7.37}$$

We evaluate (7.36) for all components. Since the four-dimensional curl is antisymmetric, all diagonal elements must vanish:

$$F_{00} = F_{11} = F_{22} = F_{33} = 0. \tag{7.38a}$$

Next, we consider all space and time components. From (7.10) follows

$$F_{i0} = \partial_i A_0 - \partial_0 A_i = -\frac{1}{c}\frac{\partial \varphi}{\partial x^i} - \frac{1}{c}\dot{A}_i = \frac{1}{c}E_i = -F_{0i}. \tag{7.38b}$$

For the pure spatial components this yields

$$F_{12} = \frac{\partial A_y}{\partial x^1} - \frac{\partial A_x}{\partial x^2} = B_z = -F_{21}, \tag{7.38c}$$

$$F_{13} = -B_y = -F_{31}, \tag{7.38d}$$

$$F_{23} = B_x = -F_{32}. \tag{7.38e}$$

Therefore, written in matrix form, we have

$$F_{\mu\nu} = \begin{pmatrix} 0 & -E_x/c & -E_y/c & -E_z/c \\ E_x/c & 0 & B_z & -B_y \\ E_y/c & -B_z & 0 & B_x \\ E_z/c & B_y & -B_x & 0 \end{pmatrix}. \tag{7.39}$$

The *contravariant* field strength tensor can be defined using the general relation between covariant and contravariant tensors:

$$F^{\mu\nu} = \eta^{\mu\alpha}\eta^{\nu\beta} F_{\alpha\beta} = \begin{pmatrix} 0 & E_x/c & E_y/c & E_z/c \\ -E_x/c & 0 & B_z & -B_y \\ -E_y/c & -B_z & 0 & B_x \\ -E_z/c & B_y & -B_x & 0 \end{pmatrix}. \tag{7.40}$$

Now that we know both the covariant and contravariant field strength tensor, we can obtain another Lorentz scalar by contraction. The result is

$$F_{\mu\nu}F^{\mu\nu} = \frac{2}{c^2}\left(c^2 B^2 - E^2\right). \tag{7.41}$$

7.3.2　Dual Field Strength Tensor

To formulate Maxwell's equations in a form as compact as possible, we introduce another tensor. We first define the completely antisymmetric *Levi-Civita tensor:*[3]

$$
\varepsilon^{\kappa\lambda\mu\nu} = \begin{cases} 1 & \text{if } \{\kappa\lambda\mu\nu\} \text{ is an even permutation of } 1,\,2,\,3,\,4, \\ -1 & \text{if } \{\kappa\lambda\mu\nu\} \text{ is an odd permutation of } 1,\,2,\,3,\,4, \\ 0 & \text{otherwise.} \end{cases} \tag{7.42}
$$

The Levi-Civita tensor is a *pseudotensor* of fourth rank, which means that it changes sign under the parity transformation. Thus, a change of coordinates results in

$$
\Lambda^{\alpha}{}_{\kappa}\Lambda^{\beta}{}_{\lambda}\Lambda^{\gamma}{}_{\mu}\Lambda^{\delta}{}_{\nu}\varepsilon^{\kappa\lambda\mu\nu} = \varepsilon^{\alpha\beta\gamma\delta}. \tag{7.43}
$$

The Levi-Civita tensor helps us to define the *dual field strength tensor*

$$
\widehat{F}^{\mu\nu} = \frac{1}{2}\varepsilon^{\mu\nu\alpha\beta}F_{\alpha\beta} = \begin{pmatrix} 0 & B_x & B_y & B_z \\ -B_x & 0 & -E_z/c & E_y/c \\ -B_y & E_z/c & 0 & -E_x/c \\ -B_z & -E_y/c & E_x/c & 0 \end{pmatrix}. \tag{7.44}
$$

The tensor $\widehat{F}^{\mu\nu}$ is also a pseudotensor. Moreover, by contraction of $\widehat{F}^{\mu\nu}$ with $F_{\mu\nu}$ we can construct the *pseudoscalar:*

$$
F_{\mu\nu}\widehat{F}^{\mu\nu} = -\frac{1}{4c}\boldsymbol{B}\cdot\boldsymbol{E}. \tag{7.45}
$$

7.3.3　First Conclusions

From the Lorentz scalar (7.41) and the pseudoscalar (7.45) we can already gather important information on how the electric and magnetic field transform under Lorentz transformations. We exploit the fact that in all inertial frames of reference Lorentz scalars have the same numerical value. We can deduce the following results:

1. If in one inertial system one has $\boldsymbol{E}\cdot\boldsymbol{B} = 0$, or $\boldsymbol{E}\perp\boldsymbol{B}$, then one has $\boldsymbol{E}\cdot\boldsymbol{B} = 0$, or $\boldsymbol{E}\perp\boldsymbol{B}$, in *all* inertial systems.
2. If, in addition, $\boldsymbol{E}^2 - c^2\boldsymbol{B}^2 > 0$, then there exists an inertial system in which $\boldsymbol{B}' = \boldsymbol{0}$. In that system, only an electric field is present. The magnetic field is transformed away.

---

[3] Tullio Levi-Civita, 1873–1941, Italian mathematician.

If, on the contrary, $E^2 - c^2 B^2 < 0$, then there exists an inertial system in which $E' = 0$. In that system, only a magnetic field is present. The electric field is transformed away.

3. If one has $E \cdot B \neq 0$ in one inertial system, then this relation must apply in all inertial systems. Neither of the fields can be transformed away.

4. If $E^2 - c^2 B^2 = 0$ in one inertial system, then $|E| = c|B|$ in *all* inertial systems. Furthermore, if the scalar product $E \cdot B = 0$, then E, B and k form an orthogonal system for electromagnetic waves.

7.3.4 Covariant form of Maxwell's Equations

With all these preparations, we can now write down Maxwell's equations for the fields in a covariant form. For example, using the field strength tensor, we can write

$$\partial_\nu F^{\mu\nu} = \mu_0 j^\mu. \tag{7.46}$$

Let us verify this equation. For $\mu = 0$ we have

$$\partial_i F^{0i} = -\frac{1}{c} \nabla \cdot E = \mu_0 j^0 = \mu_0 c \rho_{\text{el}}, \quad \text{or} \quad \nabla \cdot E = \frac{\rho_{\text{el}}}{\varepsilon_0}. \tag{7.47a}$$

The result for $\mu = 1$ is

$$\partial_\nu F^{1\nu} = -\frac{1}{c^2} \dot{E}_x + \left(\frac{\partial B_z}{\partial y} - \frac{\partial B_y}{\partial z} \right) = -\frac{1}{c^2} \dot{E}_x + (\nabla \times B)_x, \tag{7.47b}$$

and generally for $\mu = i \neq 0$

$$\partial_\nu F^{i\nu} = -\frac{1}{c^2} \dot{E} + (\nabla \times B) = \mu_0 j, \quad \text{or} \quad \frac{1}{\mu_0} \nabla \times B - \varepsilon_0 \dot{E} = j. \tag{7.47c}$$

Thus, the equation $\partial_\nu F^{\mu\nu} = \mu_0 j^\mu$ is the covariant form of the *two* Eqs. (7.3) and (7.4), which tell us how to generate electric and magnetic fields from currents and charges.

Furthermore, Maxwell's equations involving only the fields can be written as

$$\partial_\nu \widehat{F}^{\mu\nu} = 0. \tag{7.48}$$

Let us check this equation. For $\mu = 0$ we obtain

$$\partial_\nu \widehat{F}^{0\nu} = \nabla \cdot B = 0. \tag{7.49a}$$

For $\mu = 1$ we have

$$\partial_\nu \widehat{F}^{1\nu} = -\frac{1}{c}\dot{B}_x + \frac{1}{c}\left(\frac{\partial E_y}{\partial z} - \frac{\partial E_z}{\partial y}\right) = -\frac{1}{c}\left[\dot{B}_x + (\nabla \times E)_x\right] \tag{7.49b}$$

and generally for $\mu = i \neq 0$:

$$\partial_\nu \widehat{F}^{i\nu} = -\frac{1}{c}\left(\dot{B} + \nabla \times E\right) = 0. \tag{7.49c}$$

Thus, the equation $\partial_\nu \widehat{F}^{\mu\nu} = 0$ is the covariant form of the *two* Maxwell's equations (7.1) and (7.2).

One can formulate these equations also in terms of the original field strength tensor, but then has to put up with the slightly longer expression

$$\partial_\lambda F_{\mu\nu} + \partial_\mu F_{\nu\lambda} + \partial_\nu F_{\lambda\mu} = 0. \tag{7.50}$$

7.4   Change of the Frame of Reference

The field strength tensor transforms according to the inverse Lorentz transformation:

$$F'_{\mu\nu} = \Lambda_\mu{}^\alpha \Lambda_\nu{}^\beta F_{\alpha\beta}. \tag{7.51}$$

To evaluate (7.51), it is advantageous to write it, in analogy with (5.13), in the form $F' = \Lambda F \Lambda^{\mathrm{T}}$.

When we change the frames of reference (inertial systems), the electric *and* magnetic fields transform jointly. Therefore, they cannot be considered separately.

As an example, we perform a boost from the laboratory system to a system that moves in the x-direction, and whose axes are parallel to that of the laboratory system. From (7.51) we then obtain

$$F'_{\mu\nu} =$$

$$\begin{pmatrix} 0 & -E_x/c & -\gamma(E_y/c - \beta B_z) & -\gamma(E_z/c + \beta B_y) \\ E_x/c & 0 & \gamma(-\beta E_y/c + B_z) & -\gamma(\beta E_z/c + \beta B_y) \\ \gamma(E_y/c - \beta B_z) & -\gamma(-\beta E_y/c + B_z) & 0 & B_x \\ \gamma(E_z/c + \beta B_y) & \gamma(\beta E_z/c + B_y) & -B_x & 0 \end{pmatrix}. \tag{7.52}$$

Comparing with $F_{\mu\nu}$ in (7.39) we find the relations

$$\begin{aligned} E'_x = E_x, & \quad E'_y = \gamma(E_y - c\beta B_z), & \quad E'_z = \gamma(E_z + c\beta B_y), \\ B'_x = B_x, & \quad B'_y = \gamma(\beta E_z/c + B_y), & \quad B'_z = \gamma(-\beta E_y/c + B_z). \end{aligned} \tag{7.53}$$

Evidently, the x-components of the fields remain invariant, but the y- and z-components of the fields mix. Even for a general boost, a very compact expression can be found for the transformed fields, see Exercise 7.1.

7.5 Field of a Moving Point Charge

Following the presentation in Jackson [1], we use the relations just derived to investigate, in the laboratory system S, the field of a point charge Q, which rests in the origin of a system S' that moves along the x-axis of S with velocity β. Both S and S' are assumed to be in standard configuration, i.e., the origins of S and S' coincide at $t = t' = 0$. In the rest frame of the point charge, only a static electric field is present,

$$\mathbf{E}'(\mathbf{r}', t') = \frac{Q}{4\pi \varepsilon_0} \frac{\mathbf{r}'}{r'^3}, \tag{7.54}$$

but *no* magnetic field. If we want to investigate this field from the point of view of the laboratory system S, we must bear in mind that our fixed observation point P, at rest in S, moves with velocity $-\beta$ relative to S'. Thus, even though (7.54) is a static, time-independent field, the position of the observation point in S' depends on time.

If we chose our observation point in the origin of S, the point charge would directly pass through it, producing a singularity. To avoid this, we choose the trajectory of P in S' as

$$\mathbf{r}'_P(t') = (-\beta c t', b', 0)^{\mathsf{T}}. \tag{7.55}$$

Here b' is the offset of the trajectory on both the y- and the y'-axis.

From (7.55) follows $r' = \sqrt{\beta^2 c^2 t'^2 + b'^2}$. Inserting this in (7.54) yields the components of the field $\mathbf{E}'$

$$
\begin{aligned}
E'_x &= \frac{Q}{4\pi \varepsilon_0} \frac{-\beta c t'}{r'^3}, \\
E'_y &= \frac{Q}{4\pi \varepsilon_0} \frac{b'}{r'^3}, \\
E'_z &= 0.
\end{aligned}
\tag{7.56}
$$

For symmetry reasons, the electric field in P can have no component in the z-direction in both systems. With the substitution $\beta \mapsto -\beta$ we can now use (7.53) to calculate the field components in the laboratory system.

Since only E'_x and E'_y are different from zero, we obtain

$$E_x = E'_x, \quad E_y = \gamma E'_y, \quad E_z = \gamma E'_z,$$
$$B'_x = 0, \qquad B'_y = 0, \qquad B_z = \gamma \beta E'_y / c = \beta E_y / c. \tag{7.57}$$

This is not yet our final result, since we have to remember that the components of the field E' depend on the coordinates of S'. They also must be transformed using the well-known relations

$$ct' = \gamma(ct - \beta x) = \gamma ct,$$
$$x' = \gamma(-\beta ct + x) = -\gamma \beta ct, \tag{7.58}$$
$$b = b'.$$

Therefore, $r' = \sqrt{\gamma^2 \beta^2 c^2 t^2 + b^2}$, and

$$E_x = -\frac{Q}{4\pi\varepsilon_0} \frac{\gamma \beta ct}{(\gamma^2 \beta^2 c^2 t^2 + b^2)^{3/2}},$$
$$E_y = \frac{Q}{4\pi\varepsilon_0} \frac{\gamma b}{(\gamma^2 \beta^2 c^2 t^2 + b^2)^{3/2}}, \tag{7.59}$$
$$B_z = \gamma \beta E'_y / c = \beta E_y / c.$$

As an example, Fig. 7.1 shows the electric field of a moving electron for an offset of $b = 1$ a_B and different velocities. In Fig. 7.1 the time is measured in units of $a_B/c \approx 1.77 \cdot 10^{-19}$ s. This is the time that it takes a photon to cover a distance corresponding to the *Bohr radius*[4] [2]

$$a_B := \frac{\hbar}{m_e c \alpha} = 5.29177210544(82) \cdot 10^{-11} \text{ m}. \tag{7.60}$$

We can further improve the formal representation of the electric field in a way that its physical structure becomes even more apparent. In the laboratory system, the vector from the point charge to the observation point is given by

$$\mathbf{r} = \begin{pmatrix} -\beta ct \\ b \\ 0 \end{pmatrix} \quad \text{with} \quad r = \sqrt{\beta^2 c^2 t^2 + b^2}. \tag{7.61}$$

---

[4] Niels Bohr, 1885–1962, Danish physicist. Nobel Prize in physics 1922.

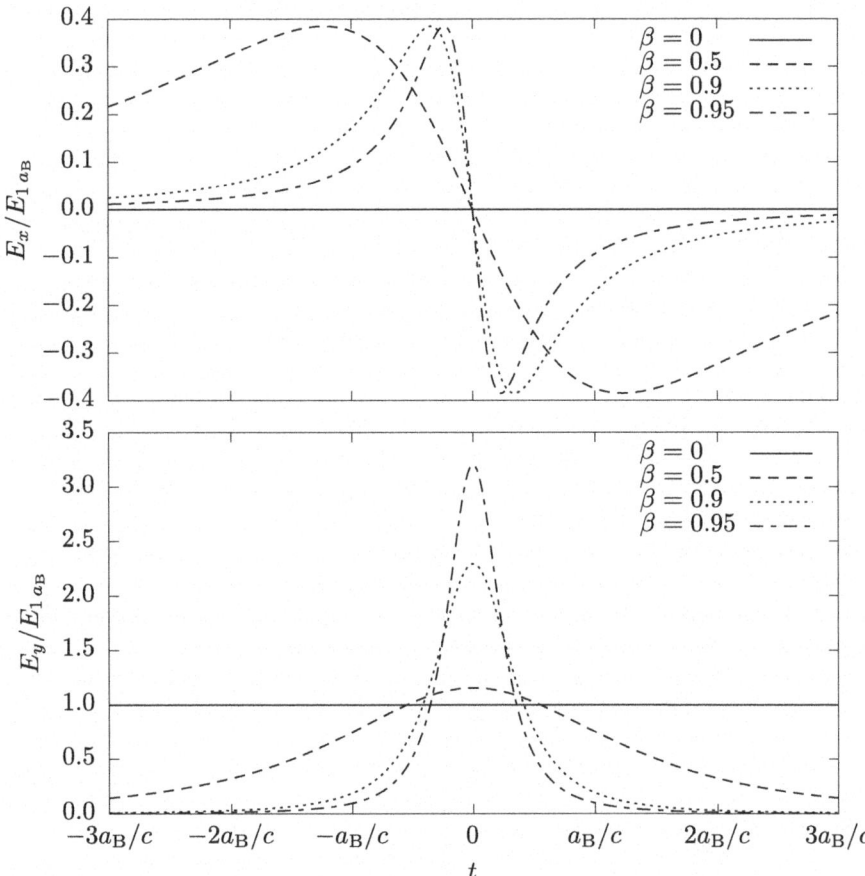

Fig. 7.1 Field strength of a moving electron for a fixed observation point at $y = b = 1a_B$ for different velocities. The field strength is given in units of $E_{1ab} = E_0/(4\pi\varepsilon_0 a_B^2)$

From this we see, first, that

$$\frac{E_x}{E_y} = \frac{r_x}{r_y}. \tag{7.62}$$

This implies that, viewed from the point charge, the electric field E always points in the direction of r, i.e., it is still a radial field. Furthermore, we have

$$b = r \sin(\varphi), \tag{7.63}$$

where φ is the angle between the direction of motion of the charge and the vector from the charge to the observation point.

Second, with the help of these relations, we can write

$$r'^2 = \gamma^2 r^2 + r^2 \sin^2(\varphi)(1 - \gamma^2) = \gamma^2 r^2 \left(1 + \frac{1 - \gamma^2}{\gamma^2} \sin^2(\varphi)\right)$$

$$= \gamma^2 r^2 \left(1 - \beta^2 \sin^2(\varphi)\right). \tag{7.64}$$

We insert this expression in the components of the E field in (7.59) and find

$$\boldsymbol{E}(\boldsymbol{r}) = -\frac{Q}{4\pi \varepsilon_0} \frac{\boldsymbol{e}_r}{r^2 \gamma^2 \left(1 - \beta^2 \sin^2(\varphi)\right)^{3/2}}. \tag{7.65}$$

This expression describes the electric field with the moving point charge at its center. We see that a dependence on the angle is superimposed on the usual decrease proportional to $1/r^2$. The sine term vanishes both in the direction of motion and opposite to it. The field then corresponds to that of a point charge at rest, attenuated by the factor $1/\gamma^2$.

In the direction perpendicular to the direction of motion, we have $(1 - \beta^2 \sin^2(\varphi))^{3/2} = 1/\gamma^3$. This means that the field is amplified by a factor of γ in comparison with the field of a charge at rest. In Fig. 7.2 the corresponding lines of constant field strength are shown for different velocities.

Let us also have a look at the magnetic field. In classical electrodynamics the magnetic field of a moving point charge is given by the *Biot-Savart law*[5],[6]

$$\boldsymbol{B} = \frac{\mu_0}{4\pi} Q \frac{\boldsymbol{v} \times \boldsymbol{r}}{r^2}. \tag{7.66}$$

In our case the z-component of the field would be

$$B_z = \frac{\mu_0}{4\pi} Q \frac{c\beta b}{(\beta^2 c^2 t^2 + b^2)^{3/2}}. \tag{7.67}$$

Apart from the relativistic corrections, this coincides with our result

$$B_z = \frac{\mu_0}{4\pi} Q \frac{c\gamma\beta b}{(\gamma^2 \beta^2 c^2 t^2 + b^2)^{3/2}}. \tag{7.68}$$

7.6   Covariant Form of the Lorentz Force

Using the field strength tensor, the Lorentz force can be written in a covariant way:

$$\frac{\mathrm{d}}{\mathrm{d}\tau} p_\mu = q F_{\mu\nu} u^\nu. \tag{7.69}$$

---

[5] Jean-Baptiste Biot, 1774–1862, French physicist and mathematician.
[6] Félix Savart, 1791–1841, French physician and physicist.

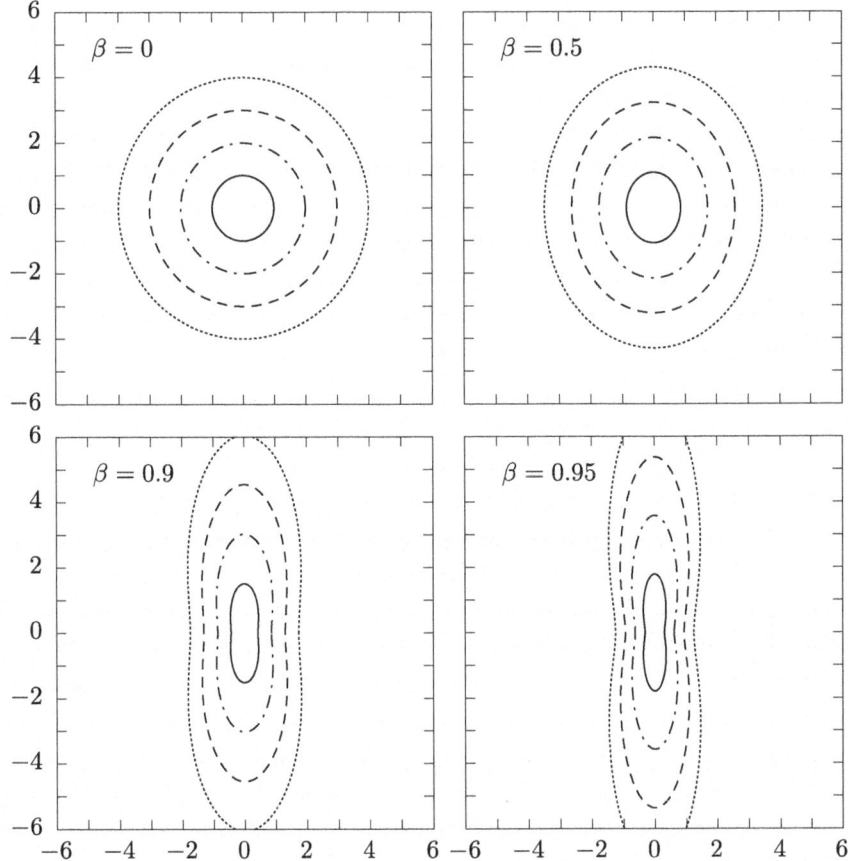

Fig. 7.2 Lines of constant field strength for an electron moving at constant velocity in x-direction. Isolines of the field strengths are shown for distances of one to four Bohr radii in the rest frame of the electron. The field is squeezed in the direction of motion, and stretched perpendicular to it. All lengths are measured in units of the Bohr radius

Let us look at this equation in detail. For $\mu = i \in \{1, 2, 3\}$ the equation yields

$$\frac{\mathrm{d}}{\mathrm{d}\tau} p_i = \gamma \frac{\mathrm{d}}{\mathrm{d}t} p_i = \gamma q (\boldsymbol{E} + \boldsymbol{v} \times \boldsymbol{B}) = \frac{\mathrm{d}}{\mathrm{d}\tau} p^i, \tag{7.70a}$$

while for $\mu = 0$ it follows

$$\frac{\mathrm{d}}{\mathrm{d}\tau} p_0 = -\frac{\mathrm{d}}{\mathrm{d}\tau} (m_0 \gamma c) = -\gamma \frac{q}{c} \boldsymbol{E} \cdot \boldsymbol{v} = -\frac{\mathrm{d}}{\mathrm{d}\tau} p^0, \tag{7.70b}$$

or

$$\frac{\mathrm{d}}{\mathrm{d}t} (\gamma m_0 c^2) = q \boldsymbol{E} \cdot \boldsymbol{v} = \frac{\mathrm{d}W}{\mathrm{d}t} = q \boldsymbol{E} \cdot \frac{\mathrm{d}\boldsymbol{s}}{\mathrm{d}t}.$$

Therefore,

$$dW = q\boldsymbol{E} \cdot d\boldsymbol{s}, \tag{7.71}$$

the increase in energy is equal to the work performed by the electric field. Working with the *contravariant* component of the four-momentum, we can also write

$$\frac{d}{d\tau} p^\mu = F^\mu = q \eta^{\mu\alpha} F_{\alpha\nu} u^\nu, \tag{7.72}$$

with the *Minkowski force* F^μ. We can define the *Minkowski force density* f^μ by the substitutions $q \mapsto \rho_{el}, qu^\nu \mapsto \rho_{el} u^\nu = j^\nu$:

$$f^\mu \equiv \eta^{\mu\alpha} F_{\alpha\nu} j^\nu = \begin{pmatrix} \frac{1}{c} \boldsymbol{j} \cdot \boldsymbol{E} \\ \rho_{el} \boldsymbol{E} + \boldsymbol{j} \times \boldsymbol{B} \end{pmatrix}. \tag{7.73}$$

7.7   Energy-Momentum Tensor of the Electromagnetic Field

From classical electrodynamics it is well known that electromagnetic fields, similar to gravitational fields, possess an energy content. To describe this energy, we now introduce the *energy-momentum tensor*.

7.7.1   Introducing the Energy-Momentum Tensor

Classical electrodynamics knows two energy-related quantities. One is the *field energy* w

$$w = \frac{1}{2} \left(\varepsilon_0 \boldsymbol{E}^2 + \frac{1}{\mu_0} \boldsymbol{B}^2 \right), \tag{7.74}$$

the other is the *Poynting vector* (energy current)

$$\boldsymbol{S} = \frac{1}{\mu_0} \boldsymbol{E} \times \boldsymbol{B}. \tag{7.75}$$

None of these two physical quantities are Lorentz covariant. However, a corresponding Lorentz covariant quantity is the *energy-momentum tensor*. It is defined as

$$T^{\mu\nu} = \frac{1}{\mu_0} \left(\eta^{\mu\beta} F_{\beta\alpha} F^{\alpha\nu} + \frac{1}{4} \eta^{\mu\nu} F_{\alpha\beta} F^{\alpha\beta} \right). \tag{7.76}$$

According to (7.41), $F_{\alpha\beta} F^{\alpha\beta} = 2 \left(\boldsymbol{B}^2 - \boldsymbol{E}^2/c^2 \right)$. We shall motivate this definition. Before, we have a closer look at the individual components of the tensor. For this purpose we evaluate the components $F_{\mu\alpha}F^{\alpha\nu}$. For $\mu = \nu = 0$ we simply have

$$F_{0\alpha}F^{\alpha 0} = \frac{E^2}{c^2}, \tag{7.77a}$$

and for $\mu = 0$, $\nu = i$

$$F_{0\alpha}F^{\alpha i} = \frac{1}{c}(E \times B)_i = \frac{\mu_0}{c}S_i, \tag{7.77b}$$

and finally for $\mu = i$, $\nu = j$

$$F_{i\alpha}F^{\alpha j} = \frac{E_i E_j}{c^2} + B_i B_j - \delta_i^j B^2. \tag{7.77c}$$

With these results, we can determine the components of $T_\mu{}^\nu$:

$$T_0{}^0 = \frac{1}{\mu_0}\left[-\frac{1}{c^2}E^2 - \frac{1}{2}\left(B^2 - \frac{E^2}{c^2}\right)\right] = -\frac{1}{2\mu_0}\left(\frac{E^2}{c^2} + B^2\right) = -w,$$
$$\tag{7.78a}$$

$$T^{0i} = -\frac{1}{c}S_i = T^{i0}, \tag{7.78b}$$

$$T^{ij} = G^{ij}. \tag{7.78c}$$

In (7.78c), G^{ij} denotes the *Maxwell stress tensor*, defined as

$$G^{ij} = \frac{E_i E_j}{c^2} + B_i B_j - \frac{1}{2}\delta^{ij}\left(B^2 + \frac{E^2}{c^2}\right). \tag{7.79}$$

The result can be summarized in matrix form:

$$T^{\mu\nu} = \begin{pmatrix} -w & -S^T/c \\ -S/c & G^{ij} \end{pmatrix}. \tag{7.80}$$

7.7.2 Interpretation of the Energy-Momentum Tensor

To elucidate the meaning of the energy-momentum tensor, we consider a small cube exposed to an electromagnetic wave traveling in the x-direction. Then the Poynting vector has only an x-component, $S = S_x e_x$. The energy passing through the cube with cross section ΔA during the time Δt is given by

$$\Delta W = S_x \Delta A \,\Delta t \stackrel{!}{=} F_x \Delta x = F_x c\Delta t, \tag{7.81a}$$

or

$$\frac{F_x}{\Delta A} = p_S = \frac{1}{c} S_x = \frac{\Delta p_x}{\Delta A \Delta t} = c \frac{\Delta p_x}{\Delta V}, \tag{7.81b}$$

where p_S denotes the *radiation pressure*. The x-component of the momentum density, Π_x, is then given by

$$\frac{\Delta p_x}{\Delta V} = \Pi_x, \tag{7.82}$$

Quite generally, the *momentum density* is defined as

$$\boldsymbol{\Pi} = \frac{1}{c^2} \boldsymbol{S}. \tag{7.83}$$

The Maxwell stress tensor G^{ij} determines the pressure an electromagnetic force exerts on a given volume element, in our case on the small cube:

$$\frac{\boldsymbol{F}}{\Delta A} = -\boldsymbol{G} \cdot \boldsymbol{n}, \tag{7.84}$$

with the normal vector $\boldsymbol{n}$ perpendicular to the surface element ΔA. Therefore,

$$\mathrm{d}\boldsymbol{F} = -\boldsymbol{G} \mathrm{d}f, \tag{7.85}$$

with $\mathrm{d}f = \boldsymbol{n}\mathrm{d}A$. The diagonal elements of G^{ij} represent pressures or tensions, the off-diagonal elements shear stresses, see Fig. 7.3. The physical dimension of $T^{\mu\nu}$ is energy per volume, or force per area, i.e., pressure. The energy-momentum tensor is closely related to the Minkowski force density. To see this, we use (7.46) to write (7.73) in the form

$$f^\mu = \frac{1}{\mu_0} \eta^{\mu\alpha} F_{\alpha\nu} \partial_\lambda F^{\nu\lambda}. \tag{7.86}$$

Next we put the derivative ∂_λ in front of $F_{\alpha\nu}$, and subtract the new term that arises from the application of the product rule:

$$f^\mu = \frac{1}{\mu_0} \eta^{\mu\alpha} \partial_\lambda \left(F_{\alpha\nu} F^{\nu\lambda} \right) - \frac{1}{\mu_0} \eta^{\mu\alpha} \partial_\lambda \left(F_{\alpha\nu} \right) F^{\nu\lambda}. \tag{7.87}$$

We can further modify the second term. We exploit the antisymmetry of the field strength tensor, and write

$$\begin{aligned}
\frac{1}{\mu_0} \eta^{\mu\alpha} \partial_\lambda \left(F_{\alpha\nu} \right) F^{\nu\lambda} &= \frac{1}{2\mu_0} \eta^{\mu\alpha} \left(F^{\nu\lambda} \partial_\lambda F_{\alpha\nu} + F^{\lambda\nu} \partial_\nu F_{\alpha\lambda} \right) \\
&= \frac{1}{2\mu_0} \eta^{\mu\alpha} F^{\nu\lambda} \left(\partial_\lambda F_{\alpha\nu} + \partial_\nu F_{\lambda\alpha} \right).
\end{aligned} \tag{7.88}$$

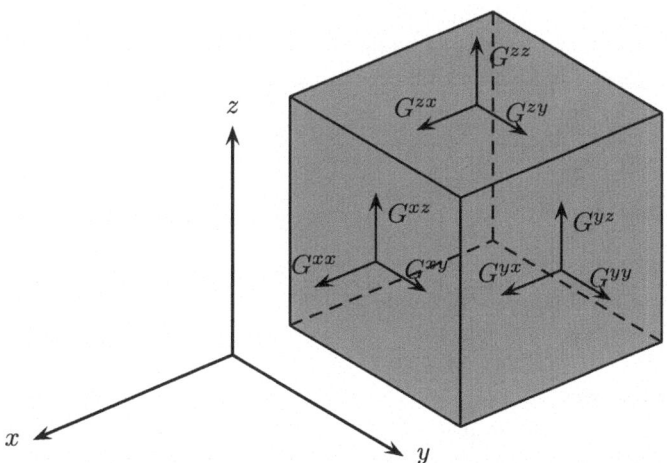

Fig. 7.3 Interpretation of the Maxwell stress tensor. G^{ij} is the stress, i.e., the force per area, in the direction $\mathbf{e}_j$, on the area with normal direction $\mathbf{e}_i$. The diagonal elements are pressure or tension, the off-diagonal elements represent shear stresses

We can replace the last term using (7.50), and obtain

$$
\begin{aligned}
\frac{1}{2\mu_0}\eta^{\mu\alpha}F^{\nu\lambda}\left(\partial_\lambda F_{\alpha\nu}+\partial_\nu F_{\lambda\alpha}\right) &= -\frac{1}{2\mu_0}\eta^{\mu\alpha}F^{\nu\lambda}\partial_\alpha F_{\nu\lambda} \\
&= -\frac{1}{4\mu_0}\eta^{\mu\alpha}\partial_\alpha\left(F^{\nu\lambda}F_{\nu\lambda}\right).
\end{aligned}
\tag{7.89}
$$

In the second step, we have again made use of the product rule. In total, we have, with $\eta^{\mu\alpha}\delta_\alpha^\lambda = \eta^{\mu\lambda}$,

$$
\begin{aligned}
f^\mu &= \frac{1}{\mu_0}\eta^{\mu\alpha}\left[\partial_\lambda\left(F_{\alpha\nu}F^{\nu\lambda}\right)+\frac{1}{4}\partial_\alpha\left(F^{\nu\lambda}F_{\nu\lambda}\right)\right] \\
&= \frac{1}{\mu_0}\eta^{\mu\alpha}\left[\partial_\lambda\left(F_{\alpha\nu}F^{\nu\lambda}\right)+\frac{1}{4}\delta_\alpha^\lambda\partial_\lambda\left(F^{\nu\kappa}F_{\nu\kappa}\right)\right] \\
&= \frac{1}{\mu_0}\partial_\lambda\left(\eta^{\mu\alpha}F_{\alpha\nu}F^{\nu\lambda}+\frac{1}{4}\eta^{\mu\lambda}F^{\nu\kappa}F_{\nu\kappa}\right).
\end{aligned}
\tag{7.90}
$$

The comparison with (7.76) reveals the relation

$$
f^\mu = \partial_\lambda T^{\mu\lambda},
\tag{7.91}
$$

with the four-gradient ∂_λ from (5.40). We see that the Minkowski force density is the divergence of the energy-momentum tensor. To understand the meaning of the relation, we look at it component by component:

$$\frac{\partial w}{\partial t} + \nabla \cdot \boldsymbol{S} = -\boldsymbol{j} \cdot \boldsymbol{E} \qquad\qquad \text{for } \mu = 0, \qquad (7.92a)$$

$$\frac{1}{c^2}\frac{\partial S_i}{\partial t} + \nabla \cdot \boldsymbol{G}_i = \rho_{\text{el}} E_i + (\boldsymbol{j} \times \boldsymbol{B})_i \qquad \text{for } \mu = i. \qquad (7.92b)$$

In the last equation, the i in $\boldsymbol{G}_i$ stands for the row number i of $\boldsymbol{G}$, and, as usual, for the i-th component of the vectors.

Results in Vacuum

In vacuum there are no currents, $j^\mu = 0$, therefore the Minkowski force density is also zero, $f^\mu = 0$, and hence $\partial_\nu T^{\mu\nu} = 0$. From (7.92a) and (7.92b) then follow the four continuity equations

$$\begin{aligned} \frac{\partial w}{\partial t} + \nabla \cdot \boldsymbol{S} &= 0, \\ \frac{1}{c^2}\frac{\partial S_i}{\partial t} + \nabla \cdot \boldsymbol{G}^i &= 0, \end{aligned} \qquad (7.93)$$

with $(1/c^2)\partial S_i/\partial t = \partial \Pi_i/\partial t$. Thus, G^{ij} also describes a momentum density current.

Outlook on General Relativity

In general relativity, the energy-momentum tensor plays a much more important role than in special relativity. However, in general relativity one does not deal with the energy-momentum tensor of electrodynamics, but with an analogous tensor for gravitation. When both electrodynamics and gravitation are important, the tensor $T^{\mu\nu}$ can be split into

$$T^{\mu\nu} = T^{\mu\nu}_{\text{em}} + T^{\mu\nu}_{\text{mat}}, \qquad (7.94)$$

where $T^{\mu\nu}_{\text{em}}$ describes only the electromagnetic fields, while $T^{\mu\nu}_{\text{mat}}$ includes the effects of matter, i.e., also of charges and currents, plus other contributions, e.g., from particle fields and gravitational fields. We will not consider all these cases, though.

However, with an outlook on general relativity, we can already notice the fundamental structure of energy-momentum tensors. They always have the form

$$T = \begin{pmatrix} \text{energy density} & \text{currents} \\ \text{currents} & \text{pressure and shear tensions} \end{pmatrix}. \qquad (7.95)$$

We mention again that pressure has the physical unit of an energy density. We will make use of this fact in several instances in the following chapters.

7.8   Exercises

7.1. Transformation of the fields in a general boost Using Eq. (3.31), show that in a general boost the transformed fields are

$$E' = \gamma(E + c\beta \times B) - \frac{\gamma^2}{\gamma + 1}\beta(\beta \cdot E) \tag{7.96a}$$

and

$$B' = \gamma(B - \beta \times E/c) - \frac{\gamma^2}{\gamma + 1}\beta(\beta \cdot B). \tag{7.96b}$$

References

1. Jackson, J.D.: Classical Electrodynamics (3rd edn.). Wiley, New York (1999)
2. Mohr, P.J., Tiesinga, E., Newell, D.B., Taylor, B.N.: Codata Internationally Reconmmended [sic!] 2022 Values of the Fundamental Physical Constants. https://physics.nist.gov/constants

Visual Effects at Relativistic Velocities

8

Contents

In special relativity, when we say we are "observing" objects, we have to define precisely what we really mean by "observing". On the one hand, "observing" can mean that we *measure* the length of a rod. This requires two individuals in the same frame of reference, with synchronized clocks. If these individuals see the two ends of the rod at the same time, they can determine the length of the rod from the difference of their coordinates, i.e., their distance. On the other hand, "observing" can also mean the way we are actually *seeing* an object. What we see of an object, at a definite moment in time, is what enters our eye, or a camera at that specific moment in time. Therefore, when and where the light started is irrelevant. What is important is that the light arrives *simultaneously* in our eye, or the camera.

In this chapter we shall concentrate on this second aspect of "seeing", and study the visual appearance of fast moving objects in more detail. Furthermore, we shall have a look at how an observer traveling at relativistic velocities would perceive the starry sky.

8.1   Relativistic Doppler Effect

The *Doppler effect*[1] for sound waves is a well-known phenomenon in everyday life. If an observer moves towards a source of sound, or the source moves toward the observer, the sound frequency increases. Conversely, when moving away from the source of sound it decreases. A corresponding effect can be described for electromagnetic waves in the framework of special relativity. In addition to the change in frequency, we shall also understand quantitatively the aberration, which we already discussed in Sect. 1.5.5.

8.1.1   Electromagnetic Waves in Vacuum

We need a covariant formulation of electromagnetic waves in order to apply the Lorentz transformation to the quantities characterizing these waves. We shall restrict ourselves to the situation where the electromagnetic wave arrives in the form of a plane wave,

$$E(r, t) = E_0 \, e^{i(k \cdot r - \omega t)}, \quad B(r, t) = B_0 \, e^{i(k \cdot r - \omega t)}. \tag{8.1}$$

For this wave, we have the relations

$$E_0 \perp B_0 \perp k. \tag{8.2}$$

Electric and magnetic field are perpendicular to the direction of propagation, given by the wavevector k, with $k = \omega/c$. With the *four-wavevector* $\underline{k} = k^\mu \partial_\mu$ and its components

$$k^\mu = \begin{pmatrix} \omega/c \\ k \end{pmatrix} = p^\mu/\hbar \quad \text{or} \quad k_\mu = \begin{pmatrix} -\omega/c \\ k \end{pmatrix}, \tag{8.3}$$

which we already know from Sect. 6.6.2, we can formulate the plane wave in (7.31) in Einstein notation in a covariant way.

8.1.2   Transformation of the Four-Wavevector

To investigate the Doppler effect quantitatively, we determine the components of the four-wavevector in a reference frame at rest and a reference frame moving with relative velocity $\underline{u}$, and compare them. The reference frame at rest has the unit vectors

$$\underline{e}_{(0)} = \frac{1}{c}\partial_t, \quad \underline{e}_{(1)} = \partial_x, \quad \underline{e}_{(2)} = \partial_y, \quad \underline{e}_{(3)} = \partial_z. \tag{8.4}$$

---

[1] Christian Doppler, 1803–1853, Austrian mathematician and physicist.

As to the mathematically precise definition of a vector, we refer to Sect. 11.1.3. For the sake of simplicity we choose the direction of the velocity along the x-axis, and obtain the unit vectors in the moving reference frame by a Lorentz boost

$$\underline{e}'_{(0)} = \gamma(\underline{e}_{(0)} + \beta \, \underline{e}_{(1)}), \quad \underline{e}'_{(1)} = \gamma(\beta \, \underline{e}_{(0)} + \underline{e}_{(1)}), \quad \underline{e}'_{(2)} = \underline{e}_{(2)}, \quad \underline{e}'_{(3)} = \underline{e}_{(3)}. \tag{8.5}$$

Now we can represent the four-wavevector from (8.3) in both reference frames. In spherical coordinates it has the form

$$\begin{aligned}
\underline{k} &= \frac{\omega}{c} \left(-\underline{e}_{(0)} + \sin(\vartheta) \cos(\varphi) \, \underline{e}_{(1)} + \sin(\vartheta) \sin(\varphi) \, \underline{e}_{(2)} + \cos(\vartheta) \, \underline{e}_{(3)} \right) \\
&= \frac{\omega'}{c} \left(-\underline{e}'_{(0)} + \sin(\vartheta') \cos(\varphi') \, \underline{e}'_{(1)} + \sin(\vartheta') \sin(\varphi') \, \underline{e}'_{(2)} + \cos(\vartheta') \, \underline{e}'_{(3)} \right).
\end{aligned} \tag{8.6}$$

In the second line we insert the definitions of the unit vectors (8.5) of the moving reference frame, and compare the coefficients. This leads to the four relations

$$\omega = \omega' \gamma [1 - \beta \sin(\vartheta') \cos(\varphi')], \tag{8.7a}$$

$$\omega \sin(\vartheta) \cos(\varphi) = \omega' \gamma [\sin(\vartheta') \cos(\varphi') - \beta], \tag{8.7b}$$

$$\omega \sin(\vartheta) \sin(\varphi) = \omega' \sin(\vartheta') \sin(\varphi'), \tag{8.7c}$$

$$\omega \cos(\vartheta) = \omega' \cos(\vartheta'). \tag{8.7d}$$

Equation (8.7a) already gives the Doppler shift in dependence on the angle of incidence. To capture the change in frequency quantitatively, the *redshift parameter* is introduced

$$z = \frac{\lambda' - \lambda}{\lambda} = \frac{\lambda'}{\lambda} - 1 = \frac{\omega - \omega'}{\omega'} = \frac{\omega}{\omega'} - 1. \tag{8.8}$$

It is of special significance in cosmology. From the definition of z we can see that signals with $z > 0$ are redshifted, while signals with $z < 0$ are blueshifted, and we always have $z > -1$.

From (8.7a) follows the redshift, or generally the *frequency shift*,

$$z(\vartheta', \varphi') = \gamma [1 - \beta \sin(\vartheta') \cos(\varphi')] - 1 \tag{8.9}$$

as a function of the angles ϑ' and φ' in the moving reference frame. We can derive the angles in the reference frame at rest as functions of the angles in the moving reference frame using (8.7a) and (8.7b)–(8.7d)

$$\cos(\vartheta) = \frac{\cos(\vartheta')}{\gamma [1 - \beta \sin(\vartheta') \cos(\varphi')]}, \quad \tan(\varphi) = \frac{\sin(\vartheta') \sin(\varphi')}{\gamma [\sin(\vartheta') \cos(\varphi') - \beta]}. \tag{8.10}$$

To obtain the inverse relationships, we write, in (8.10), instead of the unprimed angles the primed angles, and vice versa, and replace β with $-\beta$,

$$\cos(\vartheta') = \frac{\cos(\vartheta)}{\gamma[1 + \beta \sin(\vartheta)\cos(\varphi)]}, \quad \tan(\varphi') = \frac{\sin(\vartheta)\sin(\varphi)}{\gamma[\sin(\vartheta)\cos(\varphi) + \beta]}. \tag{8.11}$$

With these formulas, we can transform any viewing direction from the system at rest to the system in motion, and then use (8.9) to assess how strongly the light signal is redshifted.

8.1.3 Doppler Effect and Aberration

To demonstrate the implications of the Doppler effect, it is instructive to consider the two special cases where the light propagates either in the direction of motion, or opposite to it. This will simplify the calculations, and at the same time illustrate the symmetry with respect to the direction of motion.

Our objective is to calculate the observation angles χ and χ' of a ray of light in both reference frames. The relation between χ and the spherical angles ϑ and φ can be obtained from the scalar product of the wavevector $\underline{k}$ with the basis vector $\underline{e}_{(1)}$

$$\cos(\chi) = \frac{\underline{e}_{(1)} \cdot \underline{k}}{k} = \sin(\vartheta)\cos(\varphi), \tag{8.12}$$

see Fig. 8.1.

We can determine the relation between χ' and χ directly from the equations of the preceding section. With $\varphi = \varphi' = 0$ and $\chi = \pi/2 - \vartheta$, or $\chi' = \pi/2 - \vartheta'$, the ratio of the frequencies is,

$$\frac{\omega'}{\omega} = \frac{1}{\gamma[1 - \beta\cos(\chi')]}, \tag{8.13}$$

Fig. 8.1 The wavevector in a local coordinate system. The angles ϑ and φ correspond to the usual spherical coordinates, the angle χ is the angle between the ray of light and the direction of motion $\underline{u}$

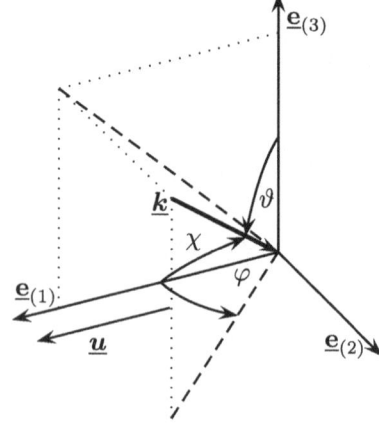

and the redshift is

$$z(\chi') = \gamma[1 - \beta \cos(\chi')] - 1. \tag{8.14}$$

We can derive the formula for the *aberration* from (8.7b) by setting $\sin(\vartheta) = \cos(\chi)$ and $\sin(\vartheta') = \cos(\chi')$, and exchanging the angles and the velocities,

$$\cos(\chi') = \frac{\cos(\chi) + \beta}{1 + \beta \cos(\chi)}. \tag{8.15}$$

Longitudinal and Transverse Doppler Effect

It is obvious that the change in frequency is maximum when the observer is moving directly towards a source of light or directly away from it. In this case we have $\chi' = 0$ or $\chi' = \pi$, and we obtain from (8.13), with the definition of γ and using the third binomial formula, for the motion towards the light source

$$\left.\frac{\omega'}{\omega}\right|_{\text{towards}} = \sqrt{\frac{1 + \beta}{1 - \beta}} \approx 1 + \beta \quad \text{or} \quad z \approx -\beta \tag{8.16}$$

and

$$\left.\frac{\omega'}{\omega}\right|_{\text{away}} = \sqrt{\frac{1 - \beta}{1 + \beta}} \approx 1 - \beta \quad \text{or} \quad z \approx \beta \tag{8.17}$$

for motion away from it. In each case the approximations are valid for velocities $\beta \ll 1$. This is the *longitudinal Doppler effect*. For $\beta \to 1$, the frequency shift becomes arbitrarily large in the first case, and in the second case the frequency converges to zero. Let us assume that a spacecraft flies directly towards a star, then the light of the star is shifted to γ-ray wavelengths for sufficiently high velocities, while signals from the Earth could be shifted down to the range of radio wavelengths, see Fig. 8.2.

Even if the observer moves perpendicularly to the connecting line to the source, a frequency shift occurs in special relativity, due to time dilation. This is the *transverse Doppler effect*, and from (8.13) we find for this case

$$\left.\frac{\omega'}{\omega}\right|_{\perp} = \sqrt{1 - \beta^2} \approx 1 - \frac{\beta^2}{2}, \quad \text{or} \quad z \approx \frac{\beta^2}{2}. \tag{8.18}$$

Obviously, the effect is one order higher in β than the longitudinal Doppler effect.

Aberration

In Sect. 1.5.5 we already derived an angle of aberration ϑ for a star in the zenith

$$\vartheta'_{\text{Newton}} = \arctan(\beta) \approx \beta - \beta^3/3. \tag{8.19}$$

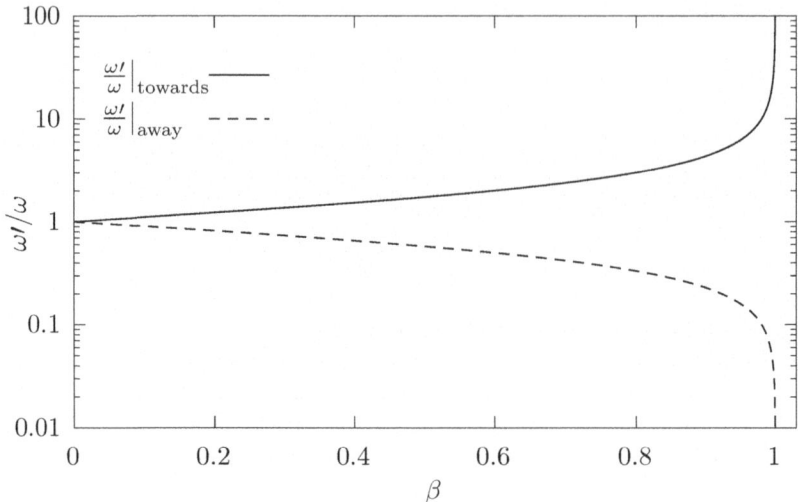

Fig. 8.2 Frequency shift for motion directly towards the source, or away from it. For motion towards the source, the frequency shift ω'/ω becomes arbitrarily large as $\beta \to 1$, for motion away from the source, the frequency tends to zero

In the limit $\beta \to 1$ this formula predicts an aberration of $\vartheta' = \pi/4$. Using (8.15) we can derive the correct angle of aberration. For $\chi = \pi/2$ we have

$$\vartheta'_{\text{SR}} = \frac{\pi}{2} - \chi' = \frac{\pi}{2} - \arccos(\beta) = \arcsin(\beta) \approx \beta + \frac{\beta^3}{6}. \tag{8.20}$$

Thus, for $\beta \to 1$ the observation angle χ' goes to zero, the star is seen almost in the direction of motion, see Fig. 8.3.

In fact this is true for all observation angles. The faster the observer is moving, the more objects appear in front of him. In the case $\beta = 0$, (8.15) yields the obvious result $\chi' = \chi$. In the limit $\beta \to 1$ follows the by no means obvious result $\chi' = 0$, independent of χ: *all* objects appear directly in front of the observer.

So far we have only taken into account the aberration. From (8.14) we can derive the observation angles $\chi'_{z=0}$ under which objects are neither redshifted nor blueshifted, i.e., objects with $z = 0$,

$$\chi'_{z=0} = \arccos\left(\frac{\gamma - 1}{\gamma \beta}\right). \tag{8.21}$$

In the limit $\beta \to 1$, this angle also goes to zero, see Fig. 8.4. Consequently, almost the entire celestial sphere is redshifted, only a very small section directly in the direction of motion is blueshifted. Figure 8.5 summarizes these results for different velocities.

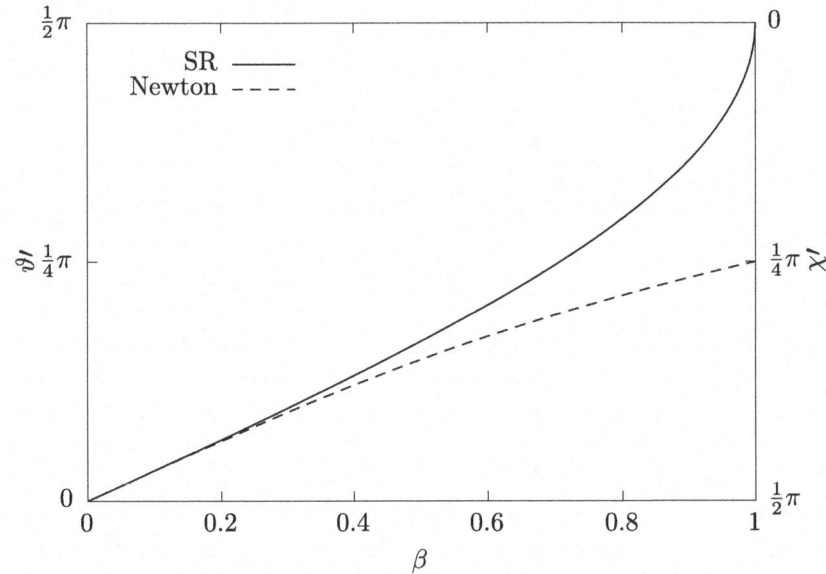

Fig. 8.3 Angle of aberration perpendicular to the direction of motion. The non-relativistic formula (8.19) yields a maximum change in angle of $\pi/4$, the relativistic result (8.20) shows that for $\beta \to 1$ the moving observer finally can see the source of light directly in front of him

Fig. 8.4 Signals arriving from the direction $\chi'_{z=0}$ at the moving observer are neither redshifted nor blueshifted. For $\beta \to 1$ this angle goes to zero

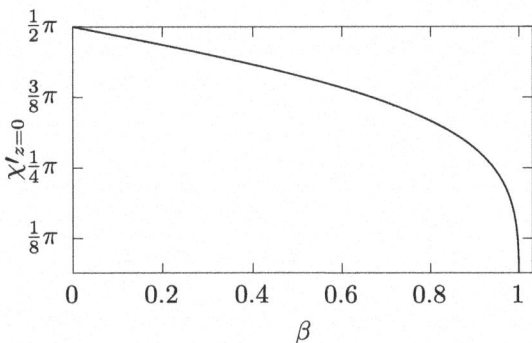

Thus, the relativistic Doppler effect has quite unusual implications. At ultrarelativistic velocities, almost the whole sky appears in front of the observer, and at the same time, all of the celestial sphere is redshifted. The situation turns even more complicated when the finite distance to an object is taken into account. In this case, during the motion, there occurs a change of the observation angles under which an observer at rest perceives an object at his respective location. During a long voyage with constant acceleration this can lead to the phenomenon that the apparent positions of stars in the sky seem to "freeze" for the space traveler.

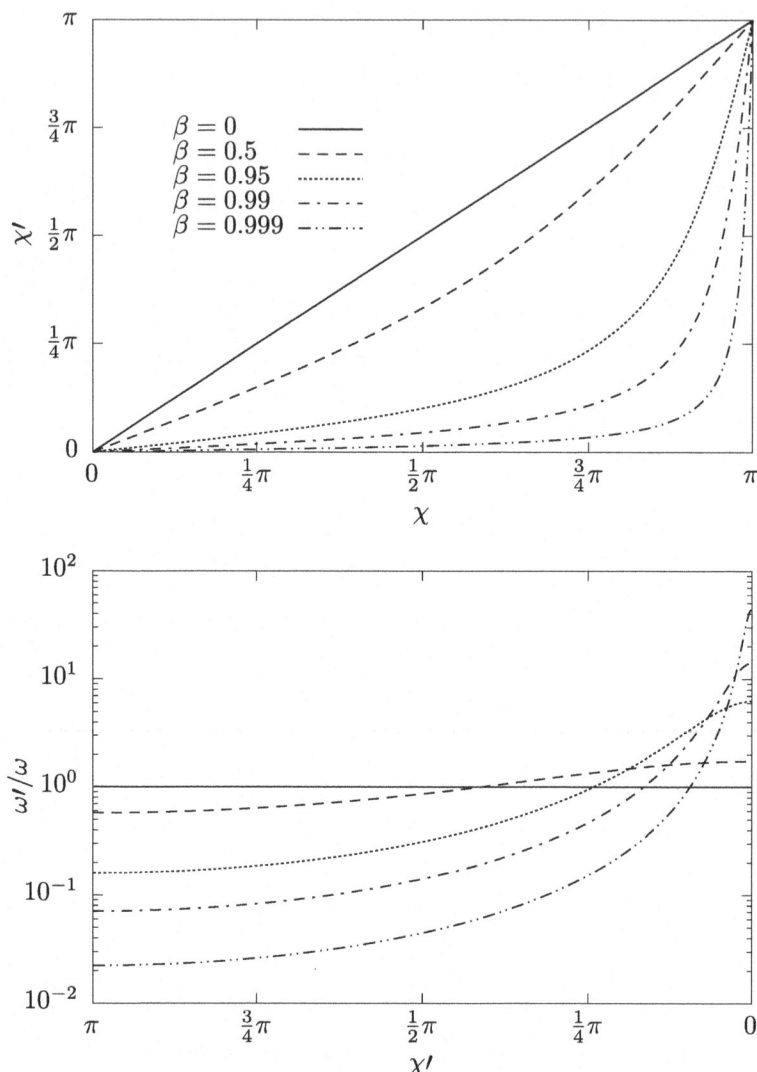

Fig. 8.5 Aberration and Doppler effect for different velocities and observation angles. *Top*: An observer at rest and an observer moving relative to him see a source of light at infinite distance according to (8.15) under different angles χ and χ' relative to the direction of motion of the moving observer. In comparison with the observer at rest, the latter sees an ever-increasing part of the celestial sphere in front of him, the faster he moves. *Bottom*: Redshift as a function of the observation angle. With increasing velocity, the signals become redshifted in a larger region of the sky

8.2 The Ever-Accelerating Rocket

To illustrate the formulas derived in the previous section, we return to the case of the rocket with constant acceleration already discussed in Sect. 6.7. But now the rocket is assumed to move away from the Earth for all times with the constant acceleration g. The Earth is the origin of the reference frame of the traveling observer. We express everything as a function of the proper time τ. Then the time $t(\tau)$ passed on Earth, the distance $x(\tau)$ of the rocket from the Earth, and the current velocity $\beta(\tau)$ of the rocket are given by

$$t(\tau) = \frac{c}{g} \sinh\left(\frac{g}{c}\tau\right), \quad x(\tau) = \frac{c^2}{g}\left[\cosh\left(\frac{g}{c}\tau\right) - 1\right], \quad \beta(\tau) = \tanh\left(\frac{g}{c}\tau\right),$$
(8.22)

see the first two branches of Eqs. (6.78), (6.79) and of (6.82). We will assume that the observer in the rocket and another observer remaining on Earth can communicate with each other by exchanging radio signals. This example will help us again to compare the views of two observers in motion relative to each other.

8.2.1 Messages from the Earth to the Space Traveler

Our first task is to calculate the proper time τ after which a signal sent from Earth at t_E reaches the space traveler. At times $t > t_E$, the signal sent at t_E is at $x = c(t - t_E)$. Therefore, we have the following condition,

$$c\left[t(\tau) - t_E\right] = x(\tau).$$
(8.23)

This equation can be easily solved with respect to the proper time, and we find

$$\tau_R(t_E) = -\frac{c}{g} \ln\left(1 - \frac{g}{c}t_E\right).$$
(8.24)

In order to compare the situations of both observers, we solve (8.24) also with respect to the terrestrial time t_E. The result is

$$t_E(\tau_R) = \frac{c}{g}\left(1 - e^{-\frac{g}{c}\tau_R}\right).$$
(8.25)

For $t_E \geq c/g$ the argument of the logarithm in (8.24) becomes zero or negative. From this we conclude that only signals sent from the Earth at times $t_E < c/g$ can still reach the space traveler, even though he always moves with a velocity smaller than the speed of light.

This, however, is not an effect of special relativity, but just due to the fact that the velocity of the space traveler asymptotically approaches the speed of light.

In Newtonian mechanics, if we consider a particle that starts from the origin at $t = 0$, with a velocity that exponentially tends to 1, $v(t) = 1 - e^{-\alpha t}$, we can easily see that a particle starting at a later time and moving at constant velocity $v = 1$, can catch up with the first particle only if it starts at a time $t < 1/\alpha$.

From (8.25) we can recognize the situation as experienced by the space traveler. It is true that for all times he does receive signals from the Earth, but at ever-increasing time intervals. If, for example, the observer on Earth sends a signal every second, according to (8.25) the time interval between two subsequent signals arriving at the space traveler becomes longer and longer.

Inserting (8.24) in $\beta(\tau)$ from Eq. (8.22), we can immediately determine the velocity of the space traveler when he receives a certain signal, and the redshift of the signal. Using the identity $\tanh(x) = (e^{2x} - 1)/(e^{2x} + 1)$ we have

$$\tanh[-\ln(y)] = \frac{1 - y^2}{1 + y^2}, \tag{8.26}$$

where, in our case, $y = 1 - gt_E/c$. Thus, we obtain the interim result

$$\beta(t_E) = \frac{1 - \left(1 - \frac{g}{c}t_E\right)^2}{1 + \left(1 - \frac{g}{c}t_E\right)^2} \quad \text{for} \quad t_E < \frac{c}{g}. \tag{8.27}$$

Of course, in this example only the longitudinal Doppler effect contributes. To determine the frequency shift of a certain signal we insert the expression (8.27) in (8.17) and obtain

$$\frac{\omega'(t_E)}{\omega} = \sqrt{\frac{1 - \beta(t_E)}{1 + \beta(t_E)}} = 1 - \frac{g}{c}t_E. \tag{8.28}$$

The frequency received therefore decreases linearly with the point in time of the transmission. At $t = t_E$ the frequency would be zero, i.e., the redshift would be infinite. This is intuitively clear, since this case corresponds to an infinite time of flight of the signal to the space traveler, and the signal reaches him asymptotically only for $\beta \to 1$. All these results are summarized in Fig. 8.6.

Finally, we derive the expression $z(t_E)$ for the redshift, as we shall use it frequently later in the discussion of cosmology. Because of $\lambda'/\lambda = \omega/\omega'$ and $z = \lambda'/\lambda - 1$, we obtain from (8.28)

$$z(t_E) = \frac{t_E}{\frac{c}{g} - t_E}. \tag{8.29}$$

As expected, the signal sent at time zero is not redshifted, as it reaches the space traveler immediately. The redshift diverges for $t_E \to c/g$ even faster than the time of reception of the signal, see Fig. 8.6.

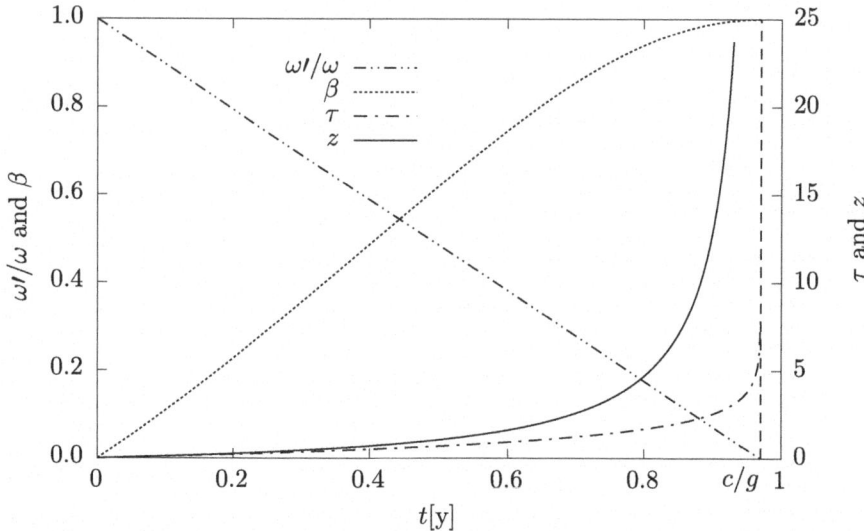

Fig. 8.6 Frequency shift, velocity of the space traveler, and time of arrival of signals in the communication with the space traveler for constant acceleration $g = 1.03$ ly y^{-2} and $c = 1$ ly y^{-1}, as measured from Earth

The presentation in Fig. 8.6 is from the point of view of the observer on Earth: He can tell when, and with what redshift, a signal is received by the space traveler, sent at a definite point in time.

8.2.2   Messages from the Space Traveler to Earth

The situation is different for signals from the space traveler to Earth. In particular, all signals reach the Earth in finite time. The situation is illustrated in Fig. 8.7. A signal sent by the space traveler at time τ_E in the direction of the Earth starts at $x(\tau_E)$ at the coordinate time $t(\tau_E)$ and is described by the equation

$$x(t) = x(\tau_E) + c\,[t(\tau_E) - t]\,. \tag{8.30}$$

The negative sign of the coordinate time indicates that the signal propagates in the direction of the Earth, i.e., in the direction of decreasing values of x. The time of arrival of the signal on Earth can be calculated from (8.30):

$$t_R(\tau_E) = \frac{c}{g}\left(e^{\frac{g}{c}\tau_E} - 1\right)\,. \tag{8.31}$$

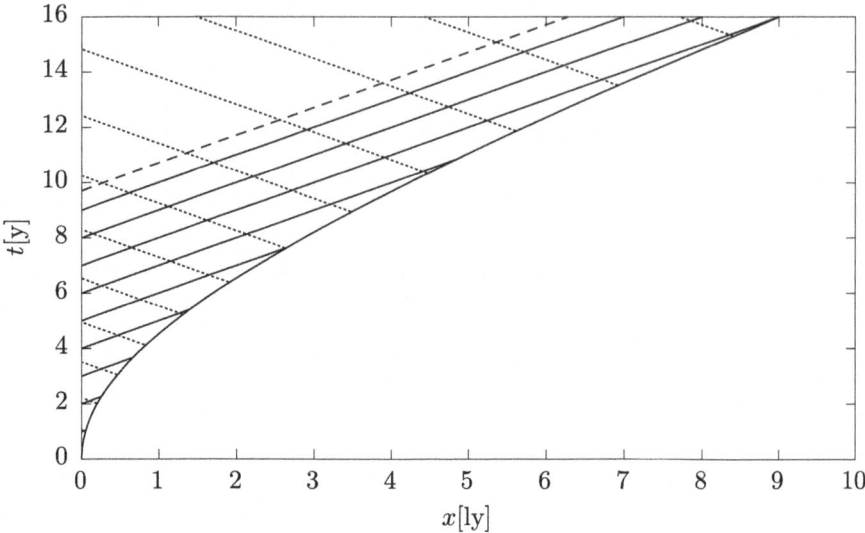

Fig. 8.7 Radio signals for the communication of the space traveler (solid curve) and an observer on Earth (fixed at $x = 0$). Both send a radio signal to the other each year. Signals from Earth are shown as solid lines, those from the space traveler as dotted lines. The dashed line corresponds to the first signal which does not reach the space traveler anymore. For the clarity of the graphical presentation, the acceleration is set to only a tenth of the gravitational acceleration of the Earth, $g = 0.103 \, \mathrm{ly \, y^{-2}}$

Obviously this time remains finite for all values of τ_E. The same holds true for the redshift, which can be obtained directly from $\beta(\tau_E)$,

$$z(\tau_E) = e^{\frac{g}{c}\tau_E}. \tag{8.32}$$

8.2.3 Appearance of the Starry Sky

For the presentation of the starry sky from the point of view of an observer moving with constant acceleration we use the stellar positions from the Hipparcos catalog [5]. The connecting lines of the constellations are drawn using the planetarium software *Stellarium* [3]. When we solve the velocity β of the observer from (8.22) with respect to the proper time τ, we can parametrize the observation time t and the observer position x by the velocity β. We obtain

$$t(\beta) = \frac{c}{g}\gamma\beta \quad \text{and} \quad x(\beta) = \frac{c^2}{g}(\gamma - 1). \tag{8.33}$$

Here, the x-position corresponds to the measured distance to the Earth in the direction of the vernal equinox, at right ascension $\alpha = 0\,\mathrm{h}$ and declination $\delta = 0°$,

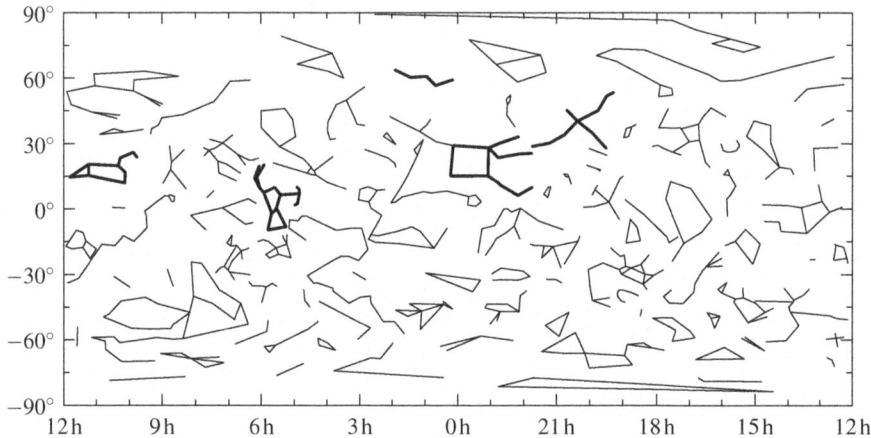

Fig. 8.8 The starry sky with constellations according to ancient mythology in rectangular projection for an observer at rest at the position of the Earth. On the abscissa, the right ascension (α) is plotted in hours, on the ordinate the declination (δ) in degrees

see Sect. 1.5.7. At time $t = 0$ the rocket of the observer is still at rest, and his view of the sky is shown in Fig. 8.8. The distortions at the top and bottom of the image are caused by the rectangular projection. Well-known constellations are drawn in bold: *Orion* ($\alpha \approx 6$ h, $\delta \approx 0°$), *Cassiopeia* ($\alpha \approx 1.5$ h, $\delta \approx 60°$), *Pegasus* ($\alpha \approx 0$ h, $\delta \approx 25°$), *Cygnus* ($\alpha \approx 20$ h, $\delta \approx 40°$), or *Leo* ($\alpha \approx 11$ h, $\delta \approx 15°$).

If the rocket accelerates constantly with $g = 9.81\,\mathrm{m\,s^{-2}} = 1.03\,\mathrm{ly\,y^{-2}}$, then it reaches half the speed of light after approximately 0.53 years. During this time it has covered a distance of roughly 0.15 light years in the direction of the vernal equinox. In comparison with the view of the starry sky at rest, we can recognize that the constellations have moved toward the direction of the vernal equinox (center of Fig. 8.9) due to aberration. After about 2.00 years, the rocket reaches 90% of the speed of light. It is now about 1.26 light-years away from Earth, and the aberration makes most constellations appear to be moving in the direction of travel, see Fig. 8.10. Because of the giant distances to the stars, the exact position of the observer is practically irrelevant. The differences between Figs. 8.8, 8.9, and 8.10 are mainly due to the different aberration at the different velocities $\beta = 0, 0.5$, and 0.9. In other words, at the position of the Earth we would have a very similar view at these velocities.

Further details on the apparent view of the starry sky for an accelerated observer, including the Doppler shift and the associated changes in the apparent magnitudes of the stars, can be found in Müller et al. [1]. There, also a derivation is given of why the apparent positions of the stars in the sky seem to be "frozen" for the space traveler.

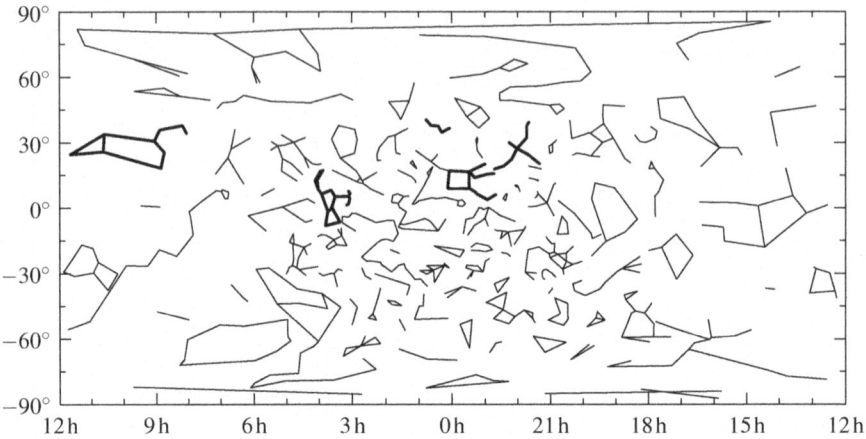

Fig. 8.9 View of the starry sky for an observer constantly accelerating with $g = 1.03$ ly y^{-2} at the time $t = 0.56$ y at the position $x = 0.15$ ly and the velocity $\beta = 0.5$

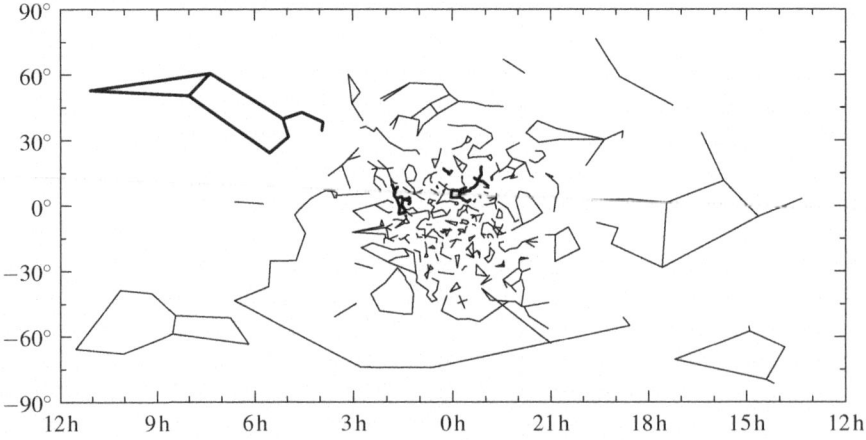

Fig. 8.10 View of the starry sky for the accelerated observer at the time $t = 2.00$ y at the position $x = 1.26$ ly and the velocity $\beta = 0.9$

8.3   Appearance of Fast Moving Objects

When an object moves at relativistic velocities, this has a decisive effect on how an observer at rest *perceives* the object. We know from Sect. 4.2 that the length of an object in the direction of motion, when *measured*, is contracted by a factor of $1/\gamma$. This, however, does not imply that an observer at rest actually *perceives* the object shortened by this factor.

We will study this phenomenon in more detail for the example of a fast moving sphere [2,4]. Because of Lorentz contraction, the sphere is squeezed into an ellipsoid

Fig. 8.11 When a fast
moving sphere is observed,
part of its back becomes
visible, because the sphere
moves away from the path of
the light emitted at the back
and passing to the observer

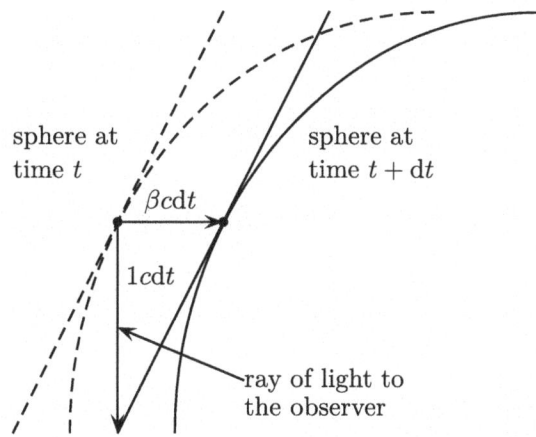

in the direction of motion. However, we shall see that, independent of its velocity,
to an observer at rest the sphere always retains its circular silhouette.

We assume that the sphere moves along the x-axis at velocity β. The first
unfamiliar phenomenon which occurs is that the observer sees parts of the sphere
not facing him, and instead does not see parts of the sphere facing him. The reason
is that the sphere is moving away from the path of the rays of light emitted at the
back of the sphere and propagating to the observer, and instead moves into the path
of rays of light from its front.

To understand this phenomenon quantitatively we consider Fig. 8.11. A ray of
light starts from the back side of the sphere in the direction of the observer. In the
time step dt it has traveled a distance

$$s_{\text{light}} = c\,dt \tag{8.34}$$

in the direction of the observer. During this time interval, the sphere moves to the
right by a distance of

$$s_{\text{sphere}} = \beta\,c\,dt. \tag{8.35}$$

In order for the ray of light to reach the observer, the sphere must make sufficient
way to the ray of light by moving to the right to keep the ray of light from hitting it.
This means that the tangent to the surface of the ellipse must have a slope of at least

$$\tan(\alpha) = m = \frac{1}{\beta}. \tag{8.36}$$

As the slope of the ellipse decreases to the right, the condition for the point farthest
on the back of the sphere from which light still reaches the observer is exactly

$$\frac{dy}{dx} = \frac{1}{\beta}. \tag{8.37}$$

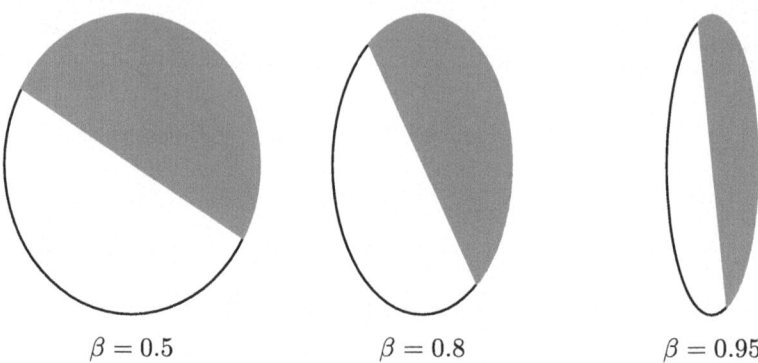

$\beta = 0.5$ $\qquad\qquad\qquad$ $\beta = 0.8$ $\qquad\qquad\qquad$ $\beta = 0.95$

Fig. 8.12 The visible region of a fast moving sphere changes with its velocity. The situation is shown for a sphere moving to the right

An ellipse with a major semi-axis of length R in the y-direction and a minor semi-axis of length R/γ in the x-direction is described by the equation

$$\frac{y^2}{R^2} + \frac{x^2}{R^2/\gamma^2} = 1. \tag{8.38}$$

Solving with respect to y yields for the top or bottom part of the ellipse, respectively,

$$y = \pm R\sqrt{1 - \frac{x^2}{R^2/\gamma^2}}. \tag{8.39}$$

From this we can calculate the slope of the ellipse

$$\frac{dy}{dx} = -\frac{x\gamma^2}{y}. \tag{8.40}$$

The condition (8.37) leads to the two points

$$x_0 = \pm\frac{R}{\gamma^2}, \quad y_0 = \mp R\beta. \tag{8.41}$$

Thus, for $\beta \to 1$, the observer at rest sees almost only the left part of the sphere, see Fig. 8.12.

Looking at Fig. 8.13 we can realize why the observer perceives the ellipse not as an ellipse but, in spite of Lorentz contraction, as a sphere, though with distorted surface.

In comparison with a point on the ellipse at $y = 0$, the light from point A must travel an additional distance $R\beta$ to arrive at the observer. During the time of flight, the sphere moves a distance $R\beta^2$. Therefore, the light from point A was emitted

Fig. 8.13 To any observer, a fast moving sphere appears in spherical shape. The effects of Lorentz contraction and the finite time of flight of the light exactly cancel. In comparison with a ray of light in the ($y = 0$) plane, a ray of light from point A must start earlier by a time $ct = R\beta$, and from point B later by the time $ct = R\beta$. During this time, the sphere has moved a distance $R\beta^2$

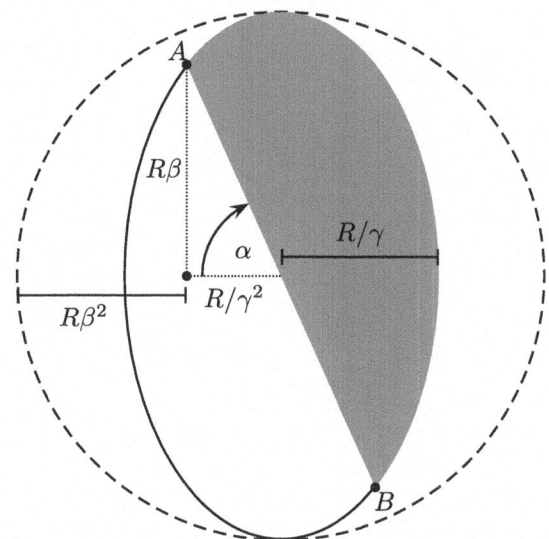

when the ellipse was further to the left by the distance $R\beta^2$. Correspondingly, the light from point B must cover a distance shorter by $R\beta$, and is emitted only when the ellipse is further to the right by the distance $R\beta^2$. Since the observer sees different points of the ellipse at different times, the ellipse seems broadened to him in the x-direction by $2R\beta^2$. For the observed radius this yields an enlargement of

$$\frac{R}{\gamma^2} + R\beta^2 = R(1 - \beta^2 + \beta^2) = R. \tag{8.42}$$

Thus, the effect of the Lorentz contraction is exactly canceled by the finite time of flight of the rays of light.

How is the fact that an observer at rest sees parts of the back of the sphere compatible with the impressions of a co-moving observer? After all, to him the sphere is at rest, and he should not see the back of the sphere. On the other hand, information about an object cannot depend on the state of motion of each observer, since the observer at rest and the observer moving at the same velocity as the sphere should receive the same rays of light from the sphere when they happen to be at the same position at the time of their observations. The solution to this problem is provided by the aberration formula (8.15). With an observation angle $\chi = \pi/2$, one finds the observation angle of the co-moving observer

$$\cos(\chi') = \beta. \tag{8.43}$$

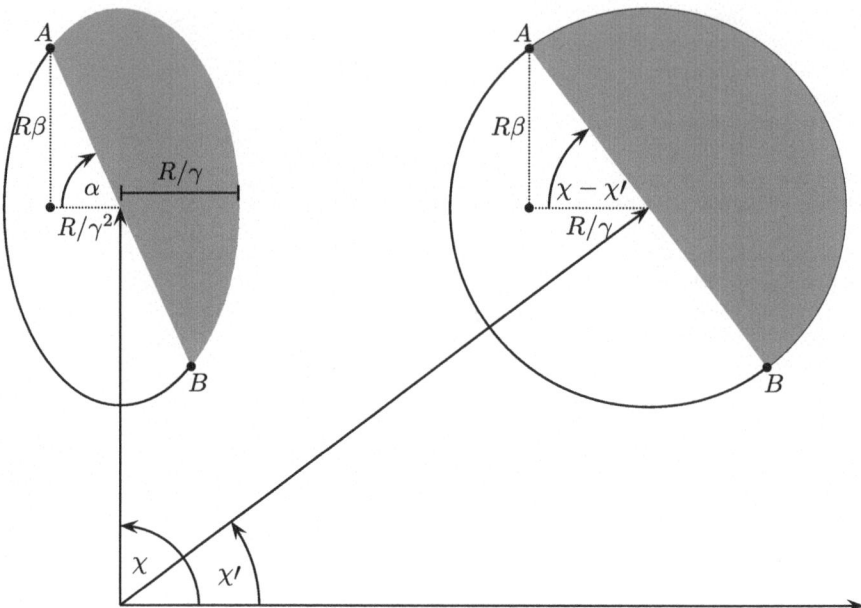

Fig. 8.14 Observation of the moving sphere from an observer at rest, and a comoving observer. The latter sees the sphere under an angle χ', the visible region of the sphere therefore changes by an angle of $\chi - \chi' = \pi/2 - \chi'$, in comparison with a sphere with $\chi' = \pi/2$

He sees the sphere in front of him, and a part of the surface rotated by an angle $\chi - \chi'$, in comparison with a sphere at $\chi' = \pi/2$, see Fig. 8.14. For this angle we find

$$\cos(\chi - \chi') = \cos[\pi/2 - \arccos(\beta)] = \sin[\arccos(\beta)] = \frac{1}{\gamma}, \qquad (8.44)$$

using the relation $\sin[\arccos(x)] = \sqrt{1 - x^2}$. The angles $\chi - \chi'$ on the sphere and α on the ellipse, with $\tan(\alpha) = y_0/x_0 = \gamma^2\beta$, are equivalent, when we take into account the contraction of the ellipse. We can conclude that both observers see the same, but interpret it in different ways.

8.4   Exercises

8.1. Doppler shift of spectral lines The identification of spectral lines plays an important role in astronomy. The Lyman α line of the hydrogen atom has a wavelength of $\lambda_{\text{Ly-}\alpha} = 121.6$ nm. What is the velocity necessary for an object moving away from the observer that this line is shifted to the Balmer α line with $\lambda_{\text{Ba-}\alpha} = 656.5$ nm?

References

1. Müller, T., King, A., Adis, D.: A trip to the end of the universe and the twin "paradox". Am. J. Phys. **76**, 360–373 (2008)
2. Penrose, R.: The apparent shape of a relativistically moving sphere. Math. Proc. Camb. Philos. Soc. **55**, 137–139 (1959)
3. Stellarium: https://stellarium.org
4. Terrell, J.: Invisibility of the Lorentz contraction. Phys. Rev. **116**(4), 1041–1045 (1959)
5. van Leeuwen, F.: Hipparcos, the new Reduction of the Raw data (2007). https://cdsarc.cds. unistra.fr/viz-bin/cat/I/311

Visualization in Special Relativity

9

Contents

In the previous chapters we have focused on the mathematical and geometrical features of special relativity. We have used the Minkowski diagram to gain a deeper understanding of the geometrical implications of special relativity, such as the length contraction or the time dilation. This type of visualization can be designated observer-independent visualization, since a complete overview of a situation is presented in space and time.

In this chapter we will deal with the question of what an observer would actually *see*, who himself moves in the Minkowski spacetime. This means, we put ourselves into the so-called *first-person view*. In doing so we must make the fundamental distinction between *seeing* and *measuring*, since the finite light travel time must be absolutely taken into account, as we have already seen in Sect. 8.3. It seems

---

Supplementary Information The online version contains supplementary material available at https://doi.org/10.1007/978-3-662-71332-7_9.

that Einstein himself did not make this distinction, and also *Gamow*[1] showed in the first edition of "Mr. Tompkins in Wonderland" [1] the cartoon of a cyclist who appeared shortened in his direction of motion because of length contraction. But already in 1924 *Lampa*[2] [2] had calculated that the silhouette of a fast moving sphere appears in a circular shape, and therefore the length contraction of the sphere remains unobservable.

9.1   Visualization Techniques

Apart from the discussion of the optical impressions, we would also like to give a brief outline of how such visualizations can be implemented technically. The visualization techniques presented in the following sections concentrate exclusively on the geometrical effects of special relativity. The Doppler effect from Sect. 8.1 and the change of the light intensity are ignored.

9.1.1   Image-Based Methods

The image-based method is the simplest visualization technique, but also the one with the biggest limitations with regard to the description of possible scenarios. It is based on the assumption that the observer moves through an otherwise static scene. Then it suffices to construct, for each observation time, a 4π panorama of what an observer at rest would see at the current position. This 4π panorama is an image of all rays of light which the moving observer would receive at the observation time. However, the motion and the accompanying aberration of the rays of light must also be taken into account.

A 4π panorama can be constructed in different ways. Nowadays, special cameras are on the market which can output a 4π panorama.[3] And even smartphones with appropriate software can produce such panoramas today. In the following, we discuss only one method, which is also applied in computer graphics. For this purpose, six pinhole cameras, each with an aperture of $90° \times 90°$, are placed at the center of a virtual cube. The field of view of each pinhole camera covers one side of the cube, and the center of the cube corresponds to the position of the observer at rest. The spatial orientation of the cube always remains the same, and is supposed to be aligned with the coordinate axes of a global coordinate system. All six pictures are then combined in a *cube map*, see Fig. 9.1.

---

[1] George Anthony Gamow, 1904–1968, Russian physicist, who lived in the US for a long time. He worked on problems in atomic physics, and on the Big Bang Theory, see Chaps. 19 and 29.

[2] Anton Lampa, 1868–1938, Austrian physicist.

[3] Usually these are called 360° panorama cameras, but this designation is not unique since it only gives information on the angles in the horizontal direction.

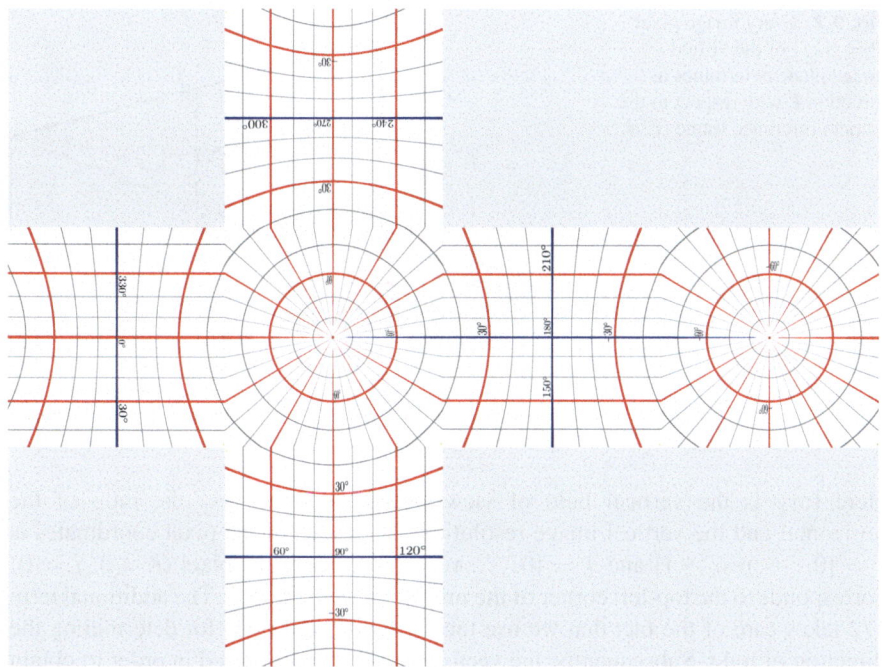

Fig. 9.1 Cube map of a grid of parallels and meridians displayed by the individual views in the six directions of the cube

To visualize what the moving observer sees, we have to replicate his eye, or rather his (pinhole) camera. For this we need a virtual image plane, represented by a CCD chip with $\mathrm{res}_h \times \mathrm{res}_v$ square pixels, and an algorithm, which for every pixel $P = (i, j)$ determines the direction $\boldsymbol{k}$ of the incident ray of light. In addition, we have to know the viewing direction $\boldsymbol{d}$ of the camera and its orientation, defined by its vertical vector $\boldsymbol{u}$ and its right vector $\boldsymbol{r} = \boldsymbol{d} \times \boldsymbol{u}$. The normalized vectors $\boldsymbol{r}$, $\boldsymbol{d}$, and $\boldsymbol{u}$ form the right-handed local coordinate system of the camera, in which we can determine the direction $\boldsymbol{k} = k_r \boldsymbol{r} + k_d \boldsymbol{d} + k_u \boldsymbol{u}$ of the incident ray of light. The viewing direction $\boldsymbol{d}$ at the same time is supposed to correspond to the direction of motion of the observer.

In the case of a pinhole camera, the individual, still unnormalized, components of the incident ray of light read, in dependence on the pixel coordinate (i, j),

$$\tilde{k}_d = 1,$$

$$\tilde{k}_r(i, j) = \rho \left(2 \frac{i + 1/2}{\mathrm{res}_h} - 1 \right) \tan \left(\frac{\mathrm{fov}_v}{2} \right),$$

$$\tilde{k}_u(i, j) = \left(1 - 2 \frac{j + 1/2}{\mathrm{res}_v} \right) \tan \left(\frac{\mathrm{fov}_v}{2} \right).$$

(9.1)

Fig. 9.2 Every image pixel $P = (i, j)$ of the virtual image plane determines a direction $\boldsymbol{k}$ with respect to the camera reference frame ($\boldsymbol{r}, \boldsymbol{d}, \boldsymbol{u}$)

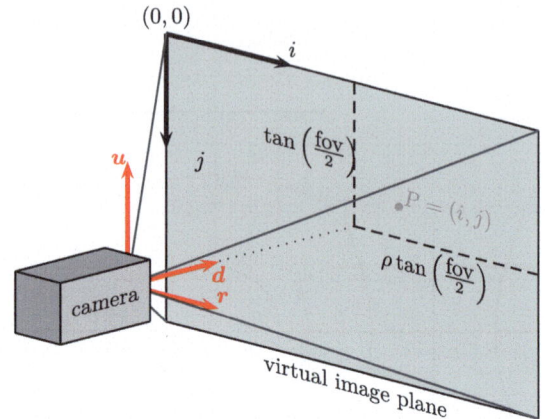

Here fov_v is the vertical field of view, and $\rho = \mathrm{res}_h/\mathrm{res}_v$ the ratio of the horizontal and the vertical image resolution. The range of the pixel coordinates is $i = \{0, \ldots, \mathrm{res}_h - 1\}$ and $j = \{0, \ldots, \mathrm{res}_v - 1\}$, where the pixel ($i = 0, j = 0$) corresponds to the top-left corner of the image, see also Fig. 9.2. The additional term $1/2$ takes care of the fact that we use the center of each pixel for determining the direction of light. Subsequently, the vector $\boldsymbol{k}$ must be normalized in order to obtain the actual components k_d, k_r, and k_u.

Instead of a pinhole camera, we can also emulate a 4π panorama camera, which leads to the display of the environment in the form of a plate-carrée map, see Fig. 9.6, top. For this purpose we first define a relation between the pixel (i, j) and the spherical angles (ϑ, φ)

$$\vartheta = \frac{j + 1/2}{\mathrm{res}_v}\pi, \quad \varphi = \left(\frac{1}{2} - \frac{i + 1/2}{\mathrm{res}_h}\right) 2\pi. \tag{9.2}$$

From this we can get the components $\boldsymbol{k}$ of the incident ray of light,

$$k_d = \sin(\vartheta)\cos(\varphi), \quad k_r = -\sin(\vartheta)\sin(\varphi), \quad k_u = \cos(\vartheta). \tag{9.3}$$

Here the viewing direction $\boldsymbol{d}$ again corresponds to the center of the image.

For the pinhole camera, too, the normalized light direction $\boldsymbol{k}$ can be written in spherical angles (θ, ϕ). We find, in analogy with the relation from (9.3), $\boldsymbol{k} = (k_d, k_r, k_u)^{\mathrm{T}} = (\sin(\theta)\cos(\phi), -\sin(\theta)\sin(\phi), \cos(\theta))^{\mathrm{T}}$. From (9.1) we obtain the two angles

$$\theta = \arccos(k_u) \quad \text{and} \quad \phi = \arctan2(-k_d, k_r). \tag{9.4}$$

As we shall see in Sect. 9.2.1, we have direct access to the color values of the 4π panorama of the cube map, via the angles (θ, ϕ), or the normalized light direction $\boldsymbol{k}$.

So far we have not yet taken the motion of the observer into account. To make up for this, we must first realize that the light direction k, derived from the pixel coordinate (i, j), is referred to the rest frame of the observer. This rest frame itself moves at the velocity β in the direction d. The 4π panorama, however, corresponds to the view of a static observer who resides at the current position of the moving observer. Therefore, we need a Lorentz transformation to transform the view of the static observer into the view of the moving observer. This transformation leads to the general formulas of the *aberration*,

$$\cos(\theta') = \frac{\cos(\theta)}{\gamma[1 - \beta\sin(\theta)\cos(\phi)]},$$

$$\cos(\phi') = \frac{\sin(\theta)\cos(\phi) - \beta}{\sin(\theta')[1 - \beta\sin(\theta)\cos(\phi)]}, \quad \sin(\phi') = \frac{\sin(\theta)\sin(\phi)}{\gamma\sin(\theta')[1 - \beta\sin(\theta)\cos(\phi)]}$$

$$(9.5)$$

(cf. (8.7)), where the unprimed angles (θ, ϕ) refer to the rest frame of the moving observer, and the primed angles (θ', ϕ') to the reference frame of the static observer. Using these angles, one can look up the color values in the 4π panorama.

The advantage of the image-based method lies in the fact that the calculations can be parallelized in a trivial way, since each pixel is independent of all the other pixels. Moreover, apart from computer-generated scenes, also real-life panorama pictures can be used to demonstrate what the world around us would look like if we could travel at relativistic velocities. To maintain the proportions in both length and time, we would have to reduce the speed of light artificially to everyday velocities.

The image-based visualization has only a limited range of applications. Apart from the restriction that only the observer can be in motion, the problem is, above all, the poor image resolution for regions opposite to the direction of motion. Because of the strong effect of aberration, large magnification factors occur which quickly cause individual pixels to become visible. This could be remedied by a cube map with a resolution that depends on the motion and its direction, which, however, is difficult to realize.

9.1.2 Polygon Rendering

The *polygon rendering method* presupposes that we can access the complete geometry of a scene. Here, we will restrict ourselves to the representation of objects by triangular meshes, known from computer graphics. To calculate the way an observer perceives an object, we have to transform all points of the triangular mesh into the reference frame of the observer using a Poincaré transformation, and taking into account the finite light travel time. The resulting, in general distorted, geometry can afterward be rendered in such a way as if both the observer and the scene were at rest. It suffices to employ well-known rendering techniques from computer graphics.

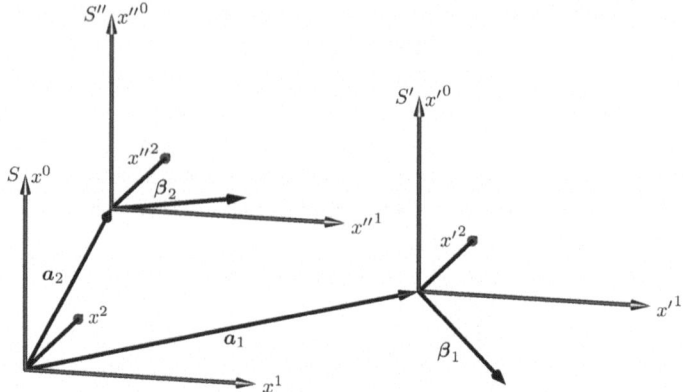

Fig. 9.3 The reference frame of the observer S' and the reference frame of the object S'' move with velocities $\boldsymbol{\beta}_1$ and $\boldsymbol{\beta}_2$ relative to the global reference frame S. The vectors $\boldsymbol{a}_1$ and $\boldsymbol{a}_2$ are the displacements of the coordinate origins relative to S at the time of clock synchronization

In the following we consider a general case of *polygon rendering*. For this, we first specify a global reference frame S with coordinates $\underline{x} = (ct, x) = (x^0, x^1, x^2, x^3)$. The reference frame of the observer S', in which he resides at the position x'_{obs}, is assumed to move at the velocity $\boldsymbol{\beta}_1$ relative to S. The origin of the coordinates in S' is displaced by $\boldsymbol{a}_1$ at time $x^0 = x'^0 = 0$, and the coordinate axes always remain parallel to each other, see Fig. 9.3.

Then the Poincaré transformations between the two systems are:

$$S' \to S: \quad x^\mu = \Lambda_1{}^\mu{}_\nu x'^\nu + a_1^\mu, \tag{9.6a}$$

$$S \to S': \quad x'^\mu = \bar{\Lambda}_1{}^\mu{}_\nu (x^\nu - a_1^\nu), \tag{9.6b}$$

where $\bar{\Lambda}_1{}^\mu{}_\nu$ is the inverse of the Lorentz transformation $\Lambda_1{}^\mu{}_\nu$.

The reference frame S'' of the object is also aligned parallel to S, and moves with velocity $\boldsymbol{\beta}_2$ relative to S. The object itself is assumed static in S''. Furthermore, the origin of S'' also is assumed to be displaced by $\boldsymbol{a}_2$ with respect to S at time $x^0 = x''^0 = 0$. Analogously, the Poincaré transformations are

$$S'' \to S: \quad x^\mu = \Lambda_2{}^\mu{}_\nu x''^\nu + a_2^\mu, \tag{9.7a}$$

$$S \to S'': \quad x''^\mu = \bar{\Lambda}_2{}^\mu{}_\nu (x^\nu - a_2^\nu). \tag{9.7b}$$

Now we must find out where each single point of the triangular mesh appears to the observer. In other words, for every single point we must find that position at which it must emit light so that it arrives at the observer at the observation time x'^0_{obs}. The simplest, and probably most intuitive method for determining this "observation position" is to transform the observer into the rest frame of that point. There the

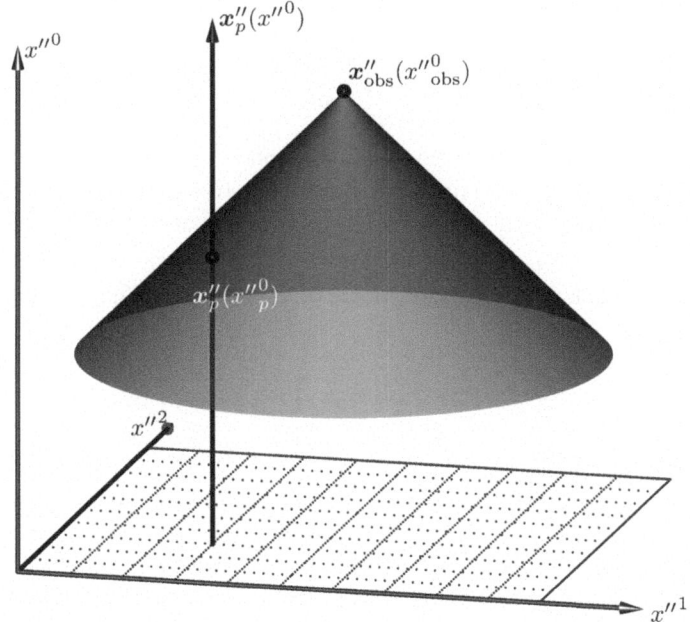

Fig. 9.4 Intersection of the backward light cone of the observer at the position x''_{obs} with the world line x''_{p} of the point at rest, displayed in a 2+1D Minkowski diagram, with the two spatial axes x''^{1} and x''^{2}, and the time axis x''^{0}

intersection of the *backward light cone* of the observer with the worldline of the point is calculated, and the time of light emission is determined, see Fig. 9.4. The backward light cone corresponds to all rays of light which can reach the observer at his observation time, see also the discussion of the Minkowski diagram in Sect. 3.5. Note that Fig. 9.4 depicts a $1+2D$ backward light cone, while Fig. 3.8 depicts $1+1D$ light cones, instead of the full $1+3D$ backward light cone that is required for the computation here.

   Afterward, the resulting spacetime event $(x''^{0}_{p}, \boldsymbol{x}''_{p})$ must be transformed into the rest frame of the observer, from which we obtain the apparent position $\boldsymbol{x}'_{p}$. In detail, we must perform the following steps: The transformation of the observer into the rest frame of the object follows from the composition of (9.6a) and (9.7b),

$$x''^{\mu}_{\text{obs}} = \bar{\Lambda}_2{}^{\mu}{}_{\nu}\left(\Lambda_1{}^{\nu}{}_{\sigma}x'^{\sigma}_{\text{obs}} + a^{\nu}_1 - a^{\nu}_2\right). \tag{9.8}$$

The intersection of the backward light cone with the wordline of the point, $\boldsymbol{x}''_{p}(x''^{0})$ leads to the emission time

$$x''^{0}_{p} = x''^{0}_{\text{obs}} - \Delta\left(\boldsymbol{x}''_{p}, \boldsymbol{x}''_{\text{obs}}\right), \tag{9.9}$$

where

$$\Delta(\boldsymbol{x}, \boldsymbol{y}) = \sqrt{(x^1 - y^1)^2 + (x^2 - y^2)^2 + (x^3 - y^3)^2} \tag{9.10}$$

is the Euclidean distance between the two points $\boldsymbol{x} = (x^1, x^2, x^3)^{\mathrm{T}}$ and $\boldsymbol{y} = (y^1, y^2, y^3)^{\mathrm{T}}$. The coordinate time $x_p''^0$ is that time (referred to the rest frame S'' of the object) at which light must leave the point in order to reach the observer at his observation time $x_{\mathrm{obs}}''^0$. Now we have to transform the spacetime point $x_p''^\mu = (x_p''^0, \boldsymbol{x}_p'')$ again back to the reference frame of the observer. With (9.7a) and (9.6b) this yields

$$x_p'^\mu = \bar{\Lambda}_1{}^\mu{}_\nu \left(\Lambda_2{}^\nu{}_\sigma x_p''^\sigma + a_2^\nu - a_1^\nu \right) \tag{9.11}$$

and thus the apparent position $\boldsymbol{x}_p'$.

If we were to describe the surface of an object $\mathcal{O}$ using a dense set of points, and to transform all these points with the method just described, then we would obtain the set of all events $\mathcal{P}_\mathcal{O}$, where the points have to emit light that arrives simultaneously in our eye. We will denote this set also by $\mathcal{P}_\mathcal{O}$, the *photo object* of $\mathcal{O}$.

As compared to the image-based method, *polygon rendering* has the advantage that there are no restrictions with regard to motion. This means that both the observer and several objects can move simultaneously in different directions. The disadvantage, however, is that only the points of the triangular mesh are transformed, edges continue to be straight lines. If, for example, the sides of a cube are constructed only by two triangles, then the edges of the cube will always appear as straight lines, and never as curved. This problem can be considerably reduced by adaptive refinement of the triangulation.

9.1.3 Ray Tracing

Ray tracing is the most general method for visualization in special and general relativity. In this method, the physical propagation of light from an object to an observer is reversed. For any viewing direction a ray of light is sent, and traced until it hits an object, or leaves the scene. This technique is well known from computer graphics, but must be extended because of the finiteness of the speed of light. Also, it must be taken into account that the ray of light is traced backwards in time. Since the rays of light are traced both in space and in time, the technique is also called *four-dimensional (4D) ray tracing*. The camera model, which we already know from the image-based method, serves as the starting point for the individual rays of light.

The advantage of the ray tracing method is the realistic visualization and the excellent quality of the resulting images due to pixel-precise calculation. The price to be paid is the long computation time. The latter can be reduced drastically by a smart combination with the polygon rendering method.

9.1.4 Other Techniques

The visualization methods discussed so far all have their advantages and disadvantages. We shall briefly sketch how these methods can be modified or combined in order that one can, at least partially, bypass the disadvantages.

In the case of polygon rendering, the triangular mesh of every single object can be changed during rendering depending on the motion. In doing so, every single triangle also can be decomposed into smaller triangles. This still does not produce round edges, indeed, but the refined decomposition leads to a better approximation to the actual appearance.

The method of *local ray tracing* [5] combines polygon rendering with ray tracing. For this purpose, first for each triangle a rectangle is estimated into which the distorted triangle will fit. This rectangle defines that region on the screen in which ray tracing is applied.

The polygon method can also be modified by considering only points and, in particular, parametrized lines instead of polygons. The resulting wireframe can be calculated very efficiently and therefore is well suited to the visualization of geometrical effects at relativistic velocities. Details can be found in Müller and Boblest [4].

9.2 Applications of Visualization Techniques

In the following, we will take a look at a few simple scenes. All examples can be treated using the ray tracing method. We make use of the *ray tracing* code GeoViS [3].

9.2.1 Image-Based Methods

If only the observer moves through an otherwise static scene, the image-based method offers the simplest and fastest possibility for the visualization of relativistic effects. Let us assume that the observer passes currently with his 4π panorama camera through the origin of a global coordinate system, and that he moves along the positive x-axis. In this case the scene only consists of an *environment map*, a static virtual sphere, with the center in the origin and an infinite radius. The sphere serves as a background panorama, and is covered by a grid of parallels and meridians. The virtual image plane of the panorama camera is given by a simple rectangle due to its representation as a plate-carrée map. As already mentioned, the calculation of the individual pixels can be performed independently of each other, which is the reason why it can be easily parallelized. Since graphics hardware was developed exactly

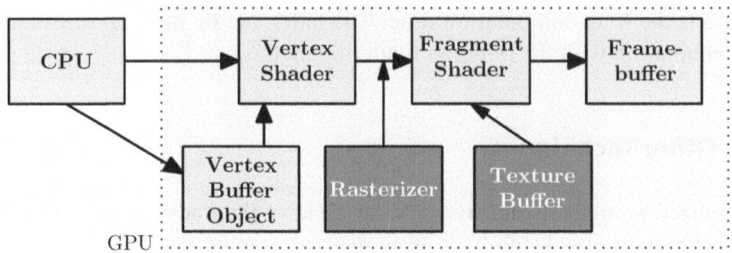

Fig. 9.5 Minimal graphics pipeline. The corners of the geometry are stored in the `vertexbufferobject`. Vertex shader and fragment shader are freely programmable. The rasterizer produces the fragments. The `texturebuffer` is accessed only from the fragment shader. The content of the frame buffer is finally displayed on the screen

for such cases, the calculation can occur on the graphics processing unit, GPU. The control of the GPU is carried out using the *Open Graphics Library* (OpenGL) and the *OpenGL Shading Language* (GLSL) [6].

Figure 9.5 shows the minimal graphics pipeline necessary. In the vertex-buffer-object, the geometry of the rectangle (i.e., the position of its four corners) is stored on the GPU. The *vertex-shader* is, in a first step, responsible for the orthographic projection of the rectangle onto the image plane. After the rasterizer, we have access to each single fragment (i, j) of the rectangle in the *fragment shader*. Using (9.2) we can calculate the corresponding direction of the incident ray of light. Since this ray of light is referred to the reference frame of the observer, we must transform the corresponding vector into the global system using the aberration equations. We can use the resulting vector directly in order to look up the pixel color of the background panorama for this direction in the *cube map*. This color is written into the *frame buffer*, and then displayed on the screen.

Figure 9.6 shows the view of an observer at rest (top), and moving in the direction of the center of the image at velocities 50% and 90% the speed of light. The observer at rest sees the grid of parallels and meridians undistorted. If he is in motion, the grid becomes more strongly distorted in the direction of motion due to the aberration effect. Regions in the direction of motion (center of the image) appear shrunk, whereas regions opposite to the direction of motion (regions in the left and right borders of the image) are strongly magnified. Moreover, the poles seem to move ever closer and closer to each other.

Unfortunately, the magnifying effect opposite to the direction of motion is the weak point of this method. As the background panorama has only a limited resolution, individual pixels become more and more visible.

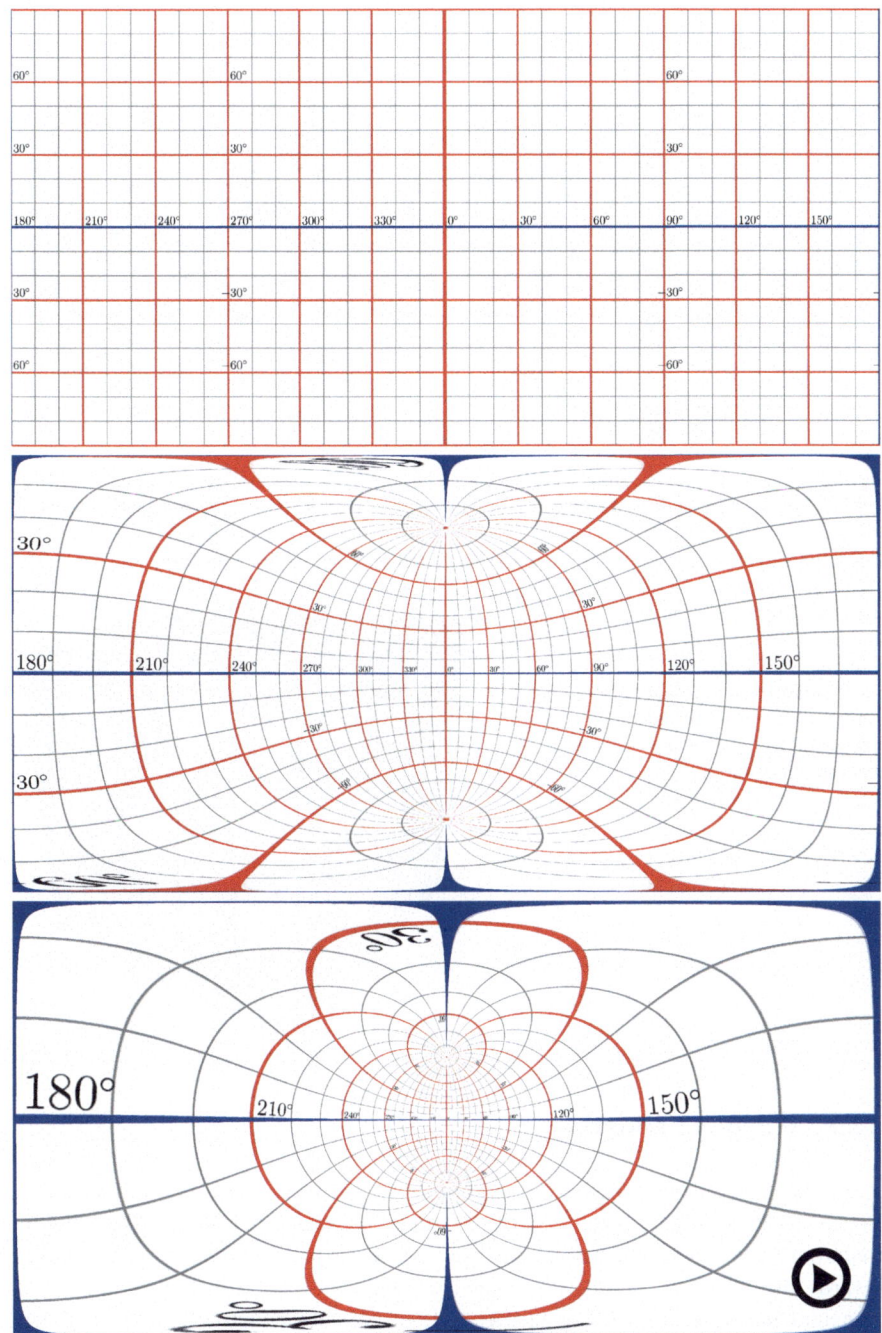

Fig. 9.6 A 4π panorama of a grid of parallels and meridians for an observer moving in the direction of the center of the image at (from *top* to *bottom*) 0%, 50%, and 90% the speed of light (▶ https://doi.org/10.1007/000-hvv)

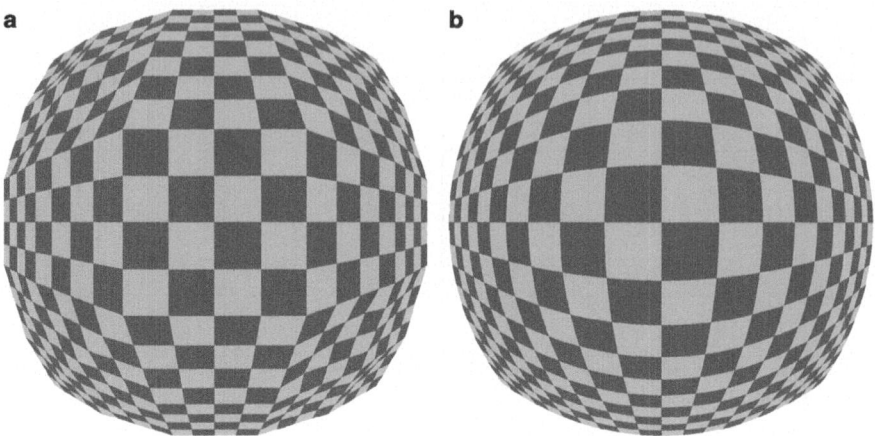

Fig. 9.7 A cube is approaching the observer face-on, see Fig. 9.8 for details about the setup. The weakness of *polygon rendering* (**a**) lies in the fact that it transforms only the corners of the underlying geometry. Ray tracing (**b**) shows what the cube really should look like. The checkerboard pattern only serves to enhance the visibility of the apparent distortions

Fig. 9.8 The cube in Fig. 9.7 consists of a triangular mesh, with each side being built by 5 × 5 squares. Each square consists of two triangles and is rendered using a 4 × 4 checkerboard pattern

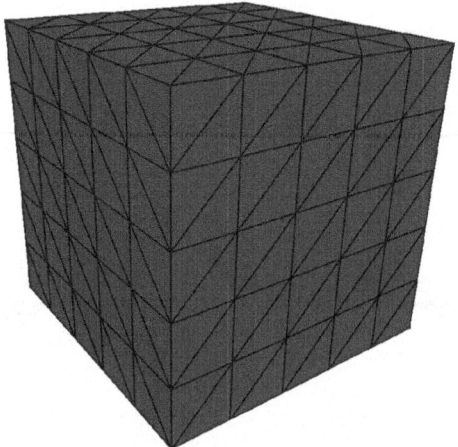

9.2.2   Polygon Rendering

Unlike the image-based method, polygon rendering is geometry-based, and therefore independent of the resolution of the image. Figure 9.7 compares the polygon rendering method with the ray tracing method, where in the example a cube moves towards the observer at 95% the speed of light. The cube is represented by a triangular mesh, see Fig. 9.8, with each side of the cube constructed by $5 \times 5 \times 2$ triangles.

In the polygon rendering method, only the corners of the triangles are transformed according to the procedure discussed in Sect. 9.1.2. Afterward, the transformed corners are again connected by straight lines. This has the consequence that

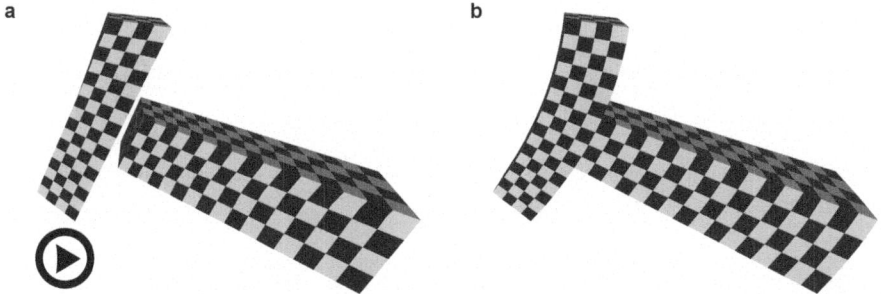

Fig. 9.9 Two cuboids, which together form the letter "T", move obliquely past the observer, at a velocity of 97% the speed of light. In the case of *polygon rendering* (**a**) the connectivity is lost, while it is preserved with *ray tracing* and *adaptive polygon rendering* (**b**) (▶ https://doi.org/10.1007/000-hvt)

the side of the cube facing the observer does not appear uniformly curved, as is the case with ray tracing. The structure of the triangular mesh of the cube can be clearly recognized. For simple objects with a very fine mesh *polygon rendering* provides a good impression of how an object would appear distorted relativistically. However, in the case where an object is constructed from several separate triangular meshes, geometric detachments can occur. As an example, we imagine the letter "T" as constructed from two cuboids, each equipped with triangular meshes. Each side of the cuboids is supposed to be represented by only two triangles. If the two cuboids now move along the direction of the T's vertical bar, they lose their connectivity in *polygon rendering*, see Fig. 9.9. As already mentioned, triangular meshes can still be subdivided during rendering. The necessary graphics pipeline resembles the one shown in Fig. 9.5. For this purpose one adds a *tessellation shader* directly behind the vertex shader, which itself consists of a *tessellation control shader* (TCS) and a *tessellation evaluation shader* (TES). The vertex shader passes the individual vertices directly to the TCS. The latter has access to all three vertices of a triangle and transforms these, as usual, according to the method discussed in Sect. 9.1.2. The TCS also determines the center points of each edge of the triangles and transforms these points as well. Subsequently, all 6 points are projected with the correct perspective onto the image plane of the observer. In this two-dimensional plane, we can now determine the distances of each transformed center point to the center point of the transformed corner points and thus define a criterion whether this edge of the triangle should be subdivided. When we denote the complete transformation of a point p to the image plane by $f_{PT}(p)$, we obtain the difference of two corner points $\delta(p_1, p_2)$

$$\delta(p_1, p_2) = \left| f_{PT}\left(\frac{1}{2}(p_1 + p_2)\right) - \frac{1}{2}\left[f_{PT}(p_1) + f_{PT}(p_2)\right]\right|. \qquad (9.12)$$

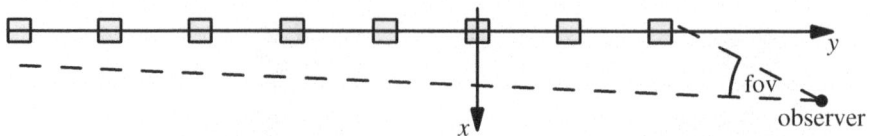

Fig. 9.10 A dice moves towards the observer along the y-axis above a chain of dice at rest. The *dashed line* limits the field of view of the observer in horizontal direction

In the first term, we have first determined and then transformed the center point, while in the second term we have first transformed the corner points, and then calculated the center point. The bigger this difference, the more subdivisions of the edge are needed in order to approximate the distortion sufficiently well in the end. In the TCS it is possible to specify, for each edge of a triangle, how often the edge shall be subdivided. Additionally, one can also specify how strong the subdivision shall be within a triangle by a number which can be calculated as the arithmetic average of the subdivisions of the edges. As a result one obtains, for each original triangle, a triangular mesh from the TCS. The TES now takes the adapted triangular mesh, and for all vertices carries out the transformation using the polygon rendering method.

9.2.3   Apparent Rotation of a Dice

We consider a dice D which moves along the y-axis above a chain of eight dice at rest at the positions $x = 0$, $y = -20 + i \cdot 4$, and $z = -0.75$, with $i = 0, \ldots, 7$. The dice have an edge length of $l = 0.5$. The observer is at the position $x_{\mathrm{obs}} = (3, 15, 1)^{\mathsf{T}}$, also at rest, see Fig. 9.10. At time $t = 0$, the dice D is supposed to pass exactly through the origin. As long as the dice D moves at a moderate velocity, the observer always sees the sides facing him with the numbers ⚀, ⚁, and ⚄. However, if the dice moves with almost the speed of light, then the observer can also see the back of the dice with the number ⚅. Figure 9.11 shows the situation for different observation times.

The dice D seems to be in front of the observer, but in fact, it is already at about the same position as the observer, or slightly to the right behind him, relative to the observer's viewing direction. The reason for this discrepancy is the finite time of flight of rays of light from their emission to the observer. The back of the dice becomes visible because it moves faster than the component of the ray of light pointing to the observer, and thus makes way for the ray of light sufficiently fast. This is the same principle as discussed in Sect. 8.3 in the context of the fast moving sphere.

Fig. 9.11 The dice D moves towards the observer above a chain of dice at rest, at a velocity of 99 % the speed of light. The observation times of the moving dice are $t_{\mathrm{obs}} = \{15.13\,\mathrm{s}, 15.33\,\mathrm{s}, 15.63\,\mathrm{s}\}$ (from behind to front) (▶ https://doi.org/10.1007/000-hvs)

9.2.4 The Silhouette of a Sphere Remains Circular

In Sect. 8.3 we have seen that a fast moving sphere turns into an ellipsoid because of length contraction. At least this is something we would measure. When observing an object, we must take the finiteness of the speed of light into account, and we have already seen that the silhouette of a sphere remains circular, no matter what velocity it has relative to the observer.

Figure 9.12a and b shows a sphere at rest, and at 93% the speed of light, with the center of the sphere always located at the same position relative to the observer. The viewing direction and the field of view are the same in both cases. The silhouette remains circular indeed, but the surface of the sphere appears distorted and rotated, as one can easily recognize from the tiling pattern. The apparent rotation has the same cause as in the example of the moving dice. As the sphere moves past the observer to the right at an angle, the observer sees the reddish back, while the yellow-green front seems to turn away to the right. Because of the motion, the sphere also seems magnified since the light from the back side can start considerably later in order to arrive at the observer, and then this part of the sphere was already closer to the observer.

Figure 9.12c and d shows a transparent wireframe model of the apparent sphere, where the color of the wireframe is plotted fainter and fainter with increasing distance to the observer. Transparency and depth information make it possible to see the apparent photo object of the sphere, which approximates a distorted ellipsoid.

Figure 9.13 shows the photo object of the sphere, moving in the positive y-direction at 93% the speed of light at the current observation time. Only the

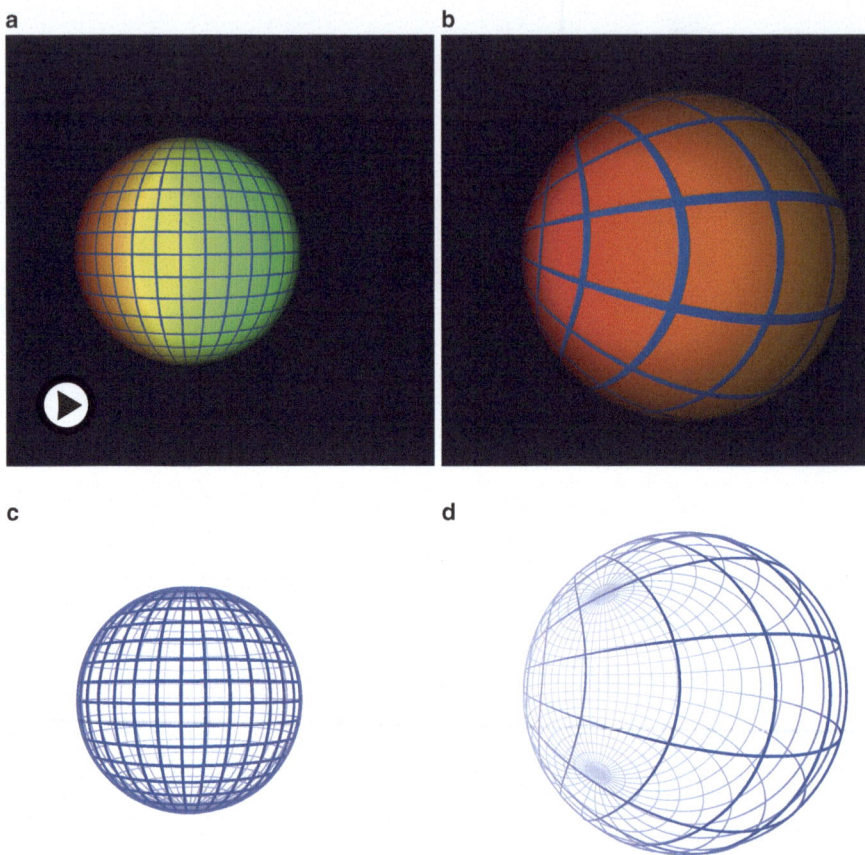

Fig. 9.12 The silhouette of a moving sphere always retains a circular shape. However, the surface seems to be distorted in itself, and the sphere rotated. The images (**a**) (sphere at rest) and (**b**) (93 % the speed of light) were produced using the ray tracing method. The wireframe model of the sphere is shown in (**c**) (sphere at rest) and (**d**) (93 % the speed of light). The color of the wires gets fainter with increasing distance to the observer (▶ https://doi.org/10.1007/000-hvw)

intersection with the xy-plane is shown. The choice of the observation time is such that the center of the moving sphere would appear in the origin. The actual position of the sphere is indicated by the dashed line. Here the length contraction in the direction of motion is taken into account, measured in the reference frame of the observer. Here again one clearly recognizes that every point on the surface of the sphere emits rays of light at different times, which then arrive at the observer simultaneously. Because of the different emission times, each point on the surface was at a different position in space at the time of emission, which explains the apparent distortion. Details on the wireframe method can be found in Müller and Boblest [4].

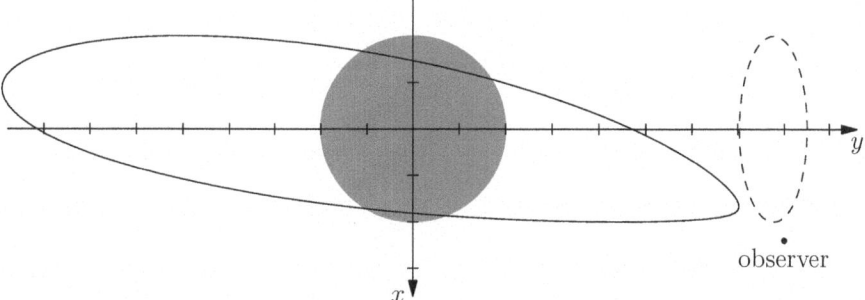

Fig. 9.13 The gray disk represents the sphere at rest, and the approximately ellipsoidal line the photo object of the sphere of Fig. 9.12b. The actual position of the sphere is indicated by the *dashed ellipse*, which takes the length contraction in the direction of motion into account

References

1. Gamow, G.: Mr. Tompkins in Wonderland. Cambridge University Press (1940)
2. Lampa, A.: Wie erscheint nach der Relativitätstheorie ein bewegter Stab einem ruhenden Beobachter? (According to the theory of relativity, how does a moving rod appear to an observer at rest?). Z. Phys. **27**, 138–148 (1924)
3. Müller, T.: GeoViS – Relativistic ray tracing in four-dimensional spacetimes. Comput. Phys. Commun. **185**(8), 2301–2308 (2014)
4. Müller, T., Boblest, S.: Visual appearance of wireframe objects in special relativity. Eur. J. Phys. **35**(6), 065025 (2014)
5. Müller, T., Grottel, S., Weiskopf, D.: Special relativistic visualization by local ray tracing. IEEE Trans. Vis. Comput. Graphics **16**(6), 1243–1250 (2010)
6. Open Graphics Library: https://www.opengl.org

Part II

General Relativity

The Equivalence Principle as the Corner Stone of General Relativity

10

Contents

General relativity is an extension of special relativity for the description of gravitation. In order to treat forces in special relativity, we had introduced the four-force $F^\mu = mb^\mu = \frac{\mathrm{d}}{\mathrm{d}\tau}p^\mu$. Evidently, if one wishes to describe gravitation within the framework of special relativity, then a covariant formulation of the gravitational force will be necessary, in analogy with that of the electromagnetic force. To achieve this, several approaches can be thought of, but they all ultimately fail due to insurmountable problems. The most obvious approach for a description of gravitation would be a Lorentz-invariant scalar field $\Phi(x^\mu)$. This scalar field should fulfill

$$F^\mu = m\frac{\mathrm{d}}{\mathrm{d}\tau}u^\mu = m\frac{\partial \Phi}{\partial x^\mu}, \quad \text{or} \quad \frac{\mathrm{d}}{\mathrm{d}\tau}u^\mu = \frac{\partial \Phi}{\partial x^\mu}. \tag{10.1}$$

© The Editor(s) (if applicable) and The Author(s), under exclusive license to
Springer-Verlag GmbH, DE, part of Springer Nature 2026
S. Boblest et al., *Special and General Relativity*,
https://doi.org/10.1007/978-3-662-71332-7_10

A problem with this approach can be seen directly when one takes the derivative of Φ along the world line $x^\mu(\tau)$, using the relation $u_\mu u^\mu = -c^2$ from Eq. (6.6) together with Eq. (10.1):

$$\frac{d\Phi}{d\tau} = \frac{\partial \Phi}{\partial x^\mu} \frac{dx^\mu}{d\tau} = \left(\frac{d}{d\tau} u_\mu \right) u^\mu = \frac{1}{2} \frac{d}{d\tau} (u_\mu u^\mu) = -\frac{1}{2} \frac{d}{d\tau} c^2 = 0. \qquad (10.2)$$

This leads to $\Phi = $ const, which makes no sense. In an analogous way, one might attempt to introduce gravitation via a four-potential, or a Lorentz tensor of second rank. But these attempts also lead to inconsistencies. Readers interested in this topic will find a comprehensive and detailed discussion in Misner et al. [2].

Since all these attempts fail, a completely new approach is required. We have already pointed out in Sect. 1.4.1 that there exist two kinds of masses, the inertial and the gravitational mass. The starting point for the development of general relativity were Einstein's considerations about the equivalence of inertial and gravitational mass. In this way, he could draw conclusions as to the properties of gravitation. Einstein's ultimate conclusion was that gravitation has an effect on the metric of spacetime. In other words: In general relativity, we no longer have the Minkowski metric $\eta_{\mu\nu}$, but a more general form of the metric, $g_{\mu\nu} = g_{\mu\nu}(x^\lambda)$, which itself depends on space and time.

10.1   Inertial Mass

Let us consider a force acting on a mass. The force will try to remove the mass from its current position. Let us think, for example, of a locomotive that we try to slow down or draw away. The mass will resist the external action of force, and try to remain in its current state of motion, or at rest. For this reason, a mass fulfilling this principle of inertia is called the *inertial mass*. In other words: The inertial mass is that property of a massive object to resist the action of an external force. The bigger the inertial mass, the bigger the force necessary to change its state of motion. Let us think again of the example of a heavy locomotive, in comparison, e.g., to a light table tennis ball. In Fig. 10.1 we consider two masses m_{i_1} and m_{i_2}. Both are attached to identical springs. We extend the springs from their position at rest by the same length Δx. Releasing the springs, the same force acts on the two masses. Therefore, we find for the accelerations $F = m_{i_1} \ddot{r}_1 = m_{i_2} \ddot{r}_2$, i.e.,

$$\ddot{r}_2 = \frac{m_{i_1}}{m_{i_2}} \ddot{r}_1. \qquad (10.3)$$

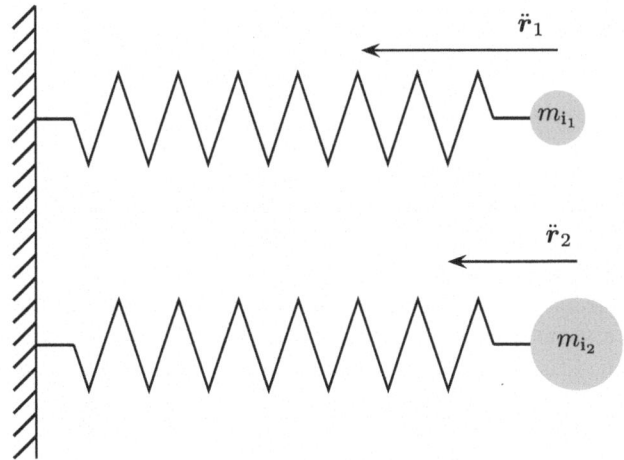

Fig. 10.1 Two identical springs are displaced from their equilibrium positions by the same length. The masses m_{i_1} and m_{i_2} are attached to the two springs and are initially at rest. Upon releasing the springs, both masses are accelerated. The ratio of the accelerations is given by $\ddot{\boldsymbol{r}}_2 = (m_{i_1}/m_{i_2})\ddot{\boldsymbol{r}}_1$

10.2   Gravitational Mass

The *gravitational mass* m_g appears in both an active and a passive form, which are closely linked. On the one hand, m_g is the property of a massive body to experience a force in the gravitational field of another massive body. This is what is meant by the passive gravitational mass. In analogy with electrodynamics, we denote this mass as the gravitational charge in the gravitational field of the gravitational charge M:

$$\boldsymbol{F}_{\text{grav}} = -\frac{m_g M}{r^2} G \, \boldsymbol{e}_r. \tag{10.4}$$

Accordingly, the active gravitational mass is that property of a massive body to generate a gravitational field.

If we place two inertial masses m_{i_1} and m_{i_2} in the gravitational field of a gravitational mass M, a force will act on these two masses. Therefore, they also possess a gravitational mass m_{g_1} and m_{g_2}.

Free fall experiments yield the result that the two masses always "fall equally fast", independent of their inertial masses, or, to put it more precisely, experience the same acceleration, $\ddot{\boldsymbol{r}}_1 = \ddot{\boldsymbol{r}}_2$, independent of the size of their inertial masses m_{i_1} and m_{i_2}. This can be summarized as follows:

$$
\begin{aligned}
m_{i_1}|\ddot{\boldsymbol{r}}_1| &= |\boldsymbol{F}_{Mm_{g_1}}| \\
m_{i_2}|\ddot{\boldsymbol{r}}_2| &= m_{i_2}|\ddot{\boldsymbol{r}}_1| = |\boldsymbol{F}_{Mm_{g_2}}|.
\end{aligned}
\tag{10.5}
$$

From this we see that

$$\frac{m_{i_1}}{m_{i_2}} = \frac{F_{Mm_{g_1}}}{F_{Mm_{g_2}}} = \frac{m_{g_1}}{m_{g_2}} \quad \text{or} \quad \frac{m_{i_1}}{m_{g_1}} = \frac{m_{i_2}}{m_{g_2}}. \tag{10.6}$$

Thus, for every object the ratio of inertial mass and gravitational mass is the same. If we choose the unit of the gravitational mass appropriately we can make this ratio equal to 1.

Because this point is so important, let us emphasize it again: In a gravitational field, objects with different inertial masses experience, for identical initial conditions, the same acceleration. For all bodies, the ratio of inertial mass and gravitational mass is the same, and with an appropriate choice of units one can set

$$\frac{\text{inertial mass}}{\text{gravitational mass}} = \frac{m_i}{m_g} = 1. \tag{10.7}$$

10.3   Free Fall Experiments

The following thought experiments date back to Einstein. We consider a test person with bathroom scales in a closed laboratory, i.e., the experimenter receives no information from outside. He has a mass of $m = 80$ kg, and stands on the scales. With this setup, we perform two types of experiments.

10.3.1 Accelerated Laboratory

In the first case, the laboratory stands (rests) in the gravitational field of the Earth with $g = 9.81$ m s^{-2}. In accordance with the mass of the test person the scales display a force of $F = 80$ kp,[1] see Fig. 10.2a. This force is calculated as

$$F = m_g g. \tag{10.8}$$

In the second case, the laboratory is located in empty space, and is accelerated upward with constant acceleration $a = g$. Again the scales display a force of 80 kp, see Fig. 10.2b. In this instance the force is calculated as

$$F = m_i g. \tag{10.9}$$

---

[1] The kilopond kp is an outdated unit for force. One has 1 kp $= 9.81$ N, i.e., it is the weight of a mass of one kilogram in the gravitational field at the surface of the Earth.

Fig. 10.2 (a) The laboratory rests in a uniform gravitational field g. (b) The laboratory is located in weightless space and is accelerated upward with constant acceleration $\ddot{r} = g$. In both cases the scales display a weight of 80 kp

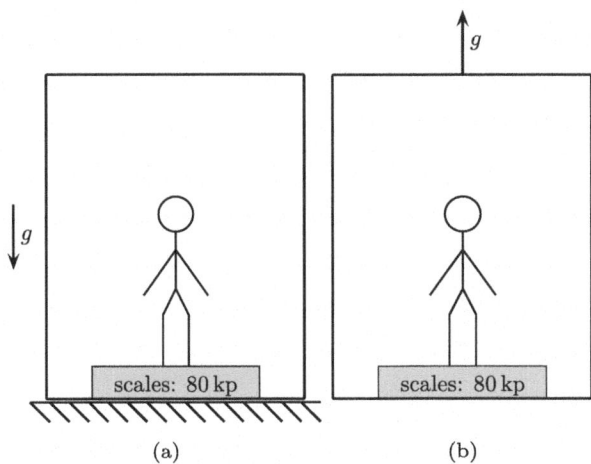

Now the crucial question is: "Can our experimenter decide by a mechanical, electrodynamic, or any other experiment, whether he rests in the gravitational field on the surface of the Earth, or whether he is accelerated upward in empty space with $a = g$?".

The answer is: No! The result of these considerations can be summarized as follows:

> The notion of a coordinate system at rest in a gravitational field is equivalent to the notion of an accordingly accelerated coordinate system without a gravitational field.

10.3.2 Laboratory at Rest

In the second type of thought experiment, we bring the laboratory into empty space where it is assumed to be at rest, or moving with constant velocity on a straight line, see Fig. 10.3a. The scales now display a force of $F = 0$ kp. Next, we set the laboratory into free fall in the uniform gravitational field of the Earth, see Fig. 10.3b. As a consequence of the equivalence principle, every object in the laboratory falls with the same velocity, there is no relative motion whatsoever. For the coordinate $x'(t)$ of an arbitrary point in the laboratory we have

$$x(t) = x_0(t) + x'(t) \quad \text{and} \quad m_{\mathrm{i}} \ddot{x} = m_{\mathrm{i}} (\ddot{x}_0 + \ddot{x}') = m_{\mathrm{g}} g. \tag{10.10}$$

Here $x_0(t)$ denotes the coordinate of the floor of the laboratory in some external inertial system. Because of $m_{\mathrm{i}} = m_{\mathrm{g}}$ and $\ddot{x}_0 = g$, we find

$$\ddot{x}' = 0. \tag{10.11}$$

Fig. 10.3 (**a**) The laboratory
rests in empty space. (**b**) The
laboratory is in free fall in the
uniform gravitational field g.
In both cases the scales
display a force 0 kp

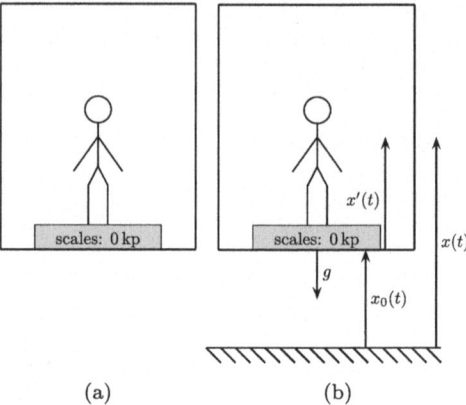

(a)   (b)

Again the scales will display a force of $F = 0$ kp. As in the previous section, we ask whether there exists any experiment which allows our experimenter to distinguish between the two situations. The answer is again: No! We can summarize this result in the *weak equivalence principle*:

> In a small laboratory, in free fall in a gravitational field, the mechanical phenomena are the same as those which can be observed in a Newtonian inertial system without a gravitational field.

In 1907 Einstein extended this statement [1] by replacing the expression "mechanical phenomena" with "laws of physics". This is the *strong equivalence principle*:

> In a small laboratory, in free fall in a gravitational field, the laws of physics are the same as those which hold in a Newtonian inertial system without a gravitational field.

This is how Einstein put it in his own words in a 1920 review article for the journal Nature, which was not printed because the editors considered it too long:

> For an observer who is in free fall from the roof of a house, there is no gravitational field— at least in his immediate vicinity. For if the falling observer drops some other bodies, then they are in a state of rest or uniform motion with respect to him. Thus, the experimentally proven independence of the gravitational acceleration is a strong argument for the fact that the relativity postulate must also be extended to coordinate systems that are in non-uniform motion relative to each other.

Einstein himself called this insight "the happiest thought of my life". We have already mentioned that, within the framework of general relativity, gravitation is described by its effect on the metric of spacetime.

If the equivalence principle were not valid, the notion of packing gravitation into a curved space-time, universal for all bodies, would not work. Therefore, an analogous approach is not possible in electrodynamics, because there the charge and the inertial mass of a particle are independent of each other.

Since, in general, gravitational fields are inhomogeneous, one must be careful to choose a laboratory, exposed to a gravitational field, so small that the deviations

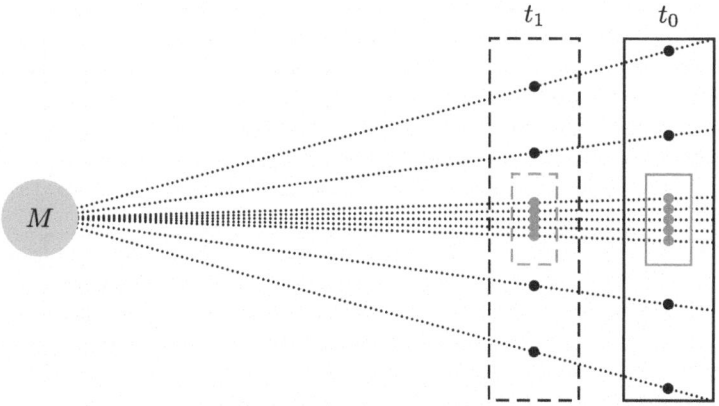

Fig. 10.4 Because of the inhomogeneity of gravitational fields, the laboratory must be so small that the inhomogeneity can be neglected. The laboratory shown in black is too big, the black balls get closer to each other. The laboratory drawn in gray is so small that the inhomogeneity is negligible in the time interval $\Delta t = t_1 - t_0$

from homogeneity can be neglected. Here "small" means a small section of spacetime, therefore we must restrict our experiment not only to a small region of space, but also to short intervals of time.

For example, if two test bodies begin to fall side by side in the gravitational field of the Earth, as time proceeds, they will get closer to each other, since both fall in the direction of the center of the Earth. On the other hand, if they start on top of each other, the acceleration of the lower body is always slightly bigger than that of the upper body, and the distance between the two bodies gradually increases, see Fig. 10.4.

If we had provided our experimenter in the examples with a very large laboratory, or conceded him very long observation times, then it would have been possible for him to distinguish between the different situations. Strictly speaking, for every point only an infinitesimally small free-falling laboratory is defined, a *local inertial system*, or free-falling frame of reference.

10.4   Light Deflection in the Gravitational Field

Using the equivalence principle alone, we can already predict that light must be deflected in a gravitational field. In Fig. 10.5 we consider a laboratory in free fall. If a laser beam is emitted in this laboratory on the left-hand side at time t_0, it arrives at the detector on the other side at time t_1 *at the same height*, because this laboratory is equivalent to a laboratory without a gravitational field. However, viewed from the outside, the laboratory has moved downward in the interval $t_1 - t_0$. Therefore, the laser beam appears deflected. On the other hand, we can also consider a laboratory with constant acceleration upward, as shown in Fig. 10.6. When the laser beam is

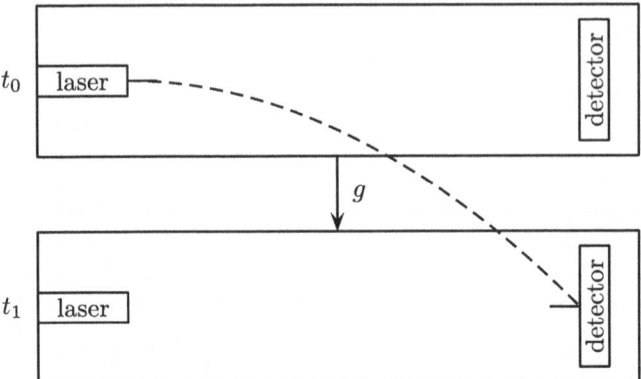

Fig. 10.5 In a laboratory in free fall, a laser beam is sent at time t_0 from the left side of the laboratory to the right side. Since a laboratory in free fall corresponds to a laboratory without a gravitational field, the laser beam arrives in the detector on the other side at time t_1 at the same height. But, viewed from the outside, the laser beam is bent

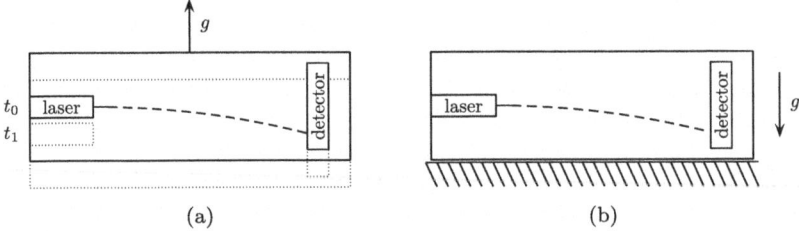

Fig. 10.6 The equivalence principle implies already the deflection of light in a gravitational field. (**a**) A laser beam is emitted in a laboratory with constant acceleration upward. Seen from the outside, the light signal propagates on a straight line. But since the laboratory has moved upward, the signal arrives at a slightly lower position on the other side. (**b**) The laboratory with constant acceleration corresponds to a laboratory at rest in a uniform gravitational field. Because of the equivalence principle, the laser beam must be bent also in this laboratory

released it will lag behind the laboratory and arrive at the other side at a slightly lower position. This can be easily understood when we remember that, viewed from the outside, the laser beam must propagate on a straight line.

The accelerated laboratory is equivalent to a laboratory at rest in a uniform gravitational field. Therefore, the ray of light must be deflected also in this laboratory.

10.5   Mathematical Formulation of the Equivalence Principle

In mathematical terms, the equivalence principle means that the spacetime *with* gravitation is *locally* Minkowskian. Let x^α be the coordinates describing the spacetime globally. Then, at any point P, there exists a coordinate transformation

$$x^\alpha \mapsto \xi^\alpha, \tag{10.12}$$

depending on x^μ, so that the metric transforms according to

$$g_{\mu\nu}(x^\alpha) \mapsto \tilde{g}_{\mu\nu}(\xi^\alpha) \quad \text{with} \quad \tilde{g}_{\mu\nu}(\xi_P^\alpha) = \eta_{\mu\nu} \tag{10.13}$$

in the neighborhood of $P = \xi_P^\alpha$, i.e.,

$$\left. \frac{\partial \tilde{g}_{\mu\nu}(\xi^\alpha)}{\partial \xi^\beta} \right|_{\xi_P^\alpha} = 0. \tag{10.14}$$

In general, higher derivatives will not vanish, which means the metric has the form

$$\tilde{g}_{\mu\nu}(\xi_P^\alpha) = \eta_{\mu\nu} + \frac{1}{2} \left(\frac{\partial^2 \tilde{g}_{\mu\nu}}{\partial \xi^\alpha \partial \xi^\beta} \xi^\alpha \xi^\beta \right). \tag{10.15}$$

In other words, in a small neighborhood around the point P there is no difference with respect to a flat spacetime.

10.6 Exercises

10.1. Inhomogeneous gravitational fields In this exercise we examine, within Newtonian gravitational theory, the consequences of the finite size of a laboratory in free fall in an inhomogeneous gravitational field.

We consider two masses with positions $r_1(t)$ and $r_2(t) = r_1(t) + \xi(t)$ falling radially towards a point mass. If the equivalence principle was fulfilled globally, their distance $\xi(t)$ would remain constant.

(a) Show that for ξ we have, in good approximation,

$$\ddot{\xi} = c^2 \frac{r_s}{r^3} \xi \tag{10.16}$$

where $r_s = 2GM/c^2$ (cf. Eq. (1.38)).

(b) Show that during a short time of observation Δt, the distance ξ between the masses changes by

$$\Delta \xi = \frac{1}{2} \left(\frac{r_s}{r} \right)^3 \left(\frac{c\Delta t}{r_s} \right)^2 \xi_0, \tag{10.17}$$

where ξ_0 is the initial distance of the masses

(c) For the laboratory to be "small", the value of $\Delta\xi$ must be smaller than the precision achievable in measurements. Calculate $\Delta\xi$ for a laboratory of the size $\xi_0 = 100$ m on the Earth's surface for observation times $\Delta t = 1$ s and 100 s. Can these laboratories still be considered as small?

10.2. Light deflection in a laboratory falling with constant velocity In this section we have derived the deflection of light in a gravitational field using the equivalence principle. But even if we observe a laboratory, falling with a *constant* velocity, from the outside, the ray of light propagating on a straight line in the laboratory must be seen on a curved trajectory. What is the crucial difference with regard to a laboratory falling with constant acceleration?

References

1. Einstein, A.: Über das Relativitätsprinzip und die aus demselben gezogenen Folgerungen. (On the principle of relativity and the conclusions drawn from it.) Jahrb. Radioakt. Elektron. **4**, 411–967 (1907)
2. Misner, C.W., Thorne, K.S., Wheeler, J.A.: Gravitation. W.H. Freeman, New York (1973)

Riemannian Geometry

11

Contents

The topic of *Riemannian geometry*[1] is the study of the curvature of spaces. In this chapter we will introduce some basic concepts of Riemannian geometry. Part of the considerations in this chapter will be a generalization of the results from Chap. 5. In comparison with the formalism of Lorentz transformations, an important extension will be the introduction of non-constant transformations, i.e., the relation between the frames of reference depends explicitly on the coordinates. This extension complicates the formalism considerably. In particular, we shall see that the concept of usual partial differentiation is no longer well suited to formulate mathematical relations in general relativity. Furthermore, we need a precise characterization of curved spaces in mathematical terms. We shall see that this task is intimately connected to the shortfall of usual partial differentiation just mentioned.

Since our main focus is on the physical discussion of general relativity, we can only give a rough outline of the mathematical foundations. For a more detailed discussion of the subject we refer the reader to more specialized literature, e.g. Nakahara [5], Wald [8], or Misner et al. [4].

11.1 Riemannian Spaces

11.1.1 Differentiable Manifolds

An n-dimensional *manifold* $\mathcal{M}^n$ is a topological space with the following properties:

1. There exists a family of pairs $\{(U_i, \Phi_i)\}_{i \in \mathcal{I}}$, where $\mathcal{I}$ denotes an index set and $\{U_i\}$ denotes a family of open sets, with

$$\bigcup_{i \in \mathcal{I}} U_i = \mathcal{M}^n \tag{11.1}$$

and $\Phi_i : U_i \to \mathbb{R}^n$ is a homeomorphism (a continuous and invertible mapping) that maps U_i onto an open subset of $\mathbb{R}^n$. Thus, each point $P \in \mathcal{M}^n$ has a neighborhood U_i, in which n real numbers $(x^1(P), \ldots, x^n(P))$ are associated with it by n functions representing the homeomorphism Φ_i. A pair (U_i, Φ_i) is called a *chart*, while the set of all pairs $\{(U_i, \Phi_i)\}_{i \in \mathcal{I}}$ is called an *atlas*.
2. For any two points there exist disjoint neighborhoods (Hausdorff space).
3. $\mathcal{M}^n$ is connected.

---

[1] Georg Friedrich Bernhard Riemann, 1826–1866, German mathematician.

$\mathcal{M}^n$ is called a *differentiable manifold*, if two overlapping coordinate systems x^i and $x^{i'}$ are connected among each other by an (r-fold) continuously differentiable coordinate transformation

$$x^{i'} = x^{i'}(x^1, \ldots, x^n), \quad i' = 1, \ldots, n \tag{11.2}$$

with non-singular functional determinant.

11.1.2 Definition of the Riemannian Space

An n-dimensional *Riemannian space* is an n-dimensional differentiable manifold $\mathcal{M}^n$, equipped with a fixed and non-singular, positive definite symmetric covariant second rank tensor field, i.e., a *metric*. If the metric of the space is not positive definite, as is the case in GR, then this is called a *pseudo-Riemannian manifold*, or also *semi-Riemannian manifold*.

11.1.3 Tangent Space and Cotangent Space

The intuitive concept of a vector as an "arrow" which connects two points cannot be generalized to manifolds, since we always remain *within* the manifold. We will define a vector $\underline{t}$ in $\mathcal{M}^n$ as a tangent, or directional derivative, of a function $f : \mathcal{M}^n \to \mathbb{R}$ along a curve $\zeta : I \subset \mathcal{M}^n$ at the point $P = \zeta(\lambda = 0)$, with $I \subset \mathbb{R}$. Let us calculate this derivative [5]

$$\frac{\mathrm{d} f(\zeta(\lambda))}{\mathrm{d}\lambda}\bigg|_{\lambda=0} = \frac{\mathrm{d}}{\mathrm{d}\lambda} \left(f \circ \Phi^{-1} \circ \Phi \circ \zeta(\lambda) \right)\bigg|_{\lambda=0}$$

$$= \frac{\partial \left(f \circ \Phi^{-1}(x^\mu) \right)}{\partial x^\nu} \frac{\mathrm{d} x^\nu(\zeta(\lambda))}{\mathrm{d}\lambda}\bigg|_{\lambda=0} \tag{11.3a}$$

$$= \frac{\partial}{\partial x^\mu} \left(f \circ \Phi^{-1}(x^\nu) \right) t^\nu. \tag{11.3b}$$

For simplicity, we have assumed that $\mathcal{M}^n$ can be represented by a single chart (U, Φ), and $f \circ \Phi^{-1}$ is the coordinate representation of the function f. Correspondingly, $\Phi \circ \zeta$ gives the coordinate representation of the curve ζ. The vector $\underline{t}$ can then be written, in shorthand notation, as the operator

$$\underline{t} \equiv t^\nu \frac{\partial}{\partial x^\nu} \equiv t^\nu \partial_\nu \tag{11.4}$$

with the *contravariant* components t^ν. If we consider curves ζ along the coordinate axes $x^\mu = \text{const}$, $\mu \neq \nu$, we arrive at the basis vectors $\{\partial_\nu \equiv \partial/\partial x^\nu\}$. These span the *tangent space* T_P at the point P, see Fig. 11.1.

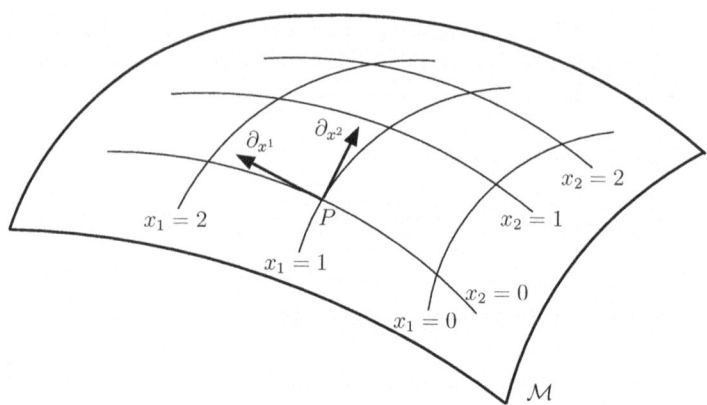

Fig. 11.1 Manifold $\mathcal{M}$ with coordinate axes x_1 and x_2 and the corresponding basis vectors ∂_{x_1} and ∂_{x_2}

We can define the dual basis $\{dx^\mu\}$ for the *cotangent space* T_P^* at the point P via the relation

$$\left\langle dx^\mu, \frac{\partial}{\partial x^\nu} \right\rangle = \frac{\partial x^\mu}{\partial x^\nu} = \delta_\nu^\mu. \tag{11.5}$$

A covector $\underline{w}$, also called a dual vector, can then be formulated as a linear combination of the dual basis with the *covariant* components w_μ, viz.

$$\underline{w} = w_\mu dx^\mu. \tag{11.6}$$

11.2   Tensor Calculus in GR

Let x^μ be the contravariant components of a vector $\underline{x}$ in an arbitrarily chosen n-dimensional basis. We will investigate the transformation behavior of different quantities under a coordinate transformation

$$x^\mu \mapsto \tilde{x}^\nu = \tilde{x}^\nu(x^\mu). \tag{11.7}$$

In special relativity we only considered constant functions $\tilde{x}^\nu(x^\mu) = \Lambda_\mu^\nu x^\mu$. The relation between the frames of reference depended only on the relative velocity β, or on several angles of rotation. This restriction must be abandoned in general relativity.

11.2.1 Contravariant and Covariant Tensors

The differentials dx^μ transform according to the usual rules of differentiation:

$$d\tilde{x}^\nu = \frac{\partial \tilde{x}^\nu}{\partial x^\mu} dx^\mu. \tag{11.8}$$

Any n-component quantity A^μ which transforms as the differentials, i.e., according to the rule

$$\tilde{A}^\nu = \frac{\partial \tilde{x}^\nu}{\partial x^\mu} A^\mu, \tag{11.9}$$

is called a *contravariant tensor of rank 1*.

To determine the transformation behavior of the derivatives $\partial/\partial x^\mu$, we consider a function $f(\tilde{x}^\nu)$. For the partial derivatives of this function we have

$$\frac{\partial}{\partial \tilde{x}^\nu} f(\tilde{x}^\nu) = \frac{\partial x^\mu}{\partial \tilde{x}^\nu} \frac{\partial}{\partial x^\mu} f(x^\mu(\tilde{x}^\nu)). \tag{11.10}$$

This yields the transformation equation for the derivatives

$$\frac{\partial}{\partial \tilde{x}^\nu} = \frac{\partial x^\mu}{\partial \tilde{x}^\nu} \frac{\partial}{\partial x^\mu}. \tag{11.11}$$

Any n-component quantity B_ν which transforms as the coordinate derivatives (n-dimensional- gradient), i.e., according to the rule

$$\tilde{B}_\nu = \frac{\partial x^\mu}{\partial \tilde{x}^\nu} B_\mu \tag{11.12}$$

is called a *covariant tensor of rank 1*. In order to simplify the notation, we introduce a shorthand notation:

$$x^{\tilde{\nu}}_{,\mu} \equiv \frac{\partial \tilde{x}^\nu}{\partial x^\mu} \quad \text{and} \quad x^\mu_{,\tilde{\nu}} \equiv \frac{\partial x^\mu}{\partial \tilde{x}^\nu}. \tag{11.13}$$

Using the new notation, we can summarize the previous results: In a coordinate transformation $x^\mu \mapsto \tilde{x}^\nu(x^\mu)$, contravariant and covariant tensors transform according to

$$\tilde{A}^\nu = x^{\tilde{\nu}}_{,\mu} A^\mu \quad \text{and} \quad \tilde{B}_\nu = x^\mu_{,\tilde{\nu}} B_\mu, \tag{11.14}$$

respectively. This definition is completely analogous to the notation of the Lorentz transformation, see Sect. 5.2. In particular one has

$$x^{\tilde{\kappa}}_{,\nu} x^\nu_{,\tilde{\beta}} = \frac{\partial \tilde{x}^\kappa}{\partial x^\nu} \frac{\partial x^\nu}{\partial \tilde{x}^\beta} = \frac{\partial \tilde{x}^\kappa}{\partial \tilde{x}^\beta} = \delta^\kappa_\beta, \tag{11.15}$$

in analogy with Eq. (5.17). Of course, this must be so, because the Lorentz transformations are a special case of the transformations considered here, namely, the special case in which all $x^{\tilde{\nu}}_{,\mu}$ are constant. For example, for a boost in the x-direction we have

$$x^{\tilde{0}}_{,0} = \gamma, \quad x^{\tilde{0}}_{,1} = -\beta\gamma, \quad x^{\tilde{1}}_{,0} = -\beta\gamma, \quad x^{\tilde{1}}_{,1} = \gamma, \quad x^{\tilde{2}}_{,2} = 1, \quad x^{\tilde{3}}_{,3} = 1.$$
(11.16)

All other derivatives vanish, see Eq. (3.20).

11.2.2 Tensors of Higher Rank

The transformation behavior of tensors of higher rank is defined in analogy with the definition introduced in special relativity for Lorentz transformations. For example, let $C^{\mu}{}_{\nu}$ be a covariant tensor of rank 1 with respect to the index μ and a contravariant tensor of rank 1 with respect to the index ν. Then this tensor transforms according to

$$\tilde{C}^{\mu}{}_{\nu} = x^{\tilde{\mu}}_{,\kappa} x^{\beta}_{,\tilde{\nu}} C^{\kappa}{}_{\beta}.$$
(11.17)

The tensor product and the tensor contraction are also defined in analogy with special relativity, e.g.,

$$D^{\mu\nu} = A^{\mu} B^{\nu},$$
(11.18)

or the scalar product

$$C = A^{\mu} B_{\mu}.$$
(11.19)

In this case, C is a *tensor of rank 0*, or a scalar.

11.2.3 Linear Forms

Let V be an n-dimensional vector space with basis $(\underline{e}_1, \ldots, \underline{e}_n)$. A mapping $f : V \to \mathbb{R}, \underline{x} \mapsto f(\underline{x})$ with $\underline{x} \in V$ is called *linear*, if it fulfills

$$f(a\underline{x} + b\underline{y}) = af(\underline{x}) + bf(\underline{y}) \quad \text{for all} \quad a, b \in \mathbb{R} \quad \text{and} \quad \underline{x}, \underline{y} \in V.$$
(11.20)

It is called a *linear form* over the vector space V. The set of all linear mappings $V \to \mathbb{R}$ is the *dual space V^**.

If $(\underline{e}_1, \ldots, \underline{e}_n)$ is a basis in V, then in its dual space there exists a uniquely determined basis $(\underline{\varepsilon}^1, \ldots, \underline{\varepsilon}^n)$ with

$$\underline{\varepsilon}^i(\underline{e}_j) = \delta^i_j. \tag{11.21}$$

A tensor of rank (p, q) (p-fold contravariant, q-fold covariant) is a mapping $(V_1^*, \ldots, V_p^*, V_{p+1}, \ldots, V_{p+q}) \to \mathbb{R}$ linear in all arguments, i.e., a *multilinear form*. With the basis

$$(\underline{e}^{j_1 \ldots j_q}_{i_1 \ldots i_p}) = (\underline{e}_{i_1} \otimes \ldots \otimes \underline{e}_{i_p} \otimes \underline{\varepsilon}^{j_1} \otimes \ldots \otimes \underline{\varepsilon}^{j_q}) \tag{11.22}$$

the components of the tensor are obtained as

$$T^{i_1 \ldots i_p}_{j_1 \ldots j_q} = T(\underline{e}^{j_1 \ldots j_q}_{i_1 \ldots i_p}). \tag{11.23}$$

Note the different positions of the indices in the components and the basis vectors.

11.2.4 Metric Tensor

A metric g defines distances and angles on a manifold $\mathcal{M}$, and can be written as a symmetric tensor of rank $(0, 2)$,

$$g = g_{\mu\nu} dx^\mu \otimes dx^\nu. \tag{11.24}$$

The flat Minkowski spacetime has the metric components[2]

$$g_{\mu\nu} = \mathrm{diag}(-c^2, 1, 1, 1).$$

The *scalar product* of two vectors $\underline{x}$ and $\underline{y}$ with respect to the metric g is given by

$$\begin{aligned}
\left\langle \underline{x}, \underline{y} \right\rangle_g &= g(\underline{x}, \underline{y}) = g(x^\rho \partial_\rho, y^\sigma \partial_\sigma) \\
&= g_{\mu\nu} x^\rho y^\sigma dx^\mu(\partial_\rho) \otimes dx^\nu(\partial_\sigma) = g_{\rho\sigma} x^\rho y^\sigma,
\end{aligned} \tag{11.25}$$

where we have used the vector definition (11.4) to represent the two vectors in their coordinate basis ∂_μ, and exploited the relation $dx^\mu(\partial_\nu) = \delta^\mu_\nu$. The length $|\underline{x}|$ of a vector follows immediately from

$$|\underline{x}|^2 = \left\langle \underline{x}, \underline{x} \right\rangle_g = g_{\mu\nu} x^\mu x^\nu. \tag{11.26}$$

---

[2] Unlike in SR, we here include the speed of light c in the metric and not in the coordinates (cf. Sect. 5.1.1, Eq. (5.5)).

Correspondingly, the inverse metric g^* is a symmetric tensor of rank $(2, 0)$,

$$g^* = g^{\mu\nu}\partial_\mu \otimes \partial_\nu. \tag{11.27}$$

The components of g and g^* fulfill

$$g_{\mu\alpha}g^{\alpha\nu} = \delta_\mu^\nu. \tag{11.28}$$

Using the metric tensor, we can, in analogy with SR, "raise" and "lower" indices. For instance, $A_\mu = g_{\mu\nu}A^\nu$ is a first rank covariant tensor. For tensors of higher rank the analogous relation applies

$$A_{\mu\nu...} = g_{\mu\alpha}g_{\nu\beta}\dots A^{\alpha\beta...}. \tag{11.29}$$

For raising indices, we need the inverse metric $g^{\mu\nu}$, and we have

$$B^{\mu\nu...} = g^{\mu\alpha}g^{\nu\beta}\dots B_{\alpha\beta...}. \tag{11.30}$$

The infinitesimal line element ds^2 is defined by the metric, similar as in special relativity, and has the form

$$ds^2 = g_{\mu\nu}dx^\mu dx^\nu. \tag{11.31}$$

In n-dimensional Minkowski coordinates, one has $g_{\mu\nu} = \eta_{\mu\nu}$, and thus

$$ds^2 = \eta_{\mu\nu}dx^\mu dx^\nu. \tag{11.32}$$

We consider a coordinate transformation $x^\mu = x^\mu(\tilde{x}^\kappa)$, $dx^\mu = x^\mu_{,\kappa}d\tilde{x}^\kappa$ from Minkowski coordinates to arbitrary other curvilinear coordinates. Then the line element is given by

$$ds^2 = \eta_{\mu\nu}x^\mu_{,\tilde\kappa}x^\nu_{,\tilde\beta}d\tilde{x}^\kappa d\tilde{x}^\beta = \tilde{g}_{\kappa\beta}d\tilde{x}^\kappa d\tilde{x}^\beta, \tag{11.33}$$

with

$$\tilde{g}_{\alpha\beta} = x^\mu_{,\tilde\alpha}x^\nu_{,\tilde\beta}\eta_{\mu\nu}. \tag{11.34}$$

For arbitrary transformations of the coordinates, the metric transforms according to

$$\tilde{g}_{\alpha\beta} = x^\mu_{,\tilde\alpha}x^\nu_{,\tilde\beta}g_{\mu\nu}. \tag{11.35}$$

This implies that $g_{\mu\nu}$ is a symmetric second rank covariant tensor.

For illustration, we consider a few examples, but will restrict ourselves to the spatial components of $g_{\mu\nu}$.

Two- and Three-Dimensional Euclidean Space

For the two-dimensional plane, we write in Cartesian coordinates $x^1 = x$, $x^2 = y$ and $ds^2 = dx^2 + dy^2$, i.e., we have

$$g_{\mu\nu} = \begin{pmatrix} 1 & 0 \\ 0 & 1 \end{pmatrix}. \tag{11.36}$$

If we use polar coordinates with $x = r\cos(\varphi)$, $y = r\sin(\varphi)$, i.e., $\tilde{x}^1 = r$, $\tilde{x}^2 = \varphi$, we obtain the individual components of the metric tensor

$$\tilde{g}_{11} = \frac{\partial x}{\partial r}\frac{\partial x}{\partial r} + \frac{\partial y}{\partial r}\frac{\partial y}{\partial r} = \cos^2(\varphi) + \sin^2(\varphi) = 1,$$

$$\tilde{g}_{22} = \frac{\partial x}{\partial\varphi}\frac{\partial x}{\partial\varphi} + \frac{\partial y}{\partial\varphi}\frac{\partial y}{\partial\varphi} = (-r\sin(\varphi))^2 + (r\cos(\varphi))^2 = r^2,$$

$$\tilde{g}_{12} = \tilde{g}_{21} = \frac{\partial x}{\partial r}\frac{\partial x}{\partial\varphi} + \frac{\partial y}{\partial r}\frac{\partial y}{\partial\varphi} \tag{11.37}$$

$$= \cos(\varphi)\,(-r\sin(\varphi)) + \sin(\varphi)\,r\cos(\varphi) = 0.$$

Therefore, the line element in polar coordinates is given by

$$ds^2 = \tilde{g}_{\mu\nu}d\tilde{x}^\mu d\tilde{x}^\nu = (d\tilde{x}^1)^2 + (\tilde{x}^1)^2(d\tilde{x}^2)^2 = dr^2 + r^2 d\varphi^2 \tag{11.38}$$

with the metric tensor

$$\tilde{g}_{\mu\nu} = \begin{pmatrix} 1 & 0 \\ 0 & (\tilde{x}^1)^2 \end{pmatrix} = \begin{pmatrix} 1 & 0 \\ 0 & r^2 \end{pmatrix}. \tag{11.39}$$

The coordinates $\tilde{x}^1$ and $\tilde{x}^2$ are not Cartesian, but nevertheless the space is flat. This is obvious since we consider a two-dimensional plane, and have only transformed the coordinates. But we can already see that one cannot tell the curvature of a space directly from the metric. Later, we will present a mathematically precise definition of whether a space is curved.

In three-dimensional Euclidean space, we can again write in Cartesian coordinates $x^1 = x$, $x^2 = y$, $x^3 = z$, $ds^2 = dx^2 + dy^2 + dz^2$ and $g_{\mu\nu} = \text{diag}\,(1, 1, 1)$. We transform to *spherical coordinates*, $x = r\sin(\vartheta)\cos(\varphi)$, $y = r\sin(\vartheta)\sin(\varphi)$, and $z = r\cos(\vartheta)$. Thus, the new coordinates are $\tilde{x}^1 = r$, $\tilde{x}^2 = \vartheta$, and $\tilde{x}^3 = \varphi$. The calculations are analogous to the planar case, and lead to the line element

$$ds^2 = (d\tilde{x}^1)^2 + (\tilde{x}^1)^2(d\tilde{x}^2)^2 + (\tilde{x}^1)^2\sin^2(\tilde{x}^2)(d\tilde{x}^3)^2$$

$$= dr^2 + r^2 d\vartheta^2 + r^2\sin^2(\vartheta)d\varphi^2 \tag{11.40}$$

with the metric tensor

$$\tilde{g}_{\mu\nu} = \begin{pmatrix} 1 & 0 & 0 \\ 0 & (\tilde{x}^1)^2 & 0 \\ 0 & 0 & (\tilde{x}^1)^2 \sin^2(\tilde{x}^2) \end{pmatrix} = \begin{pmatrix} 1 & 0 & 0 \\ 0 & r^2 & 0 \\ 0 & 0 & r^2 \sin^2(\vartheta) \end{pmatrix}. \tag{11.41}$$

Again, the coordinates $\tilde{x}^1$, $\tilde{x}^2$ and $\tilde{x}^3$ are non-Cartesian, but the underlying space still is Euclidean. The metric in (11.41) is nothing but the spatial part of $\eta_{\mu\nu}$ in spherical polar coordinates.

Surface of the Unit Sphere

As we will see, the surface of the unit sphere is a two-dimensional curved space. The description of the surface of the sphere *without* embedding it into three-dimensional space can be done by the coordinates $x^1 = \vartheta$, $x^2 = \varphi$ with the metric tensor

$$g_{\mu\nu} = \begin{pmatrix} 1 & 0 \\ 0 & \sin^2(x^1) \end{pmatrix} = \begin{pmatrix} 1 & 0 \\ 0 & \sin^2(\vartheta) \end{pmatrix}. \tag{11.42}$$

Therefore, the line element is

$$ds^2 = (dx^1)^2 + \sin^2(x^1)(dx^2)^2 = d\vartheta^2 + \sin^2(\vartheta)d\varphi^2. \tag{11.43}$$

Here appears an important difference to the two cases considered before, because for $x^1 = 0$ or $x^1 = \pi$ the metric $g_{\mu\nu}$ is not invertible!

This means that at the poles there exist *coordinate singularities*, but without these points having special properties different from all the other points on the sphere.

11.2.5 Local Tetrad

In Sect. 8.1.2 we have represented the frame of reference of a moving observer using orthonormal basis vectors. The frame of reference of an observer at a point P within a manifold $\mathcal{M}$ with metric g can also be represented by a set of four orthonormal vectors. In the literature this is called a *tetrad*, or a *vierbein*. This tetrad spans the tangential space T_P, and is valid only locally at the point P, which is why it is also called a *local tetrad*.

In a Riemannian manifold, the four tetrad vectors $\underline{e}_{(i)} = e^\mu_{(i)} \partial_\mu$ have to satisfy the orthonormality condition

$$\left\langle \underline{e}_{(i)}, \underline{e}_{(j)} \right\rangle_g = g(\underline{e}_{(i)}, \underline{e}_{(j)}) = g_{\mu\nu} e^\mu_{(i)} e^\nu_{(j)} = \delta_{ij}, \tag{11.44}$$

where $e^{\mu}_{(i)}$ are the components of $\underline{\mathbf{e}}_{(i)}$ with respect to the coordinate basis $\{\partial_{\mu}\}$. In GR, however, we are dealing with a semi-Riemannian manifold, therefore, the orthonormality relation in GR reads

$$\langle \underline{\mathbf{e}}_{(i)}, \underline{\mathbf{e}}_{(j)} \rangle_g = \eta_{ij} \tag{11.45}$$

with $\eta_{ij} = \text{diag}(-1, 1, 1, 1)$. Thus, *locally* we are always dealing with a Minkowski spacetime.

In Sect. 11.1.3 we defined the dual basis vectors dx^{μ} for the basis vectors ∂_{μ} via the relation (11.5). In an analogous way, we can define the dual local tetrad vectors $\underline{\boldsymbol{\theta}}^{(i)} = \theta^{(i)}_{\mu} dx^{\mu}$ for the local tetrad vectors. Their components then must fulfill the following condition

$$\langle \underline{\boldsymbol{\theta}}^{(i)}, \underline{\mathbf{e}}_{(j)} \rangle = \left\langle \theta^{(i)}_{\mu} dx^{\mu}, e^{\nu}_{(j)} \partial_{\nu} \right\rangle = \theta^{(i)}_{\mu} e^{\nu}_{(j)} \langle dx^{\mu}, \partial_{\nu} \rangle = \theta^{(i)}_{\mu} e^{\mu}_{(j)} = \delta^i_j. \tag{11.46}$$

A four-vector $\underline{\boldsymbol{F}} = F^{\mu} \partial_{\mu}$ can now be formulated in terms of a local tetrad,

$$\underline{\boldsymbol{F}} = F^{(i)} \underline{\mathbf{e}}_{(i)} = F^{(i)} e^{\mu}_{(i)} \partial_{\mu} = F^{\mu} \partial_{\mu}. \tag{11.47}$$

We see that, by contraction with the tetrad components $e^{\mu}_{(i)}$, we obtain the corresponding coordinate components F^{μ} from the tetrad components $F^{(i)}$. For the inverse transformation, we need the components of the dual tetrad. The result is

$$F^{(i)} = \theta^{(i)}_{\mu} F^{\mu}. \tag{11.48}$$

11.2.6 Volume Element

In Cartesian coordinates, the volume element is given by

$$dV = \prod_{i=1}^{n} dx^i. \tag{11.49}$$

Transforming the coordinates by $x^{\mu} = x^{\mu}(\tilde{x}^{\nu})$ and $dx^{\mu} = x^{\mu}_{,\tilde{\alpha}} d\tilde{x}^{\alpha}$, we obtain the volume element

$$dV = \prod_{i=1}^{n} dx^i = \det\left(x^i_{,j}\right) \prod_{k=1}^{n} d\tilde{x}^k = d\tilde{V} \text{ with the } \textit{Jacobian matrix } \mathcal{J} = \left(x^i_{,j}\right). \tag{11.50}$$

In Cartesian coordinates, $g_{\mu\nu} = \delta_{\mu\nu}$ and thus $\tilde{g}_{\alpha\beta} = x^{\tilde{\mu}}_{,\alpha} x^{\tilde{\nu}}_{,\beta} \delta_{\mu\nu}$. With the abbreviation $g \equiv \det(g_{\mu\nu})$, taking advantage of the separability of products of determinants, $\det(A \cdot B) = \det A \det B$, and using the fact that in Cartesian coordinates $\det(\delta_{\mu\nu}) = 1$, we find

$$\tilde{g} = \left[\det\left(x_{,\alpha}^{\tilde{\mu}}\right)\right]^2. \tag{11.51}$$

From (11.51) follows $d\tilde{V} = \sqrt{\tilde{g}} \prod_{i=1}^{n} d\tilde{x}^i$. Quite generally one has in curvilinear coordinates

$$d\tilde{V} = \sqrt{g} \prod_{\mu} d\tilde{x}^{\mu}. \tag{11.52}$$

If we consider, e.g., the metric (11.41), we end up with the familiar result $d\tilde{V} = r^2 \sin(\vartheta)\, dr\, d\vartheta\, d\varphi$.

11.2.7 Parallel Transport and Affine Connections

If we consider a vector $\underline{F}$ at a point P in a curved space, we have no possibility to simply compare this vector with another vector at the point Q, in the way it is always possible in Euclidean space. Only for infinitesimally neighboring points will such a comparison be possible, by "sliding" the vector parallel from P at the point x^{μ} to P' at the point $x^{\mu} + \delta x^{\mu}$. By multiple infinitesimal transports, we can move a vector along a curve from P to Q, but the result will depend on the path taken. In the following sections we will specify this concept in a precise mathematical way.

Parallel Transport in Two-Dimensional Euclidean Space

As an introductory example, we consider the two-dimensional Euclidean space. Let $\underline{\tilde{F}} = \tilde{F}^{\mu} \partial_{\mu}$ denote the vector $\underline{F} = F^{\mu} \partial_{\mu}$, transported in parallel. In Cartesian coordinates (x, y), with the line element $ds^2 = dx^2 + dy^2$ and the basis vectors $\underline{e}_x = \partial_x$, $\underline{e}_y = \partial_y$, one simply has $\tilde{F}^{\mu} = F^{\mu}$ for arbitrary displacements. We can also calculate in polar coordinates (r, φ), with the corresponding vectors ∂_r and ∂_{φ}. Then the line element is $ds^2 = dr^2 + r^2 d\varphi^2$, and the components of the vector $\underline{F} = F^r \partial_r + F^{\varphi} \partial_{\varphi}$ are given by

$$F^r = F \cos(\vartheta), \quad F^{\varphi} = F \frac{\sin(\vartheta)}{r}. \tag{11.53}$$

Here, ϑ is the angle between $\underline{F}$ and the ∂_r direction, and the modulus of the vector is $F = \sqrt{g_{\mu\nu} F^{\mu} F^{\nu}}$. The somewhat strange r-dependence of the F^{φ} component is due to the fact that the vectors ∂_r and ∂_{φ} no longer form an orthonormal basis with respect to the metric $g_{\mu\nu}$. For an orthonormal basis $\{\underline{e}_{(r)}, \underline{e}_{(\varphi)}\}$, the condition (11.44) must be satisfied. From this follows

$$\underline{e}_{(r)} = \partial_r \quad \text{and} \quad \underline{e}_{(\varphi)} = \frac{1}{r} \partial_{\varphi}. \tag{11.54}$$

The components of the vector $\underline{F} = F^{(r)} \underline{e}_{(r)} + F^{(\varphi)} \underline{e}_{(\varphi)}$ referred to this basis are $F^{(r)} = F \cos(\vartheta)$ and $F^{(\varphi)} = F \sin(\vartheta)$.

For a small displacement Δx^μ, we require that the components of the transported vector $\widetilde{\boldsymbol{F}}$ depend linearly on the components of $\boldsymbol{F}$ and the components of Δx^μ. Specifically, we now consider displacements in the r- and in the φ-direction. Transporting along r yields

$$\widetilde{F}^r = F^r, \quad \widetilde{F}^\varphi = \frac{r}{r + \delta r} F^\varphi \approx F^\varphi - \frac{\delta r}{r} F^\varphi. \tag{11.55}$$

The change of the F^φ-component during this transport is again a consequence of the non-orthonormal basis. Transporting along φ we obtain

$$\widetilde{F}^r = F \cos(\vartheta - \Delta\varphi) \simeq F \cos(\vartheta) + F \sin(\vartheta)\Delta\varphi = F^r + F^\varphi r \Delta\varphi,$$
$$\widetilde{F}^\varphi = F \frac{\sin(\vartheta - \Delta\varphi)}{r} \simeq F \frac{\sin(\vartheta)}{r} - F \frac{\cos(\vartheta)}{r} \Delta\varphi = F^\varphi - F^r \frac{\Delta\varphi}{r}, \tag{11.56}$$

see Fig. 11.2. These results can be written in the compact form

$$\widetilde{F}^\mu(x + \Delta x) = F^\mu(x) - F^\lambda \Gamma^\mu{}_{\nu\lambda}(x)\Delta x^\nu \tag{11.57}$$

with

$$\Gamma^r{}_{rr} = 0, \quad \Gamma^r{}_{r\varphi} = 0, \quad \Gamma^r{}_{\varphi r} = 0, \quad \Gamma^r{}_{\varphi\varphi} = -r, \tag{11.58a}$$
$$\Gamma^\varphi{}_{rr} = 0, \quad \Gamma^\varphi{}_{r\varphi} = \frac{1}{r}, \quad \Gamma^\varphi{}_{\varphi r} = \frac{1}{r}, \quad \Gamma^\varphi{}_{\varphi\varphi} = 0. \tag{11.58b}$$

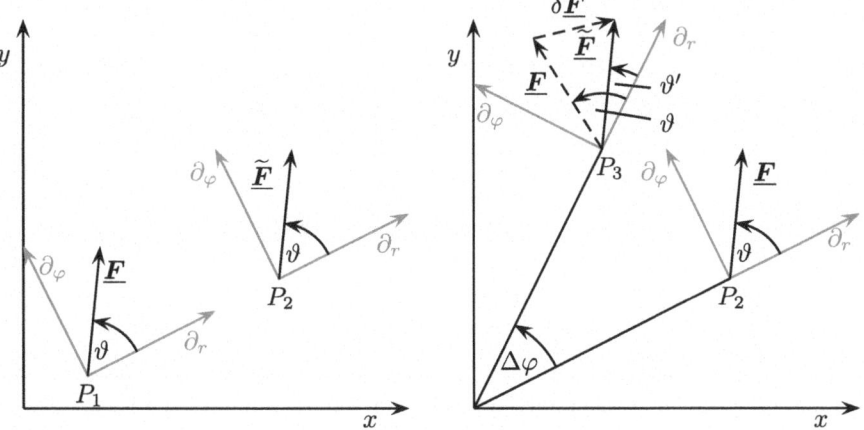

Fig. 11.2 Parallel transport of a vector in Euclidean space along the r- and the φ-coordinate. For displacements along the ∂_r-direction, the component F^r remains unchanged, the F^φ component changes, since ∂_φ is not a normalized vector. In the case of transport along the ∂_φ-direction both components change

Generalization to Arbitrary Coordinates

We consider the parallel transport of a vector $\underline{F} = F^\mu \partial_\mu$ from the point P_1 with coordinates x^β to the point P_2 with coordinates $x^\beta + \delta x^\beta$. We assume a locally Euclidean basis in P_1, and transport $F^\mu(P_1)$ parallel in this basis. Then, in P_2, the vector has the components

$$F^\mu_\parallel(P_2) = F^\mu(P_1) + \delta F^\mu(F^\alpha, \delta x^\beta). \tag{11.59}$$

It should be noted that $\underline{F}$ remains unchanged as a geometric object, it is only the projection onto the basis that changes. The deviation δF^μ is caused by the change of the direction of the coordinate lines. For small distances between P_1 and P_2, the size of $\delta F^\mu(F^\alpha, \delta x^\beta)$ again depends linearly on F^α and dx^β, i.e., it is of the form

$$\delta F^\mu = -\Gamma^\mu{}_{\alpha\beta} F^\alpha \delta x^\beta. \tag{11.60}$$

The $\Gamma^\mu{}_{\alpha\beta}$ are called *transition coefficients* or coefficients of the affine connection. $\Gamma^\mu{}_{\alpha\beta}$ is the μ-th component of the change of the basis vector $\underline{e}_\alpha$ transported parallel along a basis vector $\underline{e}_\beta$.

To calculate the transition coefficients, we consider the scalar $g_{\mu\nu} F^\mu F^\nu$, which does not change in a parallel transport. Therefore, one has

$$
\begin{aligned}
0 &= \delta\left(g_{\mu\nu} F^\mu F^\nu\right) \\
&= g_{\mu\nu,\beta} F^\mu F^\nu \delta x^\beta + g_{\mu\nu}(\delta F^\mu)F^\nu + g_{\mu\nu} F^\mu \delta F^\nu \\
&= g_{\mu\nu,\beta} F^\mu F^\nu \delta x^\beta - g_{\mu\nu}\Gamma^\mu{}_{\alpha\beta} F^\alpha F^\nu \delta x^\beta - g_{\mu\nu} F^\mu \Gamma^\nu{}_{\alpha\beta} F^\alpha \delta x^\beta \\
&= \left(g_{\mu\nu,\beta} - g_{\alpha\nu}\Gamma^\alpha{}_{\mu\beta} - g_{\mu\alpha}\Gamma^\alpha{}_{\nu\beta}\right) F^\mu F^\nu \delta x^\beta.
\end{aligned}
\tag{11.61}
$$

Since F^μ and dx^β can be chosen arbitrarily, the following system of linear equations is obtained for determining the coefficients Γ:

$$g_{\mu\nu,\beta} - g_{\alpha\nu}\Gamma^\alpha{}_{\mu\beta} - g_{\mu\alpha}\Gamma^\alpha{}_{\nu\beta} = 0. \tag{11.62}$$

The solution of this system of equation reads (see Exercise 11.1)

$$\Gamma^\alpha{}_{\mu\nu} = \frac{1}{2} g^{\alpha\sigma}\left(g_{\sigma\mu,\nu} + g_{\sigma\nu,\mu} - g_{\mu\nu,\sigma}\right). \tag{11.63}$$

In our case the transition coefficients coincide with the *Christoffel symbols of the second kind*,[3] which actually are defined by yet another property, see Sect. 11.2.8.

---

[3] Elwin Bruno Christoffel, 1829–1900, German mathematician.

We make no distinction between the terms transition coefficients and Christoffel symbols, and from now on will use only the term Christoffel symbol.[4] Because of the symmetry of the metric, $g_{\mu\nu} = g_{\nu\mu}$, it follows immediately that the Christoffel symbols are also symmetric with respect to their lower two indices. The *Christoffel symbols of the first kind* are given by

$$\Gamma_{\alpha\mu\nu} = \frac{1}{2}\left(g_{\alpha\mu,\nu} + g_{\alpha\nu,\mu} - g_{\mu\nu,\alpha}\right) \tag{11.64}$$

and are related to the Christoffel symbols of the second kind by $\Gamma^{\alpha}{}_{\mu\nu} = g^{\alpha\sigma}\Gamma_{\sigma\mu\nu}$.

Transformation Behavior of the Christoffel Symbols
At first glance, one might assume that the Christoffel symbols are tensors due to their index notation. But they are not. We will establish this point in two different ways.

An approach motivated by physics goes back to what has been said about the mathematical meaning of the equivalence principle in Sect. 10.5. In suitable coordinates, in the neighborhood of any point, the metric $g_{\mu\nu}$ can be transformed into the form $g_{\mu\nu} = \eta_{\mu\nu} + g_{\mu\nu,\alpha\beta}\xi^{\alpha}\xi^{\beta}/2$, i.e., it coincides with the Minkowski metric, up to corrections of second order. As a consequence, all derivatives vanish, which also implies that all Christoffel symbols vanish. Thus, at this point, the object $\Gamma^{\mu}{}_{\nu\alpha}$ has only entries equal to zero. From the transformation rules for tensors in Sect. 11.2.2 one can see immediately that a tensor, which in one coordinate system has only entries equal to zero, has vanishing entries in all coordinate systems. Therefore, the Christoffel symbols cannot be tensors.

The second approach is the explicit calculation of the transformation behavior. The starting point is the Christoffel symbols in transformed coordinates

$$\tilde{\Gamma}^{\mu}{}_{\alpha\beta} = \Gamma^{\tilde{\mu}}{}_{\tilde{\alpha}\tilde{\beta}} = \frac{1}{2}g^{\tilde{\mu}\tilde{\nu}}\left(g_{\tilde{\nu}\tilde{\alpha},\tilde{\beta}} + g_{\tilde{\nu}\tilde{\beta},\tilde{\alpha}} - g_{\tilde{\alpha}\tilde{\beta},\tilde{\nu}}\right). \tag{11.65}$$

---

[4] A few remarks are in place concerning the mathematical background of what just has been said: The quoted solution (11.63) of (11.62) is not unique. Because of the symmetry of $g_{\mu\nu}$ we only obtain $n^2(n+1)/2$ independent equations for the n^3 unknowns $\Gamma^{\alpha}{}_{\mu\beta}$. Our solution is obtained if one requires, additionally, that the spacetime is free of torsion. Then the Christoffel symbols and the transition coefficients are identical. To distinguish between these two quantities, the Christoffel symbols are often written in the form $\left\{{\kappa \atop \mu\nu}\right\}$ in the mathematical literature on GR. These are always symmetric in the lower two indices. The transition coefficients are then obtained as the sum of the Christoffel symbols and contributions from the torsion. However, in all spacetimes considered in GR, the torsion tensor vanishes, therefore we will not further pursue this distinction. Details on this subject can be found in Nakahara [5]. It should also be pointed out that extensions of GR with torsion were already discussed early on, and are still the subject of current research. The interested reader will find summaries of the corresponding theory in works by Hehl et al. [3], Capozziello et al. [1], and Shapiro [6].

The metric itself is a second-rank tensor, and therefore transforms according to (11.35),

$$g^{\tilde{\mu}\tilde{\nu}} = x^{\tilde{\mu}}_{,\rho} x^{\tilde{\nu}}_{,\sigma} g^{\rho\sigma}. \tag{11.66}$$

Derivatives of the metric, however, are not tensors. For the first term in the parenthesis of (11.65) we obtain

$$
\begin{aligned}
g_{\tilde{\nu}\tilde{\alpha},\tilde{\beta}} &= \frac{\partial}{\partial \tilde{x}^{\tilde{\beta}}} g_{\tilde{\nu}\tilde{\alpha}} = \frac{\partial}{\partial \tilde{x}^{\tilde{\beta}}} \left(x^{\varepsilon}_{,\tilde{\nu}} x^{\tau}_{,\tilde{\alpha}} g_{\varepsilon\tau} \right) \\
&= x^{\varepsilon}_{,\tilde{\nu}\tilde{\beta}} x^{\tau}_{,\tilde{\alpha}} g_{\varepsilon\tau} + x^{\varepsilon}_{,\tilde{\nu}} x^{\tau}_{,\tilde{\alpha}\tilde{\beta}} g_{\varepsilon\tau} + x^{\varepsilon}_{,\tilde{\nu}} x^{\tau}_{,\tilde{\alpha}} g_{\varepsilon\tau,\lambda} x^{\lambda}_{,\tilde{\beta}}.
\end{aligned}
\tag{11.67a}
$$

The other two terms are given accordingly by

$$g_{\tilde{\nu}\tilde{\beta},\tilde{\alpha}} = x^{\varepsilon}_{,\tilde{\nu}\tilde{\alpha}} x^{\tau}_{,\tilde{\beta}} g_{\varepsilon\tau} + x^{\varepsilon}_{,\tilde{\nu}} x^{\tau}_{,\tilde{\beta}\tilde{\alpha}} g_{\varepsilon\tau} + x^{\varepsilon}_{,\tilde{\nu}} x^{\tau}_{,\tilde{\beta}} g_{\varepsilon\tau,\lambda} x^{\lambda}_{,\tilde{\alpha}}, \tag{11.67b}$$

$$g_{\tilde{\alpha}\tilde{\beta},\tilde{\nu}} = x^{\varepsilon}_{,\tilde{\alpha}\tilde{\nu}} x^{\tau}_{,\tilde{\beta}} g_{\varepsilon\tau} + x^{\varepsilon}_{,\tilde{\alpha}} x^{\tau}_{,\tilde{\beta}\tilde{\nu}} g_{\varepsilon\tau} + x^{\varepsilon}_{,\tilde{\alpha}} x^{\tau}_{,\tilde{\beta}} g_{\varepsilon\tau,\lambda} x^{\lambda}_{,\tilde{\nu}}. \tag{11.67c}$$

We insert (11.67a), (11.67b), and (11.67c) into (11.65). Then, after somewhat lengthy manipulations of terms, and under the condition that second derivatives commute, we finally arrive at the transformation equation of the Christoffel symbols

$$\Gamma^{\tilde{\mu}}_{\tilde{\alpha}\tilde{\beta}} = x^{\tilde{\mu}}_{,\rho} x^{\tau}_{,\tilde{\alpha}} x^{\lambda}_{,\tilde{\beta}} \Gamma^{\rho}_{\tau\lambda} + x^{\tilde{\mu}}_{,\rho} x^{\rho}_{,\tilde{\alpha}\tilde{\beta}}. \tag{11.68}$$

The first term on the right-hand side indeed exhibits the transformation behavior of tensors, but the second term, which additionally appears, spoils the tensor property.

11.2.8 Covariant Derivative

Let $\boldsymbol{F} = F^{\mu}\partial_{\mu}$ be a contravariant vector. Then it transforms like a contravariant tensor of rank 1, $\tilde{F}^{\nu} = x^{\tilde{\nu}}_{,\mu} F^{\mu}$, see Eq. (11.9).

We now take the partial derivative of the vector with respect to x^{ν}, $F^{\mu}_{,\nu} = \partial F^{\mu}/\partial x^{\nu}$, and study its transformation behavior. Using the transformation equation for the derivatives (11.11), we obtain

$$\tilde{F}^{\mu}_{,\nu} = F^{\tilde{\mu}}_{,\tilde{\nu}} = x^{\rho}_{,\tilde{\nu}} \frac{\partial}{\partial x^{\rho}} \left(x^{\tilde{\mu}}_{,\lambda} F^{\lambda} \right) = x^{\rho}_{,\tilde{\nu}} x^{\tilde{\mu}}_{,\lambda} F^{\lambda}_{,\rho} + x^{\rho}_{,\tilde{\nu}} F^{\lambda} \underbrace{x^{\tilde{\mu}}_{,\lambda\rho}}_{Q}. \tag{11.69}$$

If $x^{\tilde{\mu}}_{,\lambda}$ is coordinate-dependent, which we will allow explicitly in GR, the term Q will not vanish. We can verify this with the help of Eq. (11.68), which we first contract with the expression $x^{\nu}_{,\tilde{\mu}}$, and subsequently permute $x^{\mu} \leftrightarrow \tilde{x}^{\mu}$. We can solve the resulting equation with respect to the second derivative, and obtain

$$x^{\tilde{\mu}}_{,\lambda\rho} = \Gamma^{\varepsilon}_{\lambda\rho} x^{\tilde{\mu}}_{,\varepsilon} - x^{\tilde{\alpha}}_{,\lambda} x^{\tilde{\beta}}_{,\rho} \Gamma^{\tilde{\mu}}_{\tilde{\alpha}\tilde{\beta}}. \tag{11.70}$$

Inserting this result into (11.69) yields

$$F^{\tilde{\mu}}_{,\tilde{\nu}} = x^{\rho}_{,\tilde{\nu}} x^{\tilde{\mu}}_{,\lambda} F^{\lambda}_{,\rho} + x^{\rho}_{,\tilde{\nu}} F^{\lambda} \left(\Gamma^{\varepsilon}_{\lambda\rho} x^{\tilde{\mu}}_{,\varepsilon} - x^{\tilde{\alpha}}_{,\lambda} x^{\tilde{\beta}}_{,\rho} \Gamma^{\tilde{\mu}}_{\tilde{\alpha}\tilde{\beta}} \right). \tag{11.71}$$

Thus, $F^{\mu}_{,\nu}$ does not transform like a tensor. This poses a major problem, because without a formalism of differentiation that exhibits tensor transformation behavior, we cannot formulate coordinate-independent differential equations in GR.

However, we can reformulate the partial derivative (11.71). If we exchange the indices $\varepsilon \leftrightarrow \lambda$ in the second term, we can combine the first two terms on the right-hand side, and arrive at

$$F^{\tilde{\mu}}_{,\tilde{\nu}} = x^{\rho}_{,\tilde{\nu}} x^{\tilde{\mu}}_{,\lambda} \left(F^{\lambda}_{,\rho} + \Gamma^{\lambda}_{\varepsilon\rho} F^{\varepsilon} \right) - \underbrace{x^{\tilde{\alpha}}_{,\lambda} \Gamma^{\tilde{\mu}}_{\tilde{\alpha}\tilde{\nu}} F^{\lambda}}_{R}. \tag{11.72}$$

In the last term R, we can apply the transformation $F^{\tilde{\alpha}} = x^{\tilde{\alpha}}_{,\lambda} F^{\lambda}$, and subsequently move it to the left-hand side of the equation. This leads to the relation

$$F^{\tilde{\mu}}_{,\tilde{\nu}} + \Gamma^{\tilde{\mu}}_{\tilde{\alpha}\tilde{\nu}} F^{\tilde{\alpha}} = x^{\rho}_{,\tilde{\nu}} x^{\tilde{\mu}}_{,\lambda} \left(F^{\lambda}_{,\rho} + \Gamma^{\lambda}_{\varepsilon\rho} F^{\varepsilon} \right), \tag{11.73}$$

which reveals a tensor transformation behavior for the expression

$$\nabla_{\rho} F^{\lambda} \equiv F^{\lambda}_{;\rho} \equiv F^{\lambda}_{,\rho} + \Gamma^{\lambda}_{\varepsilon\rho} F^{\varepsilon}. \tag{11.74}$$

Evidently, $F^{\lambda}_{;\rho}$ is a tensor of rank $(1, 1)$. It is called the *covariant derivative* of the vector $\boldsymbol{F}$.

In complete analogy, the covariant derivative of a covariant vector $\boldsymbol{G} = G_{\mu} dx^{\mu}$ is given by

$$\nabla_{\nu} G_{\mu} \equiv G_{\mu;\nu} = G_{\mu,\nu} - \Gamma^{\sigma}_{\mu\nu} G_{\sigma}. \tag{11.75}$$

The covariant derivative can also be extended to tensors of higher rank in the following way

$$\nabla_{\gamma} T^{\alpha\ldots}_{\beta\ldots} = \frac{\partial}{\partial x^{\gamma}} T^{\alpha\ldots}_{\beta\ldots} + \underbrace{\Gamma^{\alpha}_{\gamma\lambda} T^{\lambda\ldots}_{\beta\ldots}}_{\substack{\text{all contra-}\\\text{variant indices}}} - \underbrace{\Gamma^{\lambda}_{\gamma\beta} T^{\alpha\ldots}_{\lambda\ldots}}_{\substack{\text{all covariant}\\\text{indices}}}. \tag{11.76}$$

In doing so, we have to include, for each contravariant and each covariant index, a summand with a positive or negative Christoffel symbol, respectively.

For the metric tensor $g_{\mu\nu}$ from Sect. 11.2.4, we find

$$g_{\mu\nu;\alpha} = g_{\mu\nu,\alpha} - \Gamma^\lambda{}_{\mu\alpha}g_{\lambda\nu} - \Gamma^\lambda{}_{\nu\alpha}g_{\mu\lambda} = 0. \tag{11.77}$$

Therefore, the Christoffel symbols are characterized exactly by the property that the covariant derivative of the metric vanishes.

Multiple Covariant Derivatives

Multiple covariant derivatives also can be taken of a tensor. Let F^μ be a contravariant tensor of rank 1, then $F^\mu{}_{;\beta}$ is a tensor of rank $(1, 1)$, and

$$\left(F^\mu{}_{;\beta}\right)_{;\gamma} = F^\mu{}_{;\beta\gamma} \tag{11.78}$$

a tensor of rank $(1, 2)$. In taking covariant derivatives, we must be aware of the fact that multiple derivatives will not commute in general,

$$F^\mu{}_{;\alpha\beta} \neq F^\mu{}_{;\beta\alpha}, \tag{11.79}$$

whereas the usual partial derivatives of F^μ commute, $F^\mu{}_{,\alpha\beta} = F^\mu{}_{,\beta\alpha}$.

Divergence and Curl

The divergence of a vector field F^μ is defined by its covariant derivative and the subsequent contraction,

$$\nabla_\mu F^\mu = F^\mu{}_{;\mu} = F^\mu{}_{,\mu} + \Gamma^\mu{}_{\alpha\mu}F^\alpha. \tag{11.80}$$

With the abbreviation $g = \det(g_{\mu\nu})$ for the determinant of the metric $g_{\mu\nu}$, and

$$\Gamma^\mu{}_{\alpha\mu} = \frac{1}{2}g^{\mu\sigma}\left(g_{\sigma\alpha,\mu} + g_{\sigma\mu,\alpha} - g_{\alpha\mu,\sigma}\right) = \frac{1}{\sqrt{g}}\frac{\partial\sqrt{g}}{\partial x^\alpha}, \tag{11.81}$$

the divergence can also be written in the form,

$$F^\mu{}_{;\mu} = F^\mu{}_{,\mu} + \frac{1}{\sqrt{g}}\frac{\partial\sqrt{g}}{\partial x^\alpha}F^\alpha = \frac{1}{\sqrt{g}}\frac{\partial}{\partial x^\mu}\left(F^\mu\sqrt{g}\right). \tag{11.82}$$

The divergence of a contravariant tensor field $T^{\mu\nu}$ also results from its covariant derivative

$$\nabla_\lambda T^{\mu\nu} = T^{\mu\nu}{}_{;\lambda} = T^{\mu\nu}{}_{,\lambda} + \Gamma^\mu{}_{\lambda\alpha}T^{\alpha\nu} + \Gamma^\nu{}_{\lambda\alpha}T^{\mu\alpha} \tag{11.83}$$

and subsequent contraction. However, here the literature does not agree upon which index should be used for the contraction. In the case of symmetric tensor fields, $T^{\mu\nu} = T^{\nu\mu}$, this does not matter. Here we use the contraction with respect to the second index, viz.

$$\nabla_\nu T^{\mu\nu} = T^{\mu\nu}{}_{;\nu} = T^{\mu\nu}{}_{,\nu} + \Gamma^\mu{}_{\nu\alpha} T^{\alpha\nu} + \Gamma^\nu{}_{\nu\alpha} T^{\mu\alpha}. \tag{11.84}$$

The curl of a covariant vector field F_μ is defined by

$$\varphi_{\mu\nu} = F_{\nu;\mu} - F_{\mu;\nu}. \tag{11.85}$$

Thus, it is an antisymmetric covariant tensor of second rank. Moreover, for vanishing torsion, i.e., with $\Gamma^\sigma{}_{\mu\nu} = \Gamma^\sigma{}_{\nu\mu}$ the curl simplifies to

$$F_{\nu;\mu} - F_{\mu;\nu} = F_{\nu,\mu} - \Gamma^\sigma{}_{\nu\mu} F_\sigma - F_{\mu,\nu} + \Gamma^\sigma{}_{\mu\nu} F_\sigma = F_{\nu,\mu} - F_{\mu,\nu}. \tag{11.86}$$

11.3 Space Curvature

In this section we will explore how we can qualitatively and quantitatively determine whether a space is curved. This statement will not be possible directly using the metric tensor $g_{\mu\nu}$ or the affine connections $\Gamma^\alpha{}_{\mu\nu}$. Although the information about the curvature is of course incorporated in the metric, one cannot simply gather this information from its form. This is easy to see, because even for Euclidean, i.e., flat spaces, described in curvilinear coordinates such as polar coordinates, the metric and the affine connections are not trivial.

11.3.1 Curvature of Well-Known Surfaces

We investigate a few well-known surfaces with respect to their curvature. To get a feeling for the notion of curvature, we will consider whether a surface is curved by looking at the effect that parallel transport has on a vector by moving it along different paths in the surface.

Flat Spaces
As a first example, we consider flat spaces. The simplest case is a plane. Let F^μ be a vector. If F^μ is transported parallel along a closed path, the original and the final vectors match, i.e., we have $\delta F^\mu = 0$, see Fig. 11.3. Also for the case of the surface of a cylinder, the direction of the vector remains path-independent, i.e., one always has $\delta F^\mu = 0$.

Fig. 11.3 In the parallel
transport of a vector in the
plane, the original vector and
the vector parallel transported
along a closed path coincide

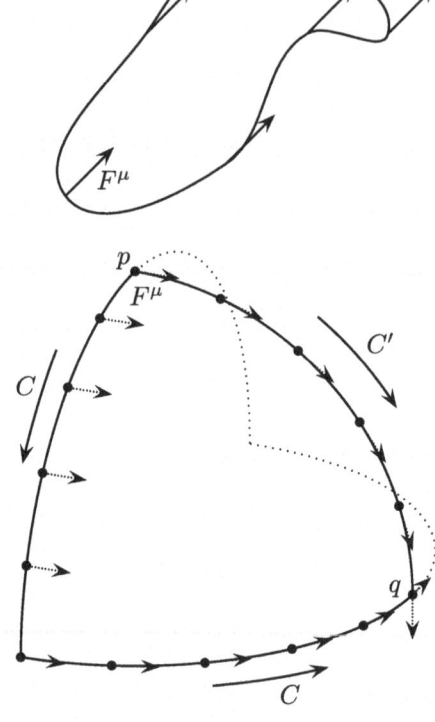

Fig. 11.4 Parallel transport
of a vector on a spherical
surface along great circles.
The direction into which F^μ
points is path-dependent

Surface of a Sphere

Now we consider a vector F^μ on the surface of a sphere, see Fig. 11.4. In this
case, the natural definition of parallel transport along a great circle is that the
angle between the vector and the great circle remains constant. If in Fig. 11.4 F^μ
undergoes parallel transport from p to q along C and C', the resulting vectors point
in different directions. Thus, the direction of the vector $F^\mu(q)$ does depend on the
path taken.

11.3.2 Curvature Tensor

In the previous examples, we have seen that the path dependence of the change
δF^μ of a vector under parallel transport differs for different spaces. In particular,
it vanishes in a flat space. Therefore, it is reasonable to assume that generally this
property can be used to characterize the curvature of the space.

Fig. 11.5 Parallel transport
of a vector F^μ from p to r

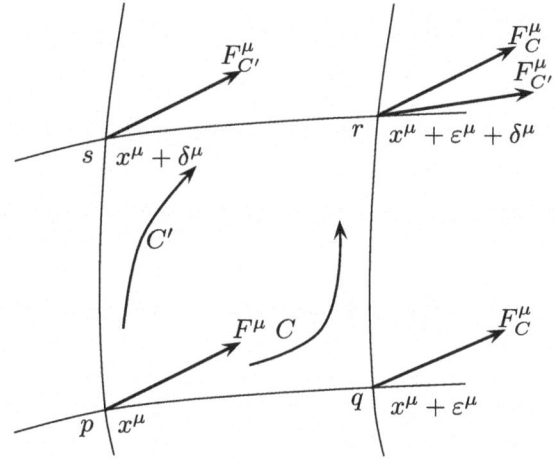

Derivation via Parallel Transport

For a rigorous treatment, we consider an infinitesimal parallelogram $pqrs$ with
coordinates x^μ, $x^\mu + \varepsilon^\mu$, $x^\mu + \varepsilon^\mu + \delta^\mu$ and $x^\mu + \delta^\mu$, see Fig. 11.5. When we
transport F^μ parallel along $C = pqr$, we obtain the vector $F_C^\mu(r)$. At q we obtain,
in linear approximation,

$$F_C^\mu(q) = F^\mu - F^\kappa \Gamma^\mu{}_{\kappa\nu}\varepsilon^\nu. \tag{11.87}$$

In the next step we go to r, and take into account terms up to second order, i.e., first
order in δ and ε, respectively:

$$
\begin{aligned}
F_C^\mu(r) =& F_C^\mu(q) - F_C^\kappa(q)\Gamma^\mu{}_{\kappa\nu}(q)\delta^\nu \\
=& F^\mu - F^\kappa \Gamma^\mu{}_{\kappa\nu}(p)\varepsilon^\nu \\
& - \left(F^\kappa - F^\rho \Gamma^\kappa{}_{\rho\xi}(p)\varepsilon^\xi\right)\left(\Gamma^\mu{}_{\kappa\nu}(p) + \Gamma^\mu{}_{\kappa\nu,\lambda}(p)\varepsilon^\lambda\right)\delta^\nu \\
\simeq& F^\mu - F^\kappa \Gamma^\mu{}_{\kappa\nu}(p)\varepsilon^\nu - F^\kappa \Gamma^\mu{}_{\kappa\nu}(p)\delta^\nu \\
& - F^\kappa \left[\Gamma^\mu{}_{\kappa\nu,\lambda}(p) - \Gamma^\rho{}_{\kappa\lambda}(p)\Gamma^\mu{}_{\rho\nu}(p)\right]\varepsilon^\lambda\delta^\nu.
\end{aligned}
\tag{11.88}
$$

For the transport along the other path $C' = psr$ we obtain in a completely analogous
way the approximation

$$
\begin{aligned}
F_{C'}^\mu(r) \simeq& F^\mu - F^\kappa \Gamma^\mu{}_{\kappa\nu}(p)\delta^\nu - F^\kappa \Gamma^\mu{}_{\kappa\nu}(p)\varepsilon^\nu \\
& - F^\kappa \left[\Gamma^\mu{}_{\kappa\lambda,\nu}(p) - \Gamma^\rho{}_{\kappa\nu}(p)\Gamma^\mu{}_{\rho\lambda}(p)\right]\varepsilon^\lambda\delta^\nu.
\end{aligned}
\tag{11.89}
$$

The difference of the two vectors finally is

$$
\begin{aligned}
F^{\mu}_{C'}(r) - F^{\mu}_{C}(r) \simeq F^{\kappa} & \left(\Gamma^{\mu}{}_{\kappa\nu,\lambda}(p) - \Gamma^{\mu}{}_{\kappa\lambda,\nu}(p) \right. \\
& \left. - \Gamma^{\rho}{}_{\kappa\lambda}(p)\Gamma^{\mu}{}_{\rho\nu}(p) + \Gamma^{\rho}{}_{\kappa\nu}(p)\Gamma^{\mu}{}_{\rho\lambda}(p) \right) \varepsilon^{\lambda}\delta^{\nu} \\
= & F^{\kappa} R^{\mu}{}_{\kappa\lambda\nu}\varepsilon^{\lambda}\delta^{\nu} .
\end{aligned}
$$

(11.90)

In this equation, the quantity

$$
R^{\mu}{}_{\kappa\lambda\nu} = \Gamma^{\mu}{}_{\kappa\nu,\lambda} - \Gamma^{\mu}{}_{\kappa\lambda,\nu} + \Gamma^{\mu}{}_{\rho\lambda}\Gamma^{\rho}{}_{\kappa\nu} - \Gamma^{\mu}{}_{\rho\nu}\Gamma^{\rho}{}_{\kappa\lambda}
$$

(11.91)

denotes the *curvature tensor* or *Riemann tensor* of rank $(1, 3)$.

Formal Definition of the Curvature Tensor

We had already seen that, in general, covariant derivatives do not commute (cf. Eq. (11.79)). Formally, the curvature tensor can be defined via the difference of two second covariant derivatives. Let A^{μ} be a contravariant tensor of first rank, then

$$
\left(\nabla_{\gamma}\nabla_{\beta} - \nabla_{\beta}\nabla_{\gamma} \right) A^{\mu} = A^{\mu}{}_{;\beta\gamma} - A^{\mu}{}_{;\gamma\beta} = -R^{\mu}{}_{\alpha\beta\gamma} A^{\alpha} .
$$

(11.92)

To see this, we explicitly evaluate the corresponding expressions, and obtain

$$
\begin{aligned}
A^{\mu}{}_{;\beta\gamma} &= \left(A^{\mu}{}_{;\beta} \right)_{,\gamma} + \Gamma^{\mu}{}_{\gamma\lambda} A^{\lambda}{}_{;\beta} - \Gamma^{\lambda}{}_{\gamma\beta} A^{\mu}{}_{;\lambda} \\
&= \left(A^{\mu}{}_{,\beta} + \Gamma^{\mu}{}_{\beta\lambda} A^{\lambda} \right)_{,\gamma} + \Gamma^{\mu}{}_{\gamma\lambda} \left(A^{\lambda}{}_{,\beta} + \Gamma^{\lambda}{}_{\beta\sigma} A^{\sigma} \right) \\
&\quad - \Gamma^{\lambda}{}_{\gamma\beta} \left(A^{\mu}{}_{,\lambda} + \Gamma^{\mu}{}_{\lambda\sigma} A^{\sigma} \right) .
\end{aligned}
$$

(11.93)

An analogous result is found for $A^{\mu}{}_{;\gamma\beta}$. Taking into account the fact that the derivatives $A^{\mu}{}_{,\alpha\beta} = A^{\mu}{}_{,\beta\alpha}$ commute, and the symmetry of the Christoffel symbols, we find

$$
\begin{aligned}
A^{\mu}{}_{;\beta\gamma} - A^{\mu}{}_{;\gamma\beta} &= - \left(\Gamma^{\mu}{}_{\alpha\beta,\gamma} - \Gamma^{\mu}{}_{\alpha\gamma,\beta} + \Gamma^{\mu}{}_{\sigma\beta}\Gamma^{\sigma}{}_{\alpha\gamma} - \Gamma^{\mu}{}_{\sigma\gamma}\Gamma^{\sigma}{}_{\alpha\beta} \right) A^{\alpha} \\
&= -R^{\mu}{}_{\alpha\beta\gamma} A^{\alpha} .
\end{aligned}
$$

(11.94)

Lowering the index μ with the help of the metric we end up with the *four-index covariant curvature tensor*

$$
R_{\mu\kappa\lambda\nu} = g_{\mu\alpha} R^{\alpha}{}_{\kappa\lambda\nu} = \Gamma_{\mu\kappa\nu,\lambda} - \Gamma_{\mu\kappa\lambda,\nu} + \Gamma_{\mu\rho\lambda}\Gamma^{\rho}{}_{\kappa\nu} - \Gamma_{\mu\rho\nu}\Gamma^{\rho}{}_{\kappa\lambda} .
$$

(11.95)

Symmetries of the Curvature Tensor

Not all of the components of the curvature tensor are independent. From the definition (11.91), or from the purely covariant form (11.95), one easily recognizes the following relations

$$R^\mu{}_{\alpha\beta\gamma} = -R^\mu{}_{\alpha\gamma\beta} \quad \text{and} \quad R^\mu{}_{\alpha\beta\gamma} + R^\mu{}_{\gamma\alpha\beta} + R^\mu{}_{\beta\gamma\alpha} = 0. \tag{11.96}$$

In an analogous way we find

$$R_{\mu\alpha\beta\gamma} = -R_{\mu\alpha\gamma\beta}, \quad R_{\mu\alpha\beta\gamma} = -R_{\alpha\mu\beta\gamma}, \quad R_{\mu\alpha\beta\gamma} = R_{\beta\gamma\mu\alpha}, \tag{11.97}$$

as well as

$$R_{\mu\alpha\beta\gamma} + R_{\mu\beta\gamma\alpha} + R_{\mu\gamma\alpha\beta} = 0. \tag{11.98}$$

As a consequence of these symmetry relations, the curvature tensor in four dimensions has only 20 independent components.

Ricci Tensor and Curvature Scalar

By contracting the curvature tensor, one obtains the *Ricci tensor*:[5]

$$R_{\kappa\nu} \equiv R^\mu{}_{\kappa\mu\nu} = \Gamma^\mu{}_{\kappa\nu,\mu} - \Gamma^\mu{}_{\kappa\mu,\nu} + \Gamma^\mu{}_{\rho\mu}\Gamma^\rho{}_{\kappa\nu} - \Gamma^\mu{}_{\rho\nu}\Gamma^\rho{}_{\kappa\mu}. \tag{11.99}$$

The Ricci tensor is a symmetric tensor of rank 2:

$$R_{\kappa\nu} = R_{\nu\kappa}. \tag{11.100}$$

To show this, we exploit the symmetry properties of the Christoffel symbols. For the second term, we resort to the relation $\Gamma^\mu{}_{\alpha\mu} = (\partial\sqrt{g}/\partial x^\alpha)/\sqrt{g}$ from (11.81), and take the partial derivative with respect to the x^β:

$$\Gamma^\mu{}_{\alpha\mu,\beta} = \frac{\partial}{\partial x^\beta}\left(\frac{1}{\sqrt{g}}\frac{\partial\sqrt{g}}{\partial x^\alpha}\right) = -\frac{1}{2\sqrt{g}^3}\frac{\partial\sqrt{g}}{\partial x^\beta}\frac{\partial\sqrt{g}}{\partial x^\alpha} + \frac{1}{\sqrt{g}}\frac{\partial^2\sqrt{g}}{\partial x^\alpha\partial x^\beta} = \Gamma^\mu{}_{\beta\mu,\alpha}. \tag{11.101}$$

A further contraction of the Ricci tensor leads to the *curvature scalar*, or *Ricci scalar*,

$$R \equiv R^\mu{}_\mu = g^{\mu\nu}R_{\mu\nu}. \tag{11.102}$$

---

[5] Gregorio Ricci-Curbastro, 1853–1925, Italian mathematician.

From the full contraction of two Riemann tensors we can construct another scalar,

$$\mathcal{K} \equiv R_{\alpha\beta\gamma\delta} R^{\alpha\beta\gamma\delta}, \tag{11.103}$$

which is called the *Kretschmann scalar*.[6]

Bianchi Identity

The *Bianchi identity* is the relation

$$R^{\mu}{}_{\nu\alpha\beta;\gamma} + R^{\mu}{}_{\nu\gamma\alpha;\beta} + R^{\mu}{}_{\nu\beta\gamma;\alpha} = 0. \tag{11.104}$$

To prove this, we consider the equation at a given point P_0 of the manifold. By choosing suitable coordinates, we can attain that

$$\Gamma^{\mu}{}_{\alpha\beta}(P_0) = 0. \tag{11.105}$$

It then follows from the definition (11.91) that at the point P_0

$$R^{\mu}{}_{\nu\alpha\beta} = \Gamma^{\mu}{}_{\nu\beta,\alpha} - \Gamma^{\mu}{}_{\nu\alpha,\beta} \tag{11.106}$$

and thus

$$R^{\mu}{}_{\nu\alpha\beta;\gamma} = R^{\mu}{}_{\nu\alpha\beta,\gamma} = \Gamma^{\mu}{}_{\nu\beta,\alpha\gamma} - \Gamma^{\mu}{}_{\nu\alpha,\beta\gamma}. \tag{11.107}$$

The other quantities can be obtained in an analogous way. If the last three indices are permuted cyclically, these terms cancel. Because of the tensor property, the Bianchi identity must hold not only at the point P_0, but in general.

Weyl Tensor

Subtracting all trace terms from the Riemann tensor (11.91), we obtain the *Weyl tensor*.[7] With the following abbreviations for the symmetrization and antisymmetrization, respectively, of an expression

$$a_{(\mu\nu)} = \frac{1}{2}\left(a_{\mu\nu} + a_{\nu\mu}\right) \quad \text{and} \quad a_{[\mu\nu]} = \frac{1}{2}\left(a_{\mu\nu} - a_{\nu\mu}\right) \tag{11.108}$$

we can write the Weyl tensor in the form

$$C_{\mu\nu\rho\sigma} = R_{\mu\nu\rho\sigma} - \left(g_{\mu[\rho} R_{\sigma]\nu} - g_{\nu[\rho} R_{\sigma]\mu}\right) + \frac{1}{3} R g_{\mu[\rho} g_{\sigma]\nu}. \tag{11.109}$$

---

[6] Erich Kretschmann, 1887–1973, German physicist.

[7] Hermann Klaus Hugo Weyl, 1885–1955, German mathematician and physicist.

The Weyl tensor has the same symmetries as the Riemann tensor. In the case of a vacuum spacetime, in which the Ricci tensor and Ricci scalar vanish identically, the Riemann tensor reduces to the Weyl tensor.

Another property of the Weyl tensor is that it is a *conformal invariant*, which means that it does not change under conformal transformations. In this context, two metrics g and $\hat{g}$ are called *conformal* if one has

$$\hat{g} = \Omega^2 g \tag{11.110}$$

with a suitable non-vanishing function $\Omega(\underline{x})$. Angles and ratios of moduli are then preserved. Moreover, the causal structure is preserved. For more details see, for example, Hawking and Ellis [2].

11.3.3 Isometries and Killing Vectors

An *isometry* is a mapping that leaves the metric g invariant. Let us consider the infinitesimal coordinate transformation

$$x^\mu \mapsto x'^\mu = x^\mu + \varepsilon K^\mu \tag{11.111}$$

with $\varepsilon \ll 1$ and a vector field $K^\mu = K^\mu(\underline{x})$. Then, for an isometry the relation

$$\mathrm{d}s'^2 = g_{\mu\nu}(x')\mathrm{d}x'^\mu\mathrm{d}x'^\nu = g_{\mu\nu}(x)\mathrm{d}x^\mu\mathrm{d}x^\nu \tag{11.112}$$

must be satisfied. We can approximate the metric and coordinate differentials referred to the new coordinates as follows,

$$g_{\mu\nu}(x') \approx g_{\mu\nu}(x) + \varepsilon g_{\mu\nu,\lambda} K^\lambda \quad \text{and} \quad \mathrm{d}x'^\mu \approx \mathrm{d}x^\mu + \varepsilon K^\mu{}_{,\lambda}\mathrm{d}x^\lambda. \tag{11.113}$$

Inserting these expressions into (11.112) and considering only terms linear in ε, we obtain

$$g_{\mu\nu,\lambda} K^\lambda + g_{\lambda\nu} K^\lambda{}_{,\mu} + g_{\mu\lambda} K^\lambda{}_{,\nu} = 0. \tag{11.114}$$

To continue the calculation, we need the partial derivative of the covariant representation of the vector field:

$$K_{\mu,\nu} = \frac{\partial K_\mu}{\partial x^\nu} = \frac{\partial}{\partial x^\nu}\left(g_{\mu\lambda} K^\lambda\right) = g_{\mu\lambda,\nu} K^\lambda + g_{\mu\lambda} K^\lambda{}_{,\nu}. \tag{11.115}$$

We can rewrite (11.114) with the help of (11.115),

$$K_{\mu,\nu} + K_{\nu,\mu} + \left(g_{\mu\nu,\lambda} - g_{\mu\lambda,\nu} - g_{\lambda\nu,\mu}\right) K^\lambda = 0 \tag{11.116}$$

and, using the covariant derivative $K_{\mu;\nu} = K_{\mu\nu} - \Gamma^\lambda{}_{\mu\nu} K_\lambda$, finally obtain the *Killing equation*[8]

$$K_{\mu;\nu} + K_{\nu;\mu} = 0. \tag{11.117}$$

Thus, a vector field $K^\mu(\underline{x})$ generates an isometry if and only if it satisfies the Killing equation. In other words, if one moves along integral curves of this vector field, the geometry of spacetime does not change.

In particular, a spacetime is *stationary*, if there exists a Killing vector field $K^\mu(\underline{x})$, which is timelike everywhere. Thus, a time translation must leave the metric invariant, which implies that the metric components $g_{\mu\nu}$ must not depend on the time coordinate. If also the mirror symmetry applies, which means that, with respect to each time-plane, the spacetime is symmetric along the Killing vector field, then the spacetime is *static*.

11.3.4 Sylvester's Inertia Law

For the mathematical formulation of GR the inertia law of *James Sylvester*[9] for quadratic forms is of central importance: In an orthonormal basis the metric $g_{\mu\nu}$ can be represented as a diagonal matrix with entries ± 1. If the matrix has r entries $+1$ and s entries -1, then this is called a metric with inertia or *signature*(r, s). For example, the Minkowski metric has the signature $(r, s) = (1, 3)$ or $(r, s) = (3, 1)$; we use the latter signature. For general relativity, this implies that gravity can be transformed away *locally*. In a sufficiently small neighborhood of a point, there exists a system of coordinates in which force-free particles move on straight lines. This is obvious because the statement means that, in suitable coordinates, the metric $g_{\mu\nu}$ can be reduced to the form $\eta_{\mu\nu}$, i.e., to the form of the Minkowski metric of flat space. When discussing the equivalence principle we had come to these insights on the basis of purely physical arguments. In the framework of Riemannian geometry, we have now rediscovered them.

11.4   Motion in Curved Spaces

In the previous sections, we have gained a better understanding of the properties of curved spaces, and have presented the mathematical tools for treating curved spaces. Now we will take a look at the motion of light and massive particles in curved spaces.

---

[8] Wilhelm Karl Joseph Killing, 1847–1923, German mathematician.

[9] James Joseph Sylvester, 1814–1897, English mathematician.

11.4.1 Geodesic Equation

In this section, we shall derive the geodesic equation for curved spaces. It is the generalization of the force-free motion in SR, see Sect. 6.5. The treatment in this section is valid independently of physical assumptions.

In Sect. 11.2.4 we have already seen that $ds^2 = g_{\mu\nu}dx^\mu dx^\nu$. As in SR, a particle is supposed to move on a trajectory such that the variation

$$\delta \int ds = 0 \tag{11.118}$$

vanishes. In the following calculation we restrict ourselves to timelike trajectories with $ds^2 < 0$. For lightlike or spacelike geodesics a similar calculation can be performed, and the present results can be transferred to these cases.

For timelike trajectories the integral in (11.118) can be written in the form

$$\int ds = \int \sqrt{-g_{\mu\nu}dx^\mu dx^\nu}\frac{ds}{ds}. \tag{11.119}$$

If we move the "ds" in the denominator of the fraction under the root, we can assign a functional of the form $\mathcal{L}(x^\alpha, dx^\alpha/ds)$ to the expression under the integral. This leads to

$$\delta \int ds = \delta \int \sqrt{-g_{\mu\nu}\frac{dx^\mu}{ds}\frac{dx^\nu}{ds}}ds = \delta \int \mathcal{L}\left(x^\alpha, \frac{dx^\alpha}{ds}\right)ds. \tag{11.120}$$

Here the function $\mathcal{L}$ is equal to 1 along the path, as can be seen immediately from the comparison with the definition of the line element. From the *Euler-Lagrange equation*[10,11] corresponding to the variation

$$\delta \int \mathcal{L}\left(x^\alpha, \frac{dx^\alpha}{ds}\right)ds = 0, \quad \text{i.e.,} \quad \frac{d}{ds}\left(\frac{\partial \mathcal{L}}{\partial\left(\frac{dx^\alpha}{ds}\right)}\right) - \frac{\partial \mathcal{L}}{\partial x^\alpha} = 0 \tag{11.121}$$

follows, with

$$\frac{\partial \mathcal{L}}{\partial\left(\frac{dx^\alpha}{ds}\right)} = -\frac{1}{2\mathcal{L}}\left(g_{\alpha\nu}\frac{dx^\nu}{ds} + g_{\mu\alpha}\frac{dx^\mu}{ds}\right) = -\frac{1}{\mathcal{L}}g_{\alpha\nu}\frac{dx^\nu}{ds} \tag{11.122}$$

and

$$\frac{\partial \mathcal{L}}{\partial x^\alpha} = -\frac{1}{2\mathcal{L}}\frac{\partial g_{\mu\nu}}{\partial x^\alpha}\frac{dx^\mu}{ds}\frac{dx^\nu}{ds}, \tag{11.123}$$

---

[10] Leonhard Euler, 1707–1783, Swiss mathematician and physicist.
[11] Joseph-Louis Lagrange, 1736–1813, Italian mathematician and astronomer.

the equation

$$\frac{d}{ds}\left[\frac{1}{\mathcal{L}}g_{\alpha\nu}\frac{dx^\nu}{ds}\right] - \frac{1}{2\mathcal{L}}\frac{\partial g_{\mu\nu}}{\partial x^\alpha}\frac{dx^\mu}{ds}\frac{dx^\nu}{ds} = 0. \tag{11.124}$$

Taking advantage of

$$\frac{d}{ds}g_{\alpha\nu} = g_{\alpha\nu,\mu}\frac{dx^\mu}{ds} \tag{11.125}$$

one finally obtains from (11.124)

$$-\frac{1}{\mathcal{L}^2}\frac{d\mathcal{L}}{ds}g_{\alpha\nu}\frac{dx^\nu}{ds} + \frac{1}{\mathcal{L}}g_{\alpha\nu,\mu}\frac{dx^\mu}{ds}\frac{dx^\nu}{ds} + \frac{1}{\mathcal{L}}g_{\alpha\nu}\frac{d^2x^\nu}{ds^2} - \frac{1}{2\mathcal{L}}g_{\mu\nu,\alpha}\frac{dx^\mu}{ds}\frac{dx^\nu}{ds} = 0. \tag{11.126}$$

Since $\mathcal{L} = 1$ along the way, we have $d\mathcal{L}/ds = 0$, i.e., the first term on the left-hand side of (11.126) vanishes, and we can omit $\mathcal{L}$ in the remaining terms. Then we have

$$g_{\alpha\nu,\mu}\frac{dx^\mu}{ds}\frac{dx^\nu}{ds} + g_{\alpha\nu}\frac{d^2x^\nu}{ds^2} - \frac{1}{2}g_{\mu\nu,\alpha}\frac{dx^\mu}{ds}\frac{dx^\nu}{ds} = 0. \tag{11.127}$$

In the next step, we exploit the fact that, due to the symmetry of $g_{\mu\nu}$, we can also write

$$g_{\alpha\nu,\mu} = \frac{1}{2}\left(g_{\alpha\nu,\mu} + g_{\nu\alpha,\mu}\right). \tag{11.128}$$

Since the summation extends over μ and ν, we can also commute these indices in the second term of (11.128), and arrive at

$$g_{\alpha\nu}\frac{d^2x^\nu}{ds^2} + \frac{1}{2}\left(g_{\alpha\nu,\mu} + g_{\mu\alpha,\nu} - g_{\mu\nu,\alpha}\right)\frac{dx^\mu}{ds}\frac{dx^\nu}{ds} = 0. \tag{11.129}$$

Multiplying by $g^{\sigma\alpha}$ and taking into account that $g^{\sigma\alpha}g_{\alpha\nu} = \delta^\sigma_\nu$, finally yields the *geodesic equation* of GR

$$\frac{d^2x^\sigma}{ds^2} + \Gamma^\sigma{}_{\mu\nu}\frac{dx^\mu}{ds}\frac{dx^\nu}{ds} = 0. \tag{11.130}$$

Thus, in the geodesic equation, the effect of gravitation is incorporated by the fact that the Christoffel symbols are defined via the metric tensor.

Geodesics can be easily defined using the covariant derivative. Let the four-velocity $u^\mu = dx^\mu/ds$ be given by the derivative with respect to the arc length. Then in analogy with classical mechanics (velocity remains constant) a geodesic can be defined via the requirement that the covariant derivative of the velocity is zero:

$$\frac{Du^\sigma}{ds} = \frac{du^\sigma}{ds} + \Gamma^\sigma{}_{\mu\nu}u^\mu\frac{dx^\nu}{ds} = \frac{d^2x^\sigma}{ds^2} + \Gamma^\sigma{}_{\mu\nu}\frac{dx^\mu}{ds}\frac{dx^\nu}{ds} = 0. \tag{11.131}$$

Therefore, the compact representation of the geodesic equation reads simply

$$\frac{Du^\sigma}{ds} = 0. \tag{11.132}$$

11.4.2 Euler-Lagrange Formalism

In the previous section we have used the Euler-Lagrange equations to derive the geodesic equation (11.130). In general, we can also use the formalism for the qualitative study of geodesics in a spacetime. For this purpose, as done in the previous section, we first set up the *Lagrange function* $\mathcal{L}$. In general, it can be obtained from the line element of a metric directly by replacing the coordinate differentials with the derivatives with respect to an affine parameter. From $ds^2 = g_{\mu\nu}dx^\mu dx^\nu$ we obtain

$$\mathcal{L} = g_{\mu\nu}\dot{x}^\mu\dot{x}^\nu = \kappa c^2, \tag{11.133}$$

where $\dot{x}^\mu = dx^\mu/d\lambda$ denotes the derivative with respect to the affine parameter λ, and the dimensionless parameter κ indicates, whether we are dealing with a timelike ($\kappa = -1$), a lightlike ($\kappa = 0$), or a spacelike ($\kappa = 1$) geodesic. From the *Euler-Lagrange equation*

$$\frac{d}{d\lambda}\frac{\partial\mathcal{L}}{\partial\dot{x}^\mu} - \frac{\partial\mathcal{L}}{\partial x^\mu} = 0 \tag{11.134}$$

we obtain again the geodesic equations for the coordinates x^μ.

One can understand Eq. (11.133) also as a *constraint*, or normalization condition, on the geodesic, which must be satisfied everywhere. Let us consider, e.g., the initial conditions of a geodesic referred to a local tetrad (cf. Sect. 11.2.5), $\underline{y} = y^\mu\partial_\mu = \dot{x}^\mu(\lambda = 0)\partial_\mu$

$$\underline{y} = y^{(i)}\underline{e}_{(i)} = y^{(i)}e^\mu_{(i)}\partial_\mu = y^\mu\partial_\mu, \tag{11.135}$$

then, due to (11.133) and the orthonormality of the tetrad $\{\underline{e}_{(i)}\}$, see (11.45), the relation

$$g_{\mu\nu}y^\mu y^\nu = \eta_{ij}y^{(i)}y^{(j)} = \kappa c^2 \tag{11.136}$$

applies.

11.4.3 Parallel and Fermi-Walker Transport

A point-like particle exposed to no external force, possessing no internal propulsion nor any other properties, follows a timelike geodesic, as discussed in Sect. 11.4.1. If one wishes to study the motion of an extended object which, like a point particle, continues to behave only as a test particle, with no back-reaction on the spacetime, one must also carry its local frame of reference along the geodesic. This means that each basis vector of the local frame of reference must be transported parallel along the geodesic. In general, for a vector $\underline{F} = F^\mu \partial_\mu$ the equation of the *parallel transport*

$$\frac{dF^\mu}{d\tau} + \Gamma^\mu{}_{\rho\sigma} u^\rho F^\sigma = 0 \tag{11.137}$$

applies, where τ is the proper time, and u^ρ is the four-velocity along the geodesic. Replacing the general vector $\underline{F}$ with the four-velocity $\underline{u} = u^\mu \partial_\mu$, we again obtain the geodesic equation (11.130) from (11.137). Thus, the tangential vector to the geodesic is transported parallel to itself.

However, if the extended object does not move along a geodesic, we also have to consider possible accelerations. Along a general world-line $x^\mu(\tau)$ with the four-velocity $u^\mu(\tau)$, the vector $\underline{F}$ must undergo a *Fermi-Walker transport*[12,13] [7].

The differential equation to be integrated then reads

$$\frac{dF^\mu}{d\tau} + \Gamma^\mu{}_{\rho\sigma} u^\rho F^\sigma + \frac{1}{c^2} \left(u^\sigma a^\mu - u^\mu a^\sigma \right) g_{\rho\sigma} F^\rho = 0. \tag{11.138}$$

The four-acceleration $\underline{a} = a^\mu \partial_\mu$ follows from

$$a^\mu = \frac{du^\mu}{d\tau} + \Gamma^\mu{}_{\rho\sigma} u^\rho u^\sigma. \tag{11.139}$$

For vanishing acceleration, the Fermi-Walker transport turns into the familiar parallel transport.

11.5 Exercises

11.1. Determining the transition coefficients Prove that the Christoffel symbols of the second kind in (11.63) satisfy the condition (11.62).

---

[12] Enrico Fermi, 1901–1954, U.S. American nuclear physicist of Italian descent, Nobel Prize in 1938. Well known for his contributions to nuclear physics. Among other things, the Fermi energy is named after him (cf. Chap. 18).

[13] Arthur Geoffrey Walker, 1909–2001, British mathematician. He also co-named the Friedmann-Lemaître-Robertson-Walker metric, which is very important in Cosmology, see Chap. 24.

11.2. Circular path in SR Using the Fermi-Walker transport, discuss the behavior of vectors moving on a circular path with radius R and velocity β, see Sect. 6.8.

11.3. Acceleration in Schwarzschild spacetime In Chap. 13 we shall get to know the Schwarzschild spacetime, which is the geometry of the exterior of a spherically symmetric mass distribution. If one sat down at any position in this spacetime, without any propulsion, one would fall into the singularity. Calculate the acceleration necessary to stay in the same place. Use the Schwarzschild metric (13.16).

References

1. Capozziello, S., Lambiase, G., Storniolo, C.: Geometric classification of the torsion tensor of spacetime. Ann. Phys. **10**(8), 713–727 (2001)
2. Hawking, S.W., Ellis, G.F.R.: The Large Scale Structure of Spacetime. Cambridge University Press (1999)
3. Hehl, F.W., Heyde, P., Kerlick, G.D., Nester, J.M.: General relativity with spin and torsion: foundations and prospects. Rev. Mod. Phys. **48**(3), 393–416 (1976)
4. Misner, C.W., Thorne, K.S., Wheeler, J.A.: Gravitation. W.H. Freeman, New York (1973)
5. Nakahara, M.: Geometry, Topology and Physics, 2nd edn. Taylor & Francis (2003)
6. Shapiro, I.L.: Physical aspects of the spacetime torsion. Phys. Rep. **357**, 113–213 (2002)
7. Stephani, H., Stewart, J.: General Relativity: An Introduction to the Theory of Gravitational Field. Cambridge University Press (1990)
8. Wald, R.M.: General Relativity. University of Chicago Press (1984)

Einstein's Field Equations

12

Contents

The main objective of general relativity is to determine the metric generated by a given distribution of masses and energy, and then to explore the physics of this spacetime, e.g., by studying the motion of objects, or the propagation of light. An ingenious summary of this problem is due to *Wheeler*:

> Matter tells space how to curve and spacetime tells matter how to move!

This requires an equation that links the metric of spacetime to the mass and energy distributions. Before we come to the formulation of this equation, we examine the non-relativistic approximation of the equations of motion in GR. The results will be useful in the subsequent analysis. From the quote just given we can already note an important property of gravitation, namely, the way it must be treated in general relativity. The statement that matter curves spacetime, and that the curvature in turn determines the motion of matter, which again has a back-reaction on spacetime, indicates that the field equations will be non-linear differential equations, which can be solved exactly only in a few cases and even numerical solutions are often hard to obtain.

12.1   Non-relativistic Limit

A necessary constraint on general relativity is that Newtonian mechanics is recovered in the non-relativistic limit of weak gravitational fields and small particle velocities $v \ll c$. This means that in the non-relativistic limit the Poisson equation of Newtonian gravitational theory

$$\Delta \phi_m(x) = 4\pi G \rho_m(x) \tag{12.1}$$

must emerge from the field equations.

In Newtonian mechanics, the equations of motion, $\ddot{x} = -\nabla \phi_m$, are obtained from a variational principle, *Hamilton's variational principle*.[1] Let $\mathcal{L} = T - V - mc^2$ be the Lagrangian of the system, then the variational principle states that the variation of the integral of the Lagrangian from point P_1 to point P_2 should vanish,

$$\delta \int_{P_1}^{P_2} \mathcal{L} \, dt = 0. \tag{12.2}$$

In (12.2), we expand the kinetic energy by including the rest energy mc^2 of the particle. The Lagrangian then takes the explicit form

$$\mathcal{L} = mc^2 \left(-1 + \frac{\dot{x}^2}{2c^2} - \frac{1}{c^2} \phi_m \right) \approx -mc^2 \sqrt{1 - \frac{\dot{x}^2}{c^2} + \frac{2\phi_m}{c^2}}, \tag{12.3}$$

with the gravitational potential $\phi_m = V/m$, and the approximation $\sqrt{1+x} \approx 1 + x/2$ for $x \ll 1$. Thus, in writing down (12.3), we have interpreted the Lagrangian of classical mechanics as an approximation to the square-root term on the right-hand side of the equation. We shall work with this approximation. The physical meaning of the condition

$$\frac{1}{c^2} \left(\frac{\dot{x}^2}{2} - \phi_m \right) \ll 1 \tag{12.4}$$

stands for the assumption of weak gravitational fields with small potentials ϕ_m. The derivatives with respect to the spatial coordinates also are assumed to be small, as well as the derivatives with respect to time. Therefore, we shall neglect products of ϕ_m and its derivatives as terms of higher order. Moreover, we consider only particles with non-relativistic velocities, $|\dot{x}| \ll c$, i.e., exactly the domain where

---

[1] William Rowan Hamilton, 1805–1865, Irish mathematician and physicist.

Newtonian mechanics is valid. All of these assumptions constitute the weak-field approximation.

Inserting (12.3) into the variational principle yields

$$
\delta \int_{P_1}^{P_2} \mathcal{L}\, dt = -mc\, \delta \int_{P_1}^{P_2} \sqrt{c^2 dt^2 \left(1 + \frac{2\phi_m}{c^2} \right) - \dot{x}^2 dt^2}
$$

$$
= -mc\, \delta \int_{P_1}^{P_2} \sqrt{(cdt)^2 \left(1 + \frac{2\phi_m}{c^2} \right) - dx^2 - dy^2 - dz^2} = 0,
$$

(12.5)

because of

$$
\dot{x}^2 dt^2 = \left[\left(\frac{dx}{dt} \right)^2 + \left(\frac{dy}{dt} \right)^2 + \left(\frac{dz}{dt} \right)^2 \right] dt^2 = dx^2 + dy^2 + dz^2.
\qquad (12.6)
$$

In general relativity, the geodesic equation also follows from a variational principle:

$$
\delta \int_{P_1}^{P_2} ds = \delta \int_{P_1}^{P_2} \sqrt{-g_{\mu\nu} dx^\mu dx^\nu} = 0.
\qquad (12.7)
$$

Comparing the two formulas we see that

$$
ds^2 = -\left(c^2 + 2\phi_m \right) dt^2 + dx^2 + dy^2 + dz^2 = 0,
\qquad (12.8)
$$

or formulated as a metric tensor

$$
g_{\mu\nu} = \begin{pmatrix} -\left(c^2 + 2\phi_m \right) & 0\;0\;0 \\ 0 & 1\;0\;0 \\ 0 & 0\;1\;0 \\ 0 & 0\;0\;1 \end{pmatrix} = \eta_{\mu\nu} + h_{\mu\nu}.
\qquad (12.9)
$$

In other words, gravitation is incorporated in the small perturbation

$$
h_{\mu\nu} = \begin{pmatrix} -2\phi_m & 0\;0\;0 \\ 0 & 0\;0\;0 \\ 0 & 0\;0\;0 \\ 0 & 0\;0\;0 \end{pmatrix}.
\qquad (12.10)
$$

Since, in general relativity, we include the factor c of the time coordinate in the metric, from now on we have

$$\eta_{\mu\nu} = \text{diag}(-c^2, 1, 1, 1) \quad \text{and} \quad \eta^{\mu\nu} = \text{diag}(-1/c^2, 1, 1, 1). \tag{12.11}$$

12.2 Formulation of the Field Equations

In this section we will "derive" the field equations. Since these are supposed to link the metric of the spacetime to the mass and energy distributions, we first need a quantity describing these quantities. This leads to the energy-momentum tensor, which we have already encountered in special relativity for the electromagnetic field in Sect. 7.7.

12.2.1 Energy-Momentum Tensor

The gravitational potential $\phi_m(r)$ appears on the right-hand side of the Poisson equation (12.1). Let us recall the structure of the energy-momentum tensor of the electromagnetic field which we had found:

$$T = \begin{pmatrix} \text{energy density} & \text{currents} \\ \text{currents} & \text{pressure and strains} \end{pmatrix}. \tag{7.95}$$

The tt-component is given by the energy density. The Poisson equation contains the mass density. If we multiply the mass density by c^2, we also obtain an energy density

$$\varepsilon_m = \rho_m c^2, \tag{12.12}$$

the *rest energy density* of matter. We can already guess that in the non-relativistic limit the Poisson equation follows from identifying the tt-component of the energy-momentum tensor with the tt-component of some other tensor. A second-rank tensor, which contains information on the curvature of spacetime, is the covariant Ricci tensor from (11.99). The tt-component of the fully covariant energy-momentum tensor is obtained from $T_t{}^t = -\rho_m c^2$ by lowering the contravariant index via (12.9)

$$T_{tt} \simeq c^4 \rho_m. \tag{12.13}$$

12.2.2 Ricci Tensor in Weak-Field Approximation

Since products of Christoffel symbols only consist of products of ϕ_m and its derivatives, which we neglect in the weak-field approximation, the expression for the Ricci tensor simplifies to

$$R_{\kappa\nu} \simeq \Gamma^{\mu}{}_{\kappa\nu,\mu} - \Gamma^{\mu}{}_{\kappa\mu,\nu}. \tag{12.14}$$

For the tt-components then only remains

$$R_{tt} \simeq \Gamma^{\mu}{}_{tt,\mu}, \tag{12.15}$$

since we also neglect derivatives with respect to time. For the metric (12.9), the Christoffel symbols then are

$$\Gamma^{t}{}_{it} \simeq \frac{\phi_{m,i}}{c^2} \quad \text{and} \quad \Gamma^{i}{}_{tt} \simeq \phi_{m,i}. \tag{12.16}$$

For the tt-component of the Ricci tensor we find

$$R_{tt} = \Gamma^{i}{}_{tt,i} \simeq \Delta\phi_m. \tag{12.17}$$

At this point we seem to have reached our goal, because we see that the tt-components of the energy-momentum tensor and the Ricci tensor yield exactly (apart from a numerical factor) the Poisson equation. We would simply have to generalize this connection to all tensor components.

We are facing, however, a problem, since from the energy-momentum tensor of electromagnetic fields follow the four continuity equations (7.93). These equations correspond to the requirement that the divergence of $T^{\mu\nu}$ vanishes, i.e., $T^{\mu\nu}{}_{,\nu} = 0$. This should also apply for the energy-momentum tensor of matter, with the modification that we now have to use the covariant instead of the normal derivative:

$$\nabla_{\nu}T^{\mu\nu} = T^{\mu\nu}{}_{;\nu} = 0. \tag{12.18}$$

However, in general $R^{\mu\nu}{}_{;\nu} \neq 0$, which means that we cannot simply use the Ricci tensor as the left-hand side of the field equations. Rather, we have to include additional terms.

12.2.3 Derivation of the Field Equations

The field-equations should meet the following five requirements:

1. The left-hand side is a symmetric tensor of rank 2 like $T_{\mu\nu}$.
2. No higher than second derivatives of $g_{\mu\nu}$ should appear on the left-hand side.

3. Second derivatives should occur only linearly.
4. The left-hand side must have zero divergence, just like $T^{\mu\nu}$.
5. In the Minkowski spacetime, the left-hand side must vanish identically, since in empty space $T_{\mu\nu} = 0$.

Weyl [2] showed that the requirements 1–3 imply this form of the field equations

$$R_{\mu\nu} + a g_{\mu\nu} R + \Lambda g_{\mu\nu} = \kappa T_{\mu\nu}, \tag{12.19}$$

where a, Λ and κ are free parameters, which still have to be determined, and R is the Ricci scalar from (11.102). We now proceed in a way different from the one used by Einstein in deriving the field equations. He made the assumption $\Lambda = 0$, which is indeed reasonable since, as we will see, the left-hand side of the field equations only vanishes in the limit of vanishing matter and energy density, if $\Lambda = 0$. The parameter Λ is called the *cosmological constant*. We shall return to the significance of this parameter in cosmology from Chap. 26 onward, but shall already retain it in the field equations.

The requirement 4 yields

$$R^{\mu\nu}{}_{;\nu} + a \left(g^{\mu\nu} R \right)_{;\nu} + \Lambda g^{\mu\nu}{}_{;\nu} \stackrel{!}{=} 0. \tag{12.20}$$

Since the metric has zero divergence, i.e. $g^{\mu\nu}{}_{;\nu} = 0$, Λ cannot be determined from this condition. But we can show that the condition leads to $a = -1/2$, and, using this value, can determine the other constants.

The Bianchi identity (11.104) results from the contraction of the Riemann tensor, and the symmetry properties of its components, see (11.96),

$$R^{\mu}{}_{\nu\mu\beta;\gamma} + R^{\mu}{}_{\nu\gamma\mu;\beta} + R^{\mu}{}_{\nu\beta\gamma;\mu} = R_{\nu\beta;\gamma} - R_{\nu\gamma;\beta} + R^{\mu}{}_{\nu\beta\gamma;\mu} = 0. \tag{12.21}$$

We multiply this equation by $g^{\nu\beta}$ and obtain, using the definition (11.102) of the Ricci scalar,

$$R_{;\gamma} - R^{\beta}{}_{\gamma;\beta} + g^{\nu\beta} \delta^{\mu}_{\alpha} R^{\alpha}{}_{\nu\beta\gamma;\mu} = 0. \tag{12.22}$$

In the third term, we have inserted the factor δ^{μ}_{α}, which we now rewrite in the form $\delta^{\mu}_{\alpha} = g^{\mu\lambda} g_{\lambda\alpha}$. The third term then can be transformed to

$$\begin{aligned}
g^{\nu\beta} g^{\mu\lambda} g_{\lambda\alpha} R^{\alpha}{}_{\nu\beta\gamma;\mu} &= g^{\nu\beta} g^{\mu\lambda} R_{\lambda\nu\beta\gamma;\mu} = -g^{\nu\beta} g^{\mu\lambda} R_{\nu\lambda\beta\gamma;\mu} \\
&= -g^{\mu\lambda} R^{\beta}{}_{\lambda\beta\gamma;\mu} = -g^{\mu\lambda} R_{\lambda\gamma;\mu} = -R^{\mu}{}_{\gamma;\mu}.
\end{aligned} \tag{12.23}$$

When we insert this relation in (12.22), and rename the summation index μ by β, we obtain $R_{;\gamma} - 2 R^{\beta}{}_{\gamma;\beta} = 0$, or

$$R^{\beta}{}_{\gamma;\beta} - \frac{1}{2}\left(\delta^{\beta}_{\gamma}R\right)_{;\beta} = 0. \tag{12.24}$$

As a final step, we raise the index γ, and get $R^{\beta\gamma}{}_{;\beta} - \frac{1}{2}\left(g^{\beta\gamma}R\right)_{;\beta} = 0$. Choosing $a = -1/2$, we have found an expression for the left-hand side of the field equations (12.19) that has zero divergence,

$$\left(R^{\mu\nu} - \frac{1}{2}Rg^{\mu\nu}\right)_{;\nu} = 0. \tag{12.25}$$

With these results, we now have

$$R_{\mu\nu} - \frac{1}{2}g_{\mu\nu}R + \Lambda g_{\mu\nu} = \kappa T_{\mu\nu}. \tag{12.26}$$

Multiplication by $g^{\mu\nu}$ and using $g^{\mu\nu}g_{\mu\nu} = 4$ leads to

$$-R + 4\Lambda = \kappa T \quad \text{or} \quad R = 4\Lambda - \kappa T \tag{12.27}$$

with the trace $T = g^{\mu\nu}T_{\mu\nu}$ of the energy-momentum tensor. We insert this expression for R in (12.26), and obtain

$$R_{\mu\nu} - \Lambda g_{\mu\nu} = \kappa T^{*}_{\mu\nu} \quad \text{with} \quad T^{*}_{\mu\nu} = T_{\mu\nu} - \frac{1}{2}Tg_{\mu\nu}. \tag{12.28}$$

To determine Λ and κ, we look again at the non-relativistic limit, and choose the parameters such that, in this limit, we end up with the Poisson equation.

In the previous section, we have already seen that in the non-relativistic limit the g_{tt}-component of the metric becomes $g_{tt} = -(c^2 + 2\phi_m)$, with the Newtonian potential ϕ_m, see (12.9). But we have to pay attention to an important subtlety: Because we assumed $v \ll c$ in (12.5), the second term of the expression $-dt^2(c^2 + 2\phi_m) + \dot{x}^2 dt^2$ in (12.5) is much smaller than the first term. This means that small perturbations of the components of the spatial part $dl^2 = dx^2 + dy^2 + dz^2$, similar to those of the tt-component of the metric, would be negligible. Therefore, we cannot use the metric (12.9) to calculate, e.g., the propagation of light in weak gravitational fields, since obviously the condition $|\dot{x}| \ll c$ is not satisfied. Furthermore, this form of the metric also will not follow as the non-relativistic limit of the exact field equations. Up to now, this was not important since we used this metric only to analyze the tt-component of the Ricci tensor, which helped us in finding an appropriate form for the field equations.

Generalizing this result, we now choose the components of the metric tensor as

$$g_{\mu\nu} = \eta_{\mu\nu} + k_{\mu\nu}, \quad \text{with} \quad k_{\mu\nu} \ll 1 \quad \text{and} \quad k_{tt} = 2\phi_m. \tag{12.29}$$

In the non-relativistic limit, velocities, and thus also currents, pressures, and strains, are negligibly small compared to the rest energy density. Then only the tt-component of the energy-momentum tensor, $T_{tt} \simeq \rho_\mathrm{m} c^4$, is significantly different from zero. Furthermore, with the metric just introduced one has

$$T = g^{\mu\nu} T_{\mu\nu} = \eta^{tt} T_{tt} = -\rho_\mathrm{m} c^2. \tag{12.30}$$

All corrections to this term are of the form $\phi_\mathrm{m} \rho_\mathrm{m}$, and therefore negligible. This is the reason why we shall use the Minkowski metric $\eta^{\mu\nu}$ instead of $g^{\mu\nu}$ for raising and lowering indices here. When we insert these results in (12.28), and make use of our approximation, $T^*_{\mu\nu} = T_{\mu\nu} - T\eta_{\mu\nu}/2$, we obtain the conditions

$$R_{tt} - \Lambda g_{tt} = \frac{1}{2}\kappa \rho_\mathrm{m} c^4, \tag{12.31a}$$

$$R_{ii} - \Lambda g_{ii} = \frac{1}{2}\kappa \rho_\mathrm{m} c^2, \tag{12.31b}$$

$$R_{\mu\nu} = 0, \quad \text{for} \quad \mu \neq \nu. \tag{12.31c}$$

To meet the fifth requirement stated at the beginning of this section, evidently we must have $\Lambda = 0$, since in the Minkowskian limit $\rho_\mathrm{m} \to 0$ we have $R_{\mu\nu} = T_{\mu\nu} = 0$. On the other hand, one might argue, on physical grounds, that Λ could be only very tiny. Then the deviations due to Λ would be important only on cosmological scales, and neglecting Λ would be compatible with the requirement 5 on smaller length scales.

Since all diagonal components of the Ricci tensor lead to the same equation (apart from the factor c^2), and all off-diagonal terms must vanish, it is reasonable to set

$$k_{\mu\nu} = \begin{cases} -\phi_\mathrm{m} & \text{for} \quad \mu = \nu = t, \\ -\phi_\mathrm{m}/c^2 & \text{for} \quad \mu = \nu = 1, 2, 3, \\ 0 & \text{for} \quad \mu \neq \nu. \end{cases} \tag{12.32}$$

Then we obtain the metric

$$g_{\mu\nu} = \begin{pmatrix} -c^2 - 2\phi_\mathrm{m} & 0 & 0 & 0 \\ 0 & 1 - \frac{2\phi_\mathrm{m}}{c^2} & 0 & 0 \\ 0 & 0 & 1 - \frac{2\phi_\mathrm{m}}{c^2} & 0 \\ 0 & 0 & 0 & 1 - \frac{2\phi_\mathrm{m}}{c^2} \end{pmatrix}. \tag{12.33}$$

To test this form, we calculate the corresponding Ricci tensor. For this purpose we need the Christoffel symbols, and consider again only terms linear in ϕ_m/c^2. The result is

$$\Gamma^i{}_{tt} = \phi_{\mathrm{m},i}, \quad \Gamma^i{}_{jj} = \Gamma^i{}_{ii} = \Gamma^t{}_{ti} = -\Gamma^j{}_{ji} = \frac{\phi_{\mathrm{m},i}}{c^2}, \quad \text{for} \quad i \neq j. \tag{12.34}$$

Next, we calculate the linearized Ricci tensor in the weak-field approximation (12.14). Inserting the Christoffel symbols (12.34), we find indeed

$$R_{\mu\nu} = \begin{cases} \Delta\phi_m & \text{for} \quad \mu = \nu = t, \\ \Delta\phi_m/c^2 & \text{for} \quad \mu = \nu = 1, 2, 3, \\ 0 & \text{for} \quad \mu \neq \nu. \end{cases} \qquad (12.35)$$

Furthermore, with the definition of the Ricci scalar in linear approximation $R = \eta^{\mu\nu} R_{\mu\nu}$, we find

$$R = -2\frac{\Delta\phi_m}{c^2}. \qquad (12.36)$$

Now we can determine the remaining free parameter κ. Comparing for example (12.31a) with the Poisson equation (12.1), we see that κ must be

$$\kappa = \frac{8\pi G}{c^4}. \qquad (12.37)$$

The numerical value of κ is very small. In SI units it reads [1]

$$\kappa = 2.076\,647(47) \cdot 10^{-43}\,\text{s}^2\,\text{m}^{-1}\,\text{kg}^{-1}. \qquad (12.38)$$

This means that only very large masses will be able to curve space noticeably. This
does not come as a surprise, because otherwise effects of general relativity would
also be important in our everyday life.

Since $R_{\mu\nu}$ and R contain second derivatives, and squares of first derivatives of the
metric tensor, the field equations are second-order non-linear differential equations.
Therefore, the superposition principle does not apply. As a consequence of their
intricate mathematical structure, Einstein's field equations can be solved analytically
only in a few special instances, and even the numerical solution is difficult in many
cases.

12.2.4 Final Form of the Field Equations

To write down the field equations in a compact form, we had introduced the
expression $T^*_{\mu\nu} = T_{\mu\nu} - Tg_{\mu\nu}/2$ in (12.28), instead of the original energy-
momentum tensor $T_{\mu\nu}$. An alternative is to define a tensor derived from the Ricci
tensor:

$$G_{\mu\nu} \equiv R_{\mu\nu} - \frac{1}{2}Rg_{\mu\nu}. \qquad (12.39)$$

This tensor is called *Einstein tensor*. Now we have two possibilities of formulating the field equations in a compact form, either

$$G_{\mu\nu} + \Lambda g_{\mu\nu} = \kappa T_{\mu\nu} \qquad (12.40a)$$

or

$$R_{\mu\nu} + \Lambda g_{\mu\nu} = \kappa T^*_{\mu\nu}. \qquad (12.40b)$$

Using (12.35) and (12.36), we find, in non-relativistic approximation,

$$G_{tt} = 2\Delta\phi_{\mathrm{m}} \qquad (12.41)$$

as the only non-vanishing component of the Einstein tensor. With the only non-vanishing component of the energy-momentum tensor, $T_{tt} = \rho_{\mathrm{m}}c^4$, this reproduces the Poisson equation.

References

1. Mohr, P.J., Tiesinga, E., Newell, D.B., Taylor, B.N.: Codata Internationally Recommended [sic!] 2022 Values of the Fundamental Physical Constants. https://physics.nist.gov/constants
2. Weyl, H.: Raum. Zeit. Materie. Springer (1919), English translation: Space. Time. Matter. Dover Publications (1952)

Schwarzschild Metric

13

Contents

One of the most important special cases in Newtonian mechanics is the gravitational field generated by a spherically symmetric mass distribution. This allows, in very good approximation, the description, e.g., of the motion of celestial bodies in the solar system. The spherically symmetric case is also one of the few for which Einstein's field equations can be solved analytically. In this chapter we will consider the region outside the spherical mass distribution, i.e., regions inside the mass distribution are exempt from the discussion. In Sect. 21.4.1 we shall consider the inner region in the context of neutron stars.

13.1   Derivation of the Schwarzschild Metric

In 1916, Karl Schwarzschild already succeeded in deriving the metric for the outer region of a spherically symmetric mass distribution [28], which is the reason why the metric is named after him.

In Newtonian mechanics, the gravitational field in the outer region of a spherically symmetric mass distribution with total mass M is given by the potential

$$\phi_{\mathrm{m}}(r) = -G\frac{M}{r}. \tag{13.1}$$

We now look for a spherically symmetric solution of Einstein's field equations (12.40) with vanishing cosmological constant ($\Lambda = 0$). In the limit of small masses or large distances, this solution must reproduce the results of Newton's theory. Since we restrict ourselves to the outer region, where neither matter nor energy is present, the energy-momentum tensor $T_{\mu\nu}$ obviously is equal to zero. The field equations (12.40) for the vacuum in the outer region therefore simplify to

$$R_{\mu\nu} = 0 \qquad \forall \mu, \nu \in \{0, 1, 2, 3\}. \tag{13.2}$$

In writing down this equation, we have taken into account that, because of $T_{\mu\nu} = 0$, we also have $T = 0$, and due to $R = -\kappa T$ from (12.27) also $R = 0$. This condition, however, does not mean that the space is flat, as one might assume at first sight.

For the spherically symmetric metric we choose the form

$$ds^2 = -e^{2\Phi(r)}c^2 dt^2 + e^{2\Psi(r)}dr^2 + r^2 d\Omega^2. \tag{13.3}$$

Here $\Phi(r)$ and $\Psi(r)$ are unknown functions, which depend solely on the radial coordinate $r \geq 0$, and which we have to determine. The choice of the expressions $e^{2\Phi(r)}$ and $e^{2\Psi(r)}$, instead of directly using $\Phi(r)$ and $\Psi(r)$, may seem arbitrary, but is often used in the literature, and we shall adopt this convention. The differential $d\Omega^2 = d\vartheta^2 + \sin^2(\vartheta)d\varphi^2$ depends on the spherical angle coordinates $\vartheta \in [0, \pi]$ and $\varphi \in [0, 2\pi)$, and describes the familiar surface element on the unit sphere.

When we compare the form (13.3) with the metric of a flat Euclidean space

$$dl^2 = dr^2 + r^2 d\Omega^2, \tag{13.4}$$

we can already see that the radial coordinate r cannot have the same meaning in both cases. We shall discuss this difference in more detail in Sect. 13.2.

In what follows, we shall concentrate on static metrics, therefore in (13.3) we have no explicit dependence on the time coordinate t (cf. Sect. 11.3.3).

Equation (13.3) leads to the following non-vanishing Christoffel symbols

$$\Gamma^t{}_{tr} = \Phi_{,r}, \tag{13.5a}$$

$$\Gamma^r{}_{tt} = c^2 \Phi_{,r}\, e^{2(\Phi - \Psi)}, \qquad \Gamma^r{}_{rr} = \Psi_{,r}, \tag{13.5b}$$

$$\Gamma^r{}_{\vartheta\vartheta} = -r e^{-2\Psi}, \qquad \Gamma^r{}_{\varphi\varphi} = -r \sin^2(\vartheta)\, e^{-2\Psi}, \tag{13.5c}$$

$$\Gamma^\vartheta{}_{r\vartheta} = \frac{1}{r}, \qquad \Gamma^\vartheta{}_{\varphi\varphi} = -\sin(\vartheta)\cos(\vartheta), \tag{13.5d}$$

$$\Gamma^\varphi{}_{r\varphi} = \frac{1}{r}, \qquad \Gamma^\varphi{}_{\vartheta\varphi} = \cot(\vartheta), \tag{13.5e}$$

with $\Phi_{,r} = d\Phi/dr$, and analogously, $\Psi_{,r} = d\Psi/dr$. From the Christoffel symbols we can calculate the Ricci tensor, whose non-vanishing components are

$$R_{tt} = c^2 \frac{e^{2(\Phi - \Psi)}}{r}\left(\Phi_{,r}^2 r - \Psi_{,r}\Phi_{,r} r + \Phi_{,rr} r + 2\Phi_{,r}\right), \tag{13.6a}$$

$$R_{rr} = \frac{1}{r}\left(2\Psi_{,r} - \Phi_{,rr} r + \Psi_{,r}\Phi_{,r} r - \Phi_{,r}^2 r\right), \tag{13.6b}$$

$$R_{\vartheta\vartheta} = e^{-2\Psi}\left(r\Psi_{,r} + e^{2\Psi} - 1 - r\Phi_{,r}\right), \tag{13.6c}$$

$$R_{\varphi\varphi} = \sin^2(\vartheta)\, R_{\vartheta\vartheta}. \tag{13.6d}$$

We multiply (13.6a) by $r e^{2(\Psi - \Phi)}/c^2$ and (13.6b) by r, and then form the sum (13.6b) + (13.6a), using (13.2), with the result

$$\Phi_{,r} + \Psi_{,r} = 0. \tag{13.7}$$

Inserting this result into $e^{2\Psi} R_{\vartheta\vartheta} = 0$ leads to the ordinary differential equation

$$2r\Psi_{,r} + e^{2\Psi} - 1 = 0. \tag{13.8}$$

By separating variables, we arrive at

$$\int \frac{d\Psi}{1 - e^{2\Psi}} = \int \frac{dr}{2r}. \tag{13.9}$$

For the integral on the right-hand side we immediately find a primitive. Introducing the substitution $x = e^{2\Psi}$ and using a partial fraction decomposition, we can easily solve the integral on the left-hand side, and end up with

$$\mathcal{A} + \Psi - \frac{1}{2}\ln\left(1 - e^{2\Psi}\right) = \frac{1}{2}\ln(r), \tag{13.10}$$

where we have hidden all integration constants in $\mathcal{A}$. Taking the exponent, and with the abbreviation $\mathcal{B} := e^{2\mathcal{A}}$, we have

$$\mathcal{B}\frac{e^{2\Psi}}{1 - e^{2\Psi}} = r \quad \text{or} \quad e^{2\Psi} = \frac{r}{\mathcal{B} + r} = \frac{1}{1 + \mathcal{B}/r}. \tag{13.11}$$

This leads to

$$\Psi_{,r} = \frac{\mathcal{B}}{2r(\mathcal{B} + r)}. \tag{13.12}$$

We insert (13.12) in (13.7), make use of another partial fraction decomposition, and integrate to obtain

$$e^{2\Phi} = \mathcal{C}\left(1 + \frac{\mathcal{B}}{r}\right). \tag{13.13}$$

The constants $\mathcal{B}$ and $\mathcal{C}$ can be determined by looking at the non-relativistic limit $r \to \infty$. For $r \to \infty$ the metric tensor $g_{\mu\nu}$ should turn into the Minkowski metric, since infinitely far away, the mass distribution can no longer have an effect on the metric. This leads to the condition

$$\lim_{r \to \infty} e^{2\Phi(r)} \stackrel{!}{=} 1 \quad \text{and thus} \quad \mathcal{C} = 1. \tag{13.14}$$

We also know from (12.33) that in the Newtonian limit $g_{tt} = -(c^2 + 2\phi_m)$, with $\phi_m = -GM/r$. From this follows

$$1 + \frac{\mathcal{B}}{r} \stackrel{!}{=} 1 - 2\frac{GM}{c^2 r} \quad \text{and thus} \quad \mathcal{B} = -2\frac{GM}{c^2}. \tag{13.15}$$

We know this constant already from Sect. 1.4.2, where we had introduced the *Schwarzschild radius*

$$r_s = 2\frac{GM}{c^2} \tag{1.38}$$

as a characteristic length scale of the gravitational interaction of a given mass M.

Summarizing the results, we finally can write down the *Schwarzschild metric*

$$ds^2 = -\left(1 - \frac{r_s}{r}\right)c^2 dt^2 + \frac{dr^2}{1 - r_s/r} + r^2 d\Omega^2 \tag{13.16}$$

with $d\Omega^2 = d\vartheta^2 + \sin^2(\vartheta)\, d\varphi^2$.

The requirement that the functions $\Phi(r)$ and $\Psi(r)$ should only depend on the radial coordinate r is, in fact, too strict. *Jebsen*[1] could already show in 1921 [19] that a more general form, in which the two functions may also depend on the time coordinate t, also leads to the Schwarzschild metric. Therefore, it represents the unique solution of the vacuum field equations with spherical symmetry and vanishing cosmological constant. The theorem that the static Schwarzschild metric is the solution for any spherically symmetric mass distribution was attributed for a long time solely to *Birkhoff*,[2] who in 1923 also found the *Birkhoff theorem* [3], named after him, see also Johansen and Ravndal [20].

13.2 Properties of the Schwarzschild Metric

In this section we will consider first physical consequences of the Schwarzschild metric.

13.2.1 Singularities

Looking at the metric components g_{tt} and g_{rr} in (13.16), we immediately recognize two problematic points (singularities), namely at $r = 0$ and $r = r_s$. We have already seen in Sect. 11.2.4 for the case of the metric of the surface of the sphere that nothing unphysical necessarily happens at such points. To make a coordinate-independent statement as to what happens at these points, we need a scalar quantity, such as the Ricci scalar. The latter, however, is identical to zero, by construction. We therefore use the *Kretschmann scalar* $\mathcal{K}$ from (11.103) to characterize these seeming singularities. For the Schwarzschild metric we have

$$\mathcal{K} = 12\frac{r_s^2}{r^6}. \tag{13.17}$$

Obviously, at $r = r_s$ this curvature scalar behaves regularly. Therefore, we only have to deal with a *coordinate singularity*, which can be removed by a proper coordinate transformation (cf. Sect. 13.3.2). Nevertheless, the surface of the sphere with radius $r = r_s$ does have a physical meaning, as can already be seen from the change in sign of the metric components g_{tt} and g_{rr}. Here the coordinates t and r evidently exchange their spacelike and timelike character. For the Schwarzschild metric, the surface of the sphere with radius $r = r_s$ is called the *event horizon*, or shortly, *horizon*, since it completely shields the inner region from the outer region. Whatever happens below the horizon cannot be observed from outside.

---

[1] Jørg Tofte Jebsen, 1888–1922, Norwegian physicist.

[2] George David Birkhoff, 1884–1944, US-American mathematician.

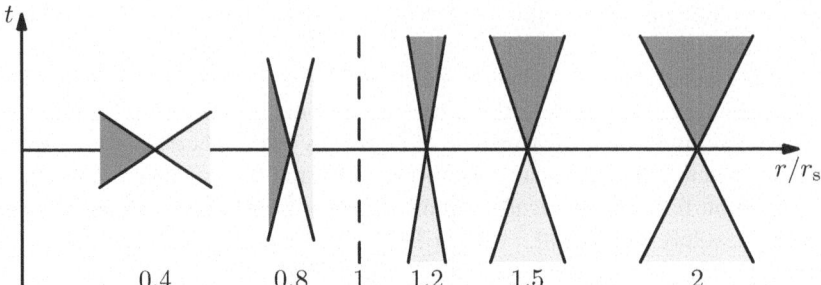

Fig. 13.1 The light cones become narrower and narrower as one approaches the Schwarzschild radius r_s. At the Schwarzschild radius, they reduce to a single line. Below the Schwarzschild radius they open again, but along the spatial axis. Regions in dark gray represent the future light cones

We can illustrate the exchange of the spacelike and timelike character of the coordinates t and r with the help of radial light cones. With $d\vartheta = d\varphi = 0$ Eq. (13.16) reduces, for lightlike geodesics, to

$$ds^2 = -\left(1 - \frac{r_s}{r}\right)c^2 dt^2 + \frac{dr^2}{1 - r_s/r} = 0. \tag{13.18}$$

From this equation the slopes

$$\frac{dt}{dr} = \pm \frac{1}{c\,(1 - r_s/r)} \tag{13.19}$$

of the radial light cones in an r-t diagram can be obtained, see Fig. 13.1.

The light cones become narrower and narrower as one approaches the Schwarzschild radius. For $r > r_s$ they are open along the time axis. For $r < r_s$ the light cones open along the spatial axis, which means that r becomes a timelike and t a spacelike coordinate. Below the horizon everything must inevitably move towards the singularity at $r = 0$.

13.2.2 Measurement of the Radial Coordinate

Because of the form of the metric component $g_{rr} = (1 - r_s/r)^{-1}$, the radial coordinate r of the Schwarzschild metric (13.16) cannot be interpreted as the actual distance from the origin, as would be the case in a flat spacetime. However, we can attribute a meaning to it by the following measuring instruction. Let us consider, at fixed time t, all points with a given, yet unknown, radial coordinate. For these points we have

$$dr = 0, \quad dt = 0, \quad d\vartheta \neq 0, \quad d\varphi \neq 0. \tag{13.20}$$

By measuring the force exerted by the mass distribution we can ascertain that all of these points, or observers at rest at these points, possess the identical radial coordinate. The latter can then be determined by measuring the surface of the sphere with radial coordinate r. From (13.16), with (13.20), we obtain the surface element

$$dF = r^2 \sin(\vartheta) \, d\vartheta \, d\varphi. \tag{13.21}$$

The corresponding surface of the sphere is $F = 4\pi r^2$. Thus, by measuring the area of the surface on which all these observers are sitting, we can identify the radial coordinate. Alternatively we can measure the perimeter of all points lying, e.g., in the plane $\vartheta = \pi/2$. This yields a length of $P = 2\pi r$. Thus, by measuring the surface or the perimeter we can, at least in principle, determine the radial coordinate.

13.2.3 Radial Distance of Points

The actually *measurable* distance of points with different radial coordinates can be determined from the line element (13.16). With the differentials

$$dr \neq 0, \quad dt = 0, \quad d\vartheta = 0, \quad d\varphi = 0 \tag{13.22}$$

we obtain the differential *proper radial distance*

$$ds_r = \frac{1}{\sqrt{1 - r_s/r}} dr \geq dr. \tag{13.23}$$

Because of the square-root expression in the denominator, we can determine radial distances only outside the Schwarzschild radius r_s. Obviously, for $r > r_s$ and $r_2 \geq r_1$ we have

$$\Delta s = \int_{r_1}^{r_2} ds_r > r_2 - r_1. \tag{13.24}$$

Therefore, the proper radial distance Δs is always bigger than the radial distance, which one would infer from the difference of the radial coordinates.

The distance to the *event horizon* is obtained as

$$\Delta s = r_s \sqrt{\frac{r}{r_s}\left(\frac{r}{r_s} - 1\right)} + \frac{r_s}{2} \ln\left[2\frac{r}{r_s} - 1 + 2\sqrt{\frac{r}{r_s}\left(\frac{r}{r_s} - 1\right)}\right]. \tag{13.25}$$

Figure 13.2 shows the radial distance Δs from (13.25) to the event horizon as a function of the radial coordinate r, normalized by the Schwarzschild radius r_s.

Fig. 13.2 Radial distance (13.24) Δs to the event horizon of a black hole as a function of the radial coordinate r, normalized with respect to the Schwarzschild radius r_{s}. For orientation, the position of the photon orbit $r = 3r_{\mathrm{s}}/2$ is also shown, see Sect. 13.2.6. There, the light is bent so strongly as to force it on a circular orbit around the black hole

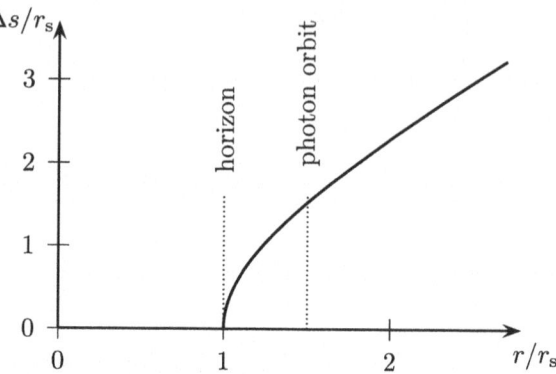

13.2.4 Meaning of the Coordinate Time

From the line element (13.16) of the Schwarzschild metric, we can also establish a relation between the time coordinate t and the actually measurable *proper time* τ of an observer. Let us consider an observer at rest at some fixed point in space. For him, we have

$$dr = d\vartheta = d\varphi = 0. \tag{13.26}$$

Inserting this into the line element (13.16) yields directly

$$ds^2 = -c^2 \left(1 - \frac{r_{\mathrm{s}}}{r}\right) dt^2 \equiv -c^2 \, d\tau^2 \quad \text{or} \quad d\tau = \sqrt{1 - \frac{r_{\mathrm{s}}}{r}} \, dt < dt. \tag{13.27}$$

For a very distant observer, $r \to \infty$, the coordinate time evidently coincides with his proper time. But the closer we get to the Schwarzschild radius r_{s}, the more the ratio between the proper time and the coordinate time tends to zero. This implies that for an observer close to the Schwarzschild radius the proper time elapses more slowly than for an observer more distant away. This has led to the popular saying "clocks in the gravitational field run more slowly". In the limit $r = r_{\mathrm{s}}$, the proper time would even come to a halt, compared to the clocks in the rest of the universe. However, the condition that an observer can keep himself at rest at $r = r_{\mathrm{s}}$, for example by rocket propulsion, requires an infinitely large acceleration (cf. Sect. 16.2.2, Eq. (16.17)).

13.2.5 Radial Lightlike Geodesics

The time $\Delta t(r_1, r_2)$ it takes for light to travel between two points in space $P_1 = (r_1, \vartheta, \varphi)$ and $P_2 = (r_2, \vartheta, \varphi)$ can be determined directly from the Schwarzschild metric (13.16). With $d\vartheta = d\varphi = 0$ and $ds^2 = 0$ we find for lightlike geodesics

$$c\Delta t(r_1, r_2) = \left| \int_{r_1}^{r_2} \frac{dr'}{1 - r_s/r'} \right| = \left| r_2 - r_1 + r_s \ln\left(\frac{r_2 - r_s}{r_1 - r_s}\right) \right|. \tag{13.28}$$

In the limit $r_1 \to r_s$ or $r_2 \to r_s$ this time interval diverges. This means, for an observer at position r_2 and a light-emitting object at $r_1 < r_2$, that light takes the more time the closer it is emitted near the event horizon. Light emitted from an object below the event horizon therefore will *never ever* reach an observer outside.

13.2.6 Photon Orbits in the Schwarzschild Metric

The geodesic equation (11.130) for the Schwarzschild metric with the metric tensor from (13.16), the spherical coordinates $x^\mu = (t, r, \vartheta, \varphi)$ for the position and $u^\mu = (u^t, u^r, u^\vartheta, u^\varphi)$ for the four-velocities has the form

$$\frac{d^2 t}{d\lambda^2} = -\frac{r_s}{r(r - r_s)} u^t u^r, \tag{13.29a}$$

$$\frac{d^2 r}{d\lambda^2} = -\frac{c^2 r_s(r - r_s)}{2r^3} u^t u^t + \frac{r_s}{2r(r - r_s)} u^r u^r$$
$$+ (r - r_s) \left[u^\vartheta u^\vartheta + \sin^2(\vartheta) u^\varphi u^\varphi \right], \tag{13.29b}$$

$$\frac{d^2 \vartheta}{d\lambda^2} = -\frac{2}{r} u^r u^\vartheta + \sin(\vartheta) \cos(\vartheta) u^\varphi u^\varphi, \tag{13.29c}$$

$$\frac{d^2 \varphi}{d\lambda^2} = -\frac{2}{r} u^r u^\varphi - 2 \cot(\vartheta) u^\vartheta u^\varphi. \tag{13.29d}$$

The special case $r = 3r_s/2$ is of particular interest. This distance characterizes the *photon orbit*: rays of light can turn around the black hole in a circular orbit.

We can derive the existence of circular orbits in the Schwarzschild metric in the following way. Because of the spherical symmetry, we can confine ourselves to orbits in the $\vartheta = \pi/2$ plane. Let the starting point of a ray of light be given by the point $x^\mu = (t_0, r, \pi/2, \varphi_0)$. For the ray of light to remain on a circular orbit, the four-velocity u^μ always must be of the form

$$u^\mu = (u^t, 0, 0, \omega). \tag{13.30}$$

From the constraint $g_{\mu\nu}u^{\mu}u^{\nu} = 0$ for lightlike geodesics, which immediately follows from the definition of the Lagrangian (11.133), we obtain

$$u^t = \frac{r\omega}{c\sqrt{1 - r_s/r}}. \tag{13.31}$$

The radial acceleration (13.29b) must also vanish. The requirement $\mathrm{d}^2r/\mathrm{d}\lambda^2 = 0$ yields

$$\left(u^t\right)^2 = \frac{2r^3}{c^2 r_s} \left(u^{\varphi}\right)^2 = \frac{2r^3}{c^2 r_s}\omega^2. \tag{13.32}$$

Inserting (13.31) into (13.32), we immediately obtain the radial coordinate

$$r_{\mathrm{po}} = \frac{3}{2}r_s \tag{13.33}$$

of the photon orbit.

13.2.7 Qualitative Behavior of Geodesics

We can study the qualitative behavior of lightlike and timelike geodesics in the Schwarzschild metric in a very straightforward way using the Euler-Lagrange formalism (cf. Sect. 11.4.2). Because of the spherical symmetry, we can again restrict ourselves to geodesics in the equatorial plane, $\vartheta = \pi/2$. In this case the *Lagrangian* takes the form

$$\mathcal{L} = -\left(1 - \frac{r_s}{r}\right)c^2\dot{t}^2 + \frac{\dot{r}^2}{1 - r_s/r} + r^2\dot{\varphi}^2 = \kappa c^2. \tag{13.34}$$

Obviously t and φ are cyclic variables since they do not appear explicitly in $\mathcal{L}$. The Euler-Lagrange equations (11.134) then imply $\partial\mathcal{L}/\partial\dot{t} = \mathrm{const}$ and $\partial\mathcal{L}/\partial\dot{\varphi} = \mathrm{const}$ and thus

$$\left(1 - \frac{r_s}{r}\right)c^2\dot{t} = k, \quad r^2\dot{\varphi} = h, \tag{13.35a}$$

and

$$\frac{1}{2}\dot{r}^2 + V_{\mathrm{eff}} = \frac{1}{2}\frac{k^2}{c^2} \tag{13.35b}$$

with the effective potential

$$V_{\mathrm{eff}} = \frac{1}{2}\left(1 - \frac{r_s}{r}\right)\left(\frac{h^2}{r^2} - \kappa c^2\right) \tag{13.36}$$

and the constants of motion k and h. In the case of timelike geodesics, these correspond to the energy and angular momentum per unit mass, respectively, in asymptotically flat space. In the case of lightlike geodesics, it is only the ratio $\varepsilon = ch/k$ which is important. Analogously to classical mechanics the relation (13.35b) is referred to as the *energy balance equation*.

If we want to specify the constants of motion from the point of view of an observer at rest, we still need his local frame of reference, which can be described, e.g., by the local tetrad

$$\underline{e}_{(t)} = \frac{1}{c\sqrt{1 - r_s/r}}\partial_t, \ \underline{e}_{(r)} = \sqrt{1 - \frac{r_s}{r}}\partial_r, \ \underline{e}_{(\vartheta)} = \frac{1}{r}\partial_\vartheta, \ \underline{e}_{(\varphi)} = \frac{1}{r\sin(\vartheta)}\partial_\varphi,$$

(13.37)

(cf. Sect. 11.2.5). Referred to this tetrad a geodesic in the equatorial plane then has the starting direction

$$\underline{y} = y^{(t)}\underline{e}_{(t)} + y^{(r)}\underline{e}_{(r)} + y^{(\varphi)}\underline{e}_{(\varphi)} = y^{(t)}\underline{e}_{(t)} + \rho\cos(\xi)\underline{e}_{(r)} + \rho\sin(\xi)\underline{e}_{(\varphi)}.$$

(13.38)

The normalization condition $\left\langle \underline{y}, \underline{y}\right\rangle_\eta = -(y^{(t)})^2 + (y^{(r)})^2 + (y^{(\varphi)})^2 = \kappa c^2$ (cf. (11.136)) leads to

$$k = c\sqrt{\rho^2 - \kappa c^2}\sqrt{1 - \frac{r_s}{r_{obs}}} \quad \text{and} \quad h = r_{obs}\,\rho\,\sin(\xi)$$

(13.39)

with the radial position $r = r_{obs}$ of the observer.

Lightlike Geodesics

If we inspect the effective potential V_{eff} for lightlike geodesics, we find only a single extremal point $r = 3r_s/2$, see Fig. 13.3. As we have already seen in Sect. 13.2.6, this position corresponds to the photon orbit. Since this position is a maximum of the effective potential, the photon orbit is unstable. For any deviation $\Delta r < 0$ ever so small the ray of light plummets into the singularity, while a positive radial displacement $\Delta r > 0$ makes the ray of light escape to infinity. With the help of the effective potential we can also determine the minimum distance r_{min} of a lightlike geodesic from the origin. For given k and h, the latter follows from the energy balance equation (13.35b) using the condition $\dot{r} = 0$ for the turning point. Solving with respect to the radius leads to the result

$$r_{min} = \frac{2\varepsilon}{\sqrt{3}}\cos\left[\frac{1}{3}\arccos\left(-\frac{3\sqrt{3}r_s}{2\varepsilon}\right)\right]$$

(13.40)

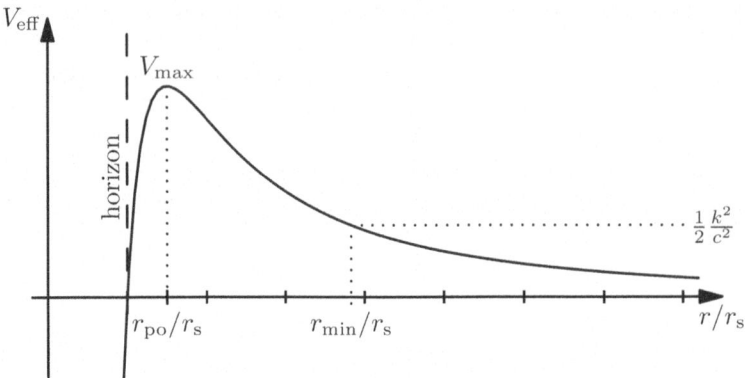

Fig. 13.3 Effective potential for a lightlike geodesic. The maximum $V_{\max}$ always lies at the position $r_{\mathrm{po}} = 3r_{\mathrm{s}}/2$ and marks the photon orbit

with $\varepsilon = ch/k = r_{\mathrm{obs}} \sin(\xi)/\sqrt{1 - r_{\mathrm{s}}/r_{\mathrm{obs}}}$, where, without loss of generality, we can choose $\rho = 1$.

For $r_{\min} = 3r_{\mathrm{s}}/2$ the geodesic asymptotically approaches the photon orbit. The angle ξ_{krit} under which the geodesic appears to an observer, results from the energy balance equation (13.35b) as

$$\sin^2(\xi_{\mathrm{krit}}) = \frac{27}{4} \frac{r_{\mathrm{s}}^2}{r_{\mathrm{obs}}^2} \left(1 - \frac{r_{\mathrm{s}}}{r_{\mathrm{obs}}}\right). \tag{13.41}$$

On the photon orbit, this relation evaluates to $\sin^2(\xi_{\mathrm{krit}}) = 1$, or $\xi_{\mathrm{krit}} = \pm 90°$. For an observer below the photon orbit, $r_{\mathrm{s}} < r_{\mathrm{obs}} < 3r_{\mathrm{s}}/2$, we must replace the angle ξ_{krit} with $180° - \xi_{\mathrm{krit}}$.

Timelike Geodesics

For timelike geodesics the effective potential can have two extremal points, located at

$$r_{\pm} = \frac{h^2 \pm h\sqrt{h^2 - 3c^2 r_{\mathrm{s}}^2}}{c^2 r_{\mathrm{s}}}. \tag{13.42}$$

The potential has a minimum at r_+, and a maximum at r_-. Obviously, if the discriminant $h^2 - 3c^2 r_{\mathrm{s}}^2$ is negative, no extremal point exists. If the discriminant is equal to zero, i.e., for $h^2 = 3c^2 r_{\mathrm{s}}^2$, one has an indifferent extremum. It is only for $h^2 > 3c^2 r_{\mathrm{s}}^2$ that two extremal points exist. This also implies that r_+ and r_- can only lie in the intervals

$$r_+ \geq 3r_{\mathrm{s}} \quad \text{and} \quad \frac{3}{2} r_{\mathrm{s}} < r_- < 3r_{\mathrm{s}}. \tag{13.43}$$

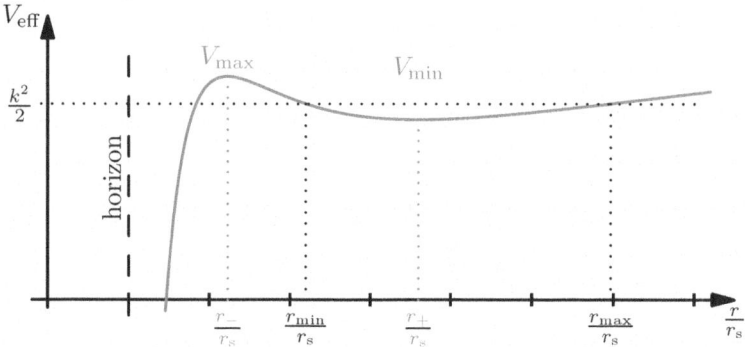

Fig. 13.4 Effective potential for a timelike geodesic with two extremal points

In Fig. 13.4 the effective potential is shown for a timelike geodesic with two extremal points. The constants of motion h and k are chosen such that the timelike geodesic always stays in the range between r_{min} and r_{max}. Here r_{min} denotes the closest approach to and r_{max} the greatest distance of the geodesic from the center, which also implies that it is a *bound orbit*. In analogy with classical mechanics, the geodesic cannot overcome the potential barrier at r_-, and therefore is kept from falling into the center. We cannot, however, tell directly from the potential whether the geodesic is following a closed orbit. Because of the effect of perihelion precession, discussed in Sect. 13.4.2, this is generally not the case. For a detailed discussion on the appearance of periodic orbits in the Schwarzschild metric we refer the reader, for example, to Levin and Perez-Giz [22].

The special case of a stable circular geodesic is obtained when we set $\dot{r} = 0$ in the Lagrangian (13.34), and choose as the radius of the circular orbit the radial coordinate r_+ at the minimum of the potential. We can then use (13.42) to find an expression for h^2 in dependence on r_+ and use it in (13.35b) together with (13.36) to also find an expression for k^2. These constants of motion then read

$$k^2 = c^4 \left(1 - \frac{r_s}{r_+}\right)^2 \left(1 - \frac{3r_s}{2r_+}\right)^{-1} \quad \text{and} \quad h^2 = \frac{c^2 r_+^2 r_s}{2r_+ - 3r_s}. \tag{13.44}$$

We can attribute an angular velocity $\omega = d\varphi/d\tau = \dot{\varphi}$ to an object moving along this geodesic path (with respect to its proper time τ), where

$$\omega^2 = \frac{h^2}{r_+^4} = \frac{c^2 r_s}{r_+^2 (2r_+ - 3r_s)}, \tag{13.45}$$

using $r^2\dot{\varphi} = h$ from (13.35a). Referred to the coordinate time, the angular velocity is $\Omega = d\varphi/dt = \dot{\varphi}/\dot{t}$, and thus

$$\Omega^2 = c^2 r_s/(2r_+^3). \tag{13.46}$$

We can also calculate the period of the complete cycle, which is

$$T_{2\pi} = \frac{2\pi}{\Omega} = \frac{2\pi r_+}{c}\sqrt{\frac{2r_+}{r_s}}.$$

(13.47)

To assign a velocity $v = c\beta$ to the object, which an observer will measure at his present position, we first have to determine the four-velocity

$$\underline{u} = \dot{t}\partial_t + \dot{\varphi}\partial_\varphi = c\gamma(\underline{e}_{(t)} + \beta\underline{e}_{(\varphi)}).$$

From this follows

$$v = c\sqrt{\frac{r_s}{2(r_+ - r_s)}}.$$

(13.48)

So far we have not taken into account the fact that the radius r_+ is bounded from below, see (13.43). For $r_+ = 3r_s$ the effective potential (13.36) has only a single (indifferent) extremal point. Any radial displacement ever so small will entail the plummeting of the object into the singularity. Notwithstanding this fact, it has become common practice to call the orbit with $r_{\text{lso}} = 3r_s$ the *last stable orbit* (lso).

From (13.48) follows, with $r_+ = 3r_s$, that the velocity of an object moving on the last stable orbit is half the speed of light, $v = c/2$. Setting $r_+ = 3r_s$ in (13.46) we obtain $\Omega = c/(3\sqrt{6}r_s)$, and the period of the last stable orbit is

$$T_{2\pi} = \frac{2\pi}{c}3r_s\sqrt{6} = 12\pi\sqrt{6}\frac{GM}{c^3}.$$

(13.49)

To obtain corresponding values in proper time, we insert $r_+ = 3r_s$ in (13.45) and obtain $\omega = c/(3\sqrt{3}r_s)$. The period of the last stable orbit measured in proper time is

$$\tau_{2\pi} = \frac{2\pi}{\omega} = 6\sqrt{3}\pi\frac{r_s}{c}.$$

(13.50)

Hence, an observer on the last stable orbit would conclude that the circumference of this orbit would be $C'_{\text{lso}} = v\tau_{2\pi} = 3\sqrt{3}\pi r_s$. The difference to the circumference $C_{\text{lso}} = 6\pi r_s$, which follows immediately from the metric (13.16), is due to the length contraction of the moving observer, where $C'_{\text{lso}} = C_{\text{lso}}/\gamma$ and $\gamma = 2/\sqrt{3}$.

13.2.8 Einstein Rings

If the center of the mass distribution is placed exactly in the line of sight of an observer to a point-like light-emitting object (in Fig. 13.5 the line of sight is the

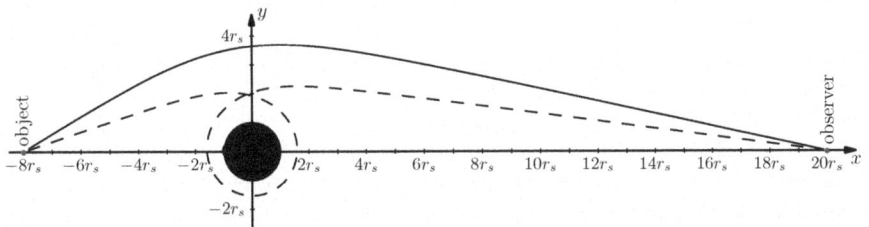

Fig. 13.5 The point-like light-emitting object at $x = -8r_s$ appears to the observer at $x = 20r_s$ as an Einstein ring with the opening angle $\xi \approx 11.9473°$ (ring of first order), or $\xi \approx 7.279793°$ (ring of second order)

x-axis), then the observer will see the object as an *Einstein ring*. The reason is the spherical symmetry of the Schwarzschild spacetime. All rays of light which arrive at the observer under the same angle ξ relative to the line of sight originate from the light-emitting object. Because of the strong deflection of light in the Schwarzschild spacetime, there exists not only a ring of first order but there exist, in principle, infinitely many rings. If a ray of light starts at the object but revolves once around the mass center before being released to the observer, this is called a ring of second order.

One can calculate the angle ξ of the Einstein ring from the positions of the observer $r = r_i$ and of the object $r = r_f$. The starting point is the energy balance equation (13.35b) with the effective potential (13.36) for $\kappa = 0$ and the constant of motion h. Then we have

$$\left(\frac{dr}{d\varphi}\right)^2 = \frac{\dot{r}^2}{\dot{\varphi}^2} = r^4 \frac{k^2}{c^2 h^2} - r^2 \left(1 - \frac{r_s}{r}\right). \tag{13.51}$$

With the substitution $x = r_s/r$ this results in

$$\left(\frac{dx}{d\varphi}\right)^2 = a^2 - x^2(1-x) \quad \text{with} \quad a^2 = \frac{k^2 r_s^2}{c^2 h^2} = \frac{x_i^2(1-x_i)}{\sin^2(\xi)} \tag{13.52}$$

and $x_i = r_s/r_i$. We can formally solve this elliptical integral, but we have to split it into two parts

$$\varphi = \int_{x_i}^{x_f} \frac{dx}{\sqrt{a^2 - x^2(1-x)}} = \int_{x_i}^{x_{min}} \frac{dx}{\sqrt{\cdots}} + \int_{x_f}^{x_{min}} \frac{dx}{\sqrt{\cdots}}. \tag{13.53}$$

depending on the minimum distance (13.40) of the geodesic from the origin

$$x_{min} = \frac{r_s}{r_{min}} = \frac{\sqrt{3}a}{2} \left\{ \cos\left[\frac{1}{3}\arccos\left(-\frac{3\sqrt{3}a}{2}\right)\right] \right\}^{-1}. \tag{13.54}$$

In (13.53) we have also taken care of the correct directions of integration and plus and minus signs. We obtain the angle ξ of the Einstein ring of first order when we choose $\varphi = \pi$ in (13.53). We cannot, however, solve the implicit equation with respect to ξ, since ξ appears not only in the integrand but also in the limits of integration. Therefore, we would have to resort to an iterative numerical procedure, in which we would have to choose the initial value and the initial limits of integration very carefully. For an Einstein ring of nth order, with $n \geq 1$, we must set $\varphi = (2n - 1)\pi$.

13.3 Black Holes

The Schwarzschild metric is valid for all spherically symmetric mass distributions with vanishing cosmological constant. If the mass distribution resides completely below the Schwarzschild radius, the metric also describes one of the most exceptional objects of General Relativity—a static black hole. As discussed in the previous Sect. 13.2, the event horizon separates the inner region completely from the outside. From this also follows that not even light can escape from the inner region to the outside, which is why this region is black. In very simple terms we have addressed this situation already in Sect. 1.4.4.

In the following sections we will investigate a few more properties of the Schwarzschild metric, which find their most extreme realization in the case of black holes.

13.3.1 Free Fall onto a Black Hole

We will study the free fall of a particle onto a black hole. What is of particular interest is the answer to the question: How does an observer, free-falling with the particle, assess the situation, and how does it appear to an observer far away? The particle is assumed to be at rest initially, and to begin its fall from the position $r = R > r_s$.

Because of the spherical symmetry, the particle can only move on a radial geodesic path. In this case the Lagrangian (13.34) reads

$$\mathcal{L} = -\left(1 - \frac{r_s}{r}\right)c^2 \dot{t}^2 + \frac{1}{1 - r_s/r}\dot{r}^2 = -c^2. \tag{13.55}$$

The dot denotes the derivative with respect to the proper time τ. We know from Sect. 13.2.7 that the time coordinate t is a cyclic variable, and the corresponding constant of motion is $k = (1 - r_s/r)c^2\dot{t}$. We replace $\dot{t}$ in the Lagrangian (13.55) using this relation and finally arrive, after a few mathematical manipulations, at the equations of motion

$$\frac{dr}{d\tau} = \pm \sqrt{\frac{k^2}{c^2} - c^2 \left(1 - \frac{r_s}{r}\right)} \tag{13.56a}$$

and

$$\frac{dt}{d\tau} = \frac{k}{c^2} \left(1 - \frac{r_s}{r}\right)^{-1} \tag{13.56b}$$

for the radial and the time coordinate in dependence on the proper time τ.

Free-Falling Observer

From (13.56a) we can derive a relation between the proper time τ of the free-falling observer and his present position r. The observer is also supposed to start at $r = R$ and to be initially at rest, $(dr/d\tau)|_{r=R} = 0$. As time proceeds, the radial position always decreases, therefore we have to choose the minus sign in (13.56a), and obtain

$$\tau(r) = - \int\limits_R^r \frac{dr'}{\sqrt{k^2/c^2 - c^2 (1 - r_s/r')}}. \tag{13.57}$$

With the help of the initial conditions and (13.56a) we can determine the constant of motion $k = c^2 \sqrt{1 - r_s/R}$. Inserting this into the integral (13.57) we arrive at the expression

$$\tau(r) = \frac{1}{c} \int\limits_r^R \frac{dr'}{\sqrt{(1 - r_s/R) - (1 - r_s/r')}} = \frac{1}{c} \int\limits_r^R \frac{dr'}{\sqrt{r_s/r' - r_s/R}}. \tag{13.58}$$

To evaluate this integral, we substitute $r' = R \sin^2(x)$, with the derivative $dr' = 2R \sin(x) \cos(x) \, dx$, and obtain, with the substituted limits of integration $x_i = \arcsin\left(\sqrt{r/R}\right)$ and $x_f = \pi/2$,

$$\tau(r) = \frac{2R^{3/2}}{c\sqrt{r_s}} \int\limits_{x_i}^{x_f} \sin^2(x) \, dx = \frac{R^{3/2}}{c\sqrt{r_s}} \left[x - \frac{1}{2} \sin(2x)\right]_{\arcsin(\sqrt{r/R})}^{\pi/2}. \tag{13.59}$$

The trigonometric relations

$$\sin(2x) = 2 \sin(x) \cos(x) \quad \text{and} \quad \arccos(x) = \frac{\pi}{2} - \arcsin(x) \tag{13.60}$$

can be used to calculate the proper time $\tau(r)$ that has elapsed for the observer since he started from R, to his present position $r \leq R$,

$$\tau(r) = \frac{R^{3/2}}{c\sqrt{r_s}} \left[\sqrt{\frac{r}{R}} \sqrt{1 - \frac{r}{R}} + \arccos\left(\sqrt{\frac{r}{R}}\right) \right]. \tag{13.61}$$

Setting $r = R$ in this expression, we obtain $\tau(R) = 0$, as of course it has to be. The free-fall time down to the Schwarzschild radius $r = r_s$ is then given by

$$\tau(r_s) = \frac{R^{3/2}}{c\sqrt{r_s}} \left[\sqrt{\frac{r_s}{R}} \sqrt{1 - \frac{r_s}{R}} + \arccos\left(\sqrt{\frac{r_s}{R}}\right) \right]. \tag{13.62}$$

Thus, the particle and the free-falling observer reach the event horizon of the black hole in *finite* proper time. Furthermore, if we set $r = 0$ in (13.61), we find, because of $\arccos(0) = \pi/2$, that the particle reaches the singularity of the black hole after the finite proper time

$$\tau(0) = \frac{\pi}{2} \frac{R^{3/2}}{c\sqrt{r_s}}. \tag{13.63}$$

We can write (13.61) in a more compact form when we introduce the cycloid coordinate η. We set

$$r = \frac{R}{2} \left[1 + \cos(\eta)\right], \tag{13.64}$$

and have $r(\eta = 0) = R$ and $r(\eta = \pi) = 0$. Therefore, we get the simple expression

$$\tau(\eta) = \frac{R^{3/2}}{2c\sqrt{r_s}} \left[\eta + \sin(\eta)\right]. \tag{13.65}$$

For $\eta \in [0, \pi]$, this is a continuous function, which evidently means that the particle passes through the event horizon without any difficulties, and that one just has to deal with a coordinate singularity.

Observer at a Great Distance

We shall see that an observer far away from the black hole will assess the free fall of the particle onto the black hole in a completely different way. We first determine the coordinate time $t(r)$ that has elapsed for him, until the particle is at its present position r. From (13.56a) and (13.56b) follows

$$\frac{dt}{dr} = \frac{\dot{t}}{\dot{r}} = -\frac{k}{c^2} \left(1 - \frac{r_s}{r}\right)^{-1} \left[\frac{k^2}{c^2} - c^2 \left(1 - \frac{r_s}{r}\right)\right]^{-1/2}. \tag{13.66}$$

Inserting the constant of motion k from the previous section, we obtain

$$t(r) = \frac{1}{2}\sqrt{1 - \frac{r_s}{R}} \int\limits_R^r \frac{dr'}{(1 - r_s/r')\sqrt{r_s/r' - r_s/R}}. \tag{13.67}$$

We immediately recognize that the factor $(1 - r_s/r')^{-1}$ behaves singularly at $r' = r_s$. For the exact evaluation of the integral we make use again of the cycloid coordinate. With the expression (13.65) already found for $\tau(\eta)$ and its derivative

$$d\tau = \frac{R^{3/2}}{2c\sqrt{r_s}}[1 + \cos(\eta)]\, d\eta, \tag{13.68}$$

Equation (13.56b) yields the relation

$$dt = \frac{R^{3/2}}{2c\sqrt{r_s}}\sqrt{1 - \frac{r_s}{R}\frac{[1 + \cos(\eta)]^2}{1 + \cos(\eta) - 2r_s/R}}\, d\eta. \tag{13.69}$$

The integration is somewhat laborious, we therefore only quote the result

$$t(\eta) = \frac{r_s}{c}\left\{\ln\left[\frac{\sqrt{R/r_s - 1} + \tan\left(\frac{\eta}{2}\right)}{\sqrt{R/r_s - 1} - \tan\left(\frac{\eta}{2}\right)}\right] + \sqrt{\frac{R}{r_s} - 1}\left[\eta + \frac{R}{2r_s}(\eta + \sin(\eta))\right]\right\}. \tag{13.70}$$

Obviously this expression diverges when the argument of the logarithm diverges. This happens for $\sqrt{R/r_s - 1} = \tan(\eta/2)$. From the trigonometric formula $\tan^2(\eta/2) = (1 - \cos(\eta))/(1 + \cos(\eta))$ then follows

$$\cos(\eta) = 2\frac{r_s}{R} - 1. \tag{13.71}$$

Inserting this expression in (13.64), we immediately see that the observer's coordinate time diverges, $t(\eta) \to \infty$, for $r \to r_s$. Since the coordinate time t of an observer at a great distance from the black hole corresponds to his own proper time, the observer would *calculate* that the particle never reaches the horizon. To find out what the observer actually would *see*, we have to consider the time $\Delta t(r, R)$ that it has taken for light to travel from the present position of the particle to the observer (cf. Sect. 13.2.5). Evidently, the latter also diverges in the limit $r \to r_s$. Taking all together, the observer sees the particle at its present position r at his own proper time $t_{obs}(r) = t(r) + \Delta t(r, R)$. Thus, for an observer far away from the black hole the particle *never* reaches the event horizon, both mathematically and visually.

13.3.2 Extensions of the Schwarzschild Metric

Looking at the form of the Schwarzschild metric (13.16) one recognizes *two* obvious singularities, one at $r = r_s$, the other at $r = 0$. As already discussed in Sect. 13.2.1, these singularities are of different types, as several authors, among them also Einstein, demonstrated in various papers from 1921 on, see, e.g., Kruskal [21].

One can best understand the fact that no physical singularity occurs at $r = r_s$, by transforming the metric to coordinates in which the singularity simply is not present. A first step in this direction can be made by introducing Eddington-Finkelstein coordinates.

Eddington-Finkelstein Coordinates

The coordinates originally found in 1924 by *Eddington*[3] [7] and rediscovered in 1958 by *Finkelstein*[4] [9, 24] are based on the inspection of photons in free radial fall onto the black hole. To this end a coordinate $\mathcal{V}$ is introduced via

$$\mathcal{V} = ct + r^* \tag{13.72}$$

with the so-called *tortoise coordinate*[5]

$$r^* = r + r_s \ln\left(\frac{r}{r_s} - 1\right) \quad \text{and} \quad dr^* = \frac{dr}{1 - r_s/r}. \tag{13.73}$$

In the new coordinates $(\mathcal{V}, r, \vartheta, \varphi)$ the metric reads

$$ds^2 = -\left(1 - \frac{r_s}{r}\right)c^2\, d\mathcal{V}^2 + 2c\, d\mathcal{V}\, dr + r^2 d\Omega^2. \tag{13.74}$$

The fact that these coordinates are well adapted to describing radially moving photons becomes evident, when one takes a look at their light cones.

From $ds^2 = 0$ and $d\vartheta = d\varphi = 0$ follow the conditions

$$\frac{d\mathcal{V}}{dr} = 0 \qquad\qquad \text{for photons moving inward and} \tag{13.75a}$$

$$\frac{d\mathcal{V}}{dr} = \frac{2}{c(1 - r_s/r)} \qquad \text{for photons moving outward.} \tag{13.75b}$$

---

[3] Arthur Stanley Eddington, 1882–1944, British astronomer, physicist and mathematician.

[4] David Finkelstein, 1929–2016, US-American physicist.

[5] The name of this coordinate traces back to one of Zeno of Elea's paradoxes on an imaginary footrace between "swift-footed" Achilles and a tortoise. The paradox states that Achilles can never catch up with a tortoise, much slower than himself, when it was given a head start and continues moving ahead. The name is used in the present context since $r^* \to -\infty$ for $r \to r_s$.

Photons falling into the black hole move on surfaces with $\mathcal{V} = $ const. However, for photons moving radially outward, there still occurs a singularity at $r = r_s$. Alternatively, one can define the coordinate $\mathcal{U} = ct - r^*$, instead of $\mathcal{V}$, for photons moving radially outward. Then a singularity occurs at $r = r_s$ for photons moving inward.

Kruskal-Szekeres Coordinates

To make the coordinate singularity at $r = r_s$ disappear, a spherically symmetric coordinate system $(v, u, \vartheta, \varphi)$ is sought in which radially moving rays of light possess the slope $dv/du = \pm 1$ everywhere, exactly as in flat space. For this purpose, *Kruskal*[6] and *Szekeres*[7] chose the form

$$ds^2 = -f^2(u, v)\left(dv^2 - du^2\right) + r^2(u, v)d\Omega^2 \tag{13.76}$$

for the line element [21,30]. Here v denotes a timelike and u a spacelike coordinate. The function f should remain finite and different from zero for $v = u = 0$ and only depend on r via u and v. The requirement that the equation

$$f^2(u, v)\left(dv^2 - du^2\right) = \left(1 - \frac{r_s}{r}\right)\left(c^2 dt^2 - dr^{*2}\right) \tag{13.77}$$

should be fulfilled, leads to the transformation equations

$$u = \pm\sqrt{\frac{r}{r_s} - 1}\, e^{r/(2r_s)} \cosh\left(\frac{ct}{2r_s}\right), \quad v = \pm\sqrt{\frac{r}{r_s} - 1}\, e^{r/(2r_s)} \sinh\left(\frac{ct}{2r_s}\right) \tag{13.78a}$$

for $r \geq r_s$, and to

$$u = \pm\sqrt{1 - \frac{r}{r_s}}\, e^{r/(2r_s)} \sinh\left(\frac{ct}{2r_s}\right), \quad v = \pm\sqrt{1 - \frac{r}{r_s}}\, e^{r/(2r_s)} \cosh\left(\frac{ct}{2r_s}\right) \tag{13.78b}$$

for $0 < r < r_s$. For the coordinates u and v this yields the relation

$$u^2 - v^2 = \left(\frac{r}{r_s} - 1\right) e^{r/r_s}. \tag{13.79}$$

Finally, the function f^2 has the form

$$f^2 = \frac{4r_s^3}{r} e^{-r/r_s}. \tag{13.80}$$

---

[6] Martin David Kruskal, 1925–2006, US-American mathematician and physicist.
[7] George Szekeres, 1911–2005, Hungarian-Australian mathematician.

This expression is positive for $r > 0$ and becomes singular only at $r = 0$. In *Kruskal coordinates* the line element then becomes

$$ds^2 = -\frac{4r_s^3}{r}e^{-r/r_s}\left(dv^2 - du^2\right) + r^2 d\Omega^2, \tag{13.81}$$

where $r = r(u, v)$ is obtained as an implicit function from (13.79). Its inverse function can be written in terms of the *Lambert W function*,[8] which is defined as the inverse of $f(x) = xe^x$,

$$r = r_s\left[\mathcal{W}\left(\frac{u^2 - v^2}{e}\right) + 1\right]. \tag{13.82}$$

The Kruskal spacetime described by the line element (13.81) extends the Schwarzschild spacetime in the following way. In the case $r \geq r_s$ one has $v/u = \tanh(ct/2r_s)$. Curves with fixed time $t = \text{const}$ fulfill $v/u = \text{const}$, and are represented by straight lines in a (u, v) diagram. Curves with fixed radial coordinate $r = \text{const}$, on the other hand, are hyperbolas in the (u, v) diagram, as can be seen directly from (13.79). For $r = r_s$, (13.79) becomes $u^2 - v^2 = 0$. Therefore, one has the asymptotic special case of straight lines, $u = \pm v$. For $r = 0$, on the other hand, one finds $u^2 - v^2 = -1$ or $v = \pm\sqrt{1 + u^2}$.

Figure 13.6 shows the four different regions of the Kruskal spacetime, and relates them to the Schwarzschild spacetime. The regions ① and ② correspond to the positive signs in Eqs. (13.78a) and (13.78b), the regions ③ and ④ to the negative signs. The region in spacetime covered by the Schwarzschild coordinates is marked in light gray. Figure 13.6b shows straight lines with constant time t. The straight lines with slope ± 1 correspond to $t = \pm\infty$. The regions drawn in dark gray would correspond to $r < 0$, and are not part of the spacetime. Figure. 13.6c shows the hyperbolas with constant radial coordinate r. In the (u, v) diagram, the straight lines $r = r_s$ lie exactly on the straight lines $t = \pm\infty$. This means that for infinite times the event horizon is mapped onto these straight lines, while for finite times all points of the spacetime with $r = r_s$ are mapped onto the point $u = v = 0$.

In Fig. 13.6d we have also drawn the light cones of three observers, A, B, and C. Observer A sits at the radial position $r_A > r_s$ in the Schwarzschild part of the spacetime. His future light cones contain world lines which always point to a larger radial coordinate r_1. By contrast, observer B resides at a radial coordinate $r_B < r_s$. All world lines proceeding from that position end in the singularity at $r = 0$. Any particle that has crossed the Schwarzschild radius to radial coordinates $(r < r_s)$ therefore will *always* end up in the singularity at $r = 0$! This is of course the reason why this region is called a *black hole*. No object entering this region will be able to leave it again. One also recognizes that all objects which want to get inside the black hole have to cross the timeline $t = \infty$. This is in agreement with the statement

---

[8] Johann Heinrich Lambert, 1728–1777, Swiss mathematician, physicist, and philosopher.

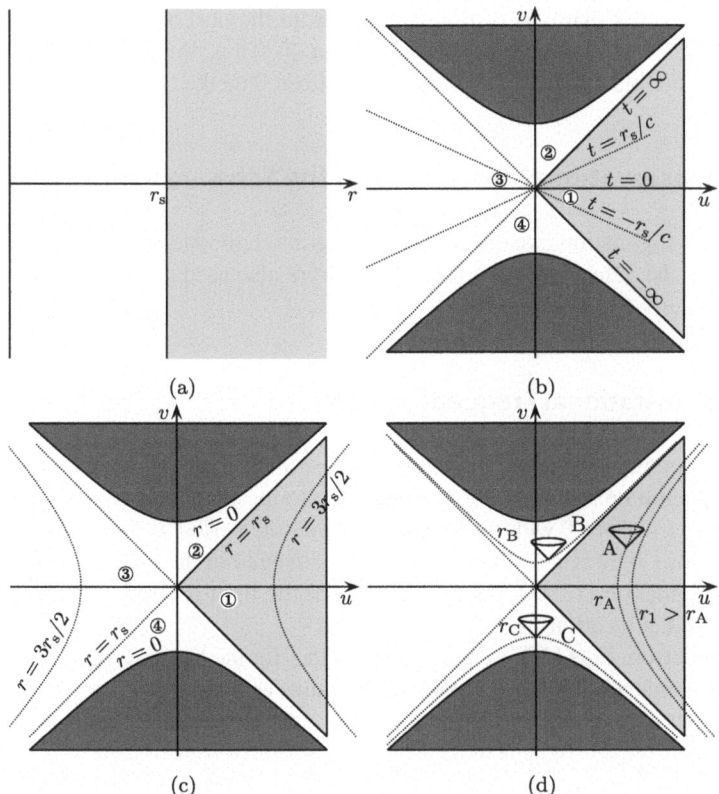

Fig. 13.6 The Schwarzschild coordinates only cover that part of spacetime marked in light gray (**a**). The Kruskal coordinates extend the Schwarzschild spacetime to three more regions. The original Schwarzschild region is also drawn in gray in the three other panels. (**b**) Lines with constant time t, (**c**) lines with constant radial coordinate r, (**d**) future light cones for different observers. Observer B resides in the black hole (region ②), C in the white hole (region ④)

found before that an observer far away from the black hole will never see the object disappearing behind the event horizon.

Finally, observer C resides at a position $r_C < r_s$ in the region ④. From there all world lines point to larger radial coordinates. This means, no object can intrude into this region, and all objects in this region are forced to move outward. In the literature this case is denoted as a *white hole*. The world lines, however, would have to cross the timeline $t = -\infty$. If white holes really were to exist then they must have been present already before the creation of the Universe! In a finite-age Universe, the initial condition for the existence of a white hole therefore can *never* be fulfilled.

Finally, let us note that the region ③ again corresponds to a Schwarzschild spacetime. The arrow of time, however, points into the wrong direction, namely in that of a *decreasing* time coordinate t, i.e., into the past. We can understand this when we consider that rays of light coming from the observer C have to cross the straight line $t = \infty$ before reaching the region ③. Thus, the world lines of objects falling into the black hole first would have to overcome the line $t = -\infty$.

13.4   Observational Tests of GR in the Schwarzschild Metric

In the following sections we discuss some general relativistic physical consequences that arise in the Schwarzschild metric. They were also used as experimental tests of General Relativity.

13.4.1 Gravitational Frequency Shift

Time measurement requires a periodic process. Such a process may be the oscillation of a pendulum, but also, for example, an atomic transition between two levels with energy $h\nu = h/T$ with the frequency ν and the period T. In such a transition, the atom emits light whose frequency will be changed in the gravitational field due to the time dilation (13.27). Let us assume that the light is emitted at location P_{em} and absorbed at location P_{abs}.

We know from Eq. (13.35a) in Sect. 13.2.7 that for the lightlike geodesic between these two points the following expression remains constant:

$$k = \left(1 - \frac{r_s}{r_{\text{em}}}\right) c^2 \frac{dt}{d\lambda}\bigg|_{P_{\text{em}}} = \left(1 - \frac{r_s}{r_{\text{abs}}}\right) c^2 \frac{dt}{d\lambda}\bigg|_{P_{\text{abs}}}. \tag{13.83}$$

If we use the same form for the four-dimensional wave vector as in (8.6), we obtain the differentials

$$\frac{dt}{d\lambda}\bigg|_{P_{\text{em}}} = \frac{\omega_{\text{em}}}{c^2\sqrt{1 - r_s/r_{\text{em}}}} \quad \text{and} \quad \frac{dt}{d\lambda}\bigg|_{P_{\text{abs}}} = \frac{\omega_{\text{abs}}}{c^2\sqrt{1 - r_s/r_{\text{abs}}}}. \tag{13.84}$$

Inserting this into (13.83), and noting that $\omega = 2\pi\nu$, we arrive at the equation for the *gravitational frequency shift*

$$\frac{\nu_{\text{abs}}}{\nu_{\text{em}}} = \sqrt{\frac{1 - r_s/r_{\text{em}}}{1 - r_s/r_{\text{abs}}}} \tag{13.85}$$

between the emitted and the absorbed frequency. The shift is determined solely by the different radial positions.

To measure the gravitational frequency shift, one needs two positions in different heights, at which the frequencies of the electromagnetic radiation are measured. Let us set $r_{\text{abs}} := r_{\text{em}} + h$, where h is the height at which the absorber is located above the emitter. If we assume h to be small compared to r_{em}, we can perform a Taylor expansion around r_{em}, which we cut off, without making a big mistake, after the term linear in h:

$$\frac{\nu_{abs}}{\nu_{em}} \approx 1 - \frac{r_s h}{2 r_{em}(r_{em} - r_s)}. \tag{13.86}$$

If one assumes the radius of the Earth $r_{em} = r_{\oplus} = 6378$ km for the location of the emitter, then one can further simplify the term due to the small Schwarzschild radius of the Earth, $r_s = 9$ mm. We then obtain the estimate for the frequency shift

$$\frac{\Delta \nu}{\nu_{em}} = \frac{\nu_{abs} - \nu_{em}}{\nu_{em}} \approx -\frac{r_s h}{2 r_{em}^2} = -\frac{g h}{c^2}. \tag{13.87}$$

Here we have replaced the Schwarzschild radius $r_s = 2GM_{\oplus}/c^2$ of the Earth by the gravitational acceleration $g = GM_{\oplus}/r_{\oplus}^2$. The negative sign shows that a frequency emitted at the Earth's surface has decreased in the height h above the emitter by this amount. A smaller frequency results in a larger wavelength, which is why this effect is also called a *gravitational redshift*. A typical distance that can be realized in an experiment is approximately $h = 30$ m. This gives us the estimate

$$\frac{\Delta \nu}{\nu_{em}} \cong -3 \cdot 10^{-15}. \tag{13.88}$$

The frequency shift was determined using *Mössbauer spectroscopy*[9] of the iron isotope ^{57}Fe. The Mössbauer effect allows measurements of nuclear transitions with an accuracy in the range of the natural line width, on the order of $z \sim 10^{-15}$ [8]. Here the *redshift parameter z* is defined as

$$z = \frac{\Delta \lambda}{\lambda} = \frac{\lambda_{abs} - \lambda_{em}}{\lambda_{em}} = \frac{\nu_{em}}{\nu_{abs}} - 1 = \sqrt{\frac{1 - r_s/r_{abs}}{1 - r_s/r_{em}}} - 1, \tag{13.89}$$

where λ denotes the wavelength of the transition (cf. Eq. (8.8)). Figure 13.7 shows a sketch of the setup of an experiment to measure the gravitational redshift. A sample of ^{57}Fe in an excited state emits γ radiation with an energy of $E = 14.4$ keV.

A sample ^{57}Fe located at height h above the emitter cannot resonantly absorb the γ radiation due to the redshift. The difference in frequency can be compensated by moving the sample at a certain velocity v_R towards the emitter, and the resulting Doppler effect. In this way, the redshift can be measured via the value of v_R. In this setup the necessary velocity was $v_R \approx 7.5 \cdot 10^{-7}$ m s^{-1}. In 1960 Pound and

---

[9] Rudolf Mößbauer, 1929–2011, German physicist. Nobel Prize 1961 for the discovery of the effect named after him.

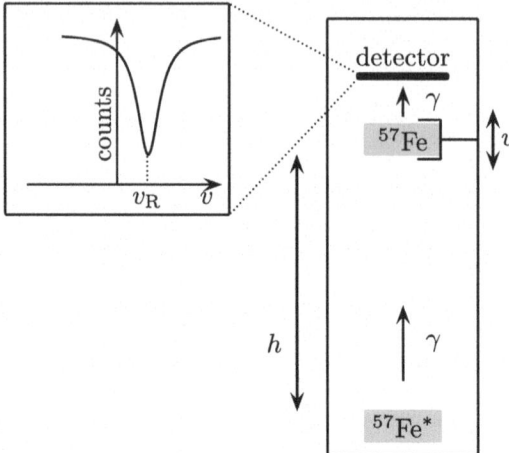

Fig. 13.7 Experimental proof of gravitational redshift: A sample of ^{57}Fe in an excited state emits γ radiation with an energy of $E = 14.4$ keV. A sample of the same isotope ^{57}Fe at a height of h above it cannot resonantly absorb the γ radiation due to the redshift. The difference in frequency can be compensated by moving the sample at a certain velocity v_R and the resulting Doppler effect. In this way, the redshift can be measured via the value of v_R. For a height of $h = 30$ m, possible in experiments, the velocity is $v_R \approx 7.5 \cdot 10^{-7}$m s^{-1}

Rebka [26] obtained, in their measurements, with a height of $h = 22.6$ m, a value of $z = (2.57 \pm 0.26) \cdot 10^{-15}$, which corresponds to a ratio of

$$\frac{\Delta \nu_{\mathrm{exp}}}{\Delta \nu_{\mathrm{theo}}} = 1.05 \pm 0.10. \tag{13.90}$$

Within the error bars, this value agrees well with the theoretical prediction. A more accurate measurement by Pound and Snider [27] in 1965 even led to a value of

$$\frac{\Delta \nu_{\mathrm{exp}}}{\Delta \nu_{\mathrm{theo}}} = 0.9990 \pm 0.0076. \tag{13.91}$$

13.4.2 Perihelion Precession

In Newtonian mechanics, the orbits of particles in the gravitational central potential $\phi_{\mathrm{m}}(r) = -GM/r$ are conic sections, e.g., Kepler ellipses. In this section we examine how the trajectories are affected by general relativity.

Setting Up the Equations of Motion

Because of the spherical symmetry, we can restrict ourselves to motions in the ($\vartheta = \pi/2$) plane. In Sect. 13.2.7 we have already written down the Lagrangian (13.34) for a general geodesic, which in the case of a timelike geodesic is

$$\mathcal{L} = -\left(1 - \frac{r_s}{r}\right) c^2 \dot{t}^2 + \frac{\dot{r}^2}{1 - r_s/r} + r^2 \dot{\varphi}^2 = -c^2. \tag{13.92}$$

The two constants of motion $k = (1 - r_s/r)c^2 \dot{t}$ and $h = r^2 \dot{\varphi}$ in (13.35a) remain unchanged. Inserting these in the Lagrangian, we obtain

$$\dot{r}^2 + \left(1 - \frac{r_s}{r}\right)\left(\frac{h^2}{r^2} + c^2\right) = \frac{k^2}{c^2}. \tag{13.93}$$

The dot in $\dot{q} = dq/d\tau$ denotes the derivative with respect to the proper time τ. Our goal is to determine the trajectory $r(\varphi)$. To this end, we first introduce the substitutions

$$r = \frac{1}{u}, \quad \frac{d\varphi}{d\tau} = hu^2, \quad \frac{dt}{d\tau} = \frac{k}{c^2(1 - r_s u)} \quad \text{and} \quad \frac{dr}{d\tau} = -h\frac{du}{d\varphi} \tag{13.94}$$

in (13.93). The expression for $dr/d\tau$ is obtained from

$$\frac{dr}{d\tau} = \frac{d}{d\tau}\frac{1}{u} = -\frac{1}{u^2}\frac{du}{d\tau} = -\frac{1}{u^2}\frac{du}{d\varphi}\frac{d\varphi}{d\tau} = -h\frac{du}{d\varphi}. \tag{13.95}$$

With the notation $u' = du/d\varphi$ the substitution leads to

$$h^2(u')^2 + (1 - r_s u)\left(h^2 u^2 + c^2\right) = \frac{k^2}{c^2}. \tag{13.96}$$

A further derivative with respect to φ yields

$$2h^2 u' u'' - 3h^2 r_s u^2 u' + 2h^2 u u' - c^2 r_s u' = 0. \tag{13.97}$$

Finally we multiply by $(-2h^2 u')^{-1}$, and obtain

$$u'' + u = \frac{c^2 r_s}{2h^2} + \underbrace{\frac{3}{2}r_s u^2}_{K}. \tag{13.98}$$

The term denoted by K is new in comparison with Newtonian mechanics.

Treatment Using Classical Perturbation Theory

For (planetary) motions, whose orbital dimension r is much larger than the Schwarzschild radius,

$$r_s u = \frac{r_s}{r} \ll 1, \tag{13.99}$$

the term K in (13.98) can be treated as a small perturbation. Therefore, we can seek a solution using classical perturbation theory.

For this purpose, we first write down the solution $u_0(\varphi)$ of the unperturbed equation of Newtonian mechanics, without the additional term of GR,

$$u_0'' + u_0 = \frac{c^2 r_s}{2h^2}. \tag{13.100}$$

The well-known solution is

$$u_0(\varphi) = \frac{c^2 r_s}{2h^2} [1 + \varepsilon \cos(\varphi)] \tag{13.101}$$

and describes, as already mentioned, conic sections. The prefactor can also be expressed in terms of the major semi-axis a and the eccentricity ε,

$$\frac{c^2 r_s}{2h^2} = \frac{1}{a\left(1 - \varepsilon^2\right)}. \tag{13.102}$$

For ellipses we had $0 \le \varepsilon < 1$, see Fig. 1.7.

We find an improved solution $u_1(\varphi)$ when we insert u_0 into the perturbing term and solve the resulting equation, which reads

$$u_1'' + u_1 \approx \frac{c^2 r_s}{2h^2} + \frac{3}{2} r_s u_0^2 = \frac{c^2 r_s}{2h^2} + \frac{3 c^4 r_s^3}{8h^4} \left[1 + 2\varepsilon \cos(\varphi) + \varepsilon^2 \cos^2(\varphi)\right]. \tag{13.103}$$

The solution of this equation is

$$u_1(\varphi) = u_0(\varphi) + \frac{3 c^4 r_s^3}{8h^4} \left\{ 1 + \underbrace{\varepsilon \varphi \sin(\varphi)}_{A} + \frac{\varepsilon^2}{2} \left[1 - \frac{1}{3} \cos(2\varphi)\right] \right\}. \tag{13.104}$$

The term denoted by A is proportional to φ, and grows with every full cycle of the planet. The other additional terms are proportional to ε^2, and therefore can be neglected for small ε. Inserting the expression for $u_0(\varphi)$, we find

$$u_1(\varphi) \approx \frac{c^2 r_s}{2h^2} \left[1 + \varepsilon \cos(\varphi) + \varepsilon \frac{3 c^2 r_s^2}{4h^2} \varphi \sin(\varphi) + \dots \right]$$

$$\approx \frac{c^2 r_s}{2h^2} \left\{ 1 + \varepsilon \cos\left[\left(1 - \frac{3 c^2 r_s^2}{4h^2}\right) \varphi\right] \right\}. \tag{13.105}$$

In the second step we have performed an "inverse Taylor expansion". Indeed, for small δ this expansion is

$$\cos[(1 - \delta)\varphi] \approx \cos(\varphi) + \sin(\varphi)\,\delta\varphi + \mathcal{O}(\delta^2). \tag{13.106}$$

Taking a look at the argument of the cosine in (13.105), we see that it becomes equal to 2π for

$$\varphi \approx 2\pi \left(1 + \frac{3c^2 r_s^2}{4h^2}\right), \tag{13.107}$$

since $(1 - \delta)^{-1} \approx 1 + \delta$. Thus, using (13.102), the result for the angle of the perihelion precession for non-relativistic velocities is

$$\Delta\varphi = 3\pi \frac{c^2 r_s^2}{2h^2} = 3\pi \frac{r_s}{a(1 - \varepsilon^2)}. \tag{13.108}$$

This angle is on the order of the ratio

$$\Delta\varphi \sim \frac{\text{Schwarzschild radius}}{\text{orbital radius}}. \tag{13.109}$$

In classical mechanics, only a pure Coulomb potential $- 1/r$ leads to *closed* periodic orbits. Any perturbation gives rise to a precession of the ellipse, and to Rosetta orbits. The perturbation of the planetary orbits caused by the interaction with the other planets was already known quantitatively in the nineteenth century. For Mercury, the calculations led to $531.5 \pm 0.3''$ per century. However, long-term observations had yielded a value of $574.3 \pm 0.4''$. Various explanations were proposed unsuccessfully to account for the difference of $42.7 \pm 0.5''$. For example, in 1859 the astronomer *Le Verrier*[10] postulated the presence of an additional planet, called Vulcan, within the orbit of Mercury, which should be responsible for the deviation.

Figure 13.8 illustrates the effect of perihelion precession. The P_i denote the consecutive points on the orbit closest to the Sun (perihelion), and the points A_i those farthest from the Sun (aphelion). Due to the reciprocal dependence on the orbital radius, the strongest perihelion precession can be expected for the planet closest to the Sun, Mercury. It has the orbital data

$$a_{\text{Mercury}} = 57.91 \cdot 10^6 \text{ km} = 0.387 \text{ AU} \quad \text{and} \quad \varepsilon_{\text{Mercury}} = 0.206, \tag{13.110}$$

with the astronomical unit from (1.57). For comparison: The orbital data for the Earth is $a_{\text{Earth}} = 149.6 \times 10^6$ km and $\varepsilon_{\text{Earth}} = 0.0167$. The perihelion precession of

---

[10] Urbain Le Verrier, 1811–1877, French mathematician and astronomer. His calculations had led to the discovery of the planet Neptune.

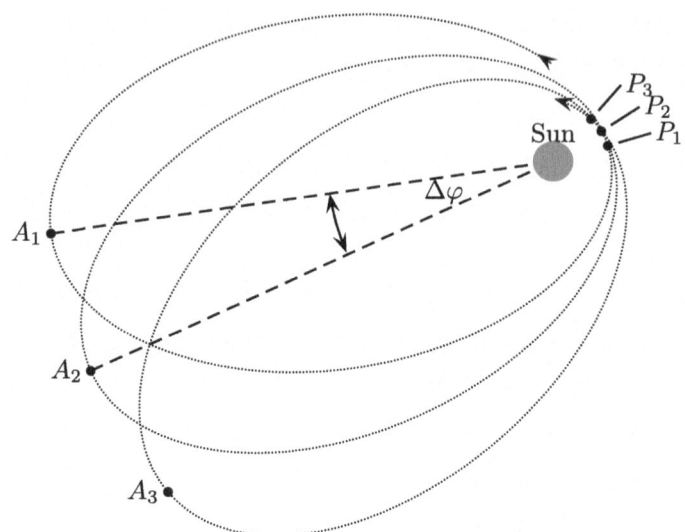

Fig. 13.8 Effect of the perihelion precession: As a consequence of deviations from a pure $1/r$-potential, the orbit of the planet is not closed. The points P_i are the consecutive points closest to the Sun (perihelion), the points A_i the farthest from the Sun (aphelion)

Mercury per century predicted by general relativity is

$$\Delta\varphi_{\text{Mercury}}\Big|_{100\,\text{years}} = 43.03''. \tag{13.111}$$

Thus the missing difference to the observations was completely explained by general relativity!

For Venus the perihelion precession per century is $8.6''$, and for the Earth only $3.8''$. Compared with these small angles, the corresponding precession of the periastron in binary pulsar systems is much larger, as we shall see in Sect. 21.5.2. In the binary pulsar system PSR B1913+16, see Table 21.1, the average periastron precession amounts to about $4.23°$ per year!

Einstein's explanation of the difference between the observed perihelion precession of Mercury, and that predicted by Newton's theory, was the first great triumph of general relativity. When Einstein made his calculations, the Schwarzschild metric was yet unknown. Einstein used an approximate metric for weak fields. He expressed his delight at this great success in a letter to *Ehrenfest*[11] in these words:[12]

> I was stunned with excitement for a few days.

---

[11] Paul Ehrenfest, 1880–1933, Austrian physicist, best known for the Ehrenfest theorem.

[12] Quote from Albert Einstein's letter of January 17, 1916 to Paul Ehrenfest, in "The Collected Papers" Vol. 8, Part A: The Berlin Years: Correspondence 1914–1917, p. 244, https://einsteinpapers.press.princeton.edu.

13.4.3 Gravitational Deflection of Light

In this section we discuss the deflection of light in a gravitational field. In analogy with the study of the perihelion precession, we first set up the equations of motion for lightlike geodesics, and solve them within the framework of perturbation theory. Then we present an alternative calculation based on a version of the Schwarzschild metric in isotropic coordinates.

Derivation Using Perturbation Theory

The starting point of our investigation is the Lagrangian (13.34) for lightlike geodesics,

$$\mathcal{L} = -\left(1 - \frac{r_s}{r}\right)c^2\dot{t}^2 + \frac{\dot{r}^2}{1 - r_s/r} + r^2\dot{\varphi}^2 = 0, \tag{13.112}$$

where the dot $\dot{q} = dq/d\lambda$ denotes the derivative with respect to the affine parameter λ. Inserting the constants of motion k and h from (13.35a) and introducing a substitution analogous to (13.95), leads to the differential equation

$$h^2(u')^2 + h^2u^2(1 - r_s u) - \frac{k^2}{c^2} = 0. \tag{13.113}$$

Another differentiation with respect to φ finally yields

$$u'' + u = \frac{3}{2}r_s u^2, \tag{13.114}$$

with the small perturbation $\frac{3}{2}r_s u^2$. Again we make use of classical perturbation theory, and first solve the unperturbed equation $u_0'' + u_0 = 0$. The solution is

$$u_0(\varphi) = \frac{1}{R}\sin(\varphi), \quad \text{or} \quad r(\varphi) = \frac{R}{\sin(\varphi)}, \tag{13.115}$$

which corresponds to a straight line. In (13.115), R is the minimal distance from the center, see Fig. 13.9. Inserting u_0 into the perturbing term leads to the equation

$$u_1'' + u_1 = \frac{3r_s}{2R^2}\sin^2(\varphi), \tag{13.116}$$

with the solution

$$u_1(\varphi) = \frac{1}{R}\sin(\varphi) + \frac{3r_s}{4R^2}\left[1 + \frac{1}{3}\cos(2\varphi)\right]. \tag{13.117}$$

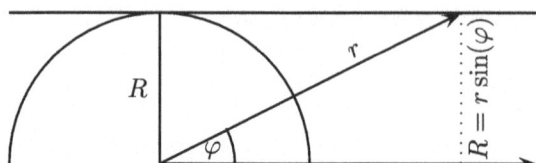

Fig. 13.9 Sketch of the deflection of light in a gravitational field: In polar coordinates a straight line in the plane ($\vartheta = \pi/2$) is described by the equation $r(\varphi) = R/\sin(\varphi)$ in spherical polar coordinates

In the asymptotic limit $r \to \infty$ we must have $u = 0$, which means that, for a ray of light coming from infinity, the condition

$$\frac{1}{R}\sin(\varphi) + \frac{3r_s}{4R^2}\left[1 + \frac{1}{3}\cos(2\varphi)\right] = 0 \qquad (13.118)$$

must be fulfilled. For $\varphi \approx 0$ we can use the approximations $\sin(\varphi) \approx \varphi$ and $\cos(2\varphi) \approx 1$, to obtain

$$\frac{\varphi}{R} + \frac{3r_s}{4R^2}\left(1 + \frac{1}{3}\right) = 0, \quad \text{i.e.} \quad \varphi_\infty \approx -\frac{r_s}{R}. \qquad (13.119)$$

The total angle of deflection is

$$\alpha = |2\varphi_\infty| = 2\frac{r_s}{R}. \qquad (13.120)$$

Result According to Newton's Theory

One can also calculate a value for the deflection of light in Newtonian mechanics by considering light as a particle carrying a momentum, and moving at the speed of light. In analogy with (13.101), the undisturbed trajectory of a ray of light is described by the equation

$$\frac{1}{r(\varphi)} = u(\varphi) = \frac{c^2 r_s}{2h^2}\left[1 + \varepsilon \sin(\varphi)\right]. \qquad (13.121)$$

Since light always moves at the speed of light c, we have the condition $R\omega = c$ at the minimal distance R from the center. Here $\omega = \mathrm{d}\varphi/\mathrm{d}t$ is the angular velocity of the "light particle" as it passes through the point of minimal distance. From this follows

$$h = R^2\frac{\mathrm{d}\varphi}{\mathrm{d}t} = R^2\omega = Rc. \qquad (13.122)$$

Inserting in (13.121) yields $R^{-1} = r_s(1 + \varepsilon)/(2R^2)$ at $\varphi = \pi/2$. Solving with respect to the excentricity ε one obtains

$$\varepsilon = 2\frac{R}{r_s} - 1 \approx 2\frac{R}{r_s}, \tag{13.123}$$

because of $R \gg r_s$. If we let go $r \to \infty$ in (13.121), we obtain

$$0 = \frac{r_s}{2R^2}[1 + \varepsilon \sin(\varphi_\infty)] \tag{13.124}$$

and, thus, for $\varepsilon \gg 1$, which is evident from (13.123),

$$\varphi_\infty = -\frac{1}{\varepsilon} = -\frac{r_s}{2R}. \tag{13.125}$$

Therefore, the total angle of deflection according to the Newtonian calculation is

$$\alpha_{\text{Newton}} = 2|\varphi_\infty| = \frac{r_s}{R}. \tag{13.126}$$

This is exactly half the value of the result of the relativistic calculation.

Isotropic Schwarzschild Metric
An alternative way for the quantitative investigation of gravitational light deflection is provided by the *isotropic Schwarzschild metric*. To introduce this metric, we define a new radial coordinate $\tilde{r}$,

$$r = \tilde{r}\left(1 + \frac{r_s}{4\tilde{r}}\right)^2 \quad \text{with} \quad dr = \left(1 + \frac{r_s}{4\tilde{r}}\right)\left(1 - \frac{r_s}{4\tilde{r}}\right)d\tilde{r}. \tag{13.127}$$

The line element assumes the form

$$ds^2 = -\left(\frac{1 - r_s/(4\tilde{r})}{1 + r_s/(4\tilde{r})}\right)^2 c^2 dt^2 + \left(1 + \frac{r_s}{4\tilde{r}}\right)^4 d\tilde{x}^2, \tag{13.128}$$

with the differential $d\tilde{x}^2 = d\tilde{r}^2 + \tilde{r}^2(d\vartheta^2 + \sin^2(\vartheta)d\varphi^2)$, and the coordinate transformation $\tilde{x} = (\tilde{r}\sin(\vartheta)\cos(\varphi), \tilde{r}\sin(\vartheta)\sin(\varphi), \tilde{r}\cos(\vartheta))^T$. For photons ($ds^2 = 0$) we then have the relation

$$\left(\frac{1 - r_s/(4\tilde{r})}{1 + r_s/(4)\tilde{r}}\right)^2 c^2 dt^2 = \left(1 + \frac{r_s}{4\tilde{r}}\right)^4 d\tilde{x}^2. \tag{13.129}$$

We rearrange this equation, and obtain for $\tilde{r} \gg r_s$, i.e., also for $\tilde{r} \approx r$,

$$\left|\frac{d\tilde{x}}{dt}\right| = \frac{1 - r_s/(4\tilde{r})}{[1 + r_s/(4\tilde{r})]^3}c \approx \left(1 - \frac{r_s}{\tilde{r}}\right)c = v_{\text{light}} < c. \tag{13.130}$$

Thus the velocity of light in the isotropic Schwarzschild metric is smaller than the speed of light in the Minkowski metric. Here one point must be clarified: The statement just given refers to a *global* property, such as measuring the transit time of light to another planet. *Locally* every observer will always measure the speed of light c!

Formally, we can take into account the fact of the globally lower velocity of light by introducing a position-dependent refractive index:

$$\frac{c}{v_{\text{light}}} = n \approx 1 + \frac{r_{\text{s}}}{r}. \tag{13.131}$$

Thus light is "diffracted" in a gravitational field, in the same way as it is diffracted by a lens. From geometrical optics we know the *eikonal equation* for the trajectories of light:

$$\frac{\mathrm{d}}{\mathrm{d}s_0}(ns_0) = \nabla n, \tag{13.132}$$

where s_0 is the tangent vector to the trajectory of the ray of light, see Fig. 13.10. If α is the angle by which the light is deflected, and R the "impact parameter" of the light relative to a scatterer (i.e., a gravitational field), we find, after a short calculation

$$\alpha = \frac{2r_{\text{s}}}{R}. \tag{13.133}$$

For the Sun we have $R = R_\odot \approx 7 \cdot 10^5$ km and $r_{\text{s}} \approx 3$ km, and therefore

$$\alpha_\odot \approx 1.75''. \tag{13.134}$$

Due to light deflection, stars in the sky that are very close to the Sun appear slightly further away from the Sun than their actual positions are, see Fig. 13.11. But since these stars are normally outshone by the Sun, this effect cannot be observed.

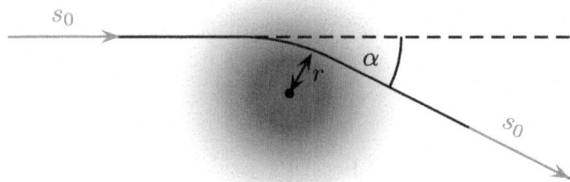

Fig. 13.10 The effect of masses on trajectories of light can be described by an apparent, position-dependent index of refraction of the spacetime. The change in the angle of the direction of the tangent vector s_0 to the trajectory is given by the eikonal equation

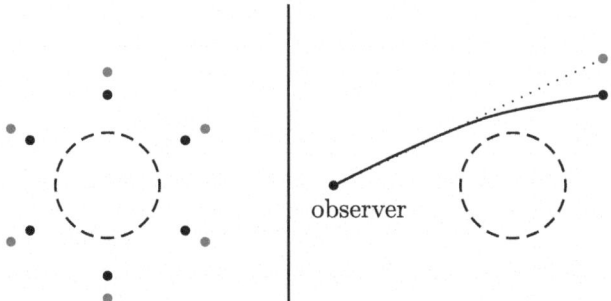

Fig. 13.11 During a solar eclipse, due to the deflection of light, stars close to the Sun appear further away from the Sun (*gray*) than their actual positions are (*black*)

However, if the Sun is covered during a solar eclipse, these stars become visible, and the apparent change of the positions of the stars can be determined.

Scientists waited for an opportunity to check whether the displacement of the positions of the stars in a solar eclipse was given by the value according to Newton's theory, or by double that value, as predicted by general relativity. The opportunity presented itself by the solar eclipse pre-calculated to happen on 29 May 1919. The eclipse was a stroke of luck in two respects: On the one hand, it happened when the Moon was close to its perigee, which meant that the Moon was big, and therefore the eclipse lasted for almost 7 minutes. On the other hand, the Sun would stand in the brightly shining star cluster of the Hyades, located in the Taurus constellation, and the individual stars close to the Sun showed up clearly. Astronomers identified the island of Principe off the coast of Spanish Guinea and the village of Sobral in northern Brazil as ideal observation sites. Half a year before the total solar eclipse, the astronomers photographed the region of the sky in which the darkened Sun would be observed on 29 May 1919. The campaign in Brazil was led by the British astronomer *Crommelin*,[13] the campaign in Principe by *Eddington*. Back in England the two groups analyzed their glass photo plates. On 6 November 1919, at a joint meeting of the Royal Society and the Royal Astronomical Society, Crommelin presented the final results: The deviation at the edge of the Sun was $1.98'' \pm 0.18''$ for one telescope, and $1.60'' \pm 0.31''$ for the other [6], confirming Einstein's result.

On 7 November 1919, The Times of London published an article with the headline

Scientific Revolution. New theory of the universe. Newton's idea overthrown.

Two days later, on 9 November 1919, the New York Times wrote:

Lights all askew in the Heavens, Stars Not Where They Seemed or Were Calculated to Be, But Nobody Need Worry –Men of science more or less agog over results of eclipse observations – Einstein Theory triumphs.

---

[13] Andrew Claude de la Cherois Crommelin, 1865–1939, British astronomer.

Because there had always been doubts about the correctness of the values, Eddington's photo plates were reanalyzed in 1979 at the Royal Greenwich Observatory using modern equipment, with the result $1.90'' \pm 0.11''$. In the solar eclipse observations, general relativity had stood another brilliant test!

The effect of light deflection is also known as *gravitational lensing*, since the massive object, in this case the Sun, acts in a way similar to an optical lens. However, there is one important difference. In an optical lens, the light is deflected more strongly when it passes through the lens further away from the optical axis. By contrast, the deflection of light in the gravitational field becomes smaller when the light passes further away from the massive object. A "gravitational lens" therefore has no focal point.

A very good and easy-to-understand treatment of the deflection of light in a gravitational field, including the historical perspective, can be found in Dominik [5]. As part of the 100th anniversary of the discovery of light deflection by the Sun, a short planetarium full-dome video was created that demonstrates light deflection, and shows what the situation would have looked like if a black hole, instead of the Sun, had passed in front of the Hyades and Pleiades star clusters [4].

Light Deflection Outside the Solar System

Using the most powerful telescopes and astronomical satellites, it is possible today to observe light deflection also outside the solar system. If light from a far distant galaxy passes by a very massive object, e.g., a cluster of galaxies, before reaching the Earth, the effect of light deflection also occurs. Because of the much bigger mass of the cluster, typically on the order of 10^{14} to 10^{15} solar masses, the deflection of light can be substantially stronger. Light emitted from the galaxy in different directions can be deflected such that it arrives in the telescopes also from different directions. In the ideal case of a spherically symmetric mass distribution, the observed object can be distorted into the shape of a ring, viz., an *Einstein ring*, see Sect. 13.2.8.

Quantitative measurements of this effect can in turn provide information on the mass of the object causing the distortion. The comparison with calculations based on the visible mass distribution reveals that much more mass is needed for the observed strength of deflection than is visible. This is one of the evidences pointing to the existence of dark matter (cf. Sect. 26.4).

13.4.4 Time Delay

Apart from light deflection, a ray of light also experiences a time delay when crossing a gravitational field. We will illustrate this effect using the example of radio waves reflected from the planets Mercury and Venus. For a maximum effect, the planets should be in conjunction. The planets and the Sun are in conjunction when, simply speaking, they lie on a straight line.

We can derive the time delay easily using the expression (13.130) for the position-dependent velocity of light in the isotropic Schwarzschild metric. We

Fig. 13.12 Sketch for the time delay in a gravitational field: To determine the delay, the expression $r(l)$ must be known, which can be derived in an elementary way

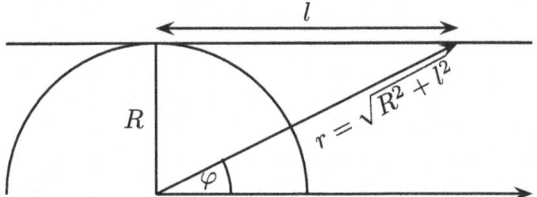

consider a ray of electromagnetic radiation that passes by the Sun at very close distance. With $dt = dl/v_{\text{light}}(r)$ we have

$$t = \int \frac{dl}{v_{\text{light}}(r(l))} = \int \frac{dl}{c\,(1 - r_{\text{s}}/r(l))}, \tag{13.135}$$

where $r(l) = \sqrt{R^2 + l^2}$, see Fig. 13.12. Because of $r \gg r_{\text{s}}$, we can expand the expression (13.135), and obtain

$$t = \int \frac{dl}{c} + \frac{r_{\text{s}}}{c} \int \frac{dl}{r} = t_{\text{Newton}} + \Delta t. \tag{13.136}$$

Thus the transit time of the ray is the sum of the transit time expected in Newton's theory, and a correction term. For an electromagnetic signal sent from the Earth to a planet and returning to Earth, we find, with the distances $l_{\oplus}$ and l_{P} of the Earth and the planet from the Sun, a time delay of

$$\Delta t = \frac{2r_{\text{s}}}{c} \left[\int_0^{l_{\text{P}}} \frac{dl}{\sqrt{R^2 + l^2}} + \int_0^{l_{\oplus}} \frac{dl}{\sqrt{R^2 + l^2}} \right] \tag{13.137}$$

$$= \frac{2r_{\text{s}}}{c} \left[\text{arsinh} \left(\frac{l_{\text{P}}}{R} \right) + \text{arsinh} \left(\frac{l_{\oplus}}{R} \right) \right] \approx \frac{2r_{\text{s}}}{c} \ln \left(\frac{4 l_{\oplus} l_{\text{P}}}{R^2} \right).$$

One can also calculate the time delay within Newton's theory for a hypothetical particle of light carrying momentum and moving at the speed of light. The two results can be summarized as follows:

$$\Delta t = (1 + \xi) \frac{r_{\text{s}}}{c} \ln \left(\frac{4 r_1 r_2}{b^2} \right), \tag{13.138}$$

with the distances r_1 and r_2 of the Earth and the planet from the Sun, and the impact parameter b. Since b changes with time T, one can finally write down a function $\Delta t(T)$, see Fig. 13.13. The calculation with general relativity leads to $\xi = 1$, the Newtonian calculation to $\xi = 0$. Therefore, we have again only half the relativistic effect in Newton's theory, as it was already the case with light deflection.

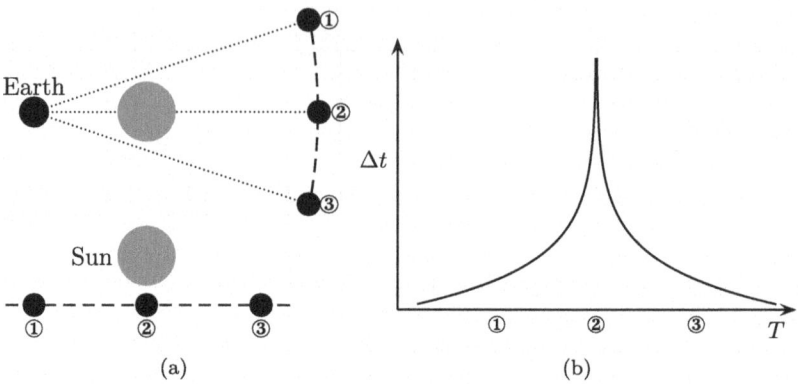

Fig. 13.13 Time delay of light when crossing the gravitational field of the Sun: (**a**) In the three constellations ①–③ the impact parameter b is changing, and therefore also the time delay (*top*: view onto the orbital plane of the Earth, *bottom*: view of the position of the planet from the Earth's perspective). (**b**) gives a schematic sketch of the time delay as a function of time as it would be expected in an actual measurement, see also Fig. 21.17

In addition to the time delay, there also occurs a Doppler shift of the signal [2]:

$$z_{\mathrm{gr}} = \frac{\Delta v}{v} = \frac{\mathrm{d}\Delta t}{\mathrm{d}t} = -2(1+\xi)\frac{r_\mathrm{s}}{c}\frac{1}{b}\frac{\mathrm{d}b}{\mathrm{d}t}. \tag{13.139}$$

In the situation② in Fig. 13.13, the transit time of the radar signal is longer than in Newton's theory because of the refractive index effect. One finds for example

$$\Delta t = 240\,\mu\mathrm{s} \quad \text{or} \quad c\,\Delta t = 36\,\mathrm{km}. \tag{13.140}$$

In an experiment carried out by Shapiro [29] in 1968 using radar waves reflected by the atmosphere of Venus, this time delay could be confirmed with an accuracy of 3 %, which corresponds to measuring the distance from Earth to Venus with an accuracy of 1 km.

Measurements Using the Cassini Spacecraft

In 2002, a much more precise measurements could be performed with the help of the Cassini spacecraft. These measurements yielded a value of

$$\xi = 1 + (2.1 \pm 2.3) \cdot 10^{-5}. \tag{13.141}$$

On its way to Saturn, the probe was in conjunction with the Sun around 6 and 7 July 2002, i.e., at maximum distance from the Earth behind the Sun, though not exactly in the orbital plane of the Earth, which meant that it was not occluded by the Sun. In contrast to the measurements with Venus, in this case the signal was not simply reflected. An electromagnetic signal was sent to the probe which actively triggered a transmitter on the probe to send back a signal. The accuracy was so high that even the Doppler shift z_{gr} from (13.139) could be measured very precisely, and thus it was possible to reach the much higher precision that is evident from the value of the parameter ξ. In fact, this constitutes the most accurate test of general relativity in the solar system so far.

13.4.5 Geodesic Precession

In Sect. 11.4.3 we have seen that, in order to describe the motion of an object along a timelike trajectory in a curved spacetime, we have to use the equation of parallel transport (11.137). In the case of a circular timelike orbit in the Schwarzschild metric, this has the consequence that after the completion of one full orbital cycle the local frame of reference no longer has the same orientation as at the start of the motion.

This *geodesic precession* is shown for the last stable orbit of the Schwarzschild metric in Fig. 13.14. The black arrow represents the $\underline{e}_{(1)}(\tau)$ direction of the

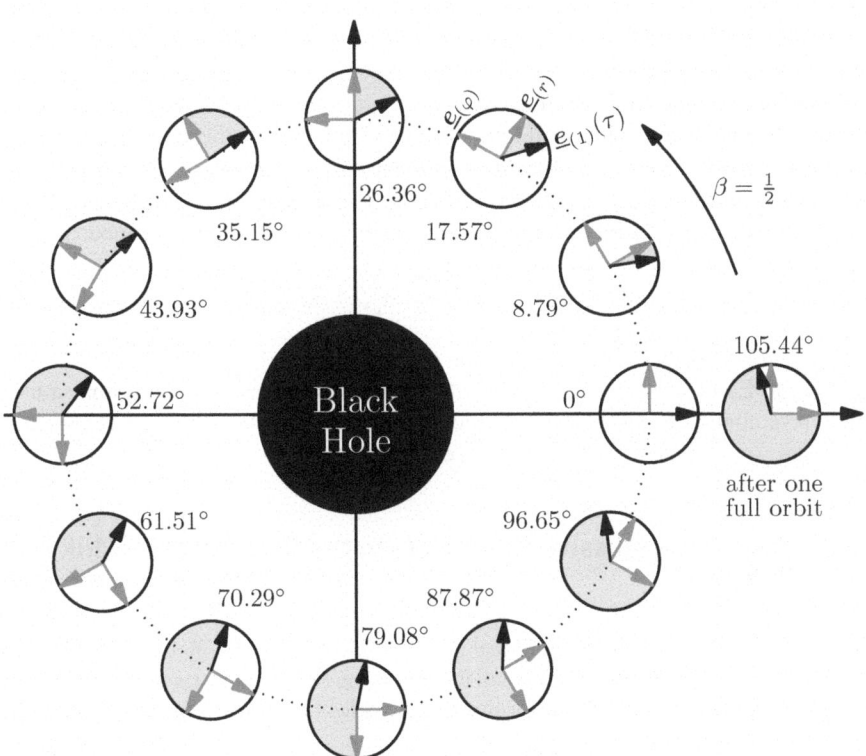

Fig. 13.14 Geodesic precession along the last stable orbit in the Schwarzschild spacetime. The angle of rotation at the current position is measured with respect to the "sky of infinitely distant fixed stars"

comoving local tetrad $\mathcal{L}_b$ until the proper time τ, whereas the gray arrows indicate the $\underline{e}_{(r)}$ and $\underline{e}_{(\varphi)}$ direction of the reference tetrad $\mathcal{L}_r$ at the current position. The current angle of rotation of $\underline{e}_{(1)}(\tau)$ is not the angle with respect to the radial direction $\underline{e}_{(r)}$ of the frame of reference, as indicated by the circular segments shaded in gray in Fig. 13.14, but is referred to "the sky of infinitely distant fixed stars". Therefore, after one full cycle the angle of rotation is $\alpha \approx 105.44°$ between the initial direction $\underline{e}_{(1)}(\tau = 0)$ and the current direction is $\underline{e}_{(1)}(\tau = \tau_{2\pi})$. The orbital period $\tau_{2\pi} = 2\pi/\omega$ then is obtained from the angular velocity ω, with respect to the proper time τ of the object, see Eq. (13.45).

13.4.6 Global Positioning System

For the operation of the *Global Positioning System* (*GPS*), effects of both special and general relativity are important. *GPS* consists of 24 satellites, out of which four satellites each circle the Earth in six orbital planes inclined at angles of 55° with respect to the equatorial plane, plus a few spare satellites. The satellites move in circular orbits in a height of approximately 20,200 km above the Earth's surface, where they orbit the Earth twice a day. Because of the big height above the Earth, and the ensuing weaker gravitational field, clocks on the satellites are ahead by roughly 45 μs per day.

However, because of the orbital velocity of approximately 3.9 km s^{-1}, the clocks are also late by 7 μs per day. Taking both effects together, the total difference in time with respect to clocks on Earth amounts to 38 μs. Since *GPS* determines the positions of users via signals of electromagnetic radiation, in a direct comparison with the clock of a user on Earth this would lead to an error of approximately

$$38 \text{ μs} \cdot 299{,}792{,}458 \text{ m s}^{-1} \approx 11.4 \text{ km} \tag{13.142}$$

per day! To avoid this, the clocks on the satellites are built in such a way that they are late on Earth by exactly this amount of time. For an exact positioning, a user must have contact to four *GPS* satellites in order to calculate the four parameters necessary: one time and three position coordinates.

13.5   The Supermassive Black Hole in the Center of the Milky Way

Two groups of astronomers, one led by *Reinhard Genzel*[14] at the Max Planck Institute for Extraterrestrial Physics in Garching [11, 12], the other by *Andrea Ghez*[15] at the University of California, Los Angeles [10], have been measuring the

---

[14] Reinhard Genzel, ∗1952, German astrophysicist.

[15] Andrea Ghez, ∗1965, US-American astronomer.

orbits of stars around the center of the Milky Way in detail since 1994. They have not only proved the existence of a massive black hole in the center of the Milky Way, but also determined its mass and its distance with high precision.

The masses obtained are consistent with each other: $(4.31 \pm 0.36) \times 10^6$ solar masses [11] and $(4.5 \pm 0.4) \times 10^6$ solar masses [10], respectively. The distance to the center of the galactic black hole could be determined with an uncertainty of only 0.3 % to $R_0 = 8178$ pc (26,670 light years) [16]. For these discoveries, Genzel and Ghez shared half of the 2020 Nobel Prize in Physics [18, 23].

The black hole at the center of the Milky Way is hidden behind dense curtains of gas and interstellar dust. This prevents its observation at *optical* wavelengths, since only about one in 10^9 photons can penetrate the dust towards the line of sight to Earth. For infrared and radio radiation, the dust is 10 times more transmissive.

Accordingly, the search for the black hole initially began with radio telescopes. In 1974, *Balick*[16] and *Brown*[17] discovered a rather intense, very compact source of non-thermal synchrotron radiation at wavelengths of 3.7 cm and 11 cm in the constellation Sagittarius, whose position coincided within a few light years with the conjectured center of the Milky Way [1]. They named the source Sagitarius A* (Sgr A*, pronounced "SAJ AY star"). The name came about because Brown presented a theory in which he interpreted Sgr A* as an excited radio source and, by analogy with labeling excited states in atomic physics, added an asterisk "*". The theory proved wrong, but the name remained.

The groups of Genzel and Ghez made their observations at near-infrared wavelengths centered around $\lambda = 2.2\,\mu m$. The GRAVITY collaboration [13] led by Genzel used the *Very Large Telescope* (VLT) of the European Southern Observatory (ESO) in the Atacama Desert in northern Chile. It consists of four identical 8.2 m telescopes that can be interconnected for interference, resulting in a "super telescope" with an effective diameter of 130 m, the *Very Large Telescope Interferometer* (VLTI). Ghez's group observed from the Keck Observatory in Hawaii with the two telescopes *Keck-I* and *Keck-II*, each with mirror diameters of 10.4 m. Interconnected for interference, this resulted in a telescope with an effective diameter of 85 m.

The high spatial resolution required for tracking stars in the vicinity of Sgr A* could be achieved using the technique of "*adaptive optics*". Turbulence in the Earth's atmosphere smears the trajectories of photons coming from stars on a timescale of less than about a second ("the twinkling of the stars"). To compensate for this, the adaptive optics technique uses either a bright reference star in the vicinity of the object to be observed, or an artificial "star" created by exciting sodium atoms in the upper atmosphere (90 km) with a laser beam. A wave front sensor measures the distortion of the wave fronts of the reference object caused by the turbulence in the atmosphere, a computer calculates the necessary reconstruction of the wave fronts, this information is passed on in a feedback loop to a deformable

---

[16] Bruce Balick, ⋆1943, US-American astronomer.

[17] Robert Lamme Brown, 1943–2014, US-American radio astronomer.

second mirror of the telescope, which is "bent" in real time in such a way that the disturbances in the image of the observed object caused by atmospheric turbulence can be completely corrected. This enabled long exposure times and delivered images as sharp as if they had been taken from space. This technical revolution also made it possible to measure the *spectra* of the stars. From the position of the spectral lines, conclusions could be drawn as to the chemical composition of the atmosphere of the stars. Furthermore, the radial velocities of the stars could be determined via the Doppler shift of the lines.

One of the stars tracked over the years (labeled S2 by Genzel's group) stands out among the observed stars. Its orbital period around Sgr A* is just under 16 years. For comparison, it takes the Sun just over 200 million years for a full orbit around the galactic center. S2 has a highly elliptical orbit with a numerical eccentricity of $\varepsilon = 0.88$. The passage of S2 through the pericenter[18] was measured twice, namely in 2002, and with extreme precision in May 2018. The distance of S2 to Sgr A* in the pericenter is only 17 light hours (or 120 AU). The orbit is inclined at an angle of $46°$ relative to the celestial plane.[19]

Figure 13.15 summarizes 26 years of observing the orbit of S2 around Sgr A*. The left figure shows the results for the orbital ellipse of S2 obtained using the ESO telescopes. The observations are accurate enough to determine the change in the position of S2 from night to night. The passage through the pericenter in May 2018 is shown enlarged in the figure at the bottom right. The best fit to the orbit is represented by the solid curve. The upper right figure shows the results of spectroscopic observations by both Keck and ESO to measure the *radial velocities* of S2 projected on the plane of the sky. It can be seen that in the pericenter S2 is approaching the Earth at a speed of over 4000 km/s, while in the apocenter it is moving away at 2000 km/s.

From the inclination of the orbital plane, conclusions can be drawn as to the three-dimensional velocities. The total velocity in the pericenter is about 7650 km/s, which corresponds to a velocity parameter of $\beta = v/c = 2.55 \times 10^{-2}$. This means that the transverse Doppler effect of special relativity must be taken into account when analyzing the observational data. As S2 approaches the pericenter, it is increasingly exposed to the strong gravitational field of the galactic black hole. This induces a gravitational redshift of the infrared radiation that also had to be taken into account in order to obtain the best fit for the orbit of S2 in the vicinity of the pericenter shown in Fig. 13.15.

Since 2016, the GRAVITY instrument on the Very Large Telescope has once again taken infrared astronomy a giant leap forward. As already mentioned, the signals of all four VLT telescopes can be combined interferometrically to form a supertelescope with an effective diameter of 130 m. This makes it possible to achieve an angular resolution that is about 16 times better than that of a single tele-

---

[18] The pericenter is the point on the elliptical orbit closest to the galactic center.

[19] The browser based app https://astro-apps.org/S2orbit makes it possible to determine the orbital parameters interactively using the measured data.

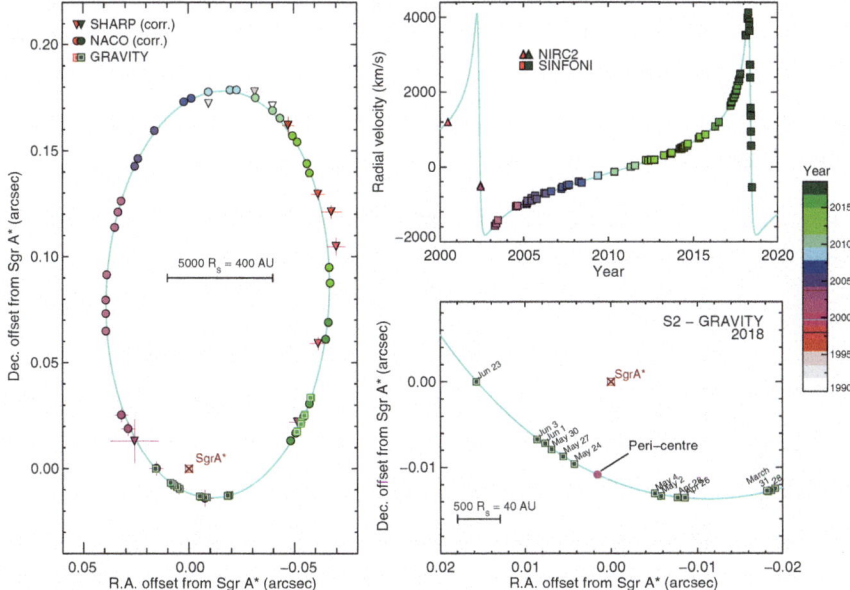

Fig. 13.15 Summary of the observational results of the orbital tracking of the star S2 around Sgr A* from 1992 to 2018 [14]. The color scale indicates the year of observation. Left: The orbit of S2 relative to the compact radio source Sgr A* (brown cross at origin) projected on the plane of the sky. The radial Doppler velocity (top right) varies between −2000 and 4000 km/s. The orbit of S2 during the passage of the pericenter in May 2018 was determined with great precision by GRAVITY (bottom right). To find the best fit for the orbit of S2 (solid curve), effects of both special and general relativity had to be taken into account. Note the length scale, which is 5000 times the Schwarzschild radius of Sgr A*, or about 400 astronomical units. Credit: GRAVITY collaboration [14], reproduced with permission ©ESO

scope. In this way, general relativistic effects such as the gravitational redshift [14] and even the Schwarzschild precession could be detected. The pericenter of S2 precesses by 12 arcseconds per orbit [17].

This is exactly the value expected in the Schwarzschild metric for a given mass of about 4 million solar masses. In addition, flares of infrared radiation originating from the accretion disk of Sgr A* [15] could be observed with GRAVITY. A belt of gas about 10 light minutes across orbits Sgr A* near the last stable orbit (known from Sect. 13.2.7). The gas whirls around the black hole at a speed of 30 % of the speed of light. Material entering below the last stable orbit falls towards the black hole and creates the flares.

A group at the University of Cologne has discovered stars, using the VLT, with an even shorter orbital period around Sgr A* [25]. The first, S62, takes 9.9 years, and the very fastest, S4711 (named after the perfume from Cologne) only 7.6 years [25]. Its velocity in the pericenter is 24,000 km/s, corresponding to $\beta = 0.08$. Another star, S4714, has an orbital period of 12 years, its orbital ellipse shows the extreme eccentricity of 0.985. These stars, with even shorter orbital periods, are other ideal candidates for observing gravitational redshift and pericenter precession in the Schwarzschild metric.

13.6   Exercises

13.1. Doppler effect in the Pound-Rebka experiment In Sect. 13.4.1 we discussed the Pound-Rebka experiment. We arrived at a velocity on the order of $v \approx 7.5 \cdot 10^{-7}$ m s^{-1}, with which the absorber must be moved to attain resonant absorption. Confirm this order of magnitude.

13.2. Time differences in GPS In Sect. 13.4.6 we gave the time difference by which a clock on the GPS-Satellite would run faster than a clock on the surface of the Earth. It is composed of contributions of time dilation in both special and general relativity. Confirm the values given in Sect. 13.4.6.

13.3. Velocity of an observer in free fall onto a black hole In Sect. 13.3.1 we discussed the case of an observer in free fall onto a black hole and determined his current position as a function of his proper time. Calculate the velocity β at his current position r referred to the local tetrad (13.37) of an observer at rest at that position.

13.4. Geodesic precession In Sect. 13.4.5 we have discussed the geodesic precession of a vector on the last stable orbit. Calculate the geodesic precession for a timelike circular orbit with arbitrary radius.

References

1. Balick, B., Brown, R. L.: Intense sub-arcsecond structure in the Galactic center, The Astrophysical Journal **194**, 265 (1974)
2. Bertotti, B., Iess, L., Tortora, P.: A test of general relativity using radio links with the Cassini spacecraft. Nature **425**, 374–376 (2003)
3. Birkhoff, G.D.: Relativity and Modern Physics. Harvard University Press, Cambridge (1923)
4. Checking up on Einstein – The Solar Eclipse of May 29, 1919, a 4 minute planetarium fulldome video https://www.eso.org/public/videos/checking_up_on_einstein_domemaster_4k/
5. Dominik, M.: The gravitational bending of light by stars: a continuing story of curiosity, scepticism, surprise, and fascination. Gen. Relativ. Gravit. **43**, 989–1006 (2011)
6. Dyson, F.W., Eddington, A.S., Davidson, C.A.: A determination of the deflection of light by the Sun's gravitational field, from observations made at the total eclipse of may 29, 1919. Phil. Trans. R. Soc. Lond. A **220**, 291–333 (1920)
7. Eddington, A.S.: A comparison of Whitehead's and Einstein's formulæ. Nature **113**, 192 (1924)
8. Eyges, L.: Physics of the Mössbauer Effect. American Journal of Physics. **33** (10), 790–802 (1965)
9. Finkelstein, D.: Past-future asymmetry of the gravitational field of a point particle. Phys. Rev. **110**(4), 965–967 (1958)

10. Ghez, A. M. et al.: Measuring distance and properties of the Milky Way's central supermassive black hole. The Astrophysical Journal **689**, 1044 (2008)
11. Gillessen, S., Eisenhauer, F., Trippe, S., Alexander, T., Genzel, R., Martins, F., Ott, T.: Monitoring stellar orbits around the massive black hole in the Galactic center. The Astrophysical Journal **692**, 1075 (2009)
12. Gillessen, S., et al.: An update on monitoring stellar orbits in the Galactic center, The Astrophysical Journal **837**, 30 (2017)
13. Gravity Collaboration, https://www.mpe.mpg.de/ir/gravity
14. GRAVITY collaboration: Detection of the gravitational redshift in the orbit of the star S2 near the Galactic center massive black hole, Astronomy & Astrophysics **615**, L15 (2018)
15. GRAVITY collaboration: Detection of orbital motions near the last stable circular orbit of the massive black hole Sgr A*. Astronomy & Astrophysics **618**, L10 (2018)
16. GRAVITY collaboration: A geometric distance measurement to the Galactic center black hole with 0.3% uncertainty. Astronomy & Astrophysics **625**, L10 (2019)
17. GRAVITY collaboration: Detection of the Schwarzschild precession in the orbit of the star S2 near the Galactic center massive black hole, Astronomy & Astrophysics **636**, L5 (2020)
18. https://www.nobelprize.org/prizes/physics/2020/press-release/
19. Jebsen, J.T.: On the general spherically symmetric solutions of Einstein's gravitational equations in vacuo. Ark. Mat. Ast. Fys. (Stockholm) **15**(18), 1–9 (1921). Reprinted in Gen. Relat. Grav. **37**, 2253–2259 (2005)
20. Johansen, N.V., Ravndal, F.: On the discovery of Birkhoff's theorem. Gen. Relat. Grav. **38**(3), 537–540 (2006)
21. Kruskal, M.D.: Maximal extension of Schwarzschild metric. Phys. Rev. **119**(5), 1743–1745 (1960)
22. Levin, J., Perez-Giz, G.: A periodic table for black hole orbits. Phys. Rev. D **77**, 103005 (2008)
23. Menten, K.: Nobelpreis - Blick ins Zentrum (Nobel Prize - A look into the center). Physik Journal **19**, Nr. 12, 28 (2020)
24. Misner, C.W., Thorne, K.S., Wheeler, J.A.: Gravitation. W.H. Freeman, New York, 1973
25. Peißker, F. et al.: S62 and S4711: Indication of a population of faint fast-moving stars inside the S2 Orbit - S4711 on a 7.6 yr orbit around Sgr A*. The Astrophysical Journal **899**, 50 (2020)
26. Pound, R.V., Rebka Jr., G.V.: Apparent weight of photons. Phys. Rev. Lett. **4**, 337–341 (1960)
27. Pound, R.V., Snider, J.L.: Effect of gravity on gamma radiation. Phys. Rev. B **140**, 788–804 (1965)
28. Schwarzschild, K.: Über das Gravitationsfeld eines Massenpunktes nach der Einstein'schen Theorie. Sitzungsberichte der Königlich-Preußischen Akademie der Wissenschaften, 189–196 (1916), English translation: On the Gravitational Field of a Mass Point according to Einstein's Theory. Gen. Rel. and Grav., **35** 5, 951–959 (2003)
29. Shapiro, I.I. et al.: Fourth test of general relativity: preliminary results. Phys. Rev. Lett. **20**(22), 1265–1269 (1968)
30. Straumann, N.: General Relativity – With Applications to Astrophysics. Springer (2004)

Kerr Metric and Detection of Two Kerr Black Holes

14

Contents

In the derivation of the Schwarzschild metric we had assumed a static spherically symmetric mass distribution. But if a spinning star collapses into a black hole, its angular momentum is conserved. Black holes characterized only by their mass and angular momentum are described by the *Kerr metric*, which *Kerr*[1] found in 1963 [10].

---

[1] Roy Kerr, ⋆1934, New Zealand mathematician.

In Boyer-Lindquist-coordinates, which are the generalization of the
Schwarzschild coordinates for the Kerr metric, the line element is [2]

$$
\begin{aligned}
\mathrm{d}s^2 = & -\left(1 - \frac{r_s r}{\Sigma}\right) c^2 \mathrm{d}t^2 - \frac{2r_s a r \sin^2(\vartheta)}{\Sigma} c\,\mathrm{d}t\,\mathrm{d}\varphi + \frac{\Sigma}{\Delta}\mathrm{d}r^2 + \Sigma\mathrm{d}\vartheta^2 \\
& + \left(r^2 + a^2 + \frac{r_s a^2 r \sin^2(\vartheta)}{\Sigma}\right)\sin^2(\vartheta)\mathrm{d}\varphi^2,
\end{aligned}
\tag{14.1}
$$

with $\Sigma = r^2 + a^2 \cos^2(\vartheta)$, $\Delta = r^2 - r_s r + a^2$, and $r_s = 2GM/c^2$. The parameter
$a = J/(Mc)$ is the angular momentum J, scaled by the mass M of the black hole
and the speed of light c. It is easy to see that for $a = 0$ the Schwarzschild metric is
recovered.

A maximally rotating Kerr black hole is obtained for $a = GM/c^2 = r_s/2$. Bigger
values of a would lead to closed timelike world lines, and thus to a violation of the
causality principle.

In the limit $r \to \infty$ the Kerr metric turns into the flat Minkowski metric.
The limit $r_s \to 0$ at constant a is not quite as obvious. By introducing the
coordinate transformation $T = t$, $Z = r\cos(\vartheta)$, $X = \sqrt{r^2 + a^2}\sin(\vartheta)\cos(\varphi)$,
$Y = \sqrt{r^2 + a^2}\sin(\vartheta)\sin(\varphi)$, one arrives again at the Minkowski metric [7].

We have seen that for the Schwarzschild metric there exists a maximal extension
in terms of the Kruskal-Szekeres coordinates. A similar maximal extension exists
for the Kerr metric, which was found by Boyer and Lindquist [3].

14.1 Horizon and Ergosphere

The *event horizon* of a Kerr black hole is given by the outer zero of Δ. There the
metric becomes singular, since $g_{rr} \to \infty$ for $\Delta \to 0$:

$$
r_+ = \frac{r_s}{2} + \sqrt{\frac{r_s^2}{4} - a^2}.
\tag{14.2}
$$

As in the case of a Schwarzschild black hole, this is a coordinate singularity.

However, before reaching the event horizon there exists a region called the
ergosphere. Its outer border is given by the outer zero of $\Sigma - r_s r$, since there the
component g_{tt} vanishes:

$$
r_{es}(\vartheta) = \frac{r_s}{2} + \sqrt{\frac{r_s^2}{4} - a^2\cos^2(\vartheta)},
\tag{14.3}
$$

see Fig. 14.1.

To understand the meaning of the ergosphere, we must first clarify the difference between a *stationary* and a *static* observer, see also Misner, Thorne, and

Fig. 14.1 The different regions of the Kerr metric for $a = 0.95\, GM/c^2$. The *gray shaded region* is the ergosphere with the static limit $r = r_{es}(\vartheta)$

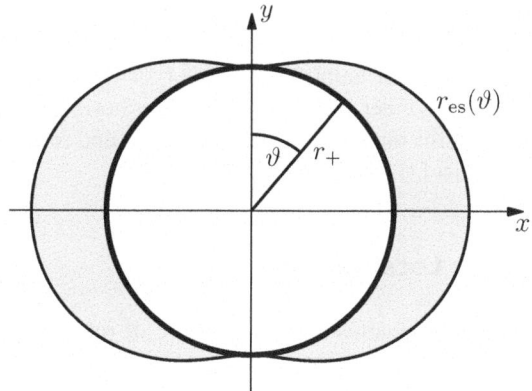

Wheeler [13]. Since the components of the Kerr metric depend explicitly on the coordinate time t and the angle φ, there exist two Killing vectors, $\underline{\xi}_t = (\partial_t)_{r,\vartheta,\varphi}$ and $\underline{\xi}_\varphi = (\partial_\varphi)_{t,r,\vartheta}$. An observer with the four-velocity

$$\underline{u} = u^t \left(\partial_t + \Omega \partial_\varphi\right) = \frac{\underline{\xi}_t + \Omega \underline{\xi}_\varphi}{|\underline{\xi}_t + \Omega \underline{\xi}_\varphi|}, \tag{14.4}$$

therefore experiences no change in the spacetime geometry in his immediate vicinity. In (14.4), $\Omega = d\varphi/dt = (d\varphi/d\tau)/(dt/d\tau) = u^\varphi/u^t$ is the angular velocity of the observer referred to the asymptotic rest frame. He is stationary relative to his local geometry. One uses the term *static* observer, when $\Omega = 0$ and the observer does not move relative to the asymptotic rest frame.

Since the four-velocity $\underline{u}$ always has to lie within the future light cone, Ω cannot assume arbitrarily large values. From $\underline{u}^2 = \langle \underline{u}, \underline{u}\rangle_g < 0$ and the metric coefficients $g_{\mu\nu}$ of the Kerr metric (14.1) follows

$$g_{tt} + 2\Omega g_{t\varphi} + \Omega^2 g_{\varphi\varphi} < 0. \tag{14.5}$$

The minimum and maximum angular velocity is obtained from (14.5) as

$$\Omega_{min} = \omega - \sqrt{\omega^2 - \frac{g_{tt}}{g_{\varphi\varphi}}} \quad \text{and} \quad \Omega_{max} = \omega + \sqrt{\omega^2 - \frac{g_{tt}}{g_{\varphi\varphi}}} \tag{14.6}$$

with $\omega = -g_{t\varphi}/g_{\varphi\varphi}$. For $r \leq r_{es}(\vartheta)$, we have $g_{tt} \geq 0$, and the minimum angular velocity $\Omega_{min} \geq 0$. Below this radius *every* observer must have a positive angular velocity, and therefore can no longer be static. This is called the *Lense-Thirring effect*[2,3] [12], or the *frame dragging*-effect. The region $r = r_{es}(\vartheta)$ is also called the *static limit*.

We will not amplify on further structures below the actual horizon, $r < r_+$, such as the inner ergosphere, the inner horizon, and the actual curvature singularity, which turns out to be a ring singularity, and refer the interested reader to Boyer and Lindquist [3] and Chandrasekhar [4].

14.2   Local Tetrads

We can now adjust the local frame of reference of the stationary observer of the previous section to the corresponding four-velocity (14.4), and so obtain the *locally non-rotating tetrad (LNRT)*

$$\underline{\mathbf{e}}_{(0)} = \sqrt{\frac{A}{\Sigma \Delta}} \frac{1}{c} \left(\partial_t + \omega \partial_\varphi \right),$$

$$\underline{\mathbf{e}}_{(1)} = \sqrt{\frac{\Delta}{\Sigma}} \partial_r, \quad \underline{\mathbf{e}}_{(2)} = \frac{1}{\sqrt{\Sigma}} \partial_\vartheta, \quad \underline{\mathbf{e}}_{(3)} = \sqrt{\frac{\Sigma}{A}} \frac{1}{\sin(\vartheta)} \partial_\varphi, \tag{14.7}$$

with $\omega = cr_s ar/A$ and $A = \Sigma \left(r^2 + a^2 \right) + r_s a^2 r \sin^2(\vartheta)$.

The statement that this tetrad is non-rotating may sound somewhat confusing, since the four-velocity in fact points along the angle φ. To understand the underlying rationale we consider two light rays which we force on circular orbits, one in the direction of rotation of the Kerr metric (prograde), and the other opposite to it (retrograde). When we measure the time it takes for the two light rays to return to the observer, we find that they both are the same [13]. The observer will conclude that he is at rest, i.e., non-rotating around the Kerr black hole, because otherwise the retrograde light ray would arrive before the prograde one. This also agrees with the statement from the previous section that the geometry does not change for him, since he moves along the Killing vectors. From the point of view of an asymptotic observer, however, the stationary observer does move around the Kerr black hole.

We can also define a frame of reference for a static observer outside the ergosphere. His local tetrad then reads

$$\underline{\mathbf{e}}_{(0)} = \frac{1}{c\sqrt{1 - r_s r/\Sigma}} \partial_t, \quad \underline{\mathbf{e}}_{(1)} = \sqrt{\frac{\Delta}{\Sigma}} \partial_r, \quad \underline{\mathbf{e}}_{(2)} = \frac{1}{\sqrt{\Sigma}} \partial_\vartheta,$$

$$\underline{\mathbf{e}}_{(3)} = \pm \frac{r_s ar \sin(\vartheta)}{c\sqrt{1 - r_s r/\Sigma}\sqrt{\Delta \Sigma}} \partial_t \mp \frac{\sqrt{1 - r_s r/\Sigma}}{\sqrt{\Delta} \sin(\vartheta)} \partial_\varphi. \tag{14.8}$$

---

[2] Josef Lense, 1890–1985, Austrian Mathematician.

[3] Hans Thirring, 1888–1976, Austrian physicist.

14.3   Qualitative Behavior of Geodesics

As in the case of the Schwarzschild metric, we will investigate the qualitative behavior of lightlike and timelike geodesics in the Kerr metric. We do this again using the Euler-Lagrange formalism. We will restrict ourselves to geodesics in the equatorial plane, $\vartheta = \pi/2$, and refer to Chandrasekhar [4] for more detailed discussions.

The *Lagrangian* for the Kerr metric (14.1) can be obtained again by replacing the differentials with the corresponding derivatives,

$$
\mathcal{L} = -\left(1 - \frac{r_s r}{\Sigma}\right)c^2 \dot{t}^2 - \frac{2r_s ar}{\Sigma}c\dot{t}\dot{\varphi} + \frac{\Sigma}{\Delta}\dot{r}^2 + \left(r^2 + a^2 + \frac{r_s a^2 r}{\Sigma}\right)\dot{\varphi}^2
$$

$$
= \kappa c^2
$$

(14.9)

with $\kappa = 0$ for lightlike and $\kappa = -1$ for timelike geodesics. Inserting (14.9) in the Euler-Lagrangian equations (11.134) leads to the equation

$$
\frac{1}{2}\dot{r}^2 + V_{\text{eff}} = 0,
$$

(14.10)

with the effective potential

$$
V_{\text{eff}} = \frac{1}{2r^3}\left\{h^2(r - r_s) + 2\frac{ahkr_s}{c} - \frac{k^2}{c^2}\left[r^3 + a^2(r + r_s)\right]\right\} - \frac{\kappa c^2 \Delta}{r^2},
$$

(14.11a)

and the constants of motion

$$
k = \left(1 - \frac{r_s}{r}\right)c^2\dot{t} + \frac{cr_s a}{r}\dot{\varphi} \quad \text{and} \quad h = \left(r^2 + a^2 + \frac{r_s a^2}{r}\right)\dot{\varphi} - \frac{cr_s a}{r}\dot{t}.
$$

(14.11b)

14.3.1 Lightlike Geodesics

From Eq. (14.10) and the condition $\dot{r} = 0$ we can determine the minimum distance r_{min} to which lightlike geodesics can approach the black hole. From the effective potential (14.11a) we obtain the condition

$$
r^3 - (\varepsilon^2 - a^2)r + r_s(\varepsilon - a)^2 = 0,
$$

(14.12)

where $\varepsilon = hc/k$. This cubic equation can be solved using the Cardanic formulas. The only physically reasonable solution is obtained as

$$r_{\min} = \sqrt{\frac{4}{3}(\varepsilon^2 - a^2)} \cos\left[\frac{1}{3}\arccos\left(-\frac{r_s(\varepsilon - a)^2}{2}\sqrt{\frac{27}{(\varepsilon^2 - a^2)^3}}\right)\right]. \qquad (14.13)$$

Clearly, this relation holds only for rays of light which pass by the black hole, and do not fall into it.

If we require, apart from the condition $\dot{r} = 0$, that the derivative of the effective potential V_{eff} with respect to r shall vanish, we obtain two circular null geodesics in the equatorial plane. From $\partial V_{\mathrm{eff}}/\partial r = 0$ follows, in addition to (14.12), the equation

$$r = \frac{3r_s}{2}\frac{\varepsilon - a}{\varepsilon + a}. \qquad (14.14)$$

Solving this equation with respect to ε and inserting the solution into (14.12) leads to the cubic equation

$$r^3 - 3r_s r^2 + \frac{9}{4}r_s^2 r - 2a^2 r_s = 0. \qquad (14.15)$$

It can be solved again using the Cardanic formulas, and taking into account the trigonometric relation $2\arccos(x) = \arccos(2x^2 - 1)$. The radii of the corresponding prograde $(-)$ and retrograde $(+)$ photon orbits then are found as

$$r_{\mathrm{po}}^{\pm} = r_s\left\{1 + \cos\left[\frac{2}{3}\arccos\left(\frac{\pm 2a}{r_s}\right)\right]\right\}, \qquad (14.16)$$

see Fig. 14.2. In the Kerr metric there exists a plethora of photon orbits lying outside the equatorial plane. A detailed discussion can be found, e.g., in Teo [15].

Using Eqs. (14.12) and (14.14), we can also calculate the two critical angles under which some ray of light asymptotically approaches the two photon orbits. Inserting (14.14) into (14.12) and solving with respect to ε yields the cubic equation

$$(\varepsilon + a)^3 - \frac{27}{4}r_s^2(\varepsilon - a) = 0, \qquad (14.17)$$

with the two physically reasonable solutions

$$\varepsilon_1 = 3r_s \cos\left[\frac{1}{3}\arccos\left(-\frac{2a}{r_s}\right)\right] - a \qquad (14.18)$$

and

$$\varepsilon_2 = -3r_s \cos\left[\frac{1}{3}\arccos\left(-\frac{2a}{r_s}\right) - \frac{\pi}{3}\right] - a. \qquad (14.19)$$

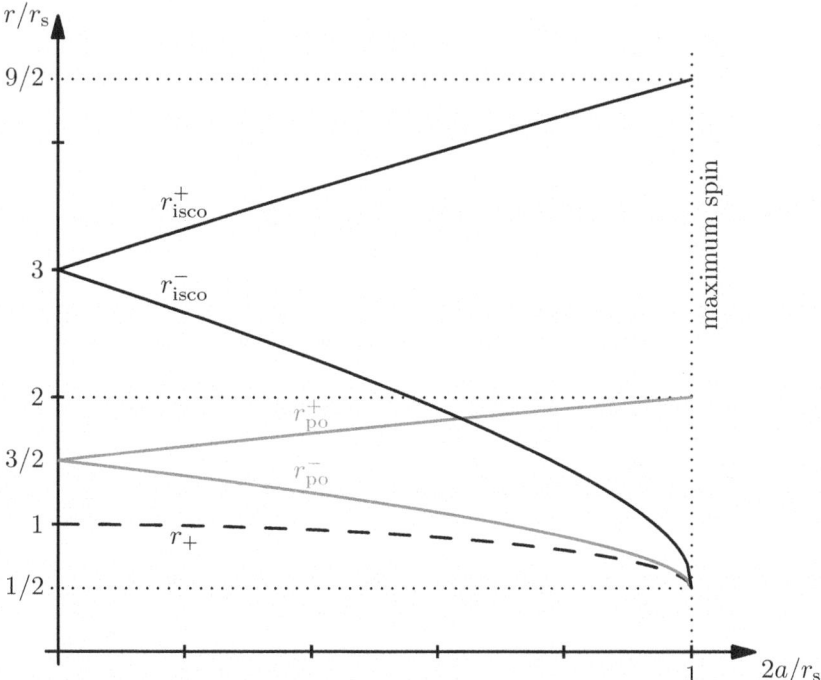

Fig. 14.2 Radii of the photon orbits $r_{po}^{\pm}$ and the innermost stable timelike circular orbits $r_{isco}^{\pm}$ as a function of the angular momentum parameter of the Kerr metric. The maximum possible value of the angular momentum parameter is $a = r_s/2$

We fix the starting direction $\underline{y}$ of a null geodesic in the equatorial plane with respect to the local tetrad (14.7) as follows:

$$\underline{y} = y^{(0)}\underline{e}_{(0)} + y^{(1)}\underline{e}_{(1)} + y^{(3)}\underline{e}_{(3)} = \pm\underline{e}_{(0)} - \cos(\xi)\underline{e}_{(1)} + \sin(\xi)\underline{e}_{(3)}. \qquad (14.20)$$

When we insert the coordinate representations of the tetrad vectors, we obtain expressions for the derivatives $\dot{t}$ and $\dot{\varphi}$, which we can use to calculate the constants of motion k and h in (14.11b),

$$k = \pm cr\sqrt{\frac{\Delta}{A}} + \frac{cr_s a}{\sqrt{A}}\sin(\xi) \quad \text{and} \quad h = \frac{\sqrt{A}}{r}\sin(\xi). \qquad (14.21)$$

Solving with respect to ξ finally yields the two critical angles

$$\xi_{1,2} = \arcsin\left(\frac{r^2\sqrt{\Delta}\varepsilon_{1,2}}{r_s ar\varepsilon_{1,2} - A}\right), \qquad (14.22)$$

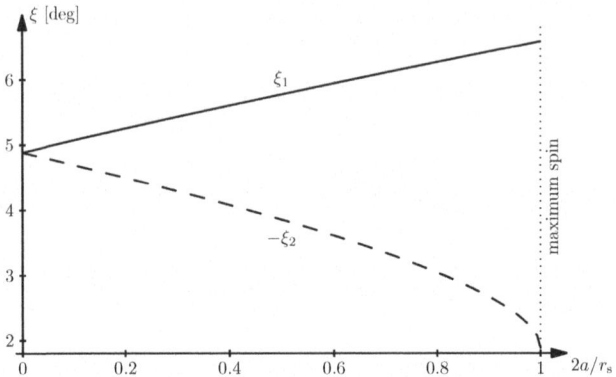

Fig. 14.3 In the Kerr metric, null geodesics with starting angles ξ_1 and ξ_2 delimit the width of the shadow of the black hole in the equatorial plane $\vartheta = \pi/2$ for an observer located at $r = 30 r_s$. In the case of maximum rotation, one finds $\xi_1 \approx 6.5853°$ and $-\xi_2 \approx 1.8779°$, whereas for $a = 0$ one has the symmetric situation with $\xi_1 = -\xi_2 \approx 4.8845°$

see Fig. 14.3. As expected, in the limit $a = 0$, the two angles reduce to the angle we had already determined for the Schwarzschild metric.

14.3.2 Timelike Circular Orbits

The calculation of timelike circular orbits from Eq. (14.10) is more difficult than for photon orbits. We therefore simply quote from Bardeen and Teukolsky [2] the result for the *innermost stable circular orbits* (isco)

$$r^\pm_{\text{isco}} = \frac{r_s}{2}\left(3 + Z_2 \pm \sqrt{(3 - Z_1)(3 + Z_1 + 2Z_2)}\right), \tag{14.23}$$

where

$$Z_1 = 1 + \left(1 - \frac{4a^2}{r_s^2}\right)^{1/3}\left[\left(1 + \frac{2a}{r_s}\right)^{1/3} + \left(1 - \frac{2a}{r_s}\right)^{1/3}\right], \tag{14.24a}$$

$$Z_2 = \sqrt{\frac{12a^2}{r_s^2} + Z_1^2}. \tag{14.24b}$$

The negative sign stands for the prograde, the positive sign for the retrograde orbit. As in the case of the Schwarzschild metric, these circular orbits are not really stable since the potential does not exhibit a minimum at these radii.

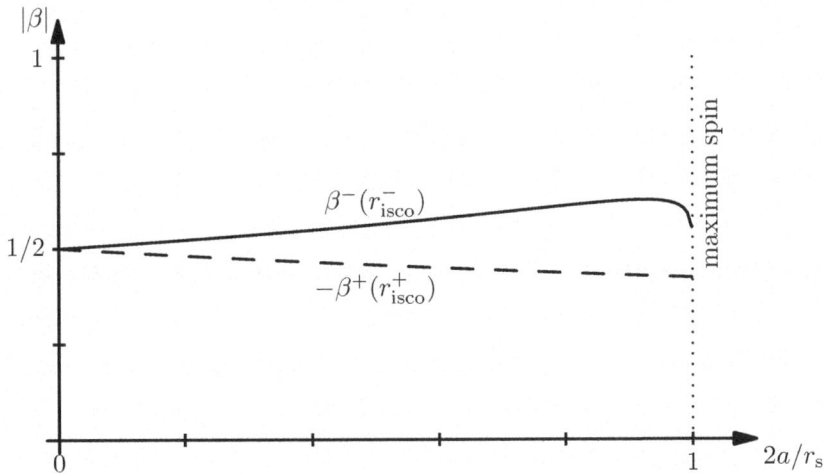

Fig. 14.4 Prograde $\beta^-(r_{\text{isco}}^-)$ and retrograde $-\beta^+(r_{\text{isco}}^+)$ velocities corresponding to the circular orbits in Fig. 14.2

The velocity necessary on the orbit $r_{\text{isco}}^\pm$, with respect to the locally non-rotating tetrad (14.7), is given by

$$\beta^\pm(r) = \frac{r_s a r^4 (3r^2 + a^2) \pm A\sqrt{2r^7 r_s}}{r^2 \sqrt{\Delta}(r_s a^2 r^2 - 2r^5)} \tag{14.25}$$

with $A = (r^2 + a^2)^2 - a^2\Delta$, see Fig. 14.4.

This relation also holds for all other timelike circular orbits in the equatorial plane as long as $|\beta| < 1$.

In Fig. 14.4, the velocity starts from its Schwarzschild value $c/2$ at $a = 0$, and slightly increases above that value for the prograde orbit, ending at $\approx 0.55\,c$ for maximum spin. For the retrograde orbit, the value slightly decreases below $c/2$, ending at $\approx 0.42\,c$ at maximum spin.

14.4   Detection of a Supermassive Kerr Black Hole

We have seen that event horizons exist both in the Schwarzschild and in the Kerr metric. These shield the inner region completely from the outside. Objects smaller than their event horizon are known as black holes. Due to their powerful gravity, no matter can leave their surroundings. Even light and other electromagnetic waves fail to escape from them, which obviously is why they are called "black".

Although there was no doubt among today's astrophysicists that the cosmos is populated by such bizarre objects, so far there had only been indirect evidence of their existence. With the aim of observing a black hole directly, scientists from twenty countries interconnected individual radio telescopes to form a large virtual telescope, the "Event Horizon Telescope" (EHT) [5]. Their efforts were crowned by success, and in spring 2019 the researchers of the EHT collaboration were able to present the first image of a black hole. The black hole has angular momentum and is therefore described by the Kerr metric.

We will discuss this discovery in some detail below.

14.4.1 The Event Horizon Telescope: Set-up and Arrangement of the Instruments

The basic principle of the Event Horizon Telescope is the radio astronomical method of "Very Long Baseline Interferometry" (VLBI, interferometry with large baselines [6]). This method allows measurements with the highest spatial resolution and positional accuracy. Wave fronts from cosmic radio sources are observed and brought to interference at two or more radio telescopes, located at large distances from each other. This creates an imaginary telescope whose resolution in radians is given by the ratio of the measured wavelength λ and the distance D between the telescopes:

$$R = \frac{\lambda}{D}. \tag{14.26}$$

During the observations in spring 2017, the EHT consisted of a network of eight radio telescopes, which were distributed over six different locations on Earth, see Fig. 14.5.

Fig. 14.5 The eight stations of the 2017 EHT campaign. Solid baselines indicate simultaneous visibility of the black hole M87* (declination $\delta = +12°$). The dashed baselines between telescopes served for calibration by observing the quasar 3C279, 5 billion light years away. From: Event Horizon Telescope collaboration [16]. © AAS. Reproduced with permission

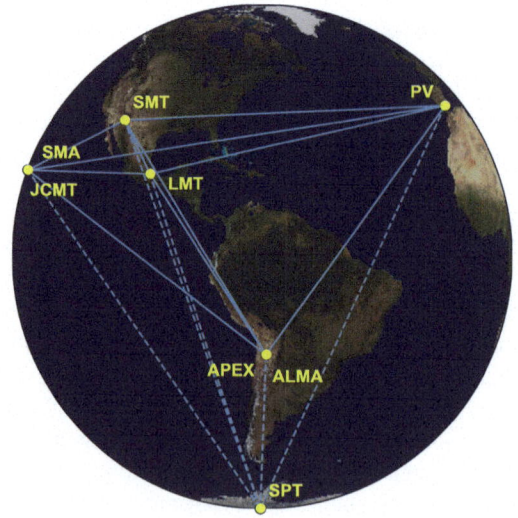

Table 14.1 The eight radio telescopes involved in the observations of the EHT in April 2017. The values given in brackets in the second column for ALMA and SMA are the number of individual telescopes at each of the sites. Adapted from Event Horizon Telescope collaboration [17]

Site	Diameter [m]	Location
ALMA	12 (54×) and 7 (12×)	Chile
APEX	12	Chile
JCMT	15	Hawaii (USA)
LMT	50	Mexico
PV 30m	30	Spain
SMA	6 (8×)	Hawaii (USA)
SMT	10	Arizona (USA)
SPT	10	Antarctica

In Table 14.1 all radio telescopes of the EHT are listed with their respective equipment. In particular, the 66 antennas from ALMA ("*Atacama Large Millimeter Array*") in Chile [1], which were integrated into the EHT network, were of prime importance for the success of the observation campaign. Baselines between pairs of facilities ranged from 160 to 10,700 km. This created a virtual telescope almost the size of the Earth.

Since only short-wave radio radiation can pass the cosmic distances between the galaxy and the observer on Earth, the researchers decided on a wavelength of $\lambda = 1.3$ mm (radio frequency $\nu = 230$ GHz). With this wavelength and the maximum baseline of $D = 10,700$ km, from (14.26) follows a maximum resolution of $R \approx 25$ µas. To illustrate: This is the angle at which a rod of length 1 m appears to an observer at a distance of 8.25 million kilometers away!

14.4.2 The Image of the Black Hole at the Center of the Galaxy M87

The giant elliptical galaxy M87 (Messier 87) is a member of the Virgo Galaxy Cluster, lying towards the constellation Virgo when looking into the night sky. M87 is about 55 million light-years away from our Milky Way. With a mass of 2400 billion solar masses, it is the most massive galaxy in the observable universe. Like our Milky Way, M87 has a supermassive black hole at its center, denoted by M87* in what follows. M87* itself cannot emit light because light cannot escape from inside the event horizon. The black hole is surrounded by a rotating accretion disk[4] made of infalling brightly glowing gas. The gas heats up due to friction effects and emits synchrotron radiation.

A characteristic feature of the galaxy M87 is a 5000 light-year unidirectional jet, a collimated beam from the poles of the black hole that ejects subatomic particles perpendicular to the accretion disk into the vast expanses of the universe at nearly the speed of light.

---

[4] In astrophysics, an accretion disk is a disk rotating around a central object that transports (accretes) matter toward the center.

Because the black hole acts as an extreme gravitational lens, the null geodesics of light near the event horizon can loop around the black hole multiple times before the light can escape. In the case of a very transparent, i.e., optically thin, plasma, almost all the radiation escapes and produces a characteristic bright ring around the shadow of M87*. The shadow marks the edge of the region in which photons can still move around the black hole on stable orbits without being caught and swallowed by the black hole.

A very instructive visualization of both the formation of the jet and the formation of the shadow using ray tracing can be found in a video by Weih and Rezzolla [14], see also Grenzebach [9].

The EHT collaboration observed M87* on April 5, 6, 10, and 11, 2017. The "brightnesses" were measured, more precisely: the two-dimensional Fourier components (spatial frequencies), of the distribution of the radio luminosity in the sky. Up to 25 scans, each lasting three to seven minutes, were recorded at each individual telescope. The observation data was digitally recorded at each telescope. In order to be able to synchronize the data later, when superimposing it in a software correlator, the recordings were time-stamped using hydrogen masers (comparable in accuracy to the most accurate atomic clocks). By the rotation of the Earth, the effective distances of the telescopes relative to the line of sight to the source changed during each observation. This meant that when the data from each pair of telescopes was brought to interference, different sub-areas in the plane of the Fourier components were covered. This resulted in considerably more information about the Fourier transform of the entire image.

Nevertheless, the plane of Fourier components of the radio luminosity distribution still showed large gaps. Therefore, complex algorithms had to be developed that enabled the researchers to reconstruct a complete picture of the luminosity distribution from the limited information. The task is similar to recognizing the complete piece of music in a song played on a piano with numerous silent keys (missing frequencies).

Four teams were formed, that worked independently of each other, using completely different reconstruction algorithms. Yet each team produced images with the same distinctive appearance: The images show a bright ring with a diameter of about 38–44 μas, with increased brightness towards the south, and the shadow of the black hole, see Fig. 14.6. Because of the good match of the images obtained with different methods, it can be ruled out, with a high degree of certainty, that the features of the images are artefacts of the reconstruction.

The brightness distribution of M87* was successfully modeled by general relativistic magnetohydrodynamic (GRMHD) simulations, see the middle panel in Fig. 14.6 [18]. The model describes a turbulent, hot magnetized disk rotating around a Kerr black hole. The simulations were able to derive a value of $M = (6.5 \pm 0.7) \times 10^9 \, M_\odot$ for the mass of the black hole.

With the assumption that the angular momentum vector of the accretion disk has a non-vanishing inclination, the asymmetric brightness of the ring can be understood. The plasma in the accretion disk moves at relativistic velocities. Therefore, due to the relativistic beaming effect, the lower side of the ring, where

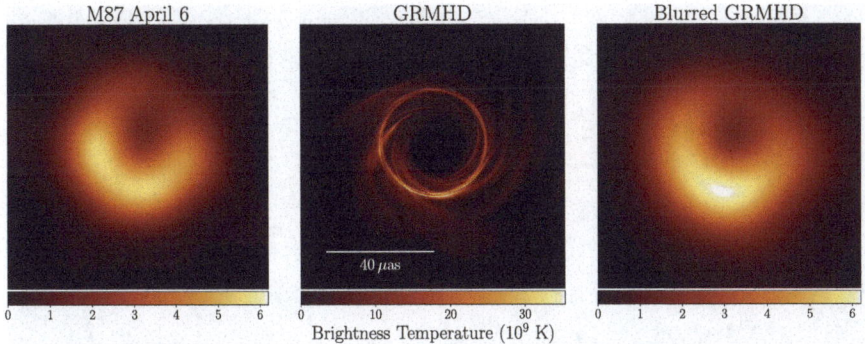

Fig. 14.6 The image of the black hole at the center of M87 (left) observed with the Event Horizon telescope agrees very well with the results of numerical simulations when folded with a Gauss kernel of 20 μas full width at half maximum (FWHM), i.e., the expected resolution of the telescope. From: Event Horizon Telescope collaboration [16]. © AAS. Reproduced with permission

the gas flows towards the Earth, appears brighter than the upper side, where it moves away from Earth.

The publication of Fig. 14.6 attracted attention not only from experts (cf. Grant [8]), but also from the general public. It also caused a wave of headlines in the press, see, e.g., Kluger [11].

14.5 The Shadow of the Black Hole in the Center of the Milky Way

The next goal of the EHT consortium was to image the black hole Sgr A* in the center of our Milky Way. Although it is 2000 times closer than M87*, imaging is no easier. The reason is that with $(4.31 \pm 0.36) \times 10^6\ M_\odot$ the black hole in the center of the Milky Way is roughly 1500 times less massive than the black hole of M87* with $M = (6.5 \pm 0.7) \times 10^9\ M_\odot$.

The difference in the masses implies a similar difference in their variability timescales. The period of the innermost stable orbit (isco), which depends on the mass and the spin of the black hole, can serve as an approximate dynamical timescale. For prograde orbits, this period ranges from $12\sqrt{6}\pi t_g$ for zero spin, see (13.49), to $4\pi t_g$ [19], where $t_g \equiv GM/c^3$. For M87*, this range extends from $\approx 3 \times 10^6$ s ≈ 35 days to $\approx 4 \times 10^5$ s ≈ 4.6 days. However, for Sgr A* the range is only ≈ 2000 s ≈ 33 minutes to ≈ 260 s ≈ 4 minutes. For this reason the source structure can evolve within a single night.

Considerable effort therefore had to be made to compensate for the effects caused by the strong variability of the accretion disk. Finally, in April 2022, the Event Horizon Collaboration was able to also present the first picture of the supermassive black hole in the center of the Milky Way, see Fig. 14.7 [19]. From the asymmetry

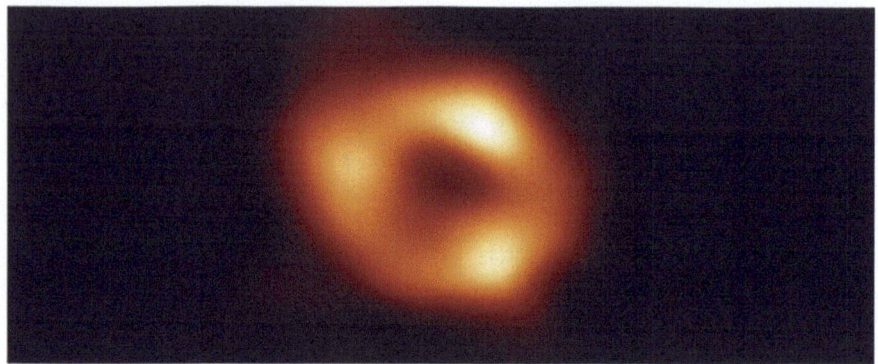

Fig. 14.7 Shadow of the black hole in the center of the Milky Way. From: Event Horizon Telescope collaboration [19]. © AAS. Reproduced with permission

Fig. 14.8 Polarization of the ring around the black hole in the center of the Milky Way. From: Event Horizon Telescope collaboration [20]. © AAS. Reproduced with permission

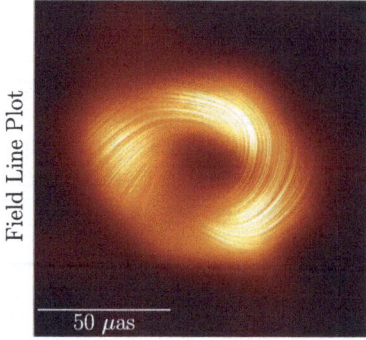

of the image we can again conclude that the black hole rotates, and is a Kerr black hole.

The Event Horizon Collaboration has also been able to measure the polarization caused by the magnetic field lines in the ring, see Fig. 14.8 [20]. Imaging the ring around Sgr A* in polarized light is not easy because, as we know, the ionized gas, or plasma, in the vicinity of the black hole orbits it in only a few minutes. Because the particles of the plasma swirl around the magnetic field lines, the magnetic field structures change rapidly during the recording of the radio waves by the EHT. Again, sophisticated instruments and computer simulations were required to capture the image of the supermassive black hole in polarized light. The image clearly demonstrates that the black hole in the center of the Milky Way is surrounded by strong, spiral-shaped magnetic fields.

14.6 Exercises

14.1. Values of the angular momentum parameter The Sun rotates with a period $P_\odot \approx 28$ d $\approx 2.5 \cdot 10^6$ s, the fastest rotating neutron star with $P \approx 1.4$ ms, see Sect. 21.2. Assume $R = 10$ km for the radius of the neutron star and $M = 1.4 M_\odot$ for its mass and determine the value of the parameter a for the neutron star and the Sun in units of the maximum value $r_s/2$. For simplicity, assume that both bodies are homogeneous solid spheres.

14.2. Time variability of the accretion disks Confirm the numerical values given in Sect. 14.5 for the different dynamical timescales of the accretion disks of M87* and Sgr A*.

References

1. ALMA-Homepage: https://www.almaobservatory.org
2. Bardeen, J.M., Press, W.H., Teukolsky, S.A.: Rotating black holes: locally nonrotating frames, energy extraction, and scalar synchroton radiation. Astrophys. J. **178**, 347–369 (1972)
3. Boyer, R.H., Lindquist, R.W.: Maximal analytic extension of the Kerr metric. J. Math. Phys. **8**, 265–281 (1967)
4. Chandrasekhar, S.: The Mathematical Theory of Black Holes. Oxford University Press, Oxford (1998)
5. Event Horizon Telescope, Homepage: https://eventhorizontelescope.org
6. Felli, M., Spencer, R.E.: Very Long Baseline Interferometry - Techniques and Applications. Springer Netherlands, Cham (1989)
7. Frolov, V.P., Zelnikov, A.: Introduction to Black Hole Physics. Oxford University Press, Oxford (2011)
8. Grant, A.: What it took to capture a black hole. Phys. Today. https://doi.org/10.1063/PT.6.1.20190411a, American Institute of Physics (2019)
9. Grenzebach, A.: The Shadow of Black Holes. An Analytic Description. Springer, Cham (2016)
10. Kerr, R.P.: Gravitational field of a spinning mass as an example of algebraically special metrics. Phys. Rev. Lett. **11**, 237–238 (1963)
11. Kluger, J.: This is the first picture of a black hole – and that's a big, even supermassive deal. https://time.com/5566225/first-black-hole-picture-photo/, Time Magazine (2019)
12. Lense, J., Thirring, H.: Über den Einfluss der Eigenrotation der Zentralkörper auf die Bewegung der Planeten und Monde nach der Einsteinschen Gravitationstheorie (On the influence of the rotation of the central bodies on the motion of the planets and moons according to Einstein's theory of gravitation). Phys. Z. **19**, 156 (1918)
13. Misner, C.W., Thorne, K.S, Wheeler, J.A.: Gravitation. Princeton University Press, Princeton (2017), ISBN 978-0-691-17779-3
14. Weih, L.R., Rezzolla, L.: Accretion flow onto a Kerr Black Hole and a visual explanation of ray tracing: https://youtu.be/jvftAadCFRI
15. Teo, E.: Spherical photon orbits around a Kerr black hole. Gen. Relativ. Gravit. **35**(11), 1909–1926 (2003)
16. The Event Horizon Telescope Collaboration: First M87 event horizon telescope results. I. The shadow of the supermassive black hole. Astrophys. J. Lett. **875**, L1 (2019)
17. The Event Horizon Telescope Collaboration: First M87 Event horizon telescope results. II. Array and instrumentation. Astrophys. J. Lett. **875**, L2 (2019)

18. The Event Horizon Telescope Collaboration: First M87 event horizon telescope results. V. Physical origin of the asymmetric ring. Astrophys. J. Lett. **875**, L5 (2019)
19. The Event Horizon Telescope Collaboration: First Sagittarius A* event horizon telescope results. I. The shadow of the supermassive black hole in the center of the Milky Way. Astrophys. J. Lett. **930**, L12 (2022)
20. The Event Horizon Telescope Collaboration: First Sagittarius A* event horizon telescope results. VII. Polarization of the ring. Astrophys. J. Lett. **964**, L25 (2022)

Gravitational Waves

15

Contents

General relativity predicts that accelerated masses emit gravitational waves, which propagate at the speed of light. The first theoretical studies of this phenomenon date back to 1916 and were performed by Einstein [9]. The observable effects of gravitational waves are tiny. Because of that they evaded direct experimental detection for almost a full century, although indirect evidence for their existence was gained in observations of binary pulsar systems much earlier, see Sect. 21.5.2.

In this chapter we will first discuss the theoretical analysis of gravitational waves using a linearized form of Einstein's field equations in Sects. 15.1–15.4. Using the results of this discussion we will then take a look at how gravitational waves could finally be detected for the first time in 2015 and how their observation in current and future experiments is a vital supplement to other astronomical methods that will help us to understand many phenomena in our universe in greater detail.

15.1   Linearization of the Field Equations

The linearization of the field equations is done in the weak-field approximation, which we have already used in establishing the connection between Newtonian theory and GR. However, in this chapter we will carry out the calculations more generally and comprehensively.

The starting point is a flat spacetime with small perturbations, i.e.,

$$g_{\mu\nu} = \eta_{\mu\nu} + h_{\mu\nu} \quad \text{with} \quad |h_{\mu\nu}| \ll 1 \quad \text{and} \quad |\partial_\lambda h_{\mu\nu}| \ll 1. \tag{15.1}$$

Here, $\eta_{\mu\nu}$ designates, as usual, the Minkowski spacetime, and $h_{\mu\nu}$ is a symmetric tensor. The Christoffel symbols are obtained in first order of the perturbation as

$$\begin{aligned}
\Gamma^\lambda{}_{\mu\nu} &= \frac{1}{2}\eta^{\lambda\kappa}\left(h_{\kappa\mu,\nu} + h_{\kappa\nu,\mu} - h_{\mu\nu,\kappa}\right) + \mathcal{O}(h^2) \\
&= \frac{1}{2}\left(h^\lambda{}_{\mu,\nu} + h^\lambda{}_{\nu,\mu} - h_{\mu\nu}{}^{,\lambda}\right),
\end{aligned} \tag{15.2}$$

where the superscript represents, in analogy with (11.13), the partial derivative $G_\mu{}^{,\lambda} = \partial^\lambda G_\mu = \partial G_\mu/\partial x_\lambda$ with respect to the covariant coordinate x_λ. In the first line of (15.2), $\eta^{\lambda\kappa}$ stands directly in front of the bracket, because perturbations in $g^{\lambda\kappa}$ drop out, since we restrict ourselves to first orders in h. Next we can calculate the Ricci tensor in first order. As in the derivation of the field equations in Sect. 12.2, a great simplification results from the fact that we can neglect all products of Christoffel symbols, since these would yield terms only in second order of h:

$$R_{\mu\nu} = \partial_\lambda \Gamma^\lambda{}_{\mu\nu} - \partial_\nu \Gamma^\lambda{}_{\lambda\mu} + \mathcal{O}(h^2). \tag{15.3}$$

We insert our results for the Christoffel symbols into this expression, and, with $2\Gamma^\lambda{}_{\lambda\mu} = \partial_\mu h^\lambda{}_\lambda + \partial_\lambda h^\lambda{}_\mu - \partial^\lambda h_{\lambda\mu}$, obtain the equation

$$\begin{aligned}
2R_{\mu\nu} &= \partial_\lambda\partial_\nu h^\lambda{}_\mu + \partial_\lambda\partial_\mu h^\lambda{}_\nu - \partial_\lambda\partial^\lambda h_{\mu\nu} - \partial_\nu\partial_\mu h^\lambda{}_\lambda - \partial_\nu\partial_\lambda h^\lambda{}_\mu + \partial_\nu\partial^\lambda h_{\lambda\mu} \\
&= \Box h_{\mu\nu} + \partial_\lambda\partial_\nu h^\lambda{}_\mu + \partial_\lambda\partial_\mu h^\lambda{}_\nu - \partial_\nu\partial_\mu h^\lambda{}_\lambda.
\end{aligned} \tag{15.4}$$

Here we encounter again the d'Alembert operator $\Box = -\partial_\mu\partial^\mu$ from (7.18), which we are already familiar with from electrodynamics. Furthermore, the last two terms in the first line eliminate each other, $-\partial_\nu\partial_\lambda h^\lambda{}_\mu + \partial_\nu\partial^\lambda h_{\lambda\mu} = 0$. We split $-\partial_\nu\partial_\mu h^\lambda{}_\lambda$

and add each half of the expression to the second and third term. We then end up with the linearized form of the field equations $R_{\mu\nu} = 8\pi G/c^4 T^*_{\mu\nu}$, viz.

$$\Box h_{\mu\nu} + \partial_\mu \left(\partial_\lambda h^\lambda{}_\nu - \frac{1}{2}\partial_\nu h^\lambda{}_\lambda \right) + \partial_\nu \left(\partial_\lambda h^\lambda{}_\mu - \frac{1}{2}\partial_\mu h^\lambda{}_\lambda \right) = \frac{16\pi G}{c^4} T^*_{\mu\nu}.$$

(15.5)

15.1.1 Transformation to Harmonic Coordinates

In the following we exploit the freedom to choose the coordinate system. All that is necessary is to retain the form $g_{\mu\nu} = \eta_{\mu\nu} + h_{\mu\nu}$. Under this restriction, we look for coordinates in which the field equations (15.5) are as simple as possible. This means, in practical terms, that we wish the two expressions in the brackets to vanish.

Therefore, we look for coordinates in which we have

$$\partial_\lambda h^\lambda{}_\nu - \frac{1}{2}\partial_\nu h^\lambda{}_\lambda = 0.$$

(15.6)

This corresponds to a gauge condition, similar to the one we already used in electrodynamics in Chap. 7. There we had chosen the four-potential in Eq. (7.21) in such a way that $\partial_\mu A^\mu = 0$ applies.

In that sense, the freedom in the choice of the coordinates in GR corresponds to the gauge freedom for the potentials in electrodynamics. For conversion into the new coordinate system, we use a transformation of the form

$$x^\mu \mapsto x'^\mu = x^\mu + s^\mu(\underline{x}) \quad \text{with} \quad |\partial_\nu s^\mu| \ll 1,$$

(15.7)

where s^μ can be arbitrarily large. This leads to

$$g'_{\mu\nu} = \eta_{\mu\nu} + h'_{\mu\nu} \quad \text{with} \quad h'_{\mu\nu} = h_{\mu\nu} - \partial_\mu s_\nu - \partial_\nu s_\mu.$$

(15.8)

Coordinates which fulfill the required gauge condition are called *harmonic coordinates*. They are characterized by the property

$$g^{\mu\nu} \Gamma^\lambda{}_{\mu\nu} \equiv \Gamma^\lambda = 0.$$

(15.9)

We evaluate this condition explicitly to demonstrate that the gauge condition indeed is fulfilled:

$$\begin{aligned} g^{\mu\nu} \Gamma^\lambda{}_{\mu\nu} &= \eta^{\mu\nu} \frac{1}{2} \left(\partial_\nu h^\lambda{}_\mu + \partial_\mu h^\lambda{}_\nu - \partial^\lambda h_{\mu\nu} \right) \\ &= \frac{1}{2} \left(\partial^\mu h^\lambda{}_\mu + \partial_\mu h^{\lambda\mu} - \partial^\lambda h^\mu{}_\mu \right) \\ &= \partial_\mu h^{\lambda\mu} - \frac{1}{2}\partial^\lambda h^\mu{}_\mu. \end{aligned}$$

(15.10)

Inserting (15.8) into (15.10) yields

$$\Gamma^\lambda = (\Gamma')^\lambda - \partial_\mu \partial^\mu s^\lambda. \tag{15.11}$$

In order to have $(\Gamma')^\lambda = 0$, the equation

$$\Box s^\lambda = \Gamma^\lambda \tag{15.12}$$

has to be satisfied in the harmonic coordinates. From the theory of differential equations we know that (15.12) always does have a solution.

15.1.2 Solution of the Linearized Field Equations

After transformation to harmonic coordinates, (15.5) simplifies to the field equations

$$\Box h_{\mu\nu} = \frac{16\pi G}{c^4} T^*_{\mu\nu}. \tag{15.13}$$

We recognize again the analogy with electrodynamics, where we arrived at the equation $\Box A^\mu = \mu_0 j^\mu$, with the four-current j^μ (cf. Sect. 7.2). Realizing this analogy, we can identify $T^*_{\mu\nu}$ as the source for the generation of gravitational waves. Since it is the d'Alembert operator that stands on the left-hand side of (15.13), it is evident that gravitational waves will propagate at the speed of light.

The *retarded solution* of the inhomogeneous equation is obtained in the form

$$h_{\mu\nu}(\boldsymbol{r}, t) = \frac{4G}{c^4} \int \frac{T^*_{\mu\nu}\left(\boldsymbol{r}, t - |\boldsymbol{r} - \boldsymbol{r}'|/c\right)}{|\boldsymbol{r} - \boldsymbol{r}'|} d^3 r'. \tag{15.14}$$

The solution is called retarded because, in evaluating the integral, the finite travel time of light $|\boldsymbol{r} - \boldsymbol{r}'|/c$ is taken into account.

The propagation of gravitational waves in vacuum is described by the homogeneous solution of (15.13), i.e., the solution of the equation

$$\Box h_{\mu\nu} = 0. \tag{15.15}$$

We choose a plane wave for the form of the homogeneous solution,

$$h_{\mu\nu}(\underline{x}) = e_{\mu\nu}\, e^{ik_\lambda x^\lambda} + e^*_{\mu\nu}\, e^{-ik_\lambda x^\lambda}. \tag{15.16}$$

Because of the tensor character of $h_{\mu\nu}$, we also have to set up an amplitude tensor $e_{\mu\nu}$. Thus, together with the four parameters of k_λ, we have 20 parameters in total in this equation. Inserting (15.16) in (15.15) we obtain

$$\Box h_{\mu\nu} = -\partial_\lambda \partial^\lambda h_{\mu\nu} = k_\lambda k^\lambda h_{\mu\nu} = 0. \tag{15.17}$$

An immediate consequence is that k_μ must be a lightlike vector, i.e., $k_\mu k^\mu = 0$.

Not all the 20 parameters in (15.16) are independent. The gauge condition leads to

$$k_\mu e^\mu{}_\nu - \frac{1}{2} k_\nu e^\mu{}_\mu = 0. \tag{15.18}$$

This corresponds to 4 conditions, the symmetry $e_{\mu\nu} = e_{\nu\mu}$ yields 6 more conditions, so that all in all we are left with 10 degrees of freedom.

In harmonic coordinates, we still have degrees of freedom in the choice of the coordinates. We choose the *TT gauge*, where the abbreviation TT stands for "transverse and traceless". We accomplish this gauge by the choice

$$e^\mu{}_\mu = 0. \tag{15.19}$$

For a gravitational wave propagating in the direction $\mathbf{e}_i$, this leads to $e^\mu{}_0 = 0$ and $e^\mu{}_i = 0$. We now consider a gravitational wave propagating in the z-direction. The *wavevector* then reads

$$k^\mu = \frac{\omega}{c} (1, 0, 0, 1)^\mathrm{T}, \quad \text{i.e.,} \quad k_\mu x^\mu = k_z z - \omega t, \tag{15.20}$$

with $k_z = \omega/c$ and $x^\mu = (ct, x, y, z)$. From this follows, in TT gauge,

$$e_{\mu\nu} = \begin{pmatrix} 0 & 0 & 0 & 0 \\ 0 & e_{11} & e_{12} & 0 \\ 0 & e_{12} & -e_{11} & 0 \\ 0 & 0 & 0 & 0 \end{pmatrix}. \tag{15.21}$$

We see that for $e_{\mu\nu}$ we are left with only two degrees of freedom. In the case of linear polarization we can split $e_{\mu\nu}$ in a linear combination of two tensors,

$$e^+_{\mu\nu} = \begin{pmatrix} 0 & 0 & 0 & 0 \\ 0 & 1 & 0 & 0 \\ 0 & 0 & -1 & 0 \\ 0 & 0 & 0 & 0 \end{pmatrix} \quad \text{and} \quad e^\times_{\mu\nu} = \begin{pmatrix} 0 & 0 & 0 & 0 \\ 0 & 0 & 1 & 0 \\ 0 & 1 & 0 & 0 \\ 0 & 0 & 0 & 0 \end{pmatrix}. \tag{15.22}$$

With these tensors, the expression for $h_{\mu\nu}$ is

$$h_{\mu\nu}(\underline{x}) = \left(A^+ e^+_{\mu\nu} + A^\times e^\times_{\mu\nu} \right) e^{i(k_z z - \omega t)} + \text{c.c.} \tag{15.23}$$

The abbreviation "c.c." stands for the complex conjugate of the preceding expression. If we set

$$p = A^+ \cos\left(\frac{\omega}{c} z - \omega t\right) \quad \text{and} \quad q = A^\times \cos\left(\frac{\omega}{c} z - \omega t\right), \tag{15.24}$$

we obtain for the line element $ds^2 = (\eta_{\mu\nu} + h_{\mu\nu})dx^\mu dx^\nu$ the form

$$ds^2 = -c^2 dt^2 + dx^2 + dy^2 + dz^2 + p\left(dx^2 - dy^2\right) + 2q\, dx\, dy. \qquad (15.25)$$

We now consider a rotation in space around an axis parallel to the direction of propagation. Apart from $z = z'$ and $t = t'$, the latter is described by

$$\begin{pmatrix} x \\ y \end{pmatrix} = \begin{pmatrix} \cos(\alpha) & \sin(\alpha) \\ -\sin(\alpha) & \cos(\alpha) \end{pmatrix} \begin{pmatrix} x' \\ y' \end{pmatrix}. \qquad (15.26)$$

The requirement of the invariance of the line element, $ds^2 = ds'^2$, means

$$p^2\left(dx^2 - dy^2\right) + 2q\, dx\, dy = p'^2\left(dx'^2 - dy'^2\right) + 2q'\, dx'\, dy'. \qquad (15.27)$$

This leads to the transformation equation

$$\begin{pmatrix} p' \\ q' \end{pmatrix} = \begin{pmatrix} \cos(2\alpha) & -\sin(2\alpha) \\ \sin(2\alpha) & \cos(2\alpha) \end{pmatrix} \begin{pmatrix} p \\ q \end{pmatrix}. \qquad (15.28)$$

It can be seen that p and q, and therefore also the polarizations, rotate twice as fast as the coordinate system. One summarizes this fact in the statement that gravitation features *helicity* 2. This also implies that the quantum of gravitation, the *graviton*, if it exists, must have spin 2.

15.2   Particles in the Field of a Gravitational Wave

We now ask: How does a particle, initially at rest, move under the action of a gravitational wave? For a particle, at rest at time $\tau = 0$, we have

$$\frac{dx^i}{d\tau} = 0. \qquad (15.29)$$

The geodesic equation for the particle then simplifies to

$$\frac{d^2 x^\mu}{d\tau^2} + \Gamma^\mu{}_{00}\frac{dx^0}{d\tau}\frac{dx^0}{d\tau} = 0. \qquad (15.30)$$

The Christoffel symbols $\Gamma^\mu{}_{00} = \frac{1}{2}\eta^{\mu\lambda}\left(2h_{\lambda 0,0} - h_{00,\lambda}\right) = 0$ are identically zero by reason of the form of the perturbation, see (15.21). From this follows that we have for a particle, initially at rest,

$$\frac{d^2 x^\mu}{d\tau^2} = 0, \quad \text{i.e.} \quad x^i(\tau) = \text{const.} \qquad (15.31)$$

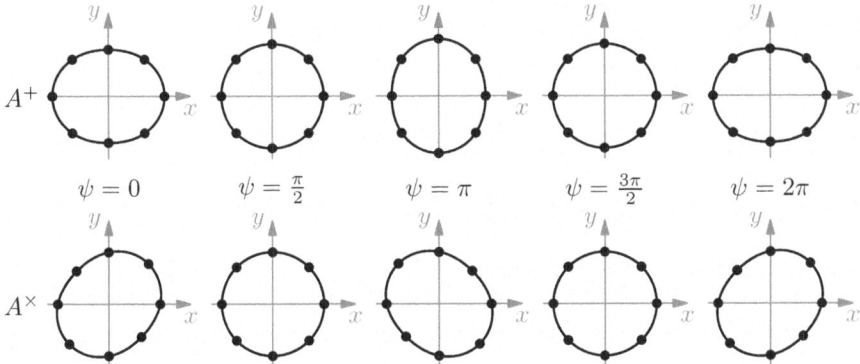

Fig. 15.1 Particles on a circle around the origin in the xy-plane under the action of a gravitational wave. Even though the coordinates of the particles remain constant, the distances from the origin change with time, in dependence of the strength and the polarization of the gravitational wave. The angular difference between the two polarizations amounts to $\varphi = \pi/4$. The angle $\psi = \omega t - k_z z$ represents the relative phase and thus also the time dependence

In other words, the coordinates of the particle remain constant! However, since coordinates alone do not incorporate the physical meaning, we have to calculate the actually measurable distances Δl between particles placed at different positions. With the line element (15.25), we find the spacelike distances in the xy-plane

$$(\Delta l)^2 = (1 + p)\Delta x^2 + (1 - p)\Delta y^2 + 2q\Delta x\Delta y. \tag{15.32}$$

As an example, we consider particles on a circle, with radius R, in the xy-plane. Because of $\Delta x = R\cos(\varphi)$ and $\Delta y = R\sin(\varphi)$, their distance from the origin is given by

$$(\Delta l)^2 = (1 + p)R^2\cos^2(\varphi) + (1 - p)R^2\sin^2(\varphi) + 2q\,R^2\sin(\varphi)\cos(\varphi)$$

$$= R^2[1 + p\cos(2\varphi) + q\sin(2\varphi)]$$

$$\tag{15.33}$$

with the time-dependent parameters p and q from (15.24). In Fig. 15.1 the temporal evolution of the distances Δl is sketched for the pure "+" and "×" polarization, respectively. For the purpose of illustration, the elongations drawn in Fig. 15.1 are strongly enlarged, since actually we have $|A^+| \ll 1$ and $|A^\times| \ll 1$.

15.3 Quadrupole Approximation

We now return to the retarded solution (15.14) of the inhomogeneous field equation. Let R be the radius of the source with barycenter $\boldsymbol{R}_S = \boldsymbol{0}$. We assume that all particles that make up the source have velocities much smaller than the speed of light.

From $v \ll c$ follows $R \ll \lambda$, where λ is the wavelength of the gravitational wave. Then $h_{\mu\nu}(\underline{x})$ can be expanded for $|\boldsymbol{r}| = r \gg \lambda \gg R$, i.e., in the far-field approximation, in terms of multipoles. When one carries out this expansion, one immediately realizes that, in analogy with electrodynamics, the monopole term vanishes. However, in contrast to electrodynamics, the dipole term vanishes, too. Therefore, the lowest non-vanishing contribution comes from the quadrupole term:

$$h_{jk}(\underline{x}) = \frac{2G}{c^6} \frac{d^2}{dt^2} T_{jk}\left(t - \frac{r}{c}\right),\tag{15.34}$$

with the reduced quadrupole moment

$$T_{jk} = \int T^{00}\left(x_j x_k - \frac{1}{3}\delta_{jk} r^2\right) d^3 x\tag{15.35}$$

of the source. The reason is that, contrary to electric charges, no negative masses exist.

Using the quadrupole approximation, we can calculate the energy radiated by gravitational waves.

15.4 Energy Radiated by Gravitational Waves

15.4.1 Radiated Power

Averaging the energy-momentum tensor $T_{\mu\nu}$ of the masses emitting a gravitational wave over all wavelengths yields

$$t_{\mu\nu} = \frac{c^4}{32\pi}\langle\partial_\mu h_{\alpha\beta}\partial_\nu h^{\alpha\beta}\rangle.$$

A derivation of this equation can be found, e.g., in Misner et al. [21].

The total energy that is radiated can be calculated via the energy flux t_{0k} passing through the surface of a sphere. The calculation leads to

$$-\frac{dE}{dt} = \frac{G}{5c^5}\langle\dddot{T}_{jk}\dddot{T}^{jk}\rangle.\tag{15.36}$$

Here T_{jk} is the reduced quadrupole tensor from Eq. (15.35).

For two masses m_1, m_2 orbiting around their common center of gravity, at a relative distance of a, the calculation yields

$$\langle \dddot{T}_{jk}\dddot{T}^{jk}\rangle = 32a^4\omega^6 m_{red}^2 \tag{15.37}$$

with the reduced mass

$$m_{red} = \frac{m_1 m_2}{m_1 + m_2} = \frac{m_1 m_2}{M} \tag{15.38}$$

and the angular velocity ω. The latter is given by Kepler's third law,

$$\omega = \left(\frac{GM}{a^3}\right)^{1/2}. \tag{15.39}$$

The result (15.37) can be understood in an intuitive way: Estimating the size of the mass quadrupole moment by $m_{red} a^2$, and taking into account the periodic time dependence of the motion $\propto \sin(\omega t)$, one obtains, apart from the numerical prefactor, the result (15.37).

Inserting Eqs. (15.37)–(15.39) in (15.36) one obtains for the total power radiated

$$-\frac{dE}{dt} = \frac{32(m_1 m_2)^2 MG^4}{5c^5 a^5}. \tag{15.40}$$

Eliminating in this equation m_1, m_2 through their Schwarzschild radii $R_{S1} = 2Gm_1/c^2$, $R_{S2} = 2Gm_2/c^2$,

$$m_1 = \frac{R_{S1}c^2}{2G}, \quad m_2 = \frac{R_{S1}c^2}{2G}, \tag{15.41}$$

one finds

$$-\frac{dE}{dt} = \frac{1}{5}\frac{c^5}{G}\left(\frac{R_{S1}}{a}\right)^2\left(\frac{R_{S2}}{a}\right)^2\frac{R_{S1}+R_{S2}}{a} \tag{15.42}$$

with the *Planck power*

$$\frac{c^5}{G} = 3.62825(82)\cdot 10^{52}\,\text{W}. \tag{15.43}$$

We will encounter the power again in the discussion of the Planck units in Chap. 30. This power is bigger by a factor of roughly 10^{26} than the luminosity of our Sun, and much higher than the total luminosity of all stars in the visible universe.

Since in general Schwarzschild radii are very much smaller than orbital radii, the actual power emitted by real systems is much less. For the Earth-Moon system, with $R_{S,Earth} = 8.9$ mm, $R_{S,Moon} = 0.11$ mm and an average distance of 384,000 km, one finds a value of 3.6×10^{-6} W. However, for two neutron stars with 1.4 solar masses each, orbiting at a distance of $700,000$ km, the value is 5.6×10^{25} W. It can be seen from Eq. (15.42) that a maximum value is reached when the orbital radii become of the order of the Schwarzschild radii, as is the case shortly before the merging of two black holes.

15.4.2 Approach of the System Partners, Chirp Mass

The energy emitted from the binary system in the form of gravitational waves is taken from the orbital energy. As a consequence, the orbital radius decreases, and the system partners come closer and closer to each other. Taking the derivative of the orbital energy

$$E = -\frac{1}{2}\frac{Gm_1 m_2}{a} \tag{15.44}$$

with respect to time

$$\frac{dE}{dt} = \frac{1}{2}\frac{Gm_1 m_2}{a^2}\frac{da}{dt}, \tag{15.45}$$

one finds for the time rate of the decrease of the distance

$$\frac{da}{dt} = 2\frac{dE}{dt}\frac{a^2}{Gm_1 m_2} = -\frac{64m_1 m_2 M G^3}{5c^5 a^3}. \tag{15.46}$$

Here, too, we can express the masses m_1 and m_2 through their Schwarzschild radii, and obtain

$$\frac{da}{dt} = -\frac{8}{5}c\frac{R_{S1}R_{S2}(R_{S1} + R_{S2})}{a^3}. \tag{15.47}$$

For the example of two neutron stars with 1.4 solar masses at a distance of $700,000$ km, the calculation leads to a decrease of the orbital radius of 6.3 m per year, see exercise 15.1.

The closer the two system partners approach, the faster they circle around each other, and the shorter is the orbital period T. As a result, the frequency of the emitted gravitational wave increases. Since the polarizations of the gravitational wave rotate twice as fast as the system, the frequency f is

$$f = \frac{2}{T} = \frac{\omega}{\pi} = \frac{1}{\pi}\left(\frac{GM}{a^3}\right)^{1/2}. \tag{15.48}$$

The increase of the frequency then is

$$\dot{f} = \frac{df}{da}\frac{da}{dt} = -\frac{3}{2\pi}\frac{\sqrt{GM}}{a^{5/2}}\left(-\frac{64m_1m_2MG^3}{5c^5a^3}\right) = \frac{96\pi^{8/3}m_1m_2G^{5/3}f^{11/3}}{5c^5M^{1/3}}.$$

$$(15.49)$$

This equation can be simplified to

$$\dot{f} = \frac{96\pi^{8/3}}{5}\left(\frac{G\mathcal{M}}{c^3}\right)^{5/3}f^{11/3},$$

$$(15.50)$$

when one introduces the so-called *chirp mass*

$$\mathcal{M} = \left(\frac{m_1^3m_2^3}{M}\right)^{1/5}.$$

$$(15.51)$$

Solving Eq. (15.50) with respect to $\mathcal{M}$ yields

$$\mathcal{M} = \frac{c^3}{G}\left(\frac{5}{96}\pi^{-8/3}f^{-11/3}\dot{f}\right)^{3/5}.$$

$$(15.52)$$

The chirp mass is of practical importance for the detection of gravitational waves, because from the frequency and the change of the frequency of the gravitational wave shortly before the merger of the two system partners, the chirp mass of the system can be directly inferred.

As an example, we consider the merger of two black holes, with k solar masses each. The result for the chirp mass is

$$\mathcal{M} = \frac{(k^2M_\odot^2)^{3/5}}{(2kM_\odot)^{1/5}} = \frac{1}{2^{1/5}}k \cdot M_\odot \approx 0.871\,k \cdot M_\odot.$$

$$(15.53)$$

Thus the chirp mass already provides a measure for the size of the two merging masses.

It should be mentioned that the chirp mass can also be derived in classical mechanics from the Kepler motion of the reduced mass in the center-of-gravity system. Solving Kepler's third law (15.39) with respect to $1/a$,

$$\frac{1}{a} = \left(\frac{\omega^2}{GM}\right)^{1/3}$$

$$(15.54)$$

and inserting this expression into Eq. (15.44) for the orbital energy, the latter can be expressed in terms of the chirp mass as

$$E = -\frac{1}{2}G^{2/3}m_1m_2M^{-1/3}\omega^{2/3} \equiv -\frac{1}{2}\left(G^2\mathcal{M}^5\right)^{1/3}\omega^{2/3}.$$

$$(15.55)$$

15.4.3 Estimate of the Amplitude of the Gravitational Wave

Next we wish to estimate the amplitude of the radiated gravitational wave. The starting point is the retarded solution (15.14), where, because of the large distance to the source ($|\mathbf{r}| \gg |\mathbf{r}'|$), we may draw the dependence on $1/r$ in the denominator in front of the integral, and use the quadrupole approximation. For two identical masses m, which revolve around their common center of gravity at a distance a, with angular velocity ω, we can estimate the energy tensor from the rotational energy as $T \sim ma^2\omega^2$. From (15.14) then follows

$$h \sim \frac{G}{c^4} \frac{ma^2\omega^2}{r}, \tag{15.56}$$

where, for simplification, in the last two steps we have suppressed the indices. When we insert ω^2 via Kepler's third law (15.39), we arrive at the estimate

$$h \sim \frac{G^2 m^2}{rac^4} \sim \frac{R_S^2}{ra} \tag{15.57}$$

with the Schwarzschild radius $R_S = 2Gm/c^2$.

The maximum of the amplitude is reached when the orbit of the two black holes has shrunk so far that their event horizons just touch. With $a = R_S$, one then has

$$h \sim \frac{R_S}{r}. \tag{15.58}$$

The amplitude is therefore given by the ratio of the Schwarzschild radii of the sources and their distance from us. For a typical value of the Schwarzschild radius of $R_S \sim 100$ km (corresponding to black holes with 33 solar masses) and for an assumed distance of one billion light years, $r \sim 10^9 \cdot 9.64 \cdot 10^{15}$ m $\sim 10^{25}$ m, the unimaginably small value of $h \sim 10^{-20}$ is obtained. The actually expected values even lie in the range of $h \sim 10^{-20} \cdots 10^{-24}$.

15.5 The Laser Interferometer Gravitational Wave Observatory (LIGO)

The first direct detection of gravitational waves was made in 2015 by the Laser Interferometer Gravitational Wave Observatory LIGO [1]. The Nobel Prize in Physics was awarded for this discovery in 2017 [15]. We will take a closer look at this observatory and the method of detection.

Fig. 15.2 Simplified
structure of the LIGO
interferometer (Credit:
Menner [CCo], via
Wikimedia Commons). The
length of each interferometer
arm is 4 km

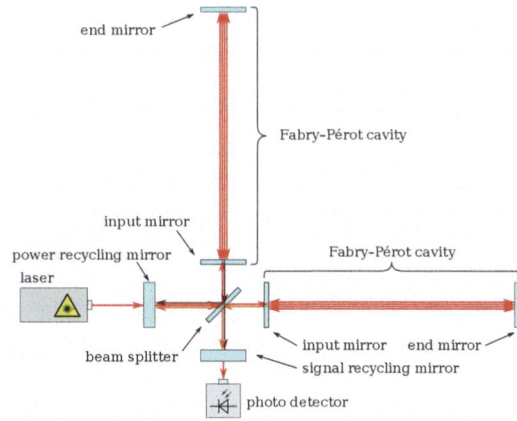

15.5.1 The LIGO Detectors

Gravitational waves are measurable by the apparent changes in length of objects
when a gravitational wave passes through them. The interferometric detection of
gravitational waves is, therefore, based on the precise measurement of the optical
phase difference between two rays of light from one common source, flying back
and forth in two interferometer arms, whose physical lengths are distorted by the
passage of the wave. The detectable signal is proportional to the gravitational strain
amplitude and the length of the interferometer arms. The challenge is that the
relative strain amplitudes that occur when a gravitational wave passes through the
Earth, are, as mentioned before, on the order of magnitude $h \sim 10^{-20} \cdots 10^{-24}$.
To get the strongest possible signal, the arm lengths of the interferometers must
be large, ideally close to a quarter of the wavelength of the gravitational wave, for
example, 750 km at 100 Hz. For operational purposes this means that the optical
path in each arm must be much longer than its physical length.

This is the principle behind the setup of the LIGO detectors at the two observing
stations in Hanford, Louisiana, USA, and Livingstone, Oregon [17]. The two
instruments are identical Michelson interferometers, enclosed in an L-shaped ultra-
high vacuum system with a pressure below 1 μPascal. The detectors are 3000 km
apart, which corresponds to a direct travel time of light of 10 ms. Figure 15.2 shows
a simplified sketch of a LIGO detector.

Each interferometer arm contains two massive (40 kg) fused silica mirrors
with a specially designed multi-layer optical coating to achieve the required high
reflectivity. Each mirror is suspended from a fourfold pendulum system, which is
mounted on an actively controlled seismic isolation platform. This also compensates
for the smallest vibrations caused by the tidal effects of the Earth's crust and
microseismic activity. Any thermal movements of the last pendulum stage are
eliminated by suspensions on fibers made of quartz glass. In each interferometer
arm, the mirrors are 4 km apart.

The laser light radiated into the interferometer with a power of 20 W is constructively superimposed with itself by a highly reflective mirror, and thus amplified to 700 W ("*power recycling*"). The light is divided by a beam splitter and sent into the two equally long arms of the interferometer. To increase sensitivity, each arm contains a Fabry-Pérot interferometer,[1] which stores the photons while the light bounces back and forth between the test mass mirrors. In this way, the effective path covered by the light in each interferometer arm increases to over 1000 km.

The light released from the Fabry-Pérot interferometers returns to the beam splitter and is guided to the exit, where the interference pattern of the two partial beams are registered by a photodetector. The output signal is also constructively superimposed by an additional high-reflecting mirror ("*signal recycling*"), which increases sensitivity. The passage of the gravitational wave changes the lengths of the interferometer arms, and thus also the interference pattern. The power of the laser light measured at the photodetector can then be converted, using a calibration method, into the shifts of the positions of the test masses, and thus into the strength of the gravitational wave signal.

15.5.2 First Direct Detection of a Gravitational Wave Event

The Advanced LIGO detectors, which measure with extremely high sensitivity, began measuring in September 2015. Already on September 14, 2015, a strong signal was registered both in Livingston and in Hanford, which could be interpreted as the signature of the passage of a gravitational wave [1]. Livingston registered the wave 6.9 ms earlier than Hanford further north, as one would expect for a source on the southern celestial hemisphere. The signal was given the designation GW150914 [1] because of the day of the observation.

Figure 15.3 summarizes the results. The signal increases in frequency and amplitude over a period of 200 ms, with a maximum at about 150 Hz. The most likely interpretation is spiraling in and eventual merging of two massive objects, which emit gravitational waves, see Fig. 15.4. From the time evolution of the frequency and its time derivative, a value of approximately 70 $M_{\odot}$ is calculated from the formula (15.50) for the chirp mass, as the lower limit for the sum of the masses of the bodies. This large mass, and the fact that the objects reach a relatively high frequency before merging, indicates that they are two orbiting black holes.

With the help of waveform models based on general relativity, further analysis of the data led to values of $35.6^{+4.8}_{-3.0}$ $M_{\odot}$ and $30.6^{+3.0}_{-4.4}$ $M_{\odot}$ for the masses of the merging black holes, and a value of $63.1^{+3.3}_{-3.0}$ $M_{\odot}$ for the remaining black hole. The energy corresponding to the mass difference of $3.1^{+0.4}_{-0.4}$ $M_{\odot}$ was thus emitted during the

---

[1] A Fabry-Pérot interferometer consists of two partially transmissive mirrors, between which the photons bounce back and forth, and at each reflection, part of the light is released through the mirrors.

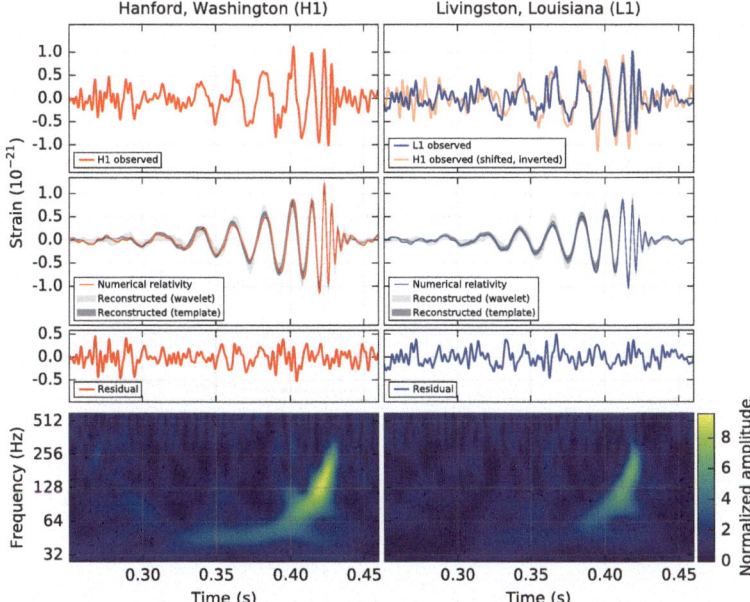

Fig. 15.3 The gravitational wave event GW150914. Above: The signals measured by LIGO Hanford and LIGO Livingston as a function of time. The signal in Livingston is superimposed on that of Hanford, shifted by the difference in arrival time of $6.9^{+0.5}_{-4}$ ms. Middle: Theoretical waveform for a system with the mass values assumed for GW150914, and the residuals remaining after subtracting the theoretical waveform from the measured signals. Bottom: Frequency of the gravitational wave as a function of time. All times are referred to the trigger time at 9:50:45 UTC on 09/14/2015. From: P.B. Abbott et al. (LIGO Scientific Collaboration and Virgo Collaboration) [1] ©APS. Reused under the terms of the Creative Commons Attribution 3.0 License

merger in the form of gravitational waves. A value of 410^{+160}_{-180} Mpc was estimated for the distance of the event.

During the first observation campaign, which lasted until January 2016, two further gravitational wave events could be detected, GW151226 and GW170104. In both cases, the observations could be interpreted as the merger of two black holes.

15.5.3 Further Gravitational-Wave Observations

After the sensitivity of the LIGO detectors had been increased again, the second observation campaign started in November 2016, and lasted until August 25, 2017. 8 more gravitational wave events were observed. A compilation of all 11 sources of gravitational waves found in the two campaigns can be found in Table 15.1, in which, in addition to the masses of the merging objects, the final mass, the energy radiated in the form of gravitational waves, and the distance and the redshift of the events are given.

Fig. 15.4 Top: The waveform calculated using numerical relativity with the mass parameters assumed for GW150914. Bottom: The distance between the two black holes in units of the Schwarzschild radius, and their velocity in units of the speed of light as a function of time. From: P.B. Abbott et al. (LIGO Scientific Collaboration and Virgo Collaboration) [1] ©APS. Reused under the terms of the Creative Commons Attribution 3.0 License

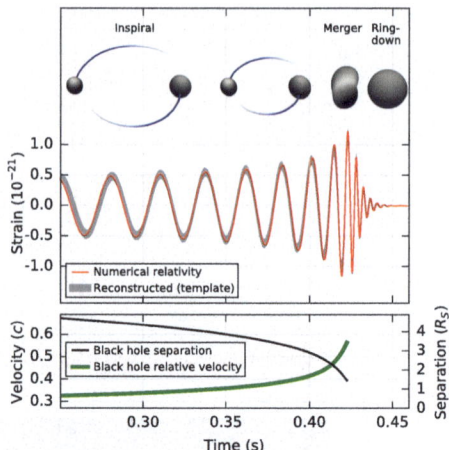

Table 15.1 List of gravitational-wave events detected by LIGO in the first two observing campaigns. Data from Abbot et al. [2]

Event	d/Mpc	z	$m_1/M_\odot$	$m_2/M_\odot$	$m_f/M_\odot$	$E_r/(M_\odot c^2)$
GW150914	410^{+160}_{-180}	$0.09^{+0.03}_{-0.04}$	$35.6^{+4.8}_{-3.0}$	$30.6^{+3.0}_{-4.4}$	$63.1^{+3.3}_{-3.0}$	$3.1^{+0.4}_{-0.4}$
GW151012	1060^{+540}_{-480}	$0.21^{+0.09}_{-0.09}$	$23.3^{+14.0}_{-5.5}$	$13.6^{+4.1}_{-4.8}$	$35.7^{+9.9}_{-3.8}$	$1.5^{+0.5}_{-0.5}$
GW151226	440^{+180}_{-190}	$0.09^{+0.04}_{-0.04}$	$13.7^{+8.8}_{-3.2}$	$8.9^{+0.3}_{-0.3}$	$20.5^{+6.4}_{-1.5}$	$1.0^{+0.1}_{-0.2}$
GW170104	960^{+430}_{-410}	$0.19^{+0.07}_{-0.08}$	$31.0^{+7.2}_{-5.6}$	$20.1^{+4.9}_{-4.5}$	$49.1^{+5.2}_{-3.9}$	$2.2^{+0.5}_{-0.5}$
GW170608	320^{+120}_{-110}	$0.07^{+0.02}_{-0.02}$	$10.9^{+5.3}_{-1.7}$	$7.6^{+1.3}_{-2.1}$	$17.8^{+3.2}_{-0.7}$	$0.9^{+0.05}_{-0.1}$
GW170729	2750^{+1350}_{-1320}	$0.48^{+0.19}_{-0.20}$	$50.6^{+16.6}_{-10.2}$	$34.3^{+9.1}_{-10.1}$	$80.3^{+14.6}_{-10.2}$	$4.8^{+1.7}_{-1.7}$
GW170809	990^{+320}_{-380}	$0.20^{+0.05}_{-0.07}$	$35.2^{+8.3}_{-6.0}$	$23.8^{+5.2}_{-5.1}$	$56.4^{+5.2}_{-3.7}$	$2.7^{+0.6}_{-0.6}$
GW170814	580^{+160}_{-210}	$0.12^{+0.03}_{-0.04}$	$30.7^{+5.7}_{-3.0}$	$25.3^{+2.9}_{-4.1}$	$53.4^{+3.2}_{-2.4}$	$2.7^{+0.4}_{-0.3}$
GW170817	40^{+10}_{-10}	$0.01^{+0.00}_{-0.00}$	$1.46^{+0.12}_{-0.10}$	$1.27^{+0.09}_{-0.09}$	≤ 2.8	≥ 0.04
GW170818	1020^{+430}_{-360}	$0.20^{+0.07}_{-0.07}$	$35.5^{+7.5}_{-4.7}$	$26.8^{+4.3}_{-5.2}$	$59.8^{+4.8}_{-3.8}$	$2.7^{+0.5}_{-0.5}$
GW170823	1850^{+840}_{-840}	$0.34^{+0.13}_{-0.14}$	$39.6^{+10.0}_{-6.6}$	$29.4^{+6.3}_{-7.1}$	$65.6^{+9.4}_{-6.6}$	$3.3^{+0.9}_{-0.8}$

15.5.4 Merger of Two Neutron Stars

The event GW170817 stands out from the list. Unlike the other events, this is not the merger of two black holes, but the merger of two *neutron stars*. The fate of the double pulsar PSR 1913+16, which will be discussed in detail in Chap. 21, namely the merging predicted in 300 million years, has already been fulfilled in the event GW170817. While for black holes the gravitational wave signal appears only for a few tenths of a second before the merger, the signal for the merger of the two neutron stars lasted over 10 s. Since the event could also be detected in the European gravitational wave detector Virgo, several thousand kilometers away near Pisa, the location of the event could be determined very precisely by triangulation, due to the long baselines, in a range of a solid angle of 16 square degrees on the southern celestial hemisphere.

What was special about this event was that, at the same time as the merger of the neutron stars occurred, the gamma satellite Fermi observed a violent burst of gamma radiation in the frequency range from 10 to 300 keV (a so-called γ ray burst) [18]. In addition to the gamma ray burst, the "Kilonova" AT2017gfo was observed, an afterglow in the infrared and in the visible range, produced by radioactive processes during the merger. The afterglow faded within 10 days. Analysis of the spectra clearly showed that the heavy element strontium had been produced [26] by rapid neutron capture. This is an example of the r-process, which we will discuss in more detail in Sect. 19.6.2.

Since this event could be measured both as a gravitational wave and in the range of electromagnetic radiation, this is seen as an example of "multi-messenger astronomy".

During the third observation campaign of the LIGO-Virgo network, on April 25, 2019, another gravitational wave, caused by the merger of two neutron stars [3], could be captured. That observing campaign, which ended in March 2020, yielded a total of 56 gravitational-wave detections. This brought the total number of gravitational-wave events detected at the time to 67. An interactive overview can be found on the Cardiff University website.[2] After over two years of upgrades, maintenance work, and construction of a new "squeezer cavity" LIGO began its fourth observing run, O4, in May 2023. The sensitivity of the detectors has been increased by 30 %. This increased sensitivity will result in a higher rate of observed gravitational-wave signals, resulting in a detection of a merger every 2 or 3 days.

15.6 Hunting Gravitational Waves Using Pulsars

Collisions, or mergers, of galaxies are supposed to have happened frequently in an early phase of the cosmos, more precisely, during the epoch of galaxy formation [25]. Such mergers even happen today, see, e.g., the montage of six mergers observed by the NASA/ESA Hubble Space Telescope [13].[3] In general, the relative angular momentum of colliding galaxies is non-zero, therefore the collisions do not happen "head-on". As a consequence, the supermassive black holes in the centers of the galaxies rotate around their common center of gravity, and emit gravitational waves at frequencies corresponding to their orbital periods.

---

[2] https://catalog.cardiffgravity.org.

[3] In fact, our galaxy is facing this fate, and is expected to merge with the Andromeda Galaxy in 4.5 billion years, to form a new galaxy nicknamed the "milkomeda".

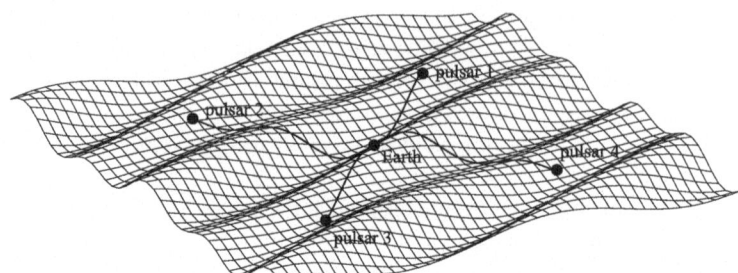

Fig. 15.5 Idea of a *Pulsar Timing Array*: Due to the gravitational wave the pulse arrival times of the radio signals change. Graphics: K. Mikić

Such binary systems of massive black holes from the epoch of galaxy formation are thought to have created a stochastic, continuous background of gravitational waves that permeates the entire universe. The frequencies of the gravitational waves would range from a few nHz to a few μHz, therefore they lie well outside the frequency range accessible to LIGO.

It was suggested early on [8] that measurements of pulsar pulse arrival times could be used to cover the frequency range up to a few μHz. The rationale is simple: Millisecond pulsars with periods shorter than 30 ms are among the most stable known rotators in the universe, and the long-term stability of their rotations over more than 10 years is comparable to that of atomic clocks. If the arrival times of the radio signals vary because they had to traverse a region of space deformed by gravitational waves, information about the gravitational waves can be extracted from this data. The pulse arrival times of sources whose connecting lines to the Earth are perpendicular to each other change in opposite directions. Some pulses arrive delayed, others earlier. Thus, the fluctuations in the arrival times are correlated with the position of the sources in the sky, see Fig. 15.5.

The *European Pulsar Timing Array* *(EPTA)* [12] has set itself the goal to detect such fluctuations, and thus to draw conclusions as to the sources of the gravitational waves that produce them. *EPTA* can detect frequencies of gravitational waves corresponding to one oscillation in 30 years, i.e., 9.47×10^8 s. The associated frequency is 1.05 nHz, and the wavelength is as large as 30 light years! On the other hand, the huge amount of timing data collected regularly by the array over more than 25 years allows for the investigation of frequencies of up to 100 oscillations per month, corresponding to a frequency of 38.5 μHz.

In a first search, 42 of the most accurate known millisecond pulsars were observed over years with the largest radio telescopes on the entire Earth [6]. No signal of gravitational waves was detected in the first search [23]. However, numerical simulations at the Max Planck Institute for Gravitational Physics (Albert Einstein Institute) had suggested that the sensitivity of the *Pulsar Timing Array* will increase with the duration of the observation period, see also Babak et al. [14], and that within the next few years gravitational waves should be detected by

EPTA—either as a dominant signal from a single source, or as a gravitational-wave background from a number of less strong sources [20].

Meanwhile, this prediction has come true. In June 2023, the *EPTA* collaboration presented the results of a second search for continuous gravitational wave signals, obtained from observations of 25 ms pulsars over 14 to 25 years [4]. The most significant candidate event from this search has a gravitational wave frequency of 4–5 nHz [5]. Most probably the signal is not due to a continuous gravitational-wave background but is generated by a supermassive black hole binary in the local universe. The detection of a gravitational wave with such a low frequency using a pulsar timing array is one of the most spectacular results of recent gravitational astronomy.

Another *Pulsar Timing Array*, the North American Nanohertz Gravitational Wave Observatory (NANOGrav) [22], reported that in data collected and analyzed over 13 years, a low-frequency signal was found that may be due to a stochastic gravitational-wave background [7].

15.7 Outlook

Other projects for the interferometric detection of gravitational waves are in progress. In Europe this is, in addition to Virgo with arm lengths of 3 km [24], GEO600 near Hanover, Germany [11] with two 600 m long arms. Both are part of the LIGO network. As the world's first detector, GEO600 uses *squeezed light*. Similar to position and momentum in quantum mechanics, the amplitude and phase of laser light obey a Heisenberg uncertainty relation, therefore they vary from measurement to measurement. If one plots the measured values in a phase-amplitude diagram, they are distributed in a circular disk with blurred borders. When the laser light is squeezed, the shape of the circle is deformed into an ellipse with the same area. As a result, the uncertainty in one variable decreases while it increases in the other. The advantage is that the shot noise of the laser light can be reduced. Essential parts of the instruments and techniques developed and tested at GEO600, were implemented in the two LIGO detectors, with which the first gravitational waves could be detected in 2015.

In Japan, the *Kamioka Gravitational Wave Detector* KAGRA is in operation [16]. Its arm length is also 3 km. What is special about this detector is that the mirrors are cooled down to 20 K to reduce thermal noise. The detection limit is $h \sim 3 \cdot 10^{-24}$ at a frequency of 100 Hz. KAGRA has joined the LIGO-Virgo network in the hunt for gravitational waves.

The European Space Agency ESA is planning the space-based project LISA (*Laser Interferometer Space Antenna*) [19]. This is a Michelson interferometer consisting of 3 satellites at a mutual distance of 2.5 million kilometers that shall follow the orbit of the Earth, but at an angular distance of 20° behind it, and with an inclination of the satellites' triangle of 60° with respect to the ecliptic plane. Together with the Earth, they will orbit around the Sun. LISA will cover the wavelength range from 0.1 to 1 Hz, which is not accessible on Earth. After a successful pathfinder mission, in which the measurement technology was tested in space, and the achieved measurement accuracy exceeded the requirements by a factor of 5, the science ministers of the nations involved in the ESA approved funding for the project in November 2019. Commissioning is planned for 2034.

Within the framework of the seventh research framework program of the European Commission, the concept of an Einstein telescope was investigated [10]. This is to be built underground to reduce seismic noise, in the shape of a triangle with three 10 km long arms, i.e., in the same geometry as LISA. Two arms each are used for two interferometers, resulting in a total of six detectors. Three of them are to be optimized for measuring low frequencies (2 to 40 Hz), three for higher frequencies. To suppress thermal noise, the optical elements are to be cooled down to 10 K. A location has not yet been selected.

Together, all these experiments should not only detect gravitational waves, but also help to better understand phenomena that cause gravitational waves. What is of central interest, is for example (as discussed in Sect. 15.6), the rotation and eventual merger of black holes in the centers of galaxies when the associated galaxies merge into one galaxy. The precise measurement of the properties of gravitational waves should also allow further tests of general relativity, as well as point to phenomena beyond existing physics, such as cosmic strings, as they are predicted by string theory [27].

A good overview of the possibilities of gravitational wave astronomy can be found on the homepage of the LISA project [19].

In summary, more than 100 years after Einstein's prediction of gravitational waves, gravitational wave astronomy has become an integral part of astrophysical research. Together with electromagnetic radiation and neutrino detection, it has opened a new era of multi-messenger astronomy.

15.8   Exercises

15.1. Decrease of orbital radius for two neutron stars Confirm the statement below Eq. (15.47) that the orbital radius of two neutron stars with $M = 1.4M_\odot$ at a distance of 700,000 km decreases by approximately 6.3 m per year.

References

1. Abbot, B.P. (LIGO Scientific Collaboration and Virgo Collaboration): Observation of gravitational waves from a binary black hole merger. Phys. Rev. Lett. **116**, 061102 (2016)
2. Abbot, B.P., et al. (LIGO Scientific Collaboration and Virgo Collaboration): GWTC-1: A gravitational-wave transient catalog of compact binary mergers by LIGO and Virgo during the first and second observing runs. Phys. Rev. X **9**, 031040 (2019)
3. Abbot, B.P., et al.: LIGO Scientific Collaboration and Virgo Collaboration: GW190425: Observation of a compact binary coalescence with total mass $\sim 3.4~M_\odot$. Astrophys. J. Lett. **892**, L3 (2020)

4. Antoniadis, J., et al.: The second data release from the European Pulsar Timing Array: I. The data set and timing analysis, arXiv:2306.16224v1 (2023)
5. Antoniadis, J., et al.: The second data release from the European Pulsar Timing Array: IV. Search for continuous gravitational wave signals, arXiv:2306.16226v1 (2023)
6. Babak, S., et al.: European Pulsar Timing Array limits on continuous gravitational waves from individual supermassive black hole binaries. Mon. Not. R. Astron. Soc. **455**, 1665 (2016)
7. De Luca, V., et al.: NANOGrav data hints at primordial black holes as dark matter. Phys. Rev. Lett. **126**, 041303 (2021)
8. Detweiler, S.: Pulsar timing measurements and the search for gravitational waves. Astrophys. J. **234**, 1100 (1979)
9. Einstein, A.: Über Gravitationswellen (About gravitational waves), pp. 154–167. Sitzungs-berichte der Königlich-Preußischen Akademie der Wissenschaften, Berlin (1916)
10. Einstein telescope Homepage: https://www.et-gw.eu
11. GEO600 Homepage: https://www.geo600.org
12. Homepage of the European Pulsar Timing Array: https://www.epta.eu.org
13. Hubble showcases six beautiful galaxy mergers: https://esahubble.org/news/heic2101/ (2021)
14. Babak, S., et al.: Forecasting the sensitivity of pulsar timing arrays to gravitational wave backgrounds. Phys. Rev. D **110**(6), 063022 (2024), DOI: https://10.1103/PhysRevD.110.063022
15. https://www.nobelprize.org/prizes/physics/2017/press-release/
16. KAGRA Homepage: https://gwcenter.icrr.u-tokyo.ac.jp/en
17. LIGO-Homepage: https://www.ligo.caltech.edu/LA
18. LIGO Scientific Collaboration and Virgo Collaboration, *Fermi* Gamma-Ray Burst Monitor, and INTEGRAL: Gravitational waves and gamma-rays from a binary neutron star merger: GW170917 and GRB 170817A. Astrophys. J. Lett. **848**, L13 (2017)
19. LISA Homepage: https://www.lisamission.org
20. Mayor, L.: Hunting gravitational waves with pulsars. Phys. World **27**(10), 38 (2014)
21. Misner, C.W., Thorne, K.S, Wheeler, J.A.: Gravitation. Princeton University Press, Princeton (2017), ISBN 978-0-691-17779-3
22. NANOGrav-Homepage: https://nanograv.org
23. Perera, B.B.P., et al.: Improving timing sensitivity in the microhertz frequency regime: limits from PSR J1713+0747 on gravitational waves produced by supermassive black hole binaries. Mon. Not. R. Astron. Soc. **478**, 218 (2018)
24. Virgo Homepage: https://www.virgo-gw.eu
25. Volonteri, M., Haardt, F., Madau, P.: The assembly and merging history of supermassive black holes in hierarchical models of galaxy formation. Astrophys. J. **582**, 559 (2003)
26. Watson, D., et al.: Identification of strontium in the merger of two neutron stars. Nature **574**, 497 (2019)
27. Zwiebach, B.: A First Course in String Theory. Cambridge University Press (2009)

Visualization in General Relativity

16

Contents

In Chap. 9 we have already dealt with visualization in special relativity. We distinguished between a more abstract representation from an outside view using Minkowski diagrams, and a representation from the first-person view. In general relativity, the prevalent focus in literature is on abstract representations. These

Supplementary Information The online version contains supplementary material available at https://doi.org/10.1007/978-3-662-71332-7_16.

include, for example, the *Penrose-Carter diagram*[1,2] for the illustration of the global structure of black-hole spacetimes, of lightlike and timelike geodesics, or of features of gravitational fields, such as quantities characterizing the curvature, or gravitational waves.

In recent years, the visualization of what an observer would actually see in a four-dimensional spacetime, has received increasing attention, and this is why we shall also focus on it.

16.1   Abstract Visualization in General Relativity

16.1.1 Lightlike and Timelike Geodesics

A first impression of the structure of a spacetime can be gained by looking at the behavior of geodesics. Figure 16.1 shows lightlike geodesics, deflected in the

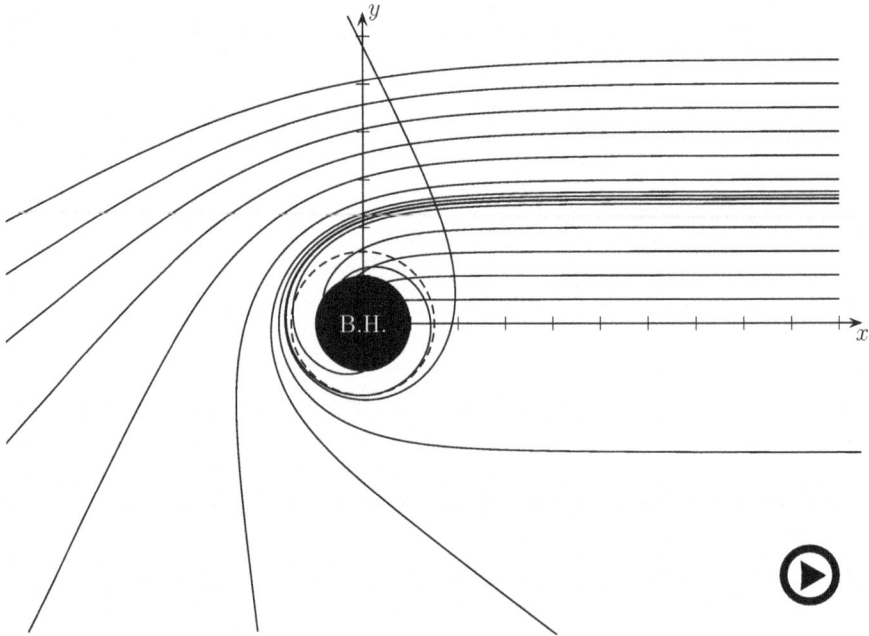

Fig. 16.1 Light deflection by a Schwarzschild black hole. The rays of light start at $x = 10r_s$ and $y = \{0, 0.5, 1, 1.5, 2, 2.5, 2.5875, 2.6, 2.67, 2.75, 3, 3.5, 4, 4.5, 5, 5.5, 6\}r_s$. The *dashed circle* corresponds to the photon orbit (▶ https://doi.org/10.1007/000-hw1)

---

[1] Roger Penrose, ⋆1931, English mathematician and theoretical physicist. Nobel Prize 2020 "for the discovery that black hole formation is a robust prediction of the general theory of relativity" (Nobel Prize Committee).

[2] Brandon Carter, ⋆1942, Australian theoretical physicist.

vicinity of a Schwarzschild black hole by the curvature of the spacetime. The deflection is the more pronounced, the closer the ray of light comes to the black hole. If a ray of light enters below the photon orbit, $r_{po} = 3r_s/2$, it must inevitably fall into the black hole. Shortly above the photon orbit, it can be deflected so strongly that it is reflected back to the observer, or winds several times around the black hole, before disappearing in some other direction.

The rays of light in Fig. 16.1 start at the position (x, y), parallel to the x-axis. The polar coordinates corresponding to this position are $r = \sqrt{x^2 + y^2}$ and $\varphi = \arctan(y, x)$. The rays of light start under an angle ξ, referred to the local tetrad (13.37), which follows from $\rho = 1$, $\vartheta = \pi/2$ and (13.38)

$$\underline{y} = y^{(t)}\underline{e}_{(t)} + \cos(\xi)\underline{e}_{(r)} + \sin(\xi)\underline{e}_{(\varphi)} \tag{16.1a}$$

$$= y^{(t)}\underline{e}_{(t)} + \cos(\xi)\sqrt{1 - \frac{r_s}{r}}\partial_r + \frac{\sin(\xi)}{r}\partial_\varphi. \tag{16.1b}$$

When we transform the directional derivatives ∂_r and ∂_φ into Cartesian coordinates, we have

$$\partial_r = \cos(\varphi)\partial_x + \sin(\varphi)\partial_y \quad \text{and} \quad \partial_\varphi = -r\sin(\varphi)\partial_x + r\cos(\varphi)\partial_y. \tag{16.2}$$

We insert these relations in (16.1b), sort according to ∂_x and ∂_y, and require $y = 0$, to obtain the expression for ξ

$$\xi = \pi + \arctan\left(-\sqrt{1 - \frac{r_s}{r}}\tan(\varphi)\right). \tag{16.3}$$

We have also taken into account that the rays of light start in the negative x-direction.

In Fig. 16.1, there exists exactly one coordinate y for the initial position $x = 10r_s$ for which the ray of light asymptotically approaches the photon orbit. Inserting (13.41) in Eq. (16.3) yields the value $y \approx 2.5875$ for the critical observation angle. At the same time, this ray of light defines the rim of the shadow of the black hole. In the limit $x \to \infty$ we obtain, with approximately twice the value for y, the apparent diameter $D \approx 5.196r_s$ of the shadow.

The interactive software (*GeodesicViewer*) for the investigation of lightlike and timelike geodesics in a spacetime is described in Müller [20].

16.1.2 Embedding Diagram

To visualize the intrinsic geometry of a curved spacetime, we shall restrict ourselves to two-dimensional sub-manifolds of the spacetime. We will embed this manifold in our familiar three-dimensional Euclidean space in such a way that the intrinsic geometry is preserved.

Fig. 16.2
Flamm's paraboloid
shows the hyperplane
(t = const, $\vartheta = \pi/2$) of the
Schwarzschild metric (13.16)
embedded in
three-dimensional
Euclidean space

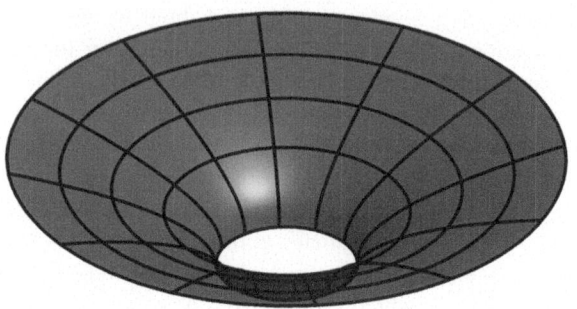

For the Schwarzschild metric we can restrict ourselves, because of its spherical symmetry, to the ($\vartheta = \pi/2$) plane at a fixed coordinate time t = const. With $dt = 0$ and $d\vartheta = 0$, the line element (13.16) reduces to

$$d\sigma_h^2 = \frac{dr^2}{1 - r_s/r} + r^2 d\varphi^2. \tag{16.4}$$

We can embed this hyperplane in Euclidean space, with the hyperplane written in cylindrical coordinates (r, φ, z), as

$$d\sigma_e^2 = \left[1 + \left(\frac{dz}{dr}\right)^2\right] dr^2 + r^2 d\varphi^2. \tag{16.5}$$

The direct comparison of $d\sigma_h^2$ with $d\sigma_e^2$ leads to the embedding function

$$z(r) = 2\sqrt{r_s}\sqrt{r - r_s}, \tag{16.6}$$

which is shown in Fig. 16.2 as a surface of revolution, and is called *Flamm's paraboloid*[3] [7].

16.1.3 Penrose-Carter Diagram

To study the causal structure and the asymptotic properties of a spacetime, we need a transformation which maps (compactifies) the complete spacetime onto a finite area, while preserving the causal structure. Penrose [29] and Carter [4] found a transformation of this type.

---

[3] Ludwig Flamm, 1885–1964, Austrian physicist.

Minkowski Spacetime

The *Penrose-Carter diagram* of the Minkowski spacetime is obtained by the transformation from spherical coordinates $(t, r, \vartheta, \varphi)$ to *conformal-compactified coordinates* $(\psi, \xi, \vartheta, \varphi)$. The transformation is defined by

$$ct + r = \tan\left(\frac{\psi + \xi}{2}\right), \quad ct - r = \tan\left(\frac{\psi - \xi}{2}\right), \tag{16.7}$$

and the inverse transformation leads to

$$\psi = \arctan(ct + r) + \arctan(ct - r) \quad \text{and}$$
$$\xi = \arctan(ct + r) - \arctan(ct - r). \tag{16.8}$$

The line element of the Minkowski metric in spherical coordinates, $ds^2 = -c^2dt^2 + dr^2 + r^2(d\vartheta^2 + \sin^2(\vartheta)d\varphi^2)$ is transformed to

$$ds^2 = \mathcal{K}^2 d\tilde{s}^2 = \mathcal{K}^2\left[-d\psi^2 + d\xi^2 + \sin^2(\xi)\left(d\vartheta^2 + \sin^2(\vartheta)d\varphi^2\right)\right], \tag{16.9}$$

with the conformal factor

$$\mathcal{K}^2 = \frac{1}{4}\left[\sec^2\left(\frac{\psi + \xi}{2}\right)\sec^2\left(\frac{\psi - \xi}{2}\right)\right]. \tag{16.10}$$

As the causal structure of a spacetime is preserved under conformal transformations, it is sufficient to investigate the compactified spacetime described by the line element $d\tilde{s}^2$. As a consequence of the transformation (16.7), the coordinates ψ and ξ are restricted to

$$-\pi < \psi + \xi < \pi, \quad -\pi < \psi - \xi < \pi, \quad \xi > 0. \tag{16.11}$$

Figure 16.3 shows the Penrose-Carter diagram for the hyperplane ($\vartheta = $ const, $\varphi = $ const). Straight lines with a slope of $\pm 45°$ correspond to rays of light (null geodesics). Timelike curves begin and end in the points $i^\pm$, whereas spacelike curves end in the point i^0. Light rays begin and end in $\mathscr{I}^\pm$. The exact structure of these boundaries represents neither points nor straight lines. A detailed discussion can be found, e.g., in Frauendiener [8].

Schwarzschild Spacetime

The Penrose-Carter diagram for the Schwarzschild spacetime can be obtained in a manner analogous to that of the Minkowski spacetime. The starting point is the Kruskal metric

$$ds^2 = -\frac{4r_s^3}{r}e^{-r/r_s}\left(dv^2 - du^2\right) + r^2d\Omega^2. \tag{13.81}$$

Fig. 16.3 Penrose-Carter
diagram of the Minkowski
spacetime. The *vertical
curves* correspond to planes
with constant radius r, and
the *dashed lines* represent
planes with constant time t.
The points $i^{\pm}$ and i^0 indicate
the timelike or spacelike
infinity; $\mathscr{I}^{\pm}$ is the lightlike
infinity. The straight line
$\xi = 0$ corresponds to $r = 0$

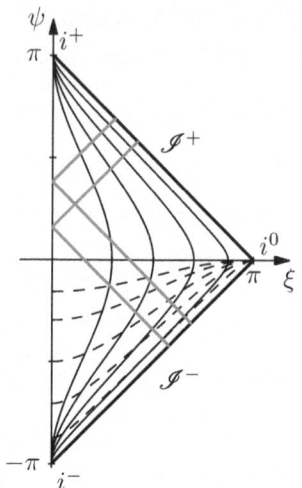

The transformation to conformal-compactified coordinates $(\psi, \xi, \vartheta, \varphi)$ reads

$$v = \frac{1}{2} \tan\left(\frac{\psi + \xi}{2}\right) + \frac{1}{2} \tan\left(\frac{\psi - \xi}{2}\right)$$

$$u = \frac{1}{2} \tan\left(\frac{\psi + \xi}{2}\right) - \frac{1}{2} \tan\left(\frac{\psi - \xi}{2}\right),$$

$$(16.12)$$

where at least $-\pi < \psi + \xi < \pi$ and $-\pi < \psi - \xi < \pi$ must be satisfied. The
line element is given by

$$ds^2 = \mathcal{K}^2 d\tilde{s}^2 = \mathcal{K}^2 \left[-\frac{4r_s^3}{r} e^{-r_s/r} \left(d\psi^2 - d\xi^2\right) + \mathcal{K}^{-2} r^2 d\Omega^2 \right], \quad (16.13)$$

with the conformal factor $\mathcal{K}^2 = [\cos(\psi) + \cos(\xi)]^{-2}$. To determine the possible
range of values of the new coordinates ψ and ξ, we insert (16.12) in (13.79), and
obtain

$$\left(\frac{r}{r_s} - 1\right) e^{r/r_s} = u^2 - v^2 = \frac{\cos(\psi) - \cos(\xi)}{\cos(\psi) + \cos(\xi)}. \quad (16.14)$$

The singularity $r = 0$ corresponds to $\psi = \pi/2$, and the horizon $r = r_s$ is
represented by $\psi = \pm \xi$, see Fig. 16.4. The region ① is the actual region outside
the black hole, and ② covers the complete region below the horizon down to the
singularity. The region ④ would correspond to a white hole, and ③ represents a
hypothetical parallel universe.

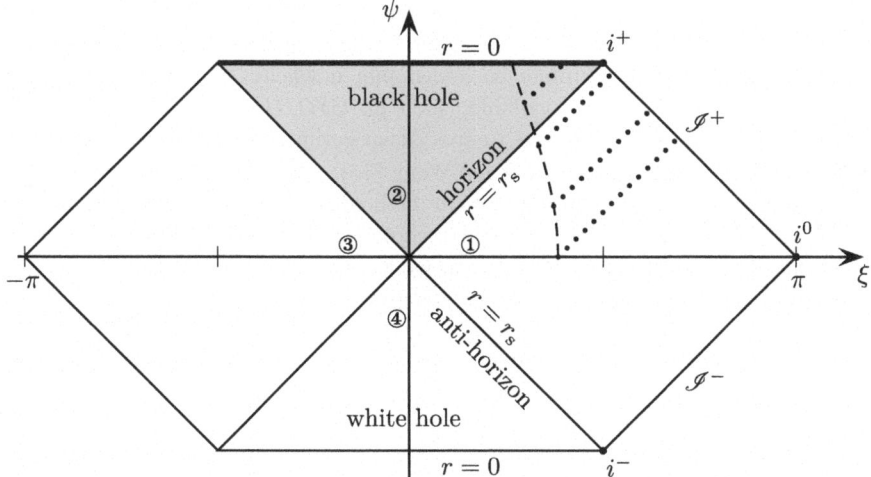

Fig. 16.4 Penrose-Carter diagram of the Schwarzschild spacetime. The *dashed line* is an arbitrary timelike geodesic. The *dotted lines* correspond to rays of light which are emitted from the current position of the timelike geodesic and propagate radially outward

---

16.2   Ray Tracing in General Relativity

Since light propagates along lightlike geodesics, the calculation of what an observer would see is based on the retracing of these geodesics from the observer to the light emitting object. The standard method for visualization in general relativity therefore is four-dimensional *ray tracing*.

The starting point of the calculation is an observer in his own frame of reference, represented by a local tetrad. Depending on the respective camera system, see Sect. 9.1.1, rays of light are started for every image pixel, and then integrated backward in time along lightlike geodesics. The integration is continued until one of the following criteria is met: Either the ray of light hits an object, or it reaches an area which is no longer of interest, or the ray of light is no longer valid. This can happen, for example, if it enters the horizon of a black hole. Since a ray of light could travel for an infinitely long time, for example along the photon orbit of a black hole, the integration should be terminated at some point. Another criterion that comes into play, especially in numerical calculations, is the validity of the ray of light per se, which can be tested by the constraint (11.133) on the geodesic.

Four-dimensional general relativistic ray tracing is very time-consuming, due to its expensive calculations of the geodesics, and the calculation of the intersections with objects in a scene. However, since a ray of light, responsible for a given image pixel, is independent of all other rays of light, the method can be trivially parallelized. This can be done using a *message passing interface* (MPI) implementation on a CPU cluster, or on the graphics hardware (GPU).

In the following examples, we shall show results obtained using the ray tracing code *GeoViS* [21] (as already in special relativity), and shall restrict ourselves to geometric distortions. It must be added that nowadays also numerous other freely accessible codes are available, such as *GYOTO* [32], *GRay* [5], *ARC-MANCER* [30], or codes adapted to special spacetimes, such as *GRTRANS* [6], Psaltis and Johannsen [31], or Yang and Wang [33].

The usually not very round numbers in the field of view of the observer are due to the fact that we have linked the image resolution to the page format $\rho = 16/9$. Therefore, when using a pinhole camera, we have to determine the horizontal and vertical viewing angles according to the equation

$$\rho = \frac{\text{res}_h}{\text{res}_v} = \frac{\tan(\text{fov}_h/2)}{\tan(\text{fov}_v/2)}, \tag{16.15}$$

see also Sect. 9.1.1, where we introduced different camera models.

16.2.1 The Shadow of a Black Hole

In the spring of 2019, the researchers of the "Event Horizon Telescope" collaboration presented the first "image" of a black hole, in the center of the galaxy M87, see Fig. 14.6 in Sect. 14.4.2. As expected, the shadow of the black hole appears as a dark disk, surrounded by a bright ring of hot gas and plasma.

In the idealized case of a Schwarzschild black hole, the apparent angular diameter 2ξ can be calculated as a function of the position of the observer r_{obs}, scaled by the Schwarzschild radius r_s, from the expression

$$\xi = \arcsin\left[\sqrt{\frac{27}{4}\frac{r_s^2}{r_{\text{obs}}^2}\left(1 - \frac{r_s}{r_{\text{obs}}}\right)}\right], \tag{16.16}$$

see (13.41). This angle corresponds exactly to that ray of light which asymptotically approaches the photon orbit. Figure 16.5 shows the apparent angular diameter for different scaled observer positions. The gray circular segments correspond to the viewing angles of those geodesics which, when followed back in time, hit the black hole. The closer the observer is to the black hole, the bigger the black hole appears to him. At the position of the photon orbit, $r_{\text{po}} = 3r_s/2$, already 50% of the entire sky is black. When the observer approaches the event horizon, the entire visible universe finally shrinks to a single point.

Figure 16.6 shows the "shadow" of two black holes with different masses in front of the background of the Milky Way panorama. The observer is at $r_{\text{obs}} = 30r_s$ (for $M = 1$) and looks in the direction of the black hole. According to (16.16), he sees the black holes with an angular diameter of $\xi \approx 4.88°$, i.e., $2\xi \approx 9.76°$, respectively. We simulate the background by a sphere of radius $R = 5000r_s$, and attach the Milky Way panorama to the inner surface of the sphere.

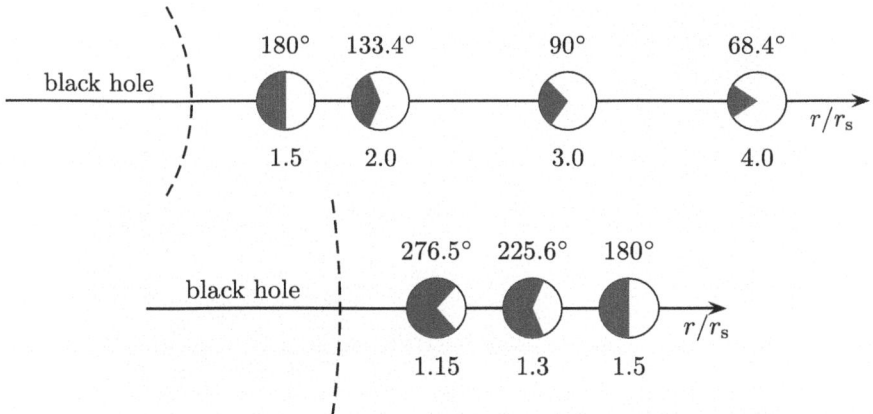

Fig. 16.5 The closer an observer resides to the event horizon, the larger the portion of the field of view that the black hole occupies, indicated by the size of the gray segments. At the event horizon, the entire visible universe finally shrinks to a single point. *Top*: Solid angle for distances outside the photon orbit, *bottom*: solid angle for distances inside the photon orbit

In ray tracing, for every single image pixel the corresponding ray of light is integrated numerically. If during integration the ray of light is "captured" by the black hole, or the integration exceeds a fixed number of integration steps, the pixel is assigned the value "black". However, if the ray of light reaches the sphere in the background, the corresponding color value of the texture is assigned to the pixel. Except for numerical inaccuracies, a pixel-precise representation of the shadow of a black hole thus emerges.

Figure 16.7 shows a similar situation, but with the Milky Way panorama replaced by a grid of parallels and meridians. Here one can clearly see from the degree lines that the deflection of light by the black hole leads to some sort of point reflection, and how the blue line of the equator is deformed to an Einstein ring, see Sect. 13.2.8.

In Fig. 16.8b we enlarge the region around the black hole, and use the background texture from Fig. 16.8a, which again represents the complete 4π view of the sky (background sphere) in the form of a rectangular projection (plate-carrée map). Note that (as already shown in Fig. 9.6), when the texture is projected onto a sphere, the upper border of the texture corresponds to the north pole, and the lower border to the south pole.

The extreme deflection of light in the vicinity of the black hole becomes again particularly evident when we take a closer look at the distorted background texture. In the direction of the observer's view, and thus behind the black hole, lies the turquoise-blue region of the texture. The reddish ring just outside the shadow is created by rays of light which have left the background texture behind the observer, and have circled the black hole once, before they arrive at the observer. But it is not only the reddish region that can be seen, but the entire color gradient of the background texture seems to be mapped on a ring around the black hole. In addition,

Fig. 16.6 Shadow of a black hole: the pinhole camera has a field of view of 79.3° × 50°. *Top*: Image of the Milky Way in flat space. *Middle*: Image of the Milky Way with a black hole (*M* = 0.25) in the foreground. *Bottom*: Image of the Milky Way with a black hole (*M* = 1) in the foreground. Panorama of the Milky Way: ESA/Gaia/DPAC, CC BY-SA 3.0 IGO (▶ https://doi.org/10.1007/000-hvy)

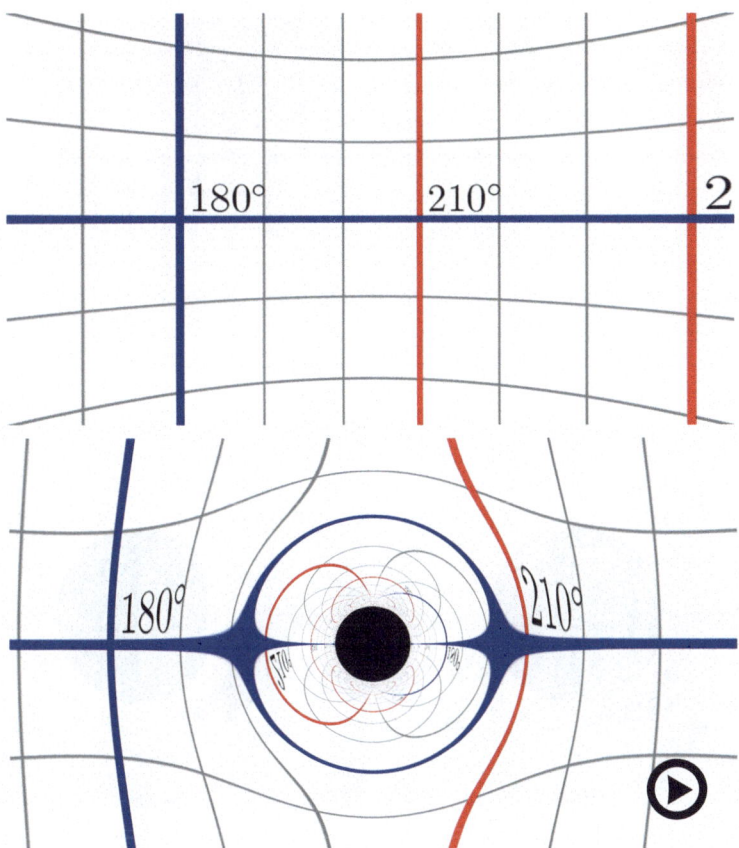

Fig. 16.7 Shadow of a black hole; the pinhole camera has a field of view of $79.3° \times 50°$. The background is a grid of parallels and meridians. *Top*: Original, *bottom:* with black hole ($M = 1$) (▶ https://doi.org/10.1007/000-hvz)

the north pole of the background sphere appears below and the south pole above the shadow.

We now proceed to the Kerr spacetime described by the metric (14.1). When we set the value of the angular momentum parameter first to $a = 0$, we obtain of course the same image as in Fig. 16.8b. When we slowly crank up the angular momentum parameter, the shadow appears increasingly more asymmetric. Qualitatively this can be easily understood: In the Kerr metric the rays of light move either along, or opposite to, the rotation axis of the spacetime, and therefore are deflected differently. In Sect. 14.3.1 we have already calculated the rays of light in the $\vartheta = \pi/2$ plane which delineate the shadow. Rays of light, however, which do not stay in that plane follow extremely complicated trajectories, which renders the calculation of the rays limiting the rim of the black hole much more difficult.

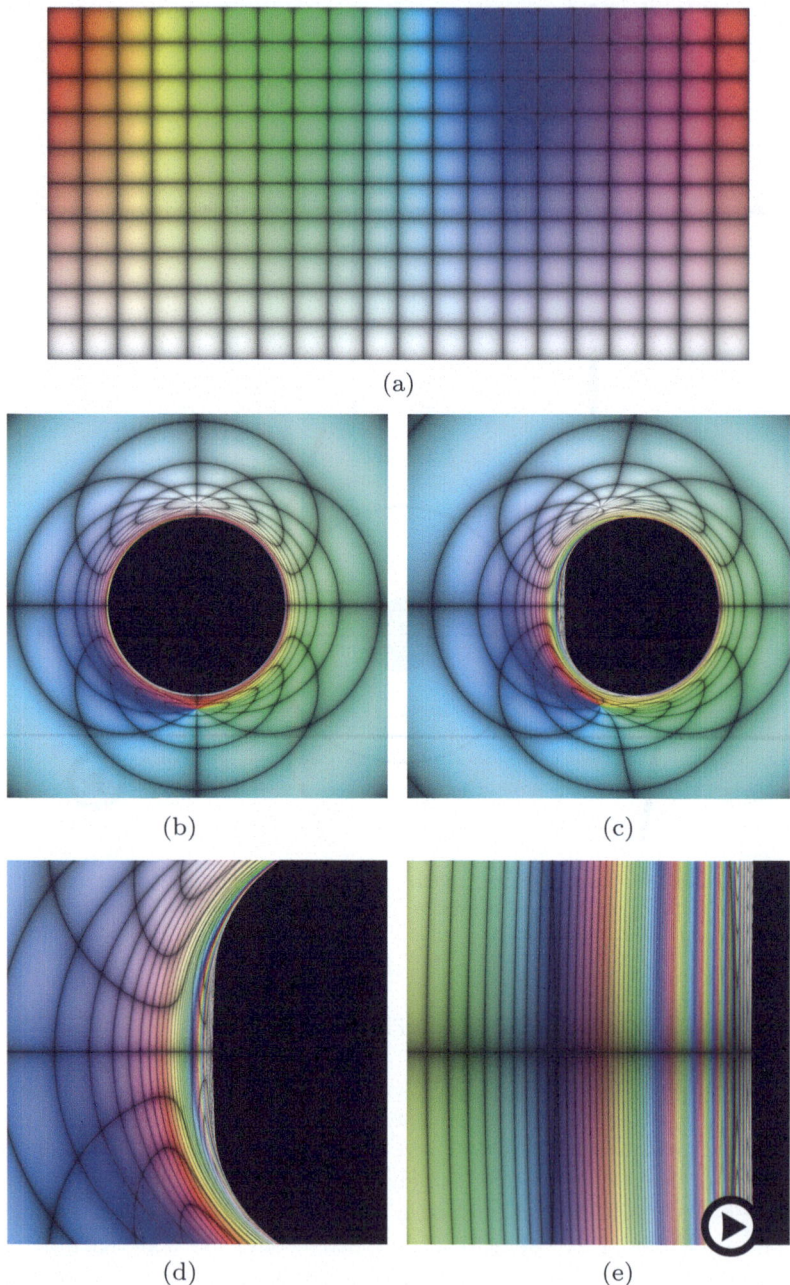

Fig. 16.8 Shadow of a Kerr black hole; (**a**) background texture, (**b**) rotation parameter $a = 0$, (**c–e**) rotation parameter $a = 1$; (**b, c**) viewing angle fov $= 25°$, (**d**) viewing angle fov $= 10°$, (**e**) viewing angle fov $= 1°$ (▶ https://doi.org/10.1007/000-hw0)

Figure 16.8c shows the distorted background texture for the angular momentum parameter $a = 1$, i.e., for a maximally rotating Kerr black hole. Here, too, the entire color gradient is distorted in the shape of a ring, yet the two poles lie no longer symmetrically above and below the shadow, but are slightly shifted to the left. In Fig. 16.8d and e the left edge of the shadow is enlarged, and it can be clearly recognized that the color gradient of the background texture is repeating itself in ever-decreasing intervals. This means, the entire 4π background shows itself in multiply repeating rings, and an observer could, in principle, see himself multiple times. The reason for the multiple images are rays of light that circle the black hole several times before reaching the observer.

16.2.2   Fall into a Black Hole

In Sect. 13.3.1 we already discussed the free fall of an observer into a Schwarzschild black hole. Here we shall visualize what the observer would see during his fall into the black hole. But first we consider the case that the observer approaches the black hole quasi-statically, which means that the velocities during the approach are much smaller than the speed of light. This is equivalent to the assumption that the observer is at rest at his current position.

Figure 16.9 shows the images of the view of the quasi-static observer for different positions r, taken with a panorama camera, with a field of view of $360° \times 90°$. The main direction of view is always to the black hole. At a distance of $r = 8r_{\rm s}$, the observer sees the black hole essentially still completely in front of himself. The closer the observer approaches the Schwarzschild radius, the bigger the black hole seems to be, and it finally occupies almost all of his field of view. The rest of the universe can then be seen only in a small region opposite to the direction to the black hole, and in the limit $r \rightarrow r_{\rm s}$ it seems to disappear completely in a single point.

We can calculate the acceleration necessary for the observer to stay at rest at his current position. We start from the four-velocity $\underline{u} = u^t \partial_t$, which possesses only a time component since the observer is at rest. From the normalization condition (11.133) we find for the time component, $u^t = 1/\sqrt{1 - r_{\rm s}/r_{\rm obs}}$. Since being at rest is a non-geodesic "motion", we have to resort to the Fermi-Walker transport from Sect. 11.4.3. It is sufficient, though, to insert the four-velocity into the four-acceleration (11.139). We obtain, as the only non-vanishing component, the radial four-acceleration $a^r = r_{\rm s}/(2r_{\rm obs}^2)$. We still have to transform it into the local reference frame of the observer, see (13.37) and (11.48), and finally obtain the acceleration which the observer must have to stay at his position,

$$a^{(r)} = \frac{c^2 r_{\rm s}}{2r_{\rm obs}^2 \sqrt{1 - r_{\rm s}/r_{\rm obs}}}, \tag{16.17}$$

In the limit $r_{\rm obs} \rightarrow r_{\rm s}$ the acceleration goes to infinity, which shows that a static observer at the horizon is impossible.

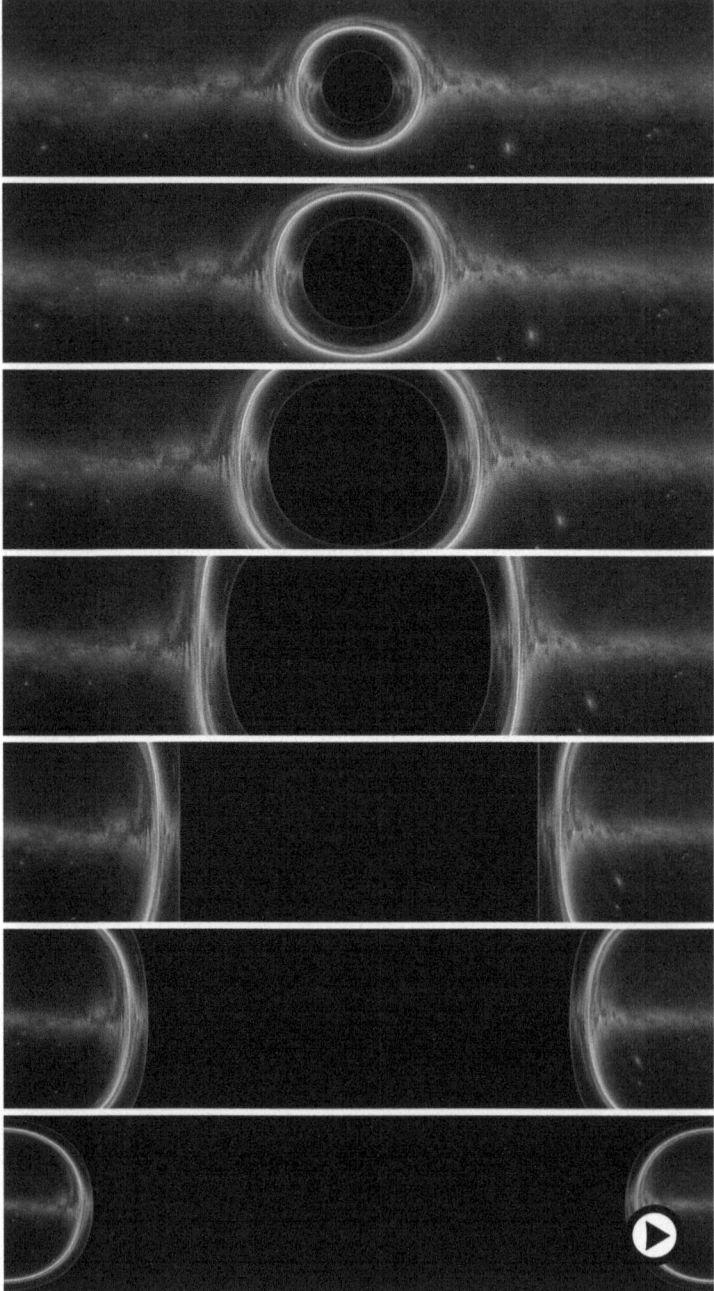

Fig. 16.9 Static approach to a black hole with the Milky Way panorama as the background; distance (from *top* to *bottom*): r_i/r_s = 8.0, 5.0, 3.0, 2.0, 1.5, 1.3, and 1.1. The field of view of the pinhole camera is 360° × 90°. Panorama of the Milky Way: ©ESO/S. Brunier (▶ https://doi.org/10.1007/000-hvx)

Figure 16.10 shows the views of the free-falling observer. His current position $r(\tau)$ at proper time τ is implicitly given by (13.61). In order to compare his views with those of the quasi-static observer, in Fig. 16.10 we have chosen the same radial positions as in Fig. 16.9. Since the two observers are at the same positions, their local geometry is always the same. Therefore, their views of the black hole only differ by their velocities, and by the special relativistic aberration caused thereby. For an observer starting at $r = R > r_{\mathrm{s}}$ out of a state of rest, his velocity β, referred to the local tetrad (13.37) of a static observer at the same position, is given by

$$\beta = \sqrt{\frac{r_{\mathrm{s}}/r - r_{\mathrm{s}}/R}{1 - r_{\mathrm{s}}/R}}, \qquad (16.18)$$

see, e.g., Müller [18], or Exercise 13.3. Regardless of the position where the observer starts, in the limit $r \to r_{\mathrm{s}}$ his velocity will always increase to the speed of light. However, this limit must be viewed with caution, since we cannot place a static observer at $r = r_{\mathrm{s}}$, and therefore are unable to measure the velocity at this radial coordinate. Nonetheless, on account of the high velocities, the strong effect of aberration almost cancels the strong deflection of light close to the horizon. Which of the two effects dominates under what physical conditions, is discussed in detail in Müller [18].

16.2.3 A Star on a Circular Orbit Around a Black Hole

The properties of geodesics in the Schwarzschild spacetime discussed in Sect. 13.2 become important especially in the observation of a star circling a black hole. The star is supposed to move on the last stable orbit, $r_\star = 3r_{\mathrm{s}}$, and the radius of the star is $R = 0.25r_{\mathrm{s}}$. We already know from Sect. 13.4.5 that the star is subject to a geodetic precession, for which reason it has turned by roughly $105.44°$ after a full cycle. The observer itself is assumed to reside at $r = 15r_{\mathrm{s}}$. A detailed discussion can be found in Müller [19].

Star at a Fixed Position

Before we allow the star to move in its orbit, we will first visualize what an observer would see from far away if the star were to stay at a fixed position. Figure 16.11 shows the star, and the trajectories of a few rays of light that reach the observer. In principle, the observer does see an infinite number of stars, with the viewing angles ξ_i to the center of the black hole turning smaller and smaller, and quickly approaching the critical angle ξ_{crit}. The latter is determined by the ray of light, emitted from the observer, that asymptotically approaches the photon orbit $r_{\mathrm{po}} = 3r_{\mathrm{s}}/2$, see (13.41). We shall call the index $i = \{1, 2, \ldots\}$ of the viewing angle the *order* of the star's image. With increasing order of the image, the angular diameter $\Delta\xi_i$ under which the star is seen, becomes very small, which is why higher orders would require an extremely high resolution of the camera.

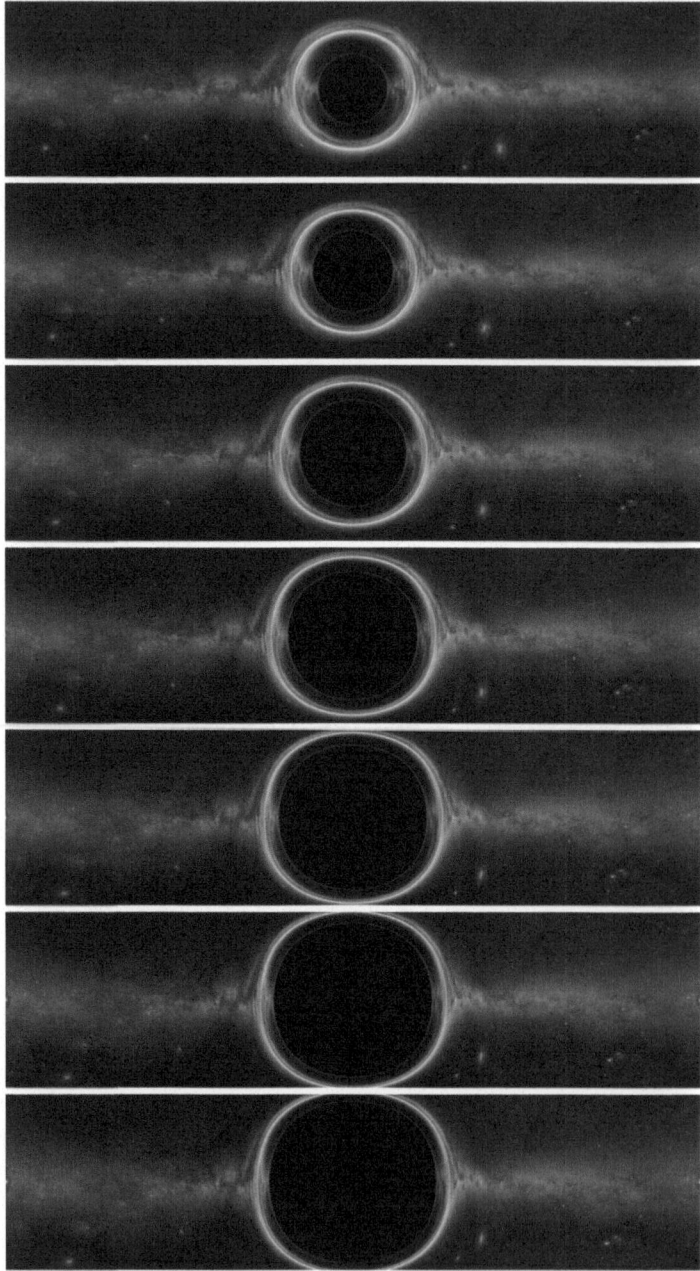

Fig. 16.10 Approach to a black hole in free fall with the Milky Way panorama as the background; distance (from *top* to *bottom*): r_i/r_s = 8.0, 5.0, 3.0, 2.0, 1.5, 1.3, and 1.1. The distances are the same as in Fig. 16.9. The images are taken with a panorama camera with a field of view of $360° × 90°$. Panorama of the Milky Way: ©ESO/S. Brunier

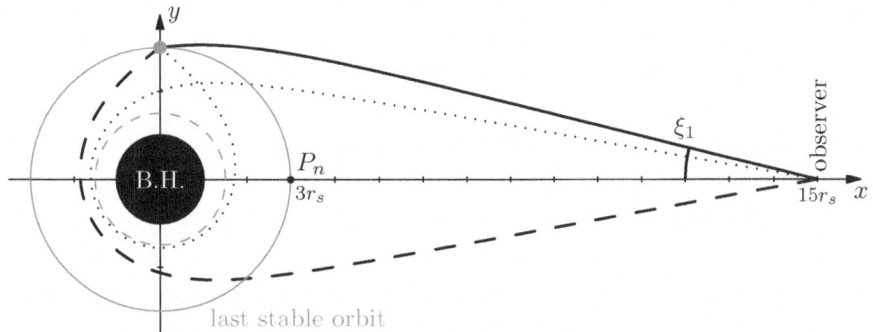

Fig. 16.11 The star is fixed at the position ($r_* = r_{\text{lso}}, \varphi = 90°$) on the last stable orbit. The observer sees the star under the angles $\xi_1 = 13.5°, \xi_2 = -10.074°$, and $\xi_3 = 9.6513°$. The critical angle is given by $\xi_{\text{crit}} \approx 9.632732°$. The *dashed inner circle* represents the photon orbit

Star in Orbit

Let us return to the originally formulated case of a star circling a black hole. To find out where the star appears to a distant observer at his time of observation, we must first calculate, for every point on the trajectory of a ray of light, the time of flight $\Delta t_\varphi(3r_s, 15r_s)$ to the observer.

We can calculate analytically the time of flight $\Delta t_{\varphi=0}(3r_s, 15r_s)$ from the point P_n on the orbit nearest to the observer directly from (13.28). For the present case we find

$$\Delta t_{\varphi=0}(3r_s, 15r_s) = (12 + \ln(7))r_s/c \approx 13.946 r_s/c.$$

Now the time of observation t_{obs} is to be chosen in such a way that the star, which passes the point nearest to the observer at time $t = 0$, is seen exactly at that position. Then we must have $t_{\text{obs}} = \Delta t_{\varphi=0}(3r_s, 15r_s)$.

We determine the time of flight to other points on the orbit by numerical integration of the geodesic equations (13.29), where we can restrict ourselves to the ($\vartheta = \pi/2$) plane. For this purpose, we use the *Motion4D* library from Müller and Grave [25]. We must take into account that rays of light can circle the black hole several times before they reach the observer. For the present example we calculate Δt_φ for $\varphi \in [-720°, 720°]$, i.e., we follow the light rays backward in time up to two cycles around the black hole in either direction. The integration starts at the position ($t = t_{\text{obs}}, r = 15r_s, \vartheta = \pi/2, \varphi = 0$) of the observer. Referred to the local tetrad (13.37) at this position, the initial directions of the rays of light are given by

$$\underline{y} = -\underline{e}_{(t)} - \cos(\xi)\underline{e}_{(r)} + \sin(\xi)\underline{e}_{(\varphi)}.$$

The intersections of the lightlike geodesics with the star's timelike geodesic yield the times of emission t_e and the corresponding positions φ_e on the orbit. The latter are recorded in a φ-t-diagram, see Fig. 16.12. The time t_e is plotted, with some

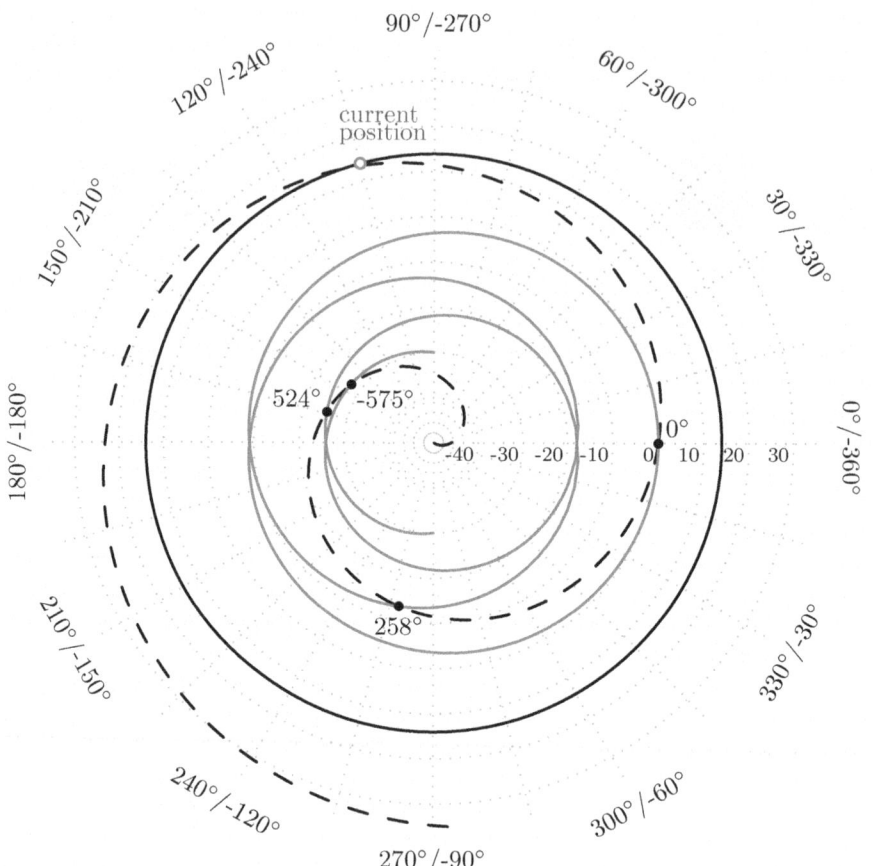

Fig. 16.12 The φ-t-diagram shows the times of emission t for the respective position φ, where light must be emitted to reach the observer at his time of observation (*gray line*). The *black circle* indicates the time of observation $t_{obs} \approx 13.946r_s/c$. The intersections of the star's timelike geodesic (*black dashed line*) with the times of emission yield the apparent positions φ_e of the star. The first image of the star is from light rays that left it at $\varphi = -575°$, the second from light rays that left it at $\varphi = 524°$, a third one from light rays that left it at $\varphi = 258°$, and the most recent one from light rays that left it at $\varphi = 0°$. The star's current position is given by the intersection of its geodesic with the circle for the time of observation at $\varphi = 105°$. See Fig. 16.13 for the corresponding lightlike geodesics

offset, as a radial coordinate in dependence on the orbital parameter, i.e., the polar angle φ (gray line). The black circle indicates the time of observation t_{obs}, and the dashed line represents the circular motion of the star. Its angular velocity is given by

$$\varphi(t) = \Omega t \quad \text{with} \quad \Omega^2 = \frac{c^2 r_s}{2r_{lso}^3} = \frac{c^2}{54r_s^2}, \tag{16.19}$$

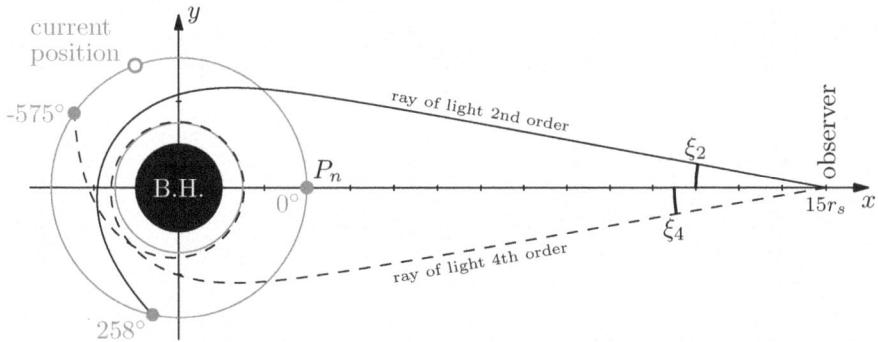

Fig. 16.13 The *filled in points* and the corresponding angles indicate the position of the star where it must emit light to reach the observer at his observation time. The "current position" of the star is its position at the time of observation. For the clarity of the graphical representation, geodesics are shown only for the observation angles ξ_2 and ξ_4 (▶ https://doi.org/10.1007/000-hw2)

cf. Eq. (13.46) in Sect. 13.2.7. The intersection of the "circle of observation time" with the orbit of the star yields the actual position of the star. The intersections of the orbit of the star and the times of emission yield the apparent positions of the star at the time of observation. In this example, the observer perceives the star not only under the angle $\xi_1 = 0°$, but also under angles $\xi_2 = 10.177°$, $\xi_3 = 9.6379°$, and $\xi_4 = -9.6356°$, see Fig. 16.13.

Figure 16.14 shows a few images out of a sequence of 370 images computed in total. The topmost image corresponds to the observation time $t_{\text{obs}} \approx 13.946r_s/c$ discussed above. By construction, the star appears directly in the line of sight with an angular diameter of $\Delta\xi_1 \approx 2.31°$. To the right, one can recognize faintly the image of second order at an observation angle of ξ_2 and an angular diameter of about $\Delta\xi_2 \approx 0.154°$. The images of third and fourth order, however, at observation angles ξ_3 and ξ_4, can no longer be recognized because of their small angular diameters of $\Delta\xi_3 \approx 0.00137°$ and $\Delta\xi_4 \approx 0.00076°$.

In the images two ($n = 144$) and three ($n = 256$), one can recognize a thin but pronounced Einstein ring, cf. Sect. 13.2.8. Einstein rings arise when the star is on the line of sight from the observer to the black hole, see Fig. 16.15. In image ($n = 300$), the star approaches the observer from the left, while on the right side it moves away. In the final image, ($n = 369$), the star has performed a full cycle, and has turned by about 105.44° due to geodetic precession.

Figure 16.15 shows the lightlike geodesics which give rise to an Einstein ring of first and second order. Because of the spherical symmetry of the Schwarzschild spacetime, the figure can be rotated around the x-axis to produce the ring structure. The corresponding observation angles are $\xi_1 \approx 11.649°$ and $\xi_2 \approx 9.72282°$.

Here, the order of the Einstein rings deviates from the definition given in Sect. 13.2.8. In the present case, we deal with an object which crosses the connecting

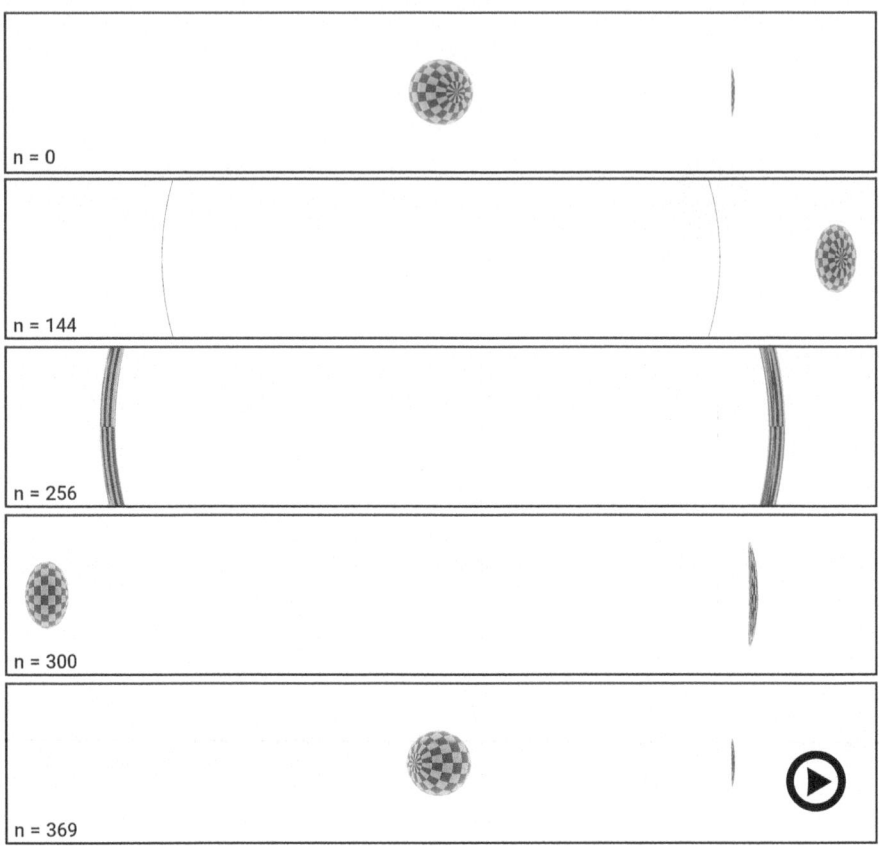

n = 0

n = 144

n = 256

n = 300

n = 369

Fig. 16.14 A star with radius $R = 0.25r_s$ is circling a black hole on the last stable orbit $r_{lso} = 3r_s$. The observer resides at the position $r_{obs} = 15r_s$. He looks with a camera in the direction of the black hole, which is always in the center of each image. His times of observation are (from *top* to *bottom*): $ct/r_s = 13.946 + 0.125 \cdot n$, with $n = \{0, 144, 256, 300, 369\}$. The *bottom image* corresponds approximately to the time of one full cycle (▶ https://doi.org/10.1007/000-hw3)

line "observer–black hole" before and behind the black hole, and these crossings have been numbered accordingly.

Müller [16] describes a method how Einstein rings can possibly be used to determine the distance to a black hole.

16.2.4 Accretion Disk

If matter falls on a black hole, an accretion disk is usually formed. Its visual appearance has already been discussed in various papers, see, e.g., Armitage and Reynolds [1], Luminet [14], Page and Thorne [28] and Fukue and Yokoyama [9].

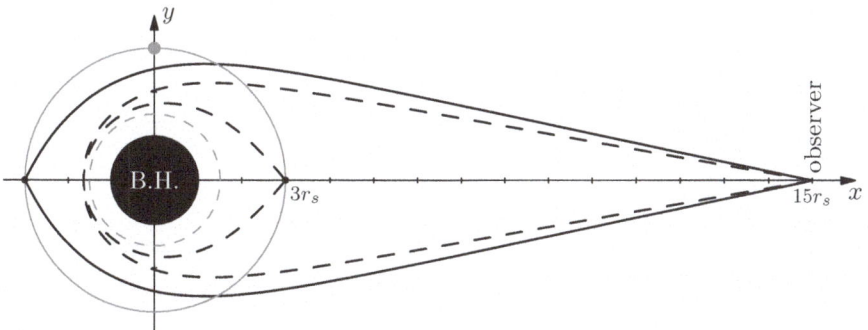

Fig. 16.15 Einstein rings are formed when an object, here the star, is located on the axis "observer–black hole". The solid lines correspond to an Einstein ring of first order (cf. Fig. 16.14, $n = 256$), the *dashed lines* give rise to an Einstein ring of second order (cf. Fig. 16.14, $n = 144$)

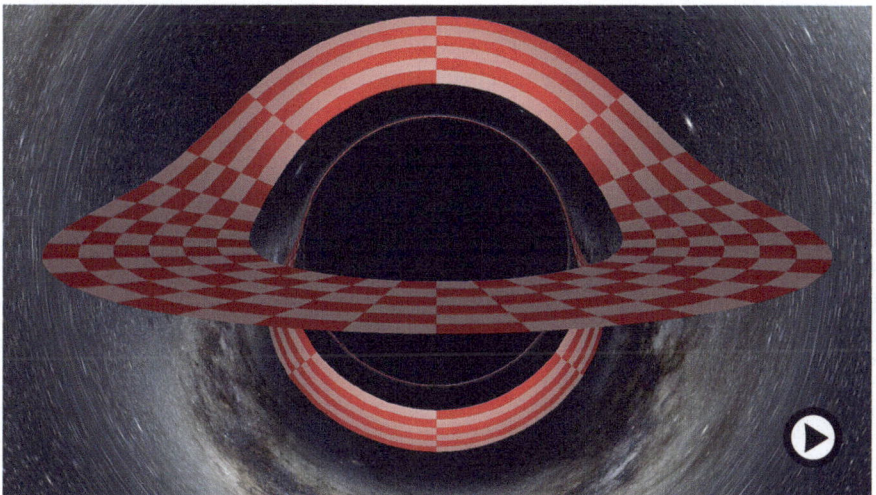

Fig. 16.16 Accretion disk from the perspective of an observer whose camera covers a field of view of $31.45° \times 18°$ (▶ https://doi.org/10.1007/000-hw4)

We place an arbitrarily thin disk around a black hole and study the effects of the deflection of light. The inner edge of the disk shall be defined by the last stable orbit $r_{\mathrm{min}} = 3r_{\mathrm{s}}$ in the Schwarzschild spacetime, and the outer edge we set arbitrarily to $r_{\mathrm{max}} = 7r_{\mathrm{s}}$. Referred to the normal vector of the ($\vartheta = \pi/2$) plane, the disk is inclined by an angle $\iota = 10°$ in the direction of the observer, who resides at $r_{\mathrm{obs}} = 30r_{\mathrm{s}}$.

Figure 16.16 shows the observer's view of the accretion disk for an inclination angle of $\iota = 10°$. Due to the deflection of light in the vicinity of the black hole, the observer sees the disk distorted, and even more than once.

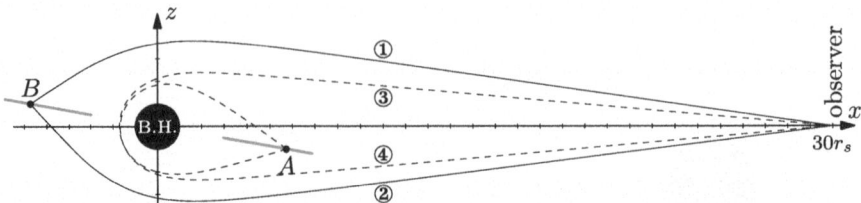

Fig. 16.17 Accretion disk around a black hole. The two gray bars indicate the cross section of the disk with the inner radius $r_{\min} = 3r_s$ and the outer radius $r_{\max} = 7r_s$, and an inclination angle $\iota = 10°$ relative to the z-axis. The two points on the bars mark a ring with radius $r_{\text{ring}} \approx 5.81r_s$

The part of the disk that lies behind the black hole, and therefore should be invisible, appears as an "arc" above the black hole. In fact, the rear side of the disk is additionally visible as an arc below the black hole. To understand this, a few rays which reach the observer are shown in Fig. 16.17. Rays of the type ① are responsible for the arc above the black hole. Rays of the type ② can explain why the arc below the black hole represents the rear bottom side of the disk. Apart from these primary rays, there are also secondary rays, which first move away from the observer, circle the black hole once, and only then reach the observer. The ray ③, for example, shows the front bottom side of the disk, and the ray ④ the corresponding top side. In Fig. 16.16, this produces the two thin arcs directly around the black hole. If the observer's camera had an even higher resolution, then he would see further arcs, which would asymptotically approach the apparent rim of the black hole. These would be produced by rays of light which circle the black hole several times before they reach the observer. In principle, there exists an infinite number of "mirror images" of the accretion disk.

Figure 16.16 shows only the geometric distortion of the accretion disk. A realistic representation would require taking into account many more effects. These include the motion of the disk and the resulting Doppler shift, the gravitational frequency shift, the gravitational lensing effect, and the radiative properties of the matter in the disk. An interactive visualization of the accretion disk where some of these effects are taken into account is described in Müller and Frauendiener [24]. Also, NASA's Goddard Space Flight Center published a commented video of an observer that falls into a black hole that is surrounded by an accretion disk [27].

16.3 Interactive Visualization in General Relativity

To gain a deeper understanding of the intrinsic geometry and the processes in a curved spacetime from the first-person view, it is helpful if you can move around freely. However, the computation of a general relativistic scene by *ray tracing* is still too time-consuming.

Here we will give a short outlook on possible techniques which could accelerate visualization computations in general relativity. This can be realized either by massive parallelization, or by using analytical solutions of the geodesic equation.

16.3.1 Massive Parallelization Using CPU or GPU Clusters

As already mentioned at the beginning of Sect. 16.2, general relativistic *ray tracing* is easily parallelized, since every ray of light can be treated separately. Today's standalone computers usually have a multi-core processor, which is already capable of running several execution threads in parallel. In spite of this, the computation of a single image with WUXGA screen resolution (1920×1200 pixels), and few objects in the scene, may take several minutes. A possibility to reduce this time is offered by parallelization on a big CPU cluster. Another possibility is to exploit the parallel compute architecture of the graphics hardware (GPU), which is specialized on computing all pixels of an image in parallel, see, e.g., Kuchelmeister et al. [13]. If several GPUs are available, one can further redistribute the work load. However, the communication between the individual execution threads also takes time, which is why a distribution which assigns to every ray of light a processor of its own, would not be reasonable.

16.3.2 Tabulation of Geodesics

In the case of a very symmetric spacetime and a very simple scene, it is worth to compute and tabulate geodesics before the actual visualization. The Schwarzschild spacetime with a background panorama projected on an "infinitely far distant" sphere, can serve as an example. This panorama could be, e.g., that of the Milky Way of Brunier [2], which is available as a rectangular projection (plate-carrée map). Due to the spherical symmetry of the Schwarzschild spacetime, we have to compute only geodesics within the ($\vartheta = \pi/2$) plane, and it is sufficient to choose the initial positions ($t = 0, r, \vartheta = \pi/2, \varphi = 0$). All the other geodesics can be obtained by an appropriate rotation around the ($\varphi = 0$) axis. However, we do not use the radial coordinate r for the pre-calculation, but rather its reciprocal value $x = r_s/r$, in the interval $(0, 1)$. Now we have to integrate the geodesic, for every initial position x and for every initial direction $\underline{y} = -\underline{e}_{(t)} + \cos(\xi)\underline{e}_{(r)} + \sin(\xi)\underline{e}_{(\varphi)}$ with $\xi \in [0, \pi]$, out to $r \to \infty$, if possible, and in this way we can obtain the angle φ_∞, where the ray of light intersects the infinitely distant sphere. Figure 16.18 shows the various regions for the pre-calculation.

The pre-calculation can be realized numerically as follows: The initial positions x and the initial angles ξ are divided in N_x and N_ξ sampling points, respectively. For each pair ($x_i = i \cdot 1/N_x, \xi_j = j \cdot \pi/N_\xi$), a lightlike geodesic is integrated according to the starting directions discussed above using a Runge-Kutta method with step control, out to a maximum radius of $R \gg 1$. In doing so, one runs the risk

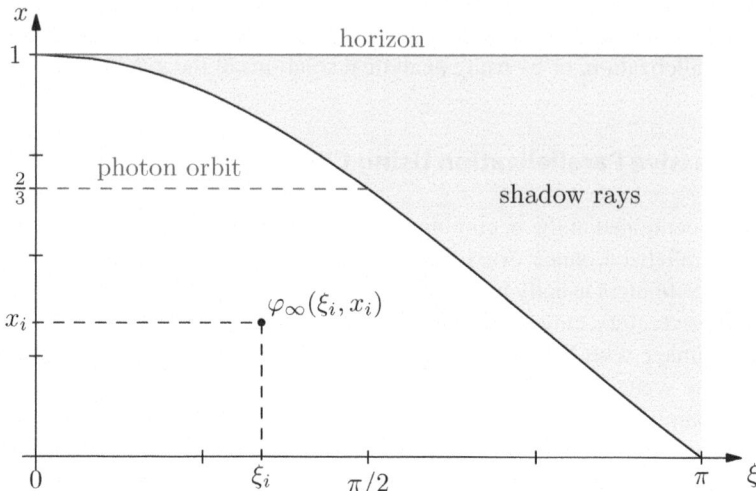

Fig. 16.18 For every ray of light within the unshaded area, starting at the position x_i in the direction ξ_i, the point of intersection φ_∞ is determined with a fictitious sphere at "infinite" distance. The thick solid line corresponds to the critical angle ξ_{crit} which defines the shadow of the black hole

that $r = r_s/x > R$ can occur, which is why it is better to switch to an analytical solution of the geodesic equation, see, e.g., Müller [15].

When the assignment table $(x, \xi) \mapsto \varphi_\infty$ has been calculated, one can use it, similarly to the image-based method of special relativity, cf. Sect. 9.1.1, to move around freely in the Schwarzschild spacetime. For this purpose, a window-filling rectangle is drawn at the current position of the observer as a representation of the camera plane. Each pixel of the image then is converted numerically into a direction referred to the local tetrad of the observer. In this process, a possible effect of aberration can also be included. Afterward, the coordinate frame must be rotated in such a way that the initial direction lies in the $(\vartheta = \pi/2)$-plane. The assignment table helps to determine the corresponding intersection point φ_∞, which, after transforming back to the original coordinates, can be used to read the color of the background image. All of this can be implemented in a fragment-shader code. In Müller [22], the image-based method is described in detail for the Schwarzschild metric, and it is also shown in which way this method can be applied, with some modifications, to the Ellis-wormhole metric. A similar approach is also used in Müller and Boblest [23] to visualize the circular motion around a black hole inside a torus. For this situation, the intersections with the torus were pre-calculated and tabulated.

16.3.3 Analytical Geodesics

For spacetimes which are less symmetric, or in the case of more complicated scenes, the pre-calculation may either be too expensive, or the amount of data calculated may be too large, or unwieldy, to store it, e.g., on a GPU. Then it may become necessary to resort to analytical solutions of the geodesic equation, see, e.g., Hackmann et al. [12], Müller [17], or Čadež and Kostić [3]. In this situation, the main stumbling block is what is called the *emitter-observer problem*. The starting point is a source of light, or a point in the geometry, and the position of the observer. The task now is to look for a geodesic which connects these two points. This is a hard problem in spacetimes with arbitrary curvature, since the solution may not be unique, because of the curved trajectories of light.

We conclude by noting that, for the Schwarzschild spacetime and the Gödel spacetime [10], which describes a rotating universe, there exist solutions which use analytical geodesics for visualization [11, 26].

References

1. Armitage, P.J., Reynolds, C.S.: The variability of accretion on to Schwarzschild black holes from turbulent magnetized discs. Mon. Not. R. Astron. Soc. **341**, 1041–1050 (2003)
2. Brunier, S.: The Milky Way Panorama (ESO). https://www.eso.org/public/images/eso0932a/
3. Čadež, A., Kostić, U.: Optics in the Schwarzschild spacetime. Phys. Rev. D **72**, 104024 (2005)
4. Carter, B.: Complete analytic extension of the symmetry axis of Kerr's solution of Einstein's equations. Phys. Rev. **141**(4), 1242–1247 (1966)
5. Chan, C., Psaltis, D., Özel, F.: GRay: A massively parallel GPU-based code for ray tracing in relativistic spacetimes. Astrophys. J. **777**(1), 13 (2013)
6. Dexter, J.: A public code for general relativistic, polarised radiative transfer around spinning black holes. Mon. Not. R. Astron. Soc. **462**, 115–136 (2016)
7. Flamm, L.: Beiträge zur Einsteinschen Gravitationstheorie (Contributions to Einstein's theory of gravitation). Phys. Z. **17**, 448–454 (1916)
8. Frauendiener, J.: Conformal Infinity. Living Rev. Relativ. **7** (2004). https://link.springer.com/article/10.12942/lrr-2004-1
9. Fukue, J., Yokoyama, T.: Color photographs of an accretion disk around a black hole. Publ. Astron. Soc. Japan **40**, 15–24 (1988)
10. Gödel, K.: An example of a new type of cosmological solutions of Einstein's field equations of gravitation. Rev. Mod. Phys. **21**, 447–450 (1949)
11. Grave, F.: The Gödel universe – physical aspects and egocentric visualizations. Dissertation, Universität Stuttgart (2010)
12. Hackmann, E., Lämmerzahl, C., Kagramanova, V., Kunz, J.: Analytical solution of the geodesic equation in Kerr-(anti-) de Sitter space-times. Phys. Rev. D **81**, 044020 (2010)
13. Kuchelmeister, D., Müller, T., Ament, M., Wunner, G., Weiskopf, D.: GPU-based four-dimensional general-relativistic ray tracing. Comput. Phys. Commun. **183**(10), 2282–2290 (2012)
14. Luminet, J.-P.: Image of a spherical black hole with thin accretion disk. Astron. Astrophys. **75**, 228–235 (1979)
15. Müller, T.: Visualisierung in der Relativitätstheorie (Visualization in the theory of relativity). Dissertation, Eberhard-Karls-Universität Tübingen (2006)
16. Müller, T.: Einstein rings as a tool for estimating distances and the mass of a Schwarzschild black hole. Phys. Rev. D **77**, 124042 (2008)

17. Müller, T.: Exact geometric optics in a Morris-Thorne wormhole spacetime. Phys. Rev. D **77**, 044043 (2008)
18. Müller, T.: Falling into a Schwarzschild black hole. Gen. Relativ. Gravit. **40**, 2185–2199 (2008)
19. Müller, T.: Analytic observation of a star orbiting a Schwarzschild black hole. Gen. Relativ. Gravit. **41**(3), 541–558 (2009)
20. Müller, T.: GeodesicViewer – A tool for exploring geodesics in the theory of relativity. Comput. Phys. Commun. **182**, 1382–1383 (2011)
21. Müller, T.: GeoViS – Relativistic ray tracing in four-dimensional spacetimes. Comput. Phys. Commun. **185**(8), 2301–2308 (2014)
22. Müller, T.: Image-based general-relativistic visualization. Eur. J. Phys. **36**, 065019 (2015)
23. Müller, T., Boblest, S.: Visualizing circular motion around a Schwarzschild black hole. Am. J. Phys. **79**(1), 63–73 (2011)
24. Müller, T., Frauendiener, J.: Interactive visualization of a thin disc around a Schwarzschild black hole. Eur. J. Phys. **33**(4), 955 (2012)
25. Müller, T., Grave, F.: Motion4D – A library for lightrays and timelike worldlines in the theory of relativity. Comput. Phys. Commun. **180**, 2355–2360 (2009)
26. Müller, T., Weiskopf, D.: Distortion of the stellar sky by a Schwarzschild black hole. Am. J. Phys. **78**, 204–214 (2010)
27. NASA's Goddard Space Flight Center/J. Schnittman and B. Powell: Plunge into a black hole explained. https://svs.gsfc.nasa.gov/14576
28. Page, D.N., Thorne, K.S.: Disk-accretion onto a black hole. Time-averaged structure of accretion disk. Astrophys. J. **191**, 499–506 (1974)
29. Penrose, R.: Zero rest-mass fields including gravitation: asymptotic behaviour. Proc. R. Soc. A **284**(1397), 159–203 (1965)
30. Pihajoki, P., Mannerkoski, M., Nättilä, J., Johansson, P.H.: General purpose ray tracing and polarized radiative transfer in general relativity. Astrophys. J. **863**(8), 1 (2018)
31. Psaltis, D., Johannsen, T.: A ray-tracing algorithm for spinning compact object spacetimes with arbitrary quadrupole moments. I. Quasi-Kerr black holes. Astrophys. J. **745**(1), 1 (2012)
32. Vincent, F.H., Paumard, T., Gourgoulhon, E., Perrin, G.: GYOTO: a new general relativistic ray-tracing code. Class. Quantum Grav. **28**(22), 225011 (2011)
33. Yang, X., Wang, J.: YNOGK: A new public code for calculating null geodesics in the Kerr spacetime. Astrophys. J. Suppl. Ser. **207**(1), 6 (2013)

Part III

Stellar Evolution

Star Formation

17

Contents

In this chapter we discuss the processes that lead to the formation of stars. The process of star formation is a central and highly topical field of research in astrophysics. Here we shall present only a simplified picture. But we can at least derive and understand the basic ideas of star formation.

Stars consist mainly of hydrogen and helium. They are formed when clouds of interstellar gas, essentially molecular hydrogen, but also helium and dust particles, are compressed under the action of their own gravitational forces. It is evident that not every cloud of gas can contract under its own self-gravitation, since we observe exactly the opposite behavior for gases under laboratory conditions. On Earth, a compressed gas, e.g., in a gas bottle, always occupies its maximum possible space, and also when the gas bottle is opened, see Fig. 17.1. The reason is the thermal motion of the gas particles. The gravitational attraction between the gas particles is completely negligible in comparison with the energy of their thermal motion. On interstellar scales, however, gravity becomes the dominant effect.

17.1   Virial Theorem

The virial theorem provides a general relation between the time-averaged kinetic and potential energy of an ensemble of particles. This relation will help us to derive a criterion for the contraction of a giant cloud of gas. To derive the virial theorem,

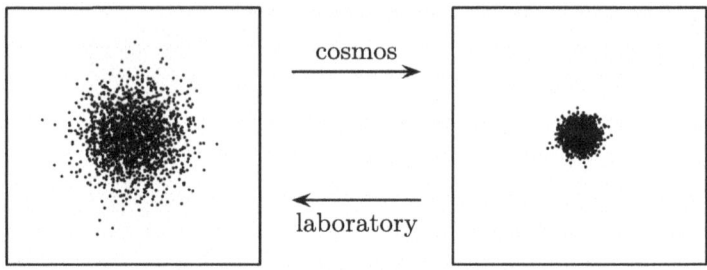

Fig. 17.1 In order for stars to form, a cloud of gas must contract under the action of its self-gravitation. In terrestrial laboratories, the opposite process is observed, since self-gravitation can be completely neglected

we first consider a particle with mass m under the action of a force F. The equation of motion then reads

$$m\ddot{r} = F. \tag{17.1}$$

We now consider the value of the quantity $F \cdot r$ averaged over a time interval $t_2 - t_1$. Integrating by parts, we find

$$\int_{t_1}^{t_2} F \cdot r \, dt = \int_{t_1}^{t_2} m\ddot{r} \cdot r \, dt = \left[m\dot{r} \cdot r \right]_{t_1}^{t_2} - \int_{t_1}^{t_2} m\dot{r} \cdot \dot{r} \, dt. \tag{17.2}$$

Thus, the intermediate result is

$$\frac{1}{t_2 - t_1} \left\{ [m\dot{r}(t_2) \cdot r(t_2) - m\dot{r}(t_1) \cdot r(t_1)] - \int_{t_1}^{t_2} m\dot{r}^2 \, dt \right\} = \frac{1}{t_2 - t_1} \int_{t_1}^{t_2} F \cdot r \, dt. \tag{17.3}$$

We can write this in the form

$$\overline{F \cdot r} + \overline{m\dot{r}^2} = \frac{1}{t_2 - t_1} [m\dot{r}(t_2) \cdot r(t_2) - m\dot{r}(t_1) \cdot r(t_1)], \tag{17.4}$$

where $\overline{X}$ denotes the time average of the quantity X.

When we make the time interval very large, the right-hand side of (17.4) goes to zero, provided the positions and velocities of the particle remain finite (bound motion), i.e., the trajectory of the particle will remain within a definite volume, and its velocity does not exceed a definite maximum velocity, which is impossible anyhow according to special relativity, since $\dot{r} < c$. Noting that $m\dot{r}^2 = 2T$ is double the kinetic energy, we arrive at the *virial theorem*

$$\overline{\boldsymbol{F} \cdot \boldsymbol{r}} + 2\overline{T} = 0. \tag{17.5}$$

In order to interpret the first term, we write the force as the negative gradient of a potential of the form $V(r) = \alpha r^n$

$$\boldsymbol{F} = -\boldsymbol{\nabla} V = -\boldsymbol{\nabla} \alpha r^n. \tag{17.6}$$

Then, with $\boldsymbol{r} = r\,\mathbf{e}_r$, the relation

$$\boldsymbol{F} \cdot \boldsymbol{r} = -[\boldsymbol{\nabla} V(r)] \cdot \boldsymbol{r} = -\alpha r^{n-1} n\, \mathbf{e}_r \cdot (r\,\mathbf{e}_r) = -n\alpha r^n = -nV(r) \tag{17.7}$$

follows. Thus, we have for the time average in (17.5)

$$2\overline{T} = n\overline{V}. \tag{17.8}$$

The important case for us is the gravitational potential with $n = -1$. In this case

$$2\overline{T} = -\overline{V} \quad \text{or} \quad \overline{T} = -\frac{1}{2}\overline{V}. \tag{17.9}$$

The relation (17.9) will help us to understand the processes occurring during the contraction of a galactic cloud of gas.

To this end, we have to additionally average the kinetic and potential energy over all particles in the cloud of gas, which we call an ensemble, at a definite point in time. We introduce the notation for the ensemble average

$$\langle T \rangle = \frac{1}{N} \sum_i T_i, \quad \langle V \rangle = \frac{1}{N} \sum_i V_i, \tag{17.10}$$

where V_i represents the potential energy, and T_i the kinetic energy of particle i.

Instead of averaging over all particles in the cloud of gas, we can also average the kinetic and potential energy of a single particle over a very long time. We further make the assumption that this time average is equal to the ensemble average, provided the average is over very many particles, or very long time intervals. This assumption is the *ergodic hypothesis*. Expressed in formulas, this means

$$\langle T \rangle = \overline{T}, \quad \langle V \rangle = \overline{V}. \tag{17.11}$$

Under the assumption of the validity of the ergodic hypothesis, the statement of the virial theorem for gravitation (17.9) leads to the relation

$$\langle T \rangle = -\frac{1}{2} \langle V \rangle \tag{17.12}$$

between the average kinetic and potential energy of the particles.

17.2 Jeans Criterion for Contraction

The density in the interstellar medium is very low. Therefore, the interactions between the particles (other than gravitation) in a contracting cloud can be neglected in good approximation. This means that a description of the cloud as an ideal gas is legitimate. Then the internal energy U of the cloud is given by

$$N\langle T \rangle = U, \tag{17.13}$$

where N is the number of molecules or atoms. It is known from thermodynamics that the thermal energy of a monoatomic gas is

$$U = \frac{3}{2}Nk_\text{B}T, \tag{17.14}$$

with the *Boltzmann constant*[1] [5]

$$k_\text{B} = 1.380\,649 \cdot 10^{-23}\,\text{J}\,\text{K}^{-1}. \tag{17.15}$$

Note that here, and in the following, T denotes temperature.

Using the virial theorem, we can express the thermal energy, which is identical to the kinetic energy of the gas particles, in terms of the potential energy as $U = -E_\text{pot}/2$. The result is

$$3k_\text{B}TN = G \int_0^{M(R)} \frac{M}{r(M)}\,\text{d}M. \tag{17.16}$$

The right-hand side follows from Eq. (1.44) with $M(r) = \frac{4}{3}\pi\rho_m r^3$ and $\text{d}M = 4\pi\rho_m r^2 \text{d}r$. During the contraction of the cloud, the modulus of the potential energy E_pot increases, therefore the thermal energy U also increases. However, only half of the released potential energy is turned into thermal energy. By energy conservation,

---

[1] Ludwig Boltzmann, 1844–1906, Austrian theoretical physicist. Delivered pioneering contributions to statistical mechanics; well known for his interpretation of entropy.

the remaining half must be released as thermal radiation. When the size of the cloud is reduced, we therefore find that 50% of the potential energy released in the contraction leads to an increase in thermal energy, and 50% is emitted as radiation.

The prerequisite for the contraction is that the modulus of the gravitational binding energy exceeds the thermal energy. The gravitational binding energy is negative, the thermal energy, corresponding to the total kinetic energy of the gas particles, is positive. Our requisite therefore corresponds to the familiar condition $E = E_{pot} + E_{kin} < 0$ for a bound mechanical system. In our present context, this is called the *Jeans criterion* for the onset of the *gravitational instability*, i.e., of the contraction. The criterion is named after *Jeans*,[2] who in 1902 investigated the propagation of density perturbations in fluids under the action of gravitation [1].

In (17.14) we replace the number of particles with the quotient of the total mass M and the average mass μ of a single molecule or atom, $N = M/\mu$,

$$U = \frac{3}{2} k_B T \frac{M}{\mu}. \qquad (17.17)$$

If we describe the cloud of gas in very simplified terms as a homogeneous solid sphere, and use the expression (1.44) for the gravitational energy, we find the inequality

$$\frac{3}{5} G \frac{M^2}{R} > \frac{3}{2} k_B T \frac{M}{\mu}. \qquad (17.18)$$

We eliminate the radius by the mass and the density, $R = [3M/(4\pi\rho_m)]^{1/3}$. Then follows from (17.18)

$$M > M_J = \left(\frac{375}{32\pi}\right)^{1/2} \left(\frac{k_B T}{\mu G}\right)^{3/2} \rho_m^{-1/2}. \qquad (17.19)$$

The quantity M_J is also called the *critical Jeans mass*. It gives a lower bound on the mass above which the cloud of gas collapses for a given density.

For a rough estimate, we consider a cloud of neutral hydrogen at a temperature $T = 100$ K and a number density of 100 atoms per cubic centimeter.

These values are of the order of magnitude typical of interstellar matter. Hydrogen has an atomic mass of $m_H \approx 1.67 \cdot 10^{-27}$ kg. We therefore have a mass density of $\rho_H = 1.67 \cdot 10^{-19}$ kg m^{-3}, and a total mass of

$$M \gtrsim 6.5 \cdot 10^{33} \text{ kg} \approx 3250 \, M_\odot. \qquad (17.20)$$

As the Jeans mass depends on the chemical composition and the temperature of the cloud, this value of course represents only a rough order of magnitude. Realistic

---

[2] Sir James Hopwood Jeans, 1877–1946, English physicist and astronomer.

$M > M_{\mathrm{J}}$

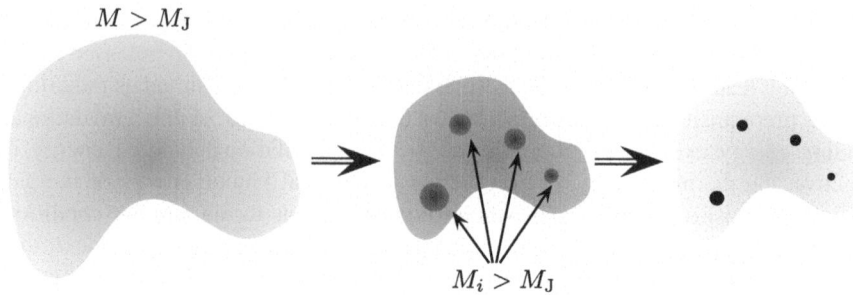

$M_i > M_{\mathrm{J}}$

Fig. 17.2 In the isothermal contraction of a cloud of gas with $M > M_{\mathrm{J}} \sim 10^4\ M_\odot$, the minimum mass necessary for contraction sinks due to $M_{\mathrm{J}} \sim 1/\sqrt{\rho_{\mathrm{m}}}$, until finally subregions can contract independently of the rest of the cloud, and individual stars are formed with masses in the range of $M \sim 1\ M_\odot$: The cloud is being fragmented

values lie rather in the range of $M_{\mathrm{J}} \sim 10^5\ M_\odot$ [6]. In our discussion, we have neglected other physical effects. Interstellar clouds are not static but dynamical objects. Internal motions of regions relative to each other, magnetic fields, and rotations greatly influence processes within a cloud of gas [3].

In any case, we recognize that the mass necessary for the contraction is much bigger than the typical mass of a star. The most massive known star is the object R136 a1 in the Large Magellanic Cloud with an assumed mass of $196^{+34}_{-27}\ M_\odot$ [2]. However, giants such as these are very rare exceptions, the absolute majority of stars has masses on the order of $M \sim 1\ M_\odot$ or less. Thus, out of a contracting interstellar cloud, not only a single star is formed, but many other stars are born, where in fact only a small fraction of the total mass actually becomes stars.

As long as the density of the cloud of gas is low, the energy which is additionally available as increased potential energy, and is converted into kinetic energy, can be dissipated by radiation. In good approximation, the contraction therefore proceeds isothermally, while the density is increasing. Because of the inverse square root dependence of the Jeans mass on the density, the minimum mass M_{J} necessary for collapse decreases. This, together with spatial mass density fluctuations, allows subregions of the cloud to collapse in the direction of their own mass centers. In these subregions, the same processes again proceed, with the result that the original cloud is subdivided in many small subregions. This sequence is called *fragmentation*, see Fig. 17.2.

Thus, individual fragments are formed from the original cloud, from which ultimately stars are born. For this reason, stars are formed in star clusters in which all stars have approximately the same age, at least in general. In the process of fragmentation, other processes probably play a role, such as supersonic motions in the cloud which lead to compressions of subregions [3].

When the density increases to such an extent that the cloud becomes optically thick, radiation from the inner regions of the cloud can escape only poorly. Then the contraction proceeds no longer isothermally, but rather adiabatically, i.e., without heat exchange with the environment. The gravitational energy which is released leads to an increase of temperature. Furthermore, radiation in the outer regions can escape more easily than radiation from the interior of the fragments. Thus, temperature in the interior increases more strongly than in the outer regions.

In this way, the mass in the center finally stabilizes in a hydrostatic equilibrium, in which the gravitational forces are compensated by the thermal pressure, see Sect. 18.1. The resulting structure is called a *protostar* with a typical mass of $M \sim 0.01\,M_{\odot}$. Subsequently, the star increases its mass by accreting the surrounding gas. The protostar is still hidden optically by the cloud of gas, or dust, and therefore visible only in the infrared. The high temperature provides sufficient energy to dissociate hydrogen molecules, and eventually for the ionization of atomic hydrogen and helium.

Readers interested in a more comprehensive account of the processes during star formation, find additional information, e.g., in Larson [3] and in McKee and Ostriker [4].

17.3 Exercises

17.1. Virial theorem for the harmonic potential In the gravitational potential, the virial theorem leads to a ratio of potential and kinetic energy of $\overline{T} = -\frac{1}{2}\overline{V}$. What is the ratio for the harmonic oscillator?

References

1. Jeans, J.H.: The stability of a spherical nebula. Phys. Trans. R. Soc. **199**, 1–53 (1902)
2. Kalari, V.M., et al.: Resolving the core of R136 in the optical. Astrophys. J. **935**(2), 162 (2022)
3. Larson, R.B.: The physics of star formation. Rep. Prog. Phys. **66**(10), 1651 (2003)
4. McKee, C.F., Ostriker, E.C.: Theory of star formation. Annu. Rev. Astron. Astrophys. **45**(1), 565–687 (2007)
5. Mohr, P.J., Tiesinga, E., Newell, D.B., Taylor, B.N.: Codata Internationally Recommended [sic!] 2022 Values of the Fundamental Physical Constants. https://physics.nist.gov/constants
6. Salaris, M., Cassisi, S.: Star formation and early evolution. In: Equation of State of the Stellar Matter, S., pp. 105–116. Wiley, Chichester (2006)

Internal Structure of Stars

18

Contents

In this chapter we will investigate the properties of matter in the interior of stars. We shall derive a set of important equations that will allow us to describe a variety of stars.

18.1 Hydrostatic Equilibrium

If an interstellar cloud collapses under the action of its own gravitational forces, and eventually forms stars, there must be a condition that causes the collapse to stop. For this to happen, a pressure gradient is necessary which supports the weight of every single shell of the star. When this state is reached, the star is stabilized by a *hydrostatic equilibrium.*

For deriving the equilibrium condition, we consider a spherical shell of thickness dr, see Fig. 18.1. Thus, we implicitly assume spherical symmetry of the problem. The assumption of spherical symmetry is expedient since then only the pressure acting in the radial direction plays a role, namely, the pressure $p(r)$ at the bottom of

Fig. 18.1 Sketch for the derivation of the condition for hydrostatic equilibrium. The gravitational pressure p_m of every spherical shell of mass δm must be compensated by the counter pressure p from inside in order that the star reaches hydrostatic equilibrium

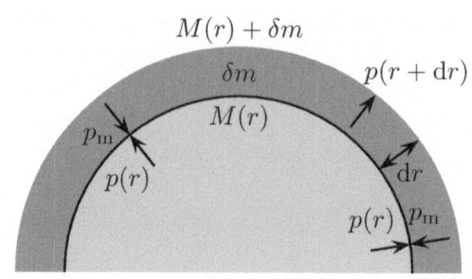

the spherical shell, and the pressure $p(r + dr)$ at the top of it. The pressure gradient must compensate the weight of the spherical shell, which is given by

$$F_m = -G\frac{M(r)\delta m}{r^2}e_r,$$ (18.1)

with the mass $M(r)$ of the sphere within the radius r, and the mass of the spherical shell $\delta m = \rho_m 4\pi r^2\, dr$, where of course $4\pi r^2 = A$ is the surface of the spherical shell. The gravitational pressure at the bottom of the shell therefore is

$$p_m = G\frac{M(r)\rho_m dr}{r^2}.$$ (18.2)

The gas pressure difference between the inner and the outer surface of the spherical shell leads to a force pointing radially outward

$$F_{\Delta p} = [p(r) - p(r + dr)]A\, e_r.$$ (18.3)

In equilibrium, the two forces must be equal, i.e.,

$$F_m + F_{\Delta p} = 0.$$ (18.4)

We linearize the expression for $p(r + dr)$, and obtain

$$p(r + dr) = p(r) + \frac{dp}{dr}\bigg|_r dr + \mathcal{O}(dr^2).$$ (18.5)

We insert this expression in (18.3), and replace $A\, dr$ with $\delta m/\rho_m$. This yields

$$F_{\Delta p} = -\frac{dp}{dr}\bigg|_r A\, dr\, e_r = -\frac{dp}{dr}\bigg|_r \frac{\delta m}{\rho_m(r)}e_r.$$ (18.6)

Inserting in (18.4) finally leads to

$$\frac{dp}{dr} = -G\frac{\rho_m(r)M(r)}{r^2}.$$ (18.7)

With (18.7) we have found a differential equation for the pressure gradient, necessary at a given radius r, in dependence of the mass density and the mass enclosed within the radius r. This pressure gradient stabilizes the star. For a spherically symmetric star, the mass is $M(r) = \int \rho_m(r) dV$ and therefore

$$\frac{dM(r)}{dr} = \rho_m(r) 4\pi r^2. \tag{18.8}$$

Using the condition for hydrostatic equilibrium, we can determine the order of magnitude of the pressure that prevails in the center of the star. Since we do not know the function $\rho_m(r)$ explicitly, we cannot calculate the mass $M(r)$ from (18.8), and therefore cannot integrate (18.7) directly.

However, we can at least estimate the pressure gradient. At the surface of the star we have $p(R) = 0$, aside from the atmospheric pressure, which we neglect. We denote the pressure in the center by p_c. For simplicity, we assume that the pressure decreases linearly with the radial coordinate, i.e.,

$$\frac{dp}{dr} = -\frac{p_c}{R}, \tag{18.9}$$

or

$$p(r) = p_c \left(1 - \frac{r}{R}\right). \tag{18.10}$$

Obviously Eq. (18.10) satisfies $p(0) = p_c$ and $p(R) = 0$. We compare the expression on the right-hand side of (18.9) with the right-hand side of (18.7). In doing so, we replace the enclosed mass $M(r)$ with the total mass M, and use an average density $\langle \rho_m \rangle$ instead of $\rho_m(r)$. The result is

$$\frac{p_c}{R} = G \frac{\langle \rho_m \rangle M}{R^2}. \tag{18.11}$$

With the estimate $\langle \rho_m \rangle = M/[(4/3)\pi R^3]$ we have

$$p_c \sim \frac{3}{4\pi} G \frac{M^2}{R^4}. \tag{18.12}$$

For the corresponding values for mass and radius of the Sun from (1.54) and (1.55), we find

$$p^\odot_{c,\,\text{estimate}} \gtrsim 2.7 \cdot 10^{14}\ \text{Pa} = 2.6 \cdot 10^9\ \text{atm}, \tag{18.13}$$

where one atmosphere (1 atm $= 101{,}325$ Pa) is approximately the air pressure at sea level. We shall see that the actual value of p_c is many times higher.

In spite of the relatively rough estimates contained in the result (18.12), we continue working with it. When we use the average density again, and express M by the Schwarzschild radius $r_s = 2GM/c^2$ from (1.38), we obtain (neglecting prefactors)

$$p_c \gtrsim c^2 \langle \rho_m \rangle \frac{r_s}{R}. \tag{18.14}$$

Therefore, we find the estimate

$$\frac{\langle p \rangle}{\langle \rho_m \rangle c^2} \approx \frac{r_s}{R}. \tag{18.15}$$

Here, $\langle p \rangle$ is the average pressure and $\langle \rho_m \rangle c^2$ the average *rest energy density*. Equation (18.15) establishes a link between the average pressure and the average density of the star.

In physics, relations between variables describing the state of matter of a given system—these variables are, in general, pressure, temperature, density, but also the chemical composition—are called *equations of state*. Equations of state, such as the one in (18.15), are frequently written in the form of a dimensionless function f of the state variables:

$$f(\rho_m, T, \dots) \equiv \frac{p(\rho_m, T, \dots)}{\rho_m c^2}. \tag{18.16}$$

Equation (18.15) is a simplified example of a dimensionless equation of state, simplified, as it links only averaged quantities.

In Sects. 18.3 and 18.4, we shall discuss equations of state in quite different physical situations. Furthermore, the relation (18.7) is only valid as long as density and pressure are so small that general relativistic effects can be neglected. For very massive objects this is no longer true. In Sect. 21.4.1 we shall treat the hydrostatic equilibrium once more in the framework of general relativity.

18.2 Physical Conditions in Stars

The physical parameters describing a star are its mass M, the luminosity L, its radius R, pressure p, and temperature T, and its chemical composition. These quantities are not independent of each other but are linked by mathematical equations. We have already seen such equations in the condition (18.7) for the hydrostatic equilibrium, and in Eq. (18.8) for the mass. Stars are, in good approximation, *black bodies*. In view of the brightness of stars, this statement may seem implausible. But black bodies are not characterized by being black, but by the physical fact that electromagnetic radiation hitting them is completely absorbed, and that the radiation emitted from them exhibits a *black body spectrum*. Stars possess these properties to a very good approximation.

The *Stefan-Boltzmann law*[1] relates the power $L = A\sigma T^4$ radiated by a black body with its surface A and temperature T. The numerical value of the *Stefan-Boltzmann constant* is [4]

$$\sigma = 5.670\,374\,419\ldots \cdot 10^{-8}\ \mathrm{W\,m^{-2}\,K^{-4}}. \tag{18.17}$$

Accordingly, for a star of radius R the law predicts a luminosity of

$$L = 4\pi R^2 \sigma T_{\mathrm{eff}}^4. \tag{18.18}$$

A black body of the same size and the same effective temperature T_{eff} would have the same luminosity as the star. The effective temperature of the Sun is $T_{\mathrm{eff}} \approx 5800$ K. The Sun maintains its temperature essentially constant, even though according to (18.18) the Sun continuously radiates energy. This is only possible when the Sun compensates these radiative losses from a reservoir of energy in its interior. Let us denote by $L(r)$ the fraction of luminosity produced within a sphere of radius r, in analogy with the enclosed mass $M(r)$. From Eq. (18.8) for the radial gradient of $M(r)$ we can infer the analogous equation for $L(r)$

$$\frac{dL(r)}{dr} = 4\pi r^2 \varepsilon(r). \tag{18.19}$$

In (18.19), $\varepsilon(r)$ denotes the *energy production rate*. In Chap. 19 we shall show how stars produce their energy by nuclear fusion processes.

We shall see that the fusion rate R, and with it $\varepsilon(r)$, depends on the density of the reacting particles and the temperature. Therefore, Eq. (18.19) is coupled to other quantities in the interior of the star. Density and temperature, and thus the energy production rate, are maximum near the center of the star. The energy produced by fusion therefore must be transported from the interior of the star to the outside.

A possible process for this heat transport is diffusion. A prerequisite for diffusion is a temperature gradient. Under the assumption of diffusive heat transport, the latter is determined by the equation

$$\frac{dT(r)}{dr} = -\frac{3\kappa(r)\rho_{\mathrm{m}}(r)}{16\sigma T^3(r)} \frac{L(r)}{4\pi r^2}. \tag{18.20}$$

The derivation of this relation can be found, e.g., in Salaris and Cassisi [6]. We recognize that the temperature gradient is negative, like the pressure gradient, which means that the temperature increases towards the center of the star. It is obvious that this is necessary for heat transport from the interior to the outside.

Another mechanism for heat transfer in gases and liquids is convection. This mechanism plays an important role also in stars, a quantitative description of heat

---

[1] Josef Stefan, 1835–1893, Austrian mathematician and physicist.

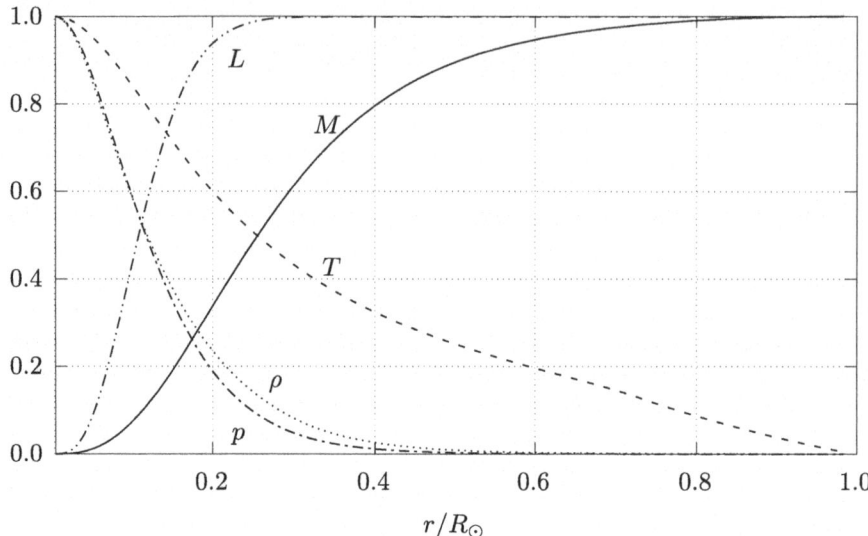

Fig. 18.2 State variables in the interior of the Sun, normalized to their maximum value, as a function of the radial position, normalized to the radius of the Sun. The curve M shows the fraction of the total mass enclosed below r, the curve L the fraction of the total luminosity produced below r [2]

transport with Eq. (18.20) alone therefore is not possible. But we see at this point how the various parameters describing stellar matter are interconnected by different equations, which must be treated jointly.

In (18.20) another quantity appears, namely the *opacity* $\kappa(r)$. It is a measure of how transparent, or opaque, the material in the star is to radiation. For low-mass stars there exists the estimate $\kappa(r) \sim \rho_{\mathrm{m}}(r)T(r)^{-3.5}$, for high-mass stars it is $\kappa(r) \sim$ const [5]. However, the detailed modeling of stars and radiative transfer requires numerical calculations of the opacity.

Figure 18.2 shows the results of quantitative computations of a model for the Sun [1], in dependence of the radial coordinate, normalized to the total radius. The different curves correspond to the pressure, the temperature, the density, the enclosed mass $M(r)$, and the fraction $L(r)$ of the luminosity produced below the radial coordinate r. In (18.13) we had found an estimate for the pressure in the center of the Sun of $p_{\mathrm{ctr}}^{\odot} \sim 3 \cdot 10^9$ atm.

The numerical calculations in the model yield the significantly higher value

$$p_{\mathrm{ctr}}^{\odot} = 2.34 \cdot 10^{16}\ \mathrm{Pa} = 2.31 \cdot 10^{11}\ \mathrm{atm}, \tag{18.21a}$$

approximately 231 billion times the air pressure on the surface of the Earth. Furthermore, the computations in the model yield the values

$$T_{\text{ctr}}^{\odot} = 1.548 \cdot 10^7 \text{ K} \tag{18.21b}$$

and

$$\rho_{\text{ctr}}^{\odot} = 1.502 \cdot 10^5 \text{ kg m}^{-3} \tag{18.21c}$$

for the temperature and the density in the center of the Sun.

It can be seen from Fig. 18.2 that pressure and temperature drop very strongly towards the outside, thus the assumption of a linear pressure gradient in (18.9) was very inaccurate. This is the reason why the estimate (18.13) yielded only a lower bound.

Since the density is much bigger in the center of the Sun than in the outer layers, a large fraction of the total mass is concentrated near the center of the Sun. For example, an inner sphere with radius $r \approx 0.25 R_{\odot}$ contains approximately 50% of the total mass. Likewise, most of the energy radiated by the Sun is produced in the center.

18.3   Equation of State for Stellar Matter

With (18.15) we have already derived a simple estimate for the equation of state of the matter that makes up a star. Now we shall consider this equation in some more detail.

Because of the high temperatures in the interior of stars, the energy due to the interaction between particles is in many cases much smaller than their kinetic energy, and can be neglected. In this case, the description of the matter which composes stars as an ideal gas is an excellent approximation. The corresponding equation of state is

$$pV = Nk_{\text{B}}T \tag{18.22}$$

where N is the number of gas particles. We divide by N and multiply by $V/N = (V/M)(M/N) = \langle m \rangle / \rho_{\text{m}}$. Here, $\langle m \rangle$ denotes the average mass per gas particle. Equation (18.22) then turns into

$$p(r) = \frac{\rho_{\text{m}}(r)}{\langle m \rangle} k_{\text{B}} T(r). \tag{18.23}$$

We have again explicitly indicated the dependence on the radius. Strictly speaking, the average particle mass $\langle m \rangle$ can differ at different locations in the star, and can also be a function of the radius. In the Sun, for example, the outer layers consist mainly of hydrogen, while in the center the fraction of helium dominates, see Fig. 19.5.

We can examine how accurately (18.23) describes the interior of the Sun, by inserting the values for pressure, temperature, and density from (18.21) into (18.23). We obtain

$$\langle m \rangle = \frac{k_B T_{\text{ctr}}^\odot \rho_{\text{ctr}}^\odot}{p_{\text{ctr}}^\odot} \approx 1.37 \cdot 10^{-27} \text{ kg.} \tag{18.24}$$

This is slightly less than the mass of a proton, and therefore gives the correct order of magnitude for a mixture of hydrogen and helium.

For a star in the hydrogen burning phase, see Sect. 19.4, it is a reasonable approximation to insert the mass of the hydrogen atom as the average particle mass. With the pressure from (18.23), the equation of state (18.16) of the foregoing section becomes

$$f[\rho_m(r), T(r)] = \frac{p(r)}{\rho_m(r)c^2} = \frac{k_B T(r)}{m_H c^2}. \tag{18.25}$$

This equation of state depends only on the temperature $T(r)$. When we insert the average temperature $\langle T \rangle$ in (18.25), and use the relation $f \approx r_s/R$ from (18.15), we obtain the estimate

$$f(T) = \frac{k_B T}{m_H c^2} \approx \frac{r_s}{R}. \tag{18.26}$$

Thus, in equilibrium, radius and temperature are related by $T \sim 1/R$. We also see that the ratio of kinetic energy $k_B T$ and rest energy $m_H c^2$ of the atoms determines the radius of the star. *Main sequence stars*, which gain their energy from the fusion of hydrogen, have typical temperatures of $T \approx 1.5 \cdot 10^7$ K, which corresponds to a thermal energy of $k_B T \approx 1.3$ keV. The rest energy of protons and neutrons is approximately $m_p c^2 \approx m_n c^2 \approx 1$ GeV. Thus, the ratio (18.26) is

$$f(T) \approx \frac{1.3 \text{ keV}}{1 \text{ GeV}} \approx \frac{10^3}{10^9} = 10^{-6}. \tag{18.27}$$

For the Sun, with $r_s^\odot \approx 3$ km and $R \approx 7 \cdot 10^5$ km, this leads to $f_\odot \approx 4.3 \cdot 10^{-6}$. Thus, for the Sun the relation (18.27) is satisfied quite well.

18.4   Degenerate Electron Gas

The second equation of state which we will discuss comes into play when the density of the matter becomes very high. We have already seen in Fig. 18.2 that in the interior of the Sun the density is much higher than in the outer layers. Although the Sun can be consistently described as an ideal gas, in other stars it is possible that this description no longer applies for the inner regions of the star. There the electrons can

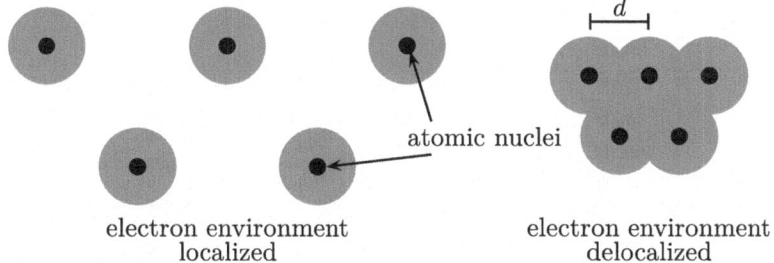

Fig. 18.3 Delocalization of electrons in a Fermi gas during the collapse of a star for very small atomic distances d by overlapping electron shells

build up a pressure due to their quantum mechanical properties. The decisive point will be that this pressure does not depend on the temperature. As a consequence, even at $T \approx 0$ K this *electron degeneracy pressure* prevents a further collapse. Before we discuss this state of matter quantitatively, we will try to understand the formation of the degeneracy pressure in an intuitive picture.

18.4.1 Intuitive Interpretation of the Degeneracy Pressure

Due to the high density, the wave functions of the electrons overlap at a certain point, see Fig. 18.3. This state corresponds to a global wave function, and resembles the state of valence electrons in a metal. The electrons become delocalized, and can move freely in the whole system. This allows for a description as a *free electron gas*, or *Fermi gas*, analogously to the situation in metals. Electrons are fermions, and obey the *Pauli principle*,[2] which says that no two fermions can occupy the same quantum state. Matter with density high enough that it is stabilized due to quantum mechanical effects is generally called *degenerate matter*.

The property of fermions to build up a pressure against gravitation can be understood qualitatively in a simple model. In a star, only the stellar volume is available as a possible location of the electrons. Therefore, as in the simple model of a potential well, the energy levels of the fermions are discrete.

This is the reason why we can derive the basic properties of degenerate matter by considering quantum mechanical particles in a three-dimensional potential well with width L, see Fig. 18.4. The way we proceed in this section is completely analogous to the way the free electron gas is treated in solid state physics, cf. Kittel [3].

In our model, the solution of the Schrödinger equation leads to the eigenfunctions

$$\Psi_k(r) = \exp(i k \cdot r), \tag{18.28}$$

---

[2] Wolfgang Ernst Pauli, 1900–1958, Austrian physicist, Nobel Prize 1945.

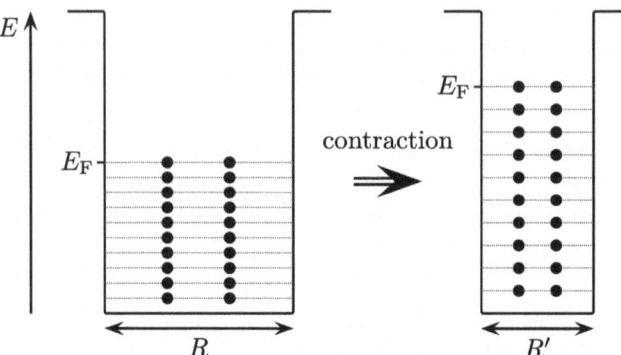

Fig. 18.4 The formation of Fermi pressure can be understood, greatly simplified, in a potential well model for the electrons (*black points*). The width R of the well corresponds to the radius of the star. When the width is reduced, then the energy levels are raised. Therefore, energy has to be expended to compress the well

that is, to plane waves. Periodic boundary conditions apply for the components of the wavevectors,

$$k_i = \frac{2\pi}{L} n, \quad \text{with} \quad n \in \mathbb{Z} \quad \text{and} \quad i \in \{x, y, z\}. \tag{18.29}$$

The boundary condition on the wavevectors guarantees that the wavefunctions vanish at the borders of the potential well. The plane waves Ψ_k are also eigenfunctions of the momentum operator $\hat{p} = -i\hbar\nabla$, with the eigenvalues, written as vectors,

$$p = \hbar k. \tag{18.30}$$

The quantum states occupied by the particles can be represented most clearly in momentum space. There, all states lie within a sphere of radius $R = |p_F|$, the *Fermi sphere*, see Fig. 18.5. The *Fermi momentum* p_F corresponds to the momentum of the highest occupied state. Due to the condition $k_i = 2\pi n/L$ in (18.29) and the relation (18.30) between wavevector and momentum, each allowed state occupies a volume $(h/L)^3$ in momentum space. Because of the Pauli principle, only two fermions with opposite spin orientation can occupy such a state. If states with a given momentum are occupied, other particles must go into states with higher momentum.

The number of states in a volume $V = 4\pi p_F^3/3$ is

$$N = 2\frac{4\pi p_F^3/3}{\left(\frac{h}{L}\right)^3} = \frac{8}{3}\pi \frac{V}{h^3} p_F^3. \tag{18.31}$$

Fig. 18.5 In momentum space, all quantum states with $p < p_F$ lie within a sphere of radius $R = p_F$. To simplify the representation, the sketch is restricted to states with $p_i \geq 0$

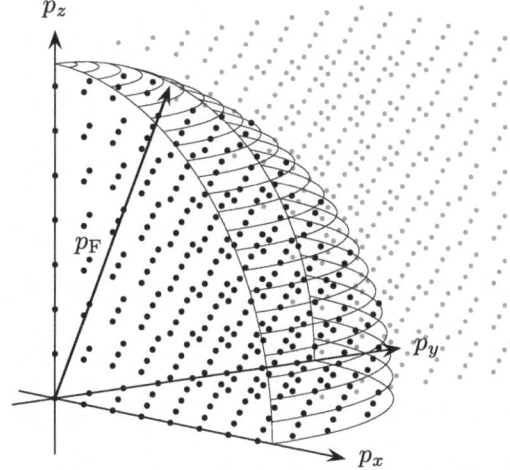

In the second step we have inserted the volume $V = L^3$ of the well. We solve with respect to the Fermi momentum, and obtain

$$p_F = \left(\frac{3N}{8\pi} \frac{h^3}{V} \right)^{1/3} .$$
(18.32)

The Fermi energy is linked to the Fermi momentum by the *relativistic energy-momentum relation*

$$E_F = \sqrt{m^2 c^4 + c^2 p_F^2}$$
(18.33)

discussed in Sect. 6.6. Thus, to further compress the degenerate electrons, energy must be expended.

18.4.2 Completely Degenerate Fermi Gas

In this section we follow the presentation in Shapiro and Teukolsky [7]. Since pressure and momentum appear simultaneously in this section, and we have used the symbol p for both quantities so far, in this section we shall always write p_e for the pressure, to avoid confusion. We continue the discussion of the previous section by considering the number density $d\mathcal{R}/(d^3 r \, d^3 p)$ of the fermions in phase space. The latter defines a distribution function $f(\mathbf{r}, \mathbf{p}, t)$ via

$$\frac{d\mathcal{R}}{d^3 r \, d^3 p} = \frac{2}{h^3} f.$$
(18.34)

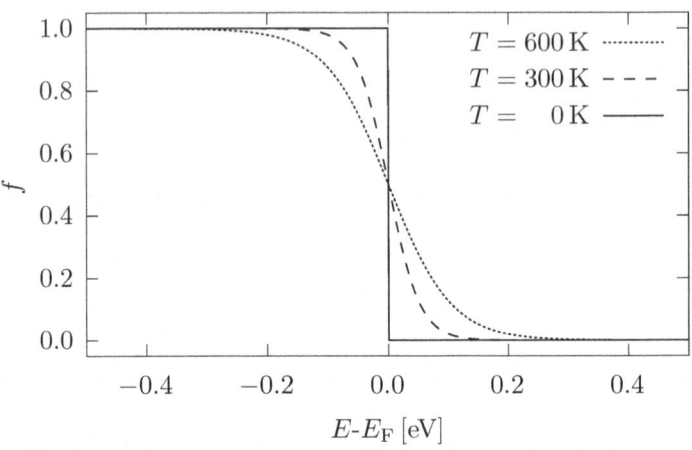

Fig. 18.6 Fermi-Dirac distribution function for different temperatures. The smaller the temperature, the more the function f tends to a step function. For $T \to 0\,\mathrm{K}$, all states with $E < E_\mathrm{F}$ are occupied, and all states with $E > E_\mathrm{F}$ are empty

Here, h^3 is the volume of one cell in phase space, introduced before, the factor 2 accounts for the spin orientations of the electrons. The expression for f is known, since for fermions *Fermi-Dirac statistics* applies, viz.

$$f(E) = \frac{1}{\exp\left(\frac{E-\mu}{k_\mathrm{B}T}\right) + 1}, \tag{18.35}$$

where μ is the chemical potential. It is the amount of energy by which the energy of the system changes when one fermion is removed from the system. For the case of degeneracy, we can therefore set $\mu \approx E_\mathrm{F}$. The smaller the temperature, the bigger the factor $1/(k_\mathrm{B}T)$ in the exponential function in (18.35). Therefore, the function f approaches the form of a step function with decreasing temperature.

In the limit of completely degenerate fermions ($T \to 0$ K, $\mu/(k_\mathrm{B}T) \to \infty$) one finally has

$$f(E) = \begin{cases} 1, & E \leq E_\mathrm{F}, \\ 0, & E > E_\mathrm{F}, \end{cases} \tag{18.36}$$

which means that all states with $E < E_\mathrm{F}$ are occupied, and all states with $E > E_\mathrm{F}$ are empty, see Fig. 18.6.

The number density of the fermions is given by

$$n = \int \frac{d\mathcal{R}}{d^3r\, d^3p} d^3p. \tag{18.37}$$

Furthermore, the pressure in a system with an isotropic momentum distribution is

$$p_e = \frac{1}{3} \int pv \frac{d\mathcal{R}}{d^3r\, d^3p} d^3p. \tag{18.38}$$

The factor 1/3 follows from the isotropy. The velocity can be written as

$$v = pc^2/E, \tag{18.39}$$

since $p = m\gamma v$ and $E = m\gamma c^2$, see Eq. (6.35).

Because of (18.36) and (18.37), the electron density is

$$n_e = \int \frac{d\mathcal{R}}{d^3r\, d^3p} d^3p = \int_0^{p_F} \frac{2}{h^3} f(E) d^3p = \int_0^{p_F} \frac{2}{h^3} 4\pi p^2 dp = \frac{8\pi}{3h^3} p_F^3. \tag{18.40}$$

The upper integration limit p_F follows from (18.36), because states with higher momentum are not occupied. When the Fermi momentum of the electrons becomes larger and larger, the behavior of the electrons finally turns relativistic. To characterize the strength of relativistic effects, a dimensionless relativity parameter is introduced:

$$x = \frac{p_F}{m_e c}. \tag{18.41}$$

It measures the Fermi momentum in units of the non-relativistic momentum of an electron which would move at the speed of light. Then

$$n_e = \frac{8\pi m_e^3 c^3}{3h^3} x^3 = \frac{1}{3\pi^2 \lambdabar_e^3} x^3 = \frac{1}{3\pi^2} n_{e,c} x^3. \tag{18.42}$$

In this equation we have introduced the reduced *Compton wavelength* of the electron [4]

$$\lambdabar_e = \frac{\hbar}{m_e c} = 3.861\,592\,6744(12) \cdot 10^{-13} \text{ m}. \tag{18.43}$$

It only differs by a factor 2π from the usual Compton wavelength, which we already know from (6.59). The quantity $\lambdabar_e$ is called reduced, because it is defined via $\hbar$ instead of h, in analogy with the two forms of Planck's quantum of action. In the following we shall simply use the term Compton wavelength. We see that we can measure the number density n_e in units of a characteristic number density

$$n_{e,c} = \lambda_e^{-3}, \tag{18.44}$$

which corresponds to one electron in a cube with edge length λ_e. Associated with $n_{e,c}$ is a corresponding characteristic mass density and energy density

$$\rho_{e,c} = m_e n_{e,c} = \frac{m_e}{\lambda_e^3} = 1.5819 \cdot 10^7 \text{ kg m}^{-3} \tag{18.45a}$$

and

$$\varepsilon_{e,c} = m_e c^2 n_{e,c} = \frac{m_e c^2}{\lambda_e^3} = 1.42178 \cdot 10^{24} \text{ J m}^{-3}. \tag{18.45b}$$

We can evaluate the pressure using the quantities just introduced. With the momentum $p = y m_e c$, (18.38), together with (18.39), yields

$$p_e = \frac{1}{3} \frac{2}{h^3} \int_0^{p_F} p \frac{p}{\sqrt{p^2 c^2 + m_e^2 c^4}} 4\pi p^2 \, dp = \frac{\varepsilon_{e,c}}{3\pi^2} \int_0^x \frac{y^4 \, dy}{\sqrt{1+y^2}}. \tag{18.46}$$

The integral appearing in the expression can be solved analytically:

$$\phi(x) \equiv \frac{1}{3\pi^2} \int_0^x \frac{y^4 \, dy}{\sqrt{1+y^2}} = \frac{1}{8\pi^2} \left[x\sqrt{1+x^2} \left(\frac{2}{3}x^2 - 1 \right) + \text{arsinh}(x) \right]. \tag{18.47}$$

We have absorbed the prefactor $1/(3\pi^2)$ in the function $\phi(x)$, since then

$$p_e = \varepsilon_{e,c} \phi(x). \tag{18.48}$$

We have not yet reached our final result, since we are looking for an expression of the form $p_e(\rho_m)$. We can express x as a function of the electron density ρ_e using (18.44) and (18.45a). Yet, in stellar matter, the electron density does not yield the dominant contribution to the total mass density. On the contrary, it contributes only an almost negligible part. Even though the pressure is almost exclusively caused by the degenerate electrons, the total mass density is dominated by the rest mass of the ions

$$\rho_{\text{total}} = \sum_i n_i m_i. \tag{18.49}$$

In this expression, n_i and m_i denote the number density and mass of ions of type i. At this point, we can already see that the properties of degenerate matter depend on the chemical composition, since the number of nucleons per ion and the ion density determine the total mass density.

To formulate (18.49) in a more elegant way, we introduce the average baryon mass

$$m_{\text{bar}} = \frac{1}{n_{\text{bar}}} \sum_i n_i m_i. \tag{18.50}$$

The baryon density

$$n_{\text{bar}} = \sum_i n_i A_i \tag{18.51}$$

is defined by the number densities n_i and the mass numbers A_i of the individual constituents of the matter. The total density then simply is

$$\rho_{\text{total}} = n_{\text{bar}} m_{\text{bar}}. \tag{18.52}$$

To link this expression again to the electron density, we introduce the average number of electrons Y_e per baryon. With $n_e = n_{\text{bar}} Y_e$ we obtain

$$\rho_{\text{total}} = \frac{n_e}{Y_e} m_{\text{bar}}. \tag{18.53}$$

With n_e from (18.42) this leads to

$$\rho_{\text{total}}(x) = \rho_c x^3, \tag{18.54}$$

where we have introduced the *critical mass density*

$$\rho_c = \frac{1}{3\pi^2} \frac{n_{e,c} m_{\text{bar}}}{Y_e} = \frac{1}{3\pi^2} \frac{m_{\text{bar}}}{\lambda_e^3 Y_e}, \tag{18.55}$$

and have inserted (18.44) in the second step. Then we can write

$$x = \left(\frac{\rho_{\text{total}}}{\rho_c} \right)^{1/3}. \tag{18.56}$$

Without making a significant error, we can insert the unified atomic mass unit [4] for the baryon mass

$$m_u = 1.660\,539\,068\,92(52) \cdot 10^{-27} \text{ kg}. \tag{18.57}$$

For any isotope of an element with nuclear charge Z and mass number A, the average electron number is given by $Y_e = Z/A$. For a great number of isotopes, e.g., ^{4}He, ^{6}C, and ^{8}O, $A = 2Z$, and therefore $Y_e = 1/2$. Counterexamples are ^{1}H, with $Y_e = 1$, the dominant constituent in main sequence stars, and ^{56}Fe with $Y_e = 0.46$, which plays an important role as a fusion end product of mass-rich stars (cf. Sect. 19.5.2). Yet the three isotopes mentioned—helium, carbon, and oxygen—are those which play the most important role in degenerate matter. Therefore, in (18.55) we can set $Y_e = 0.5$. The example of iron shows that this numerical value is also relatively accurate for other isotopes, except for hydrogen.

The result then is

$$\rho_c = 1.94786 \cdot 10^9 \text{ kg m}^{-3}. \tag{18.58}$$

To get an impression of the size of ρ_c, we can compare with the density $\rho_{ctr}^\odot$ in the center of the Sun from equation (18.21c) and find $\rho_c \simeq 1.3 \cdot 10^4 \rho_{ctr}^\odot$.

Equations (18.48) and (18.54) are the desired equations of state in parametrized form, where, of course, we can insert the expression for x from (18.56) in p_e, and then directly obtain the formula $p_e(\rho_m)$. Due to the relatively complicated form of $\phi(x)$ from (18.47), this results in a rather unwieldy expression.

For the two limiting cases $x \ll 1$, i.e., non-relativistic electrons, and $x \gg 1$, i.e., ultra-relativistic electrons, simple expressions can be obtained, when we make adequate approximations for $\phi(x)$. In the first case, we can expand $\phi(x)$ in a Taylor series around $x = 0$

$$\phi(x) \approx \frac{x^5}{15\pi^2} \quad \text{for} \quad x \ll 1. \tag{18.59}$$

To find an approximation for the ultra-relativistic limit, we exploit the fact that for x being large we have $\sqrt{1 + x^2} \approx x$. Furthermore, since for large x the function $\text{arsinh}(x)$ behaves as $\ln(2x)$, we can neglect this contribution, and find as the dominant term

$$\phi(x) \approx \frac{x^4}{12\pi^2} \quad \text{for} \quad x \gg 1. \tag{18.60}$$

In total, this leads to

$$p_e(x) = \frac{\varepsilon_{e,c}}{3\pi^2} \cdot \begin{cases} \frac{x^5}{5} & \text{for} \quad x \ll 1, \\ \frac{x^4}{4} & \text{for} \quad x \gg 1. \end{cases} \tag{18.61}$$

Finally, when we insert (18.56), we find

$$p_e(\rho_m) = \frac{\varepsilon_{e,c}}{3\pi^2} \cdot \begin{cases} \frac{1}{5}\left(\frac{\rho_m}{\rho_c}\right)^{5/3} & \text{for} \quad \rho_m \ll \rho_c, \\ \frac{1}{4}\left(\frac{\rho_m}{\rho_c}\right)^{4/3} & \text{for} \quad \rho_m \gg \rho_c. \end{cases} \tag{18.62}$$

In thermodynamics, equations of state of the form

$$p_e \sim \rho^\Gamma, \quad \text{i.e.} \quad p_e V^{-\Gamma} = \text{const}, \tag{18.63}$$

are called *polytropic equations of state*, and for this reason the term is also used for Eq. (18.62). As a matter of principle, the pressure of a star must increase with increasing density, otherwise it would be unstable. Equation (18.62) satisfies this condition, but the functional relation between pressure and density is not identical in the two limiting cases. Equations of state in which the pressure increases strongly as a function of density, are called *stiff*, and those in which the pressure increases slowly are called *soft*. The equation of state for the ultra-relativistic limit is therefore softer than in the non-relativistic limit. This statement is of course also true for the exact equation, and it plays an important role in white dwarf stars: Remnants of stars which are stabilized by the degeneracy pressure of the electrons. We shall discuss white dwarf stars in Chap. 20. For white dwarfs, the "relativistic softening" implies that these stars cannot possess arbitrarily large masses.

Expressed by numerical values, Eq. (18.62) yields

$$p_e(\rho_m) = \begin{cases} 3.16 \cdot 10^6 \rho_m^{4/3} \, [\text{kg m}^{-3}] \, \text{Pa} & \text{for} \quad \rho_m \ll \rho_c, \\ 4.93 \cdot 10^9 \rho_m^{5/3} \, [\text{kg m}^{-3}] \, \text{Pa} & \text{for} \quad \rho_m \gg \rho_c. \end{cases} \tag{18.64}$$

When we insert the value of the density in the center of the Sun from (18.21c) in the non-relativistic approximation of (18.64), we find

$$p_e(\rho_{ctr}^\odot) \approx 0.06 p_{ctr}^\odot, \tag{18.65}$$

with the pressure $p_{ctr}^\odot$ in the center of the Sun from (18.21a). This simple estimate demonstrates that the pressure in the interior of the Sun is not caused by the degeneracy pressure of electrons, but by gas pressure, according to (18.23).

Finally, we provide the expressions for the dimensionless equation of state $f = p_e/(\rho_m c^2)$ for the two approximations:

$$f(\rho_m) = \frac{1}{3\pi^2} \cdot \frac{p_{e,c}}{\rho_c} \begin{cases} \frac{1}{5} \left(\frac{\rho_m}{\rho_c}\right)^{2/3} & \text{for} \quad \rho_m \ll \rho_c, \\ \frac{1}{4} \left(\frac{\rho_m}{\rho_c}\right)^{1/3} & \text{for} \quad \rho_m \gg \rho_c, \end{cases}$$

$$\simeq 2.74 \cdot 10^{-4} \begin{cases} \frac{1}{5} \left(\frac{\rho_m}{\rho_c}\right)^{2/3} & \text{for} \quad \rho_m \ll \rho_c, \\ \frac{1}{4} \left(\frac{\rho_m}{\rho_c}\right)^{1/3} & \text{for} \quad \rho_m \gg \rho_c. \end{cases} \tag{18.66}$$

18.5   Summary

Equation (18.25) for the ideal gas, and Eq. (18.62) for the degenerate electron gas, represent completely different equations of state for stellar matter. One significant difference immediately stands out: While f in (18.25) is only a function of the temperature, Eq. (18.62) depends only on the density. This difference of course also applies when we insert the general expression for the pressure in (18.62).

The evolution of a star depends on the relative strength of the two pressures $p \sim \rho_m T$ in (18.23) and $p \sim \rho_m^{n/3}$ in (18.62). According to (18.23), a contracting non-degenerate star will increase its temperature, until fusion processes set in and stop the collapse. However, if the degeneracy pressure increases so strongly that it can stop the collapse, before the temperature is high enough for a certain fusion process to happen, the latter obviously cannot take place. It is essentially the mass of a star that decides which of the two possibilities is physically realized. We shall come back to this topic in Chap. 19, when we discuss the energy production in stars.

References

1. Bahcall, J.N., Serenelli, A.M., Basu, S.: New solar opacities, abundances, helioseismology, and neutrino fluxes. Astrophys. J. Lett. **621**(1), L85 (2005)
2. Bahcall, J.N., Serenelli, A.M., Basu, S.: BS2005 solar model [1]. https://www.sns.ias.edu/~jnb/
3. Kittel, C.: Introduction to Solid State Physics, 8th edn. John Wiley & Sons, Hoboken (2018)
4. Mohr, P.J., Tiesinga, E., Newell, D.B., Taylor, B.N.: Codata Internationally Reconmmended [sic!] 2022 Values of the Fundamental Physical Constants. https://physics.nist.gov/constants
5. Ryan, S.G., Norton, A.J.: Stellar Evolution and Nucleosynthesis. Cambridge University Press, Cambridge (2010)
6. Salaris, M., Cassisi, S.: Equation of State of the Stellar Matter, pp. 31–47. John Wiley & Sons, Chichester (2006)
7. Shapiro, S.L., Teukolsky, S.A.: Black Holes, White Dwarfs, and Neutron Stars: The Physics of Compact Objects. Wiley-Interscience, New York (1983)

Energy Production in Stars

19

Contents

In this chapter we consider the physical processes that allow stars to generate energy. In Sect. 1.5.1 we had already gained a first impression of the incredible amounts of energy released by stars. The Sun, for example, has a luminosity of $L_\odot \approx 4 \cdot 10^{26}$ W, according to (1.58). To remain in equilibrium, the Sun must produce energy with the same power.

Today we know that stars generate this energy by nuclear fusion processes which take place in their interior. A first quantitative study of the processes was carried out

© The Editor(s) (if applicable) and The Author(s), under exclusive license to
Springer-Verlag GmbH, DE, part of Springer Nature 2026
S. Boblest et al., *Special and General Relativity*,
https://doi.org/10.1007/978-3-662-71332-7_19

in the 1930s by *Weizsäcker*[1] and *Bethe*[2] [6]. In 1957 *Burbidge*,[3] *Burbidge*,[4] *Fowler*[5] and *Hoyle*[6] published a landmark paper on the origin of the chemical elements. The paper's title was "Synthesis of the Elements in Stars", but it became known as the B^2HF paper from the initials of its authors [7].

A deep knowledge of these processes is essential not only to the understanding of the evolution of stars, but also to the explanation of the formation and the abundance of the different elements occurring in nature. We begin our discussion with basic considerations concerning the conditions under which nuclear fusion processes occur, and why energy can be released, before looking at some important reactions in more detail.

19.1   Nuclear Fusion as a Source of Energy

Stars like the Sun shine for several billion years, with almost constant luminosity. Therefore, they must draw their energy from a very large reservoir. As we shall shortly see, the amounts of energy released in nuclear fusion reactions are much bigger than the energy released, e.g., in chemical reactions. For comparison: Burning 1 kg of hard coal releases $E \approx 2.9 \cdot 10^7$ J $= 1$ SKE (one hard coal equivalent). If the entire Sun consisted of hard coal, this would yield enough energy to shine for approximately 4600 years with its present luminosity, way too short compared to the age of the Sun of roughly $4.57 \cdot 10^9$ y [5].

In the nineteenth century, nuclear fusion was unknown, and therefore the question as to the source of energy of the Sun remained an unsolved problem. This becomes evident when we examine how long the Sun could draw its energy from other sources of energy known at the time. As we have seen in the previous example, chemical reactions are out of the question. Another obvious source of energy would be gravitational binding energy. In Sect. 1.4.3 we have derived the gravitational binding energy $E_m = -(3/5)GM^2/R$ for a homogeneous solid sphere with mass M and radius R, which would be released in a collapse. For the Sun, with $R_\odot \approx 7 \cdot 10^8$ km and $M_\odot = 2 \cdot 10^{30}$ kg, see Eqs. (1.55) and (1.54), this would result in an energy of

$$E_m^\odot \approx 2.29 \cdot 10^{41} \text{ J}. \tag{19.1}$$

---

[1] Carl Friedrich Freiherr von Weizsäcker, 1912–2007, German physicist and philosopher. Brother of the former President of the Federal Republic of Germany Richard von Weizsäcker.

[2] Hans Albrecht Bethe, 1906–2005, German-American physicist, Nobel Prize in physics 1967 for his calculations of the energy production in stars.

[3] Margaret Burbidge, 1919–2020, US-American astrophysicist.

[4] Geoffrey Burbidge, 1925–2010, US-American astrophysicist, husband of Margaret Burbidge.

[5] William Alfred Fowler, 1911–1995, US-American astrophysicist, Nobel Prize 1983.

[6] Fred Hoyle, 1915–2010, British astronomer and mathematician.

This amount of energy would be sufficient for the Sun to shine with its present luminosity for a period of

$$\tau_{\mathrm{KH}} = \frac{E_{\mathrm{m}}}{L_\odot} \approx 5.9 \cdot 10^{14}\,\mathrm{s} \approx 1.88 \cdot 10^7\,\mathrm{y}. \tag{19.2}$$

The ratio τ_{KH} of gravitational binding energy and luminosity is called the *Kelvin-Helmholtz time scale*.[7,8]

Interpreted as a pure gravitational effect, this would yield a lifetime of a few million years. This is much longer than it would be possible with chemical reactions, but still too small to explain the estimated age $4.57 \cdot 10^9$ y of the Sun. As we have seen, protostars indeed draw their energy from this reservoir before they become hot enough for fusion processes to set in. In the nineteenth century no bigger reservoir of energy was known from which the Sun could draw its radiation energy. The resulting age of the Sun, however, was incompatible with geological findings for the age of the Earth and Darwin's studies of evolution.

How then can energy be gained through nuclear fusion? Figure 19.1 shows binding energies per nucleon for different mass numbers A, where for every A we

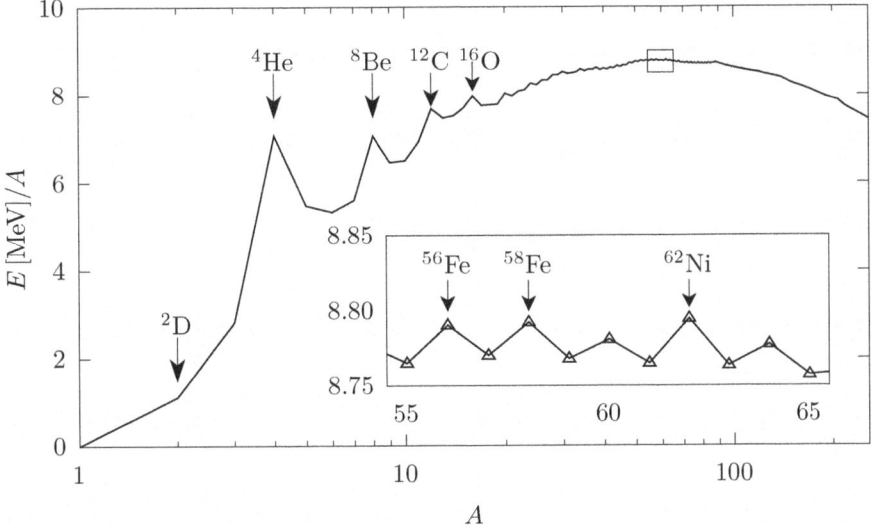

Fig. 19.1 Binding energy per nucleon for nuclides with different mass numbers A. For every mass number the nuclide with the biggest binding energy is shown. The inset shows the range from $A = 55$–65 with the most tightly bound nuclides ^{62}Ni, ^{58}Fe, and ^{56}Fe. The data in the figure is taken from Audi et al. [3]

---

[7] William Thomson, 1st Baron Kelvin, 1824–1907, British physicist.

[8] Hermann von Helmholtz, 1821–1894, German physiologist and physicist.

have chosen the most tightly bound nucleus. Since the hydrogen nucleus consists solely of a proton, its binding energy is of course zero. By contrast, the doubly magic ^{4}He nucleus has a binding energy per nucleon of roughly 7 MeV. Thus, when four protons fuse to form a helium nucleus, during which process 2 protons must be converted into neutrons, one can expect that an energy on the order of 28 MeV is released. The actual value is slightly lower since the neutrinos also produced in the fusion process do not interact with the stellar matter and so carry part of the energy away. Moreover, several reaction paths are possible for protons to produce helium nuclei, which yield slightly different power balances. For simplicity, we assume a value of 26 MeV for the total energy that is released in this fusion process.

When we assume an initial abundance of hydrogen in the Sun of 70%, and also assume a constant luminosity of $L_\odot = 3.86 \cdot 10^{26}$ W, we can estimate how long the Sun can radiate via hydrogen burning.

Converted into Joule, we obtain the energy released per helium nucleus $E_{4p \to He} = 26 \,\text{MeV} = 4.2 \cdot 10^{-12}$ J. Therefore,

$$N_s = 3.86 \cdot 10^{26} \,\text{J}/E_{4p \to He} = 9.3 \cdot 10^{37} \qquad (19.3)$$

of these reactions must take place per second to compensate for the radiative losses. The upper bound for the possible number of these reactions is approximately

$$N_{4p \to He} = \frac{0.7 M_\odot}{4 m_p} = 2.1 \cdot 10^{56}. \qquad (19.4)$$

This leads to the rough estimate

$$t_H = \frac{N_s}{N_{4p \to He}} \,\text{s} \approx 7.1 \cdot 10^{10} \,\text{y}. \qquad (19.5)$$

Thus, the Sun could meet its energy demand at the present luminosity by fusion of hydrogen for about 71 billion years. The lifetime of the Sun is actually shorter by a factor of 7 since stars fuse only roughly 10% of their hydrogen supply. The crucial factor, however, is the huge energy reservoir of nuclear fusion, which is many times bigger than that of gravitational energy.

From Fig. 19.1 we can also see that in the fusion of hydrogen to helium, by far more energy is released per nucleon, than in the fusion of heavier elements, e.g., in the fusion of helium to carbon. Such reactions happen when a star has consumed most of its hydrogen reservoir in the core. Due to the much lower energy yield, the energy supply of these reactions lasts only for much shorter periods. We will discuss these aspects in more detail in Sect. 19.5. Before we do this we would like to analyze what physical conditions must be met for fusion reactions to take place at all.

19.2   Prerequisites for Fusion Processes

The temperatures prevailing in the Sun are on the order of $T \sim 10^7$ K, see (18.21b). This corresponds to a thermal energy of $E \gtrsim 800$–900 eV, much bigger than the ionization energy of hydrogen. Thus light atoms will be present in ionized form. Protons repel each other because of the electromagnetic interaction. If one proton wishes to approach another proton, it first must overcome the *Coulomb barrier*. This means they have to get so close to each other that the short-range strong interaction can induce the nuclear reaction.

To estimate the height of the Coulomb barrier we assume a value of approximately $r_p \simeq 10^{-15}$ m for the radius of the proton, which is in good agreement with current experimental values [2].

When the protons "touch", their centers are at a distance $2r_p$. This yields a potential energy of

$$V(2r_p) = \frac{e^2}{4\pi\varepsilon_0}\frac{1}{2r_p}. \tag{19.6}$$

We compare this energy with the binding energy of the hydrogen atom. To this end, we express the proton radius by the *Bohr radius* $a_B \simeq 0.529 \cdot 10^{-10}$ m from (7.60). Therefore, $r_p \simeq 2a_B \cdot 10^{-5}$, and we have

$$V(2r_p) = \frac{e^2}{4\pi\varepsilon_0}\frac{1}{2a_B}\frac{1}{2}\cdot 10^5 = \frac{\alpha^2}{2}m_e c^2\frac{1}{2}\cdot 10^5 = \frac{1}{2}E_{Ry}\cdot 10^5 \tag{19.7}$$

with the *Rydberg energy*[9] [15]

$$E_{Ry} = \alpha^2 m_e c^2/2 = 13.605\,693\,122\,990(15)\ \text{eV} \tag{19.8}$$

and the fine structure constant α in (1.18). Thus, the height of the Coulomb barrier amounts to $V(2r_p) = 6.8 \cdot 10^5$ eV, and therefore is much bigger than the thermal energies in the range of 1 keV.

There are two reasons why fusion processes do take place nevertheless. Not all protons or nuclei have the same fixed velocity, but we have to deal with a velocity distribution. And fusion processes can be initiated by the quantum mechanical tunneling effect even if the energy of the reactants classically would be too small.

19.2.1 Velocity Distribution of the Nucleons

We have already discussed that, in good approximation, the hot matter in stars can be described as an ideal gas. Accordingly, the velocities of the particles will obey

---

[9] Johannes Rydberg, 1854–1919, Swedish physicist.

a *Maxwell-Boltzmann distribution*. The moduli of the velocities have a probability density

$$P(v)\,dv = 4\pi \left(\frac{m}{2\pi k_B T}\right)^{3/2} v^2 \exp\left(-\frac{mv^2}{2k_B T}\right) dv. \tag{19.9}$$

In fusion processes, it is not the velocities of the particles in their rest frame that is relevant, but the relative velocities in their center of mass system, which also obey a Maxwell distribution. All we have to do is to replace the mass m with the reduced mass

$$m_r = \frac{m_A m_B}{m_A + m_B}. \tag{19.10}$$

As we are dealing with atomic nuclei, it is reasonable to express the reduced mass by the atomic mass numbers of the nuclei. With $m_{A,B} = A_{A,B} m_u$, where m_u is the atomic mass unit from (18.57), we have

$$m_r = \frac{A_A A_B}{A_A + A_B} m_u = A_r m_u. \tag{19.11}$$

Furthermore, it is useful to write the Maxwell distribution as a function of energy, using the non-relativistic relation $E = p^2/2m_r$. Then we obtain

$$P(E)\,dE = 2\sqrt{\frac{E}{\pi k_B^3 T^3}} \exp\left(-\frac{E}{k_B T}\right) dE. \tag{19.12}$$

The fraction of high-energetic particles decreases exponentially with the energy, yet there always remains a small fraction of particles with very high energies. Figure 19.2 shows the Maxwell energy distribution. A short side note: The Maxwell distribution (19.9) is non-relativistic, therefore it allows also for velocities $v > c$. At temperatures not too high, the fraction of high velocities is negligibly small, and this deviation does not play a role. At high temperatures in stars, though, this assumption

Fig. 19.2 Sketch of the Maxwell energy distribution. Even though the thermal energy is much lower than the Coulomb barrier, there always exists a small fraction of high-energetic particles which can induce the fusion process

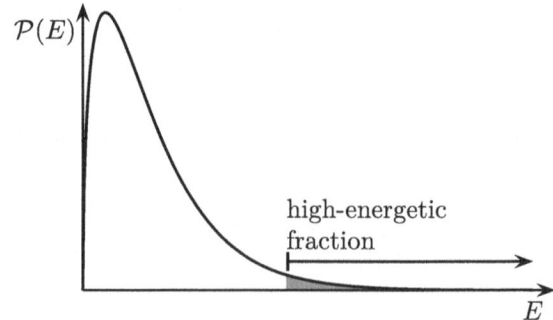

may be problematic. However, for our qualitative discussion the non-relativistic treatment certainly is sufficient. Readers interested in an introductory discussion of the unification of relativity and thermodynamics are referred to Dunkel [8] and the references therein.

19.2.2 Tunneling

In classical physics, a particle can only overcome a potential barrier of height V if its kinetic energy is bigger than V. In quantum mechanics, this restriction no longer holds: There is a small probability for the particle to tunnel through the potential well. These tunneling processes are of central importance in both nuclear decays and nuclear fusion.

The first quantitative investigation of the tunneling effect in nuclear processes was carried out by Gamow in 1928 [9]. The subject of his study was the radioactive α decay, in which a ^{4}He nucleus is emitted by another larger nucleus. We do consider fusion processes, in which exactly the opposite happens, but the corresponding results for the tunneling effect remain valid also in the reverse direction.

A simple example of a one-dimensional system in quantum mechanics is a free particle with energy E moving in a rectangular potential $V > E$. In the region of the potential, the Schrödinger equation reads

$$\frac{-\hbar^2}{2m}\psi''(x) + V\psi(x) = E\psi(x), \tag{19.13}$$

with the solution

$$\psi(x) = \psi_0 \exp\left(-\sqrt{\frac{2m}{\hbar^2}(V-E)}\,x\right). \tag{19.14}$$

Thus, the amplitude of the wave function decreases exponentially, and indeed the faster, the bigger the size of the difference $V - E$. The factor ψ_0 must be determined such that the total wave function ψ is normalized, but this is not of importance in the following considerations. What is important is that the probability for finding the particle in the interval $[x, x+dx]$ is given by $|\psi(x)|^2 dx$. If the rectangular potential extends from x_1 to x_2, the tunneling probability is

$$\mathcal{P} = \frac{|\psi(x_1)|^2}{|\psi(x_2)|^2} = \exp\left[-2\sqrt{\frac{2m}{\hbar^2}(V-E)}(x_2-x_1)\right]. \tag{19.15}$$

Thus, the tunneling probability decreases with the size of the difference $V - E$ of the potential above the energy, and with the width $x_2 - x_1$.

Fig. 19.3 Tunneling effect in fusion processes. An incident particle can tunnel through the Coulomb barrier with some probability, enter the range of the strong interaction, and undergo a fusion process with the target nucleus. In this sketch, the shape of the potential is depicted in a simplified fashion

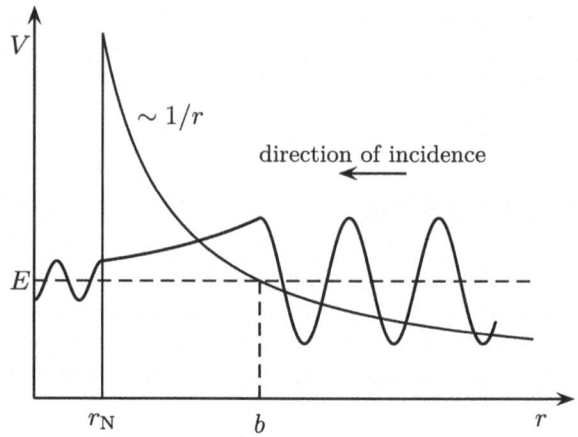

When we want to transfer this simple consideration to the fusion of two charged particles, we must replace the rectangular potential with the Coulomb potential

$$V(r) = \frac{Z_A Z_B}{4\pi\varepsilon_0} \frac{e^2}{r}, \tag{19.16}$$

see Fig. 19.3. Here, Z_A and Z_B are the charges of the two particles. Due to the non-constant potential, we cannot solve the Schrödinger equation as easily as in (19.14). In particular, we do not know the exact form of the potential close to the nucleus.

The result of the detailed calculation is [17]

$$\mathcal{P}_T \approx \exp\left(-\sqrt{E_G/E}\right), \tag{19.17}$$

with the *Gamow energy*

$$E_G = 2\pi^2\alpha^2 m_u c^2 A_r Z_A^2 Z_B^2 = 979\,\text{keV} \cdot A_r Z_A^2 Z_B^2. \tag{19.18}$$

Note that we have used the reduced mass number A_r from (19.11). In astrophysical literature frequently the notation $2\pi\eta = \sqrt{E_G/E}$ is introduced, but we shall not adopt this convention. In the fusion of two protons, $A_r = 1/2$ and $Z_A = Z_B = 1$, and one obtains $E_{G,pp} \simeq 489.6\,\text{keV}$. For the temperature $T_{ctr}^\odot$ in the center of the Sun from (18.21b), the thermal energy is roughly $E \approx k_B T_{ctr}^\odot \approx 1.3\,\text{keV}$. Thus, it is many times smaller than the Gamow energy. Therefore, $\mathcal{P}_T \approx \exp(-19.2) \approx 4.8 \cdot 10^{-9}$, and the probability of tunneling through the Coulomb barrier is extremely small.

The Gamow energy increases with the charge numbers of the reacting nuclei, therefore in this case the tunneling probability decreases even more strongly. For example, in the fusion of two ^{4}He nuclei, due to $Z_A = Z_B = 2$ and $A_r = 1/2$, the Gamow energy is already 16 times bigger than for two protons. For fusion reactions of more highly charged particles to happen, the average energy of the particles, and thus the temperature in the stars, must be much higher.

19.3 Calculation of Reaction Rates

The discussion in the two previous sections has shown that fusion processes can take place in stars. The crucial quantity, however, is the rate at which fusion reactions can happen. In the following investigation of fusion rates we follow the presentation in Ryan and Norton [17].

In physics, the probability for reactions in scattering processes is characterized by the *cross section* σ. In the present case, the latter must be proportional to the probability $\mathcal{P}_T$ in (19.17). However, the details of the specific nuclear physical processes of the reactions also play a role. These are summarized by some factor $S(E)$, which must be determined in a combination of theory and experiment, since the physical conditions prevailing in stars can hardly be realized in terrestrial laboratories. Therefore, results of laboratory experiments must be extrapolated with the help of theory.

In many cases $S(E)$ varies only weakly with energy, and therefore can be assumed as constant over wide ranges of energy. An exception to this rule are energies corresponding to excited states of the nuclei involved. Values of $S(E)$ are tabulated extensively for many different reactions [1]. This reference also includes a detailed discussion of the methods used in determining the values, and their respective uncertainties. All in all one combines these quantities in the cross section

$$\sigma(E) = \frac{S(E)}{E} \exp\left[-\left(\frac{E_G}{E}\right)^{1/2}\right]. \tag{19.19}$$

However, the rate R_{AB} of a fusion process between particles of type A and B is not solely determined by the cross section, but also by the frequency of the events in which two such particles come close to each other. This rate is the bigger, the higher the particle densities, and the faster the relative velocities of the particles, viz.

$$R_{AB} \sim n_A n_B \sigma v_r. \tag{19.20}$$

Like the velocities of the particles, their relative velocities and thus their energies obey the Maxwell distribution. Therefore, in order to obtain an expression for the reaction rate per unit volume, we have to average over the expression σv_r. As we have derived the cross section as a function of the energy, we express the velocities by the energy, $v_r = \sqrt{2E/m_r}$, and use the form of the Maxwell distribution (19.12), expressed in terms of the energy. We then have

$$\langle \sigma v_r(E) \rangle = \left(\frac{8}{\pi m_r k_B^3 T^3}\right)^{1/2} \int_0^\infty \sigma(E) E \exp\left(-\frac{E}{k_B T}\right) dE$$

$$= \left(\frac{8}{\pi m_r k_B^3 T^3}\right)^{1/2} \int_0^\infty S(E) \exp\left[-\frac{E}{k_B T} - \left(\frac{E_G}{E}\right)^{1/2}\right] dE. \tag{19.21}$$

Even if we assumed $S(E)$ as constant, we could not evaluate the integral directly. Therefore, we are looking for a suitable approximation. In the argument of the exponential function in the second line there appear two terms, the term $-E/k_B T$ from the Maxwell distribution, and the term $-(E_G/E)^{1/2}$ from the tunneling probability. For high energies, the Maxwell part diverges to minus infinity, the tunneling part tends to zero. For low energies the behavior is exactly the other way round. As a consequence, the integrand in (19.21) features relevant contributions only in a limited range of energies. This means, in physical terms, that the fraction of particles with very high energies is very small, and that, on the other hand, the tunneling probability for particles with low energy is very small.

The maximum of the exponent lies at

$$E_P = \left[E_G \left(\frac{k_B T}{2} \right)^2 \right]^{1/3},$$ (19.22)

with the value of the exponential function

$$\exp(E_P) = \exp\left[-3 \left(\frac{E_G}{4k_B T} \right)^{1/3} \right].$$ (19.23)

This maximum is called the *Gamow peak*. We expand the exponent around the maximum, and find

$$\frac{E}{k_B T} + \left(\frac{E_G}{E} \right)^{1/2} \approx 3 \left(\frac{E_G}{4k_B T} \right)^{1/3} + 3 \left(\frac{1}{16 E_G (k_B T)^5} \right)^{1/3} (E - E_P)^2.$$ (19.24)

Inserting this expression again into the exponential function yields a constant contribution, multiplied by a Gaussian, viz.

$$\langle \sigma v_r(E) \rangle \approx \left(\frac{8}{\pi m_r k_B^3 T^3} \right)^{1/2} \int_0^\infty S(E) \exp\left[-3 \left(\frac{E_G}{4k_B T} \right)^{1/3} \right]$$

$$\cdot \exp\left[-\frac{1}{2} \left(\frac{E - E_P}{\sigma_G} \right)^2 \right] dE,$$ (19.25)

where the variance is given by

$$\sigma_G = 3^{-1/2} \left[2 E_G (k_B T)^5 \right]^{1/6}.$$ (19.26)

The energy range $E \in [E_P \pm \Delta_G / 2]$ in which the exponential function has functional values greater than $1/e$ of the maximum value, is called the *Gamow window*. From

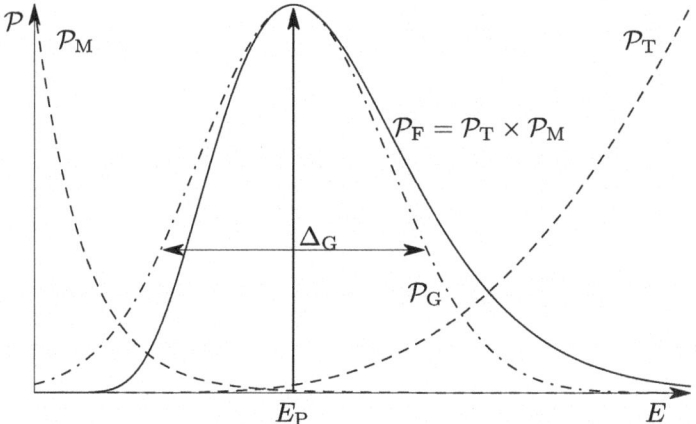

Fig. 19.4 Most fusion reactions take place in a relatively small window around the Gamow peak at $E = E_P$, since only very few high-energy particles are present due to the Maxwell distribution, and the tunneling probability $\mathcal{P}_T$ is extremely small at low energies. The quantitative evaluation is done by expanding the product $\mathcal{P}_F$ in a Gaussian around the maximum. In the sketch, $\mathcal{P}_F$ is represented as a scaled function

the definition of the Gaussian we can see that $\Delta_G = 2\sqrt{2}\sigma_G$. It is in this energy interval that the overwhelming fraction of fusion reactions takes place. Figure 19.4 illustrates the discussion of the previous paragraphs.

At this point we exploit the weak energy dependence of the S factor, and set $S(E) \approx S(E_P)$. Therefore, we can draw this term, together with the first exponential function which also does not depend on E, in front of the integral. Then the integration only extends over a pure Gaussian, with the result

$$\langle \sigma v_r(E) \rangle \approx \left(\frac{4}{m_r k_B^3 T^3} \right)^{1/2} S(E_P) \exp\left[-3 \left(\frac{E_G}{4k_B T} \right)^{1/3} \right]$$

$$\cdot \sigma_G \left[1 + \mathrm{erf}\left(\frac{E_P}{\sqrt{2}\sigma_G} \right) \right]. \tag{19.27}$$

For the fusion of two protons in the conditions prevailing in the center of the Sun, we had found $E_G \simeq 490\,\mathrm{keV}$ and $k_B T_\mathrm{ctr}^\odot \simeq 1.3\,\mathrm{keV}$. Using (19.22) and (19.26) and these values, we find the result $E_P/(\sqrt{2}\sigma_G) \approx 0.69(E_G/(k_B T_\mathrm{ctr}^\odot))^{1/6} \approx 1.85$ for the argument of the error function $\mathrm{erf}(x)$. In total, this then leads to $\mathrm{erf}(E_P/(\sqrt{2}\sigma_G)) \simeq 0.99 \approx 1$. Because of the power $1/6$ in (19.26), this ratio varies only very weakly with the ratio of Gamow energy to the thermal energy, and therefore this approximation can be used over a very wide range.

To write the rest of the expression in a more clearly arranged way, we expand by $(E_G/E_G)^{1/2}$. Together with the factor $E_G^{1/6}$ from σ_G, we then obtain the power $2/3$

for E_G as well as for $k_B T$. We combine the remaining factor $E_G^{-1/2}$ with $m_r^{-1/2}$ via $(m_r E_G)^{1/2} = \sqrt{2}\pi m_r c \alpha Z_A Z_B$, and in a next step express m_r again as $m_r = A_r m_u$. To be in a position to use tabulated values of S, we still have to change from SI units $[S] = \text{J m}^2$ to units used in nuclear physics $[S] = \text{keV b}$, where $1\,\text{b} = 1\,\text{barn} = 10^{-28}\,\text{m}^2$, i.e., $S_{SI} = 1.60 \cdot 10^{-44} S_{tab}$. Taking all numerical factors into account, including the values of the physical quantities c, m_u, and α, we arrive at the final result

$$\langle \sigma v_r(E) \rangle \simeq \frac{6.48 \cdot 10^{-24}}{A_r Z_A Z_B} \left(\frac{E_G}{4 k_B T} \right)^{2/3} \exp\left[-3 \left(\frac{E_G}{4 k_B T} \right)^{1/3} \right] \frac{S_{tab}(E_P)}{\text{keV b}} \, \text{m}^3 \, \text{s}^{-1}.$$

$$(19.28)$$

The reaction rate is given by

$$R_{AB} = \frac{n_A n_B}{1 + \delta_{AB}} \langle \sigma v_r(E) \rangle. \tag{19.29}$$

The Kronecker-delta δ_{AB} in (19.29) accounts for the case of the fusion of two identical particles. In this case, the expression $n_A n_B$ becomes $n_A^2 / 2$, since otherwise one would double count the reaction of particle ① with particle ②, as ① + ② on the one hand, and as ② + ① on the other.

To calculate the rate of a specific reaction, we need to know the particle densities of the nuclides involved, in addition to the temperature. We consider the example of the fusion of two protons in the center of the Sun. With the values previously calculated for the thermal energy and the Gamow energy, we have $E_{G,pp}/4 k_B T_{ctr}^\odot \simeq 94$. Now we need the particle density n_p in the center of the Sun. For this purpose we need to know the mass fraction of hydrogen in the center. In the astrophysical literature the convention is to denote the mass fraction of hydrogen by X, the mass fraction of helium by Y, and by Z the mass fraction of all other elements.

Since the Sun has been burning hydrogen over roughly 5 billion years, the hydrogen fraction in the center is smaller compared with the fraction at the surface.

Figure 19.5 shows the mass fractions of hydrogen and helium as a function of the radial position in the Sun, calculated in the solar model in Bahcall et al. [5]. Accordingly, the hydrogen fraction in the center of the Sun today amounts to about 36%. With the value of the density in the center of the Sun (18.21c), $\rho_{ctr}^\odot = 1.502 \cdot 10^5\,\text{kg m}^{-3}$, we find $n_p = \rho_{ctr}^\odot X_{ctr}^\odot / m_p \simeq 3.27 \cdot 10^{31}\,\text{m}^{-3}$. Furthermore, we use the tabulated value [1]

$$S_{pp} = (4.01 \pm 0.04) \cdot 10^{-22}\,\text{keV b}, \tag{19.30}$$

and find

$$R_{pp} \simeq 6.83 \cdot 10^{13}\,\text{m}^{-3}\,\text{s}^{-1}. \tag{19.31}$$

Fig. 19.5 Mass fraction of hydrogen X and helium Y at different distances from the center of the Sun. In the center, the hydrogen fraction has already declined to roughly 36%. The data in this figure is from Bahcall et al. [5]

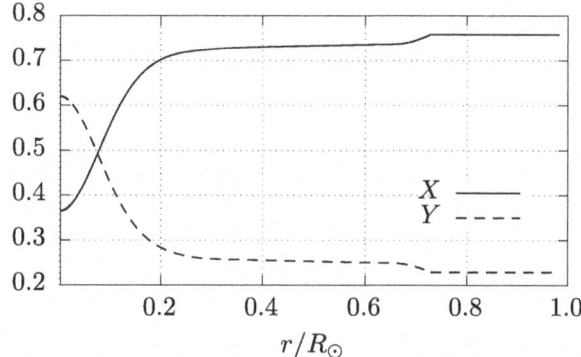

At first sight, the value of about $7 \cdot 10^{13}$ fusion reactions per cubic meter and second seems very large, but it must be compared with the roughly $3 \cdot 10^{31}$ protons per cubic meter. Therefore, the probability for a proton to be involved in a fusion reaction in a certain period of time in fact is very small. We could expect this, since we already estimated that the Sun can meet its energy demand by hydrogen burning for many billion years. When we consider that two protons are consumed per fusion reaction, the average lifetime of a proton with respect to the proton-proton reaction turns out to be

$$t_{pp} = \frac{n_p}{2R_{pp}} \simeq 7.6 \cdot 10^9 \, \text{y}. \qquad (19.32)$$

Of course, the reaction rate will not remain constant over this long period.

19.4 Fusion of Hydrogen

We already know that the fusion of hydrogen into helium can provide the energy emitted by the Sun for several billion years. In fact, all stars gain energy from the fusion of hydrogen into helium for most of their lifetime.

Already before the actual hydrogen fusion happens, another reaction takes place, *deuterium burning*, which sets in at temperatures of about $T \sim 6 \cdot 10^6 \, \text{K}$. In this reaction,

$$^2D + {}^1H \rightarrow {}^3He + \gamma, \qquad (19.33)$$

the deuterium present in the star is burnt into helium. This reaction can take place even in very small stars with masses $M \simeq 0.012 \, M_\odot$ [14], which cannot produce sufficiently high temperatures in their interior for further fusion reactions. Stars which did not get beyond the deuterium burning phase are called *brown dwarfs*. The reason why the density in brown dwarfs does not become high enough for hydrogen fusion to ignite, is the onset of electron degeneracy. The degeneracy pressure in (18.62) is independent of temperature and prevents the temperature from rising higher. The competition between degeneracy pressure and gas pressure also plays an important role on the way towards later burning phases, as we will see below.

In stars that become hot enough for the fusion of hydrogen into helium, two different processes are important: The proton-proton chain, and the CNO cycle.

19.4.1 Proton-Proton Chain

Not too massive stars like the Sun fuse hydrogen predominantly by the proton-proton chain, or short, pp chain. In this process, four protons are fused to form a helium nucleus via various intermediate stages. There are two initial reactions for this chain, the most important one is the direct fusion of two protons

$$^1H + {}^1H \ \rightarrow \ {}^2D + e^+ + \nu_e. \tag{19.34}$$

We have already discussed the rate for this reaction, in the Sun it does take place with a fraction of 99.76%. On a much smaller scale of 0.24%, another reaction takes place in which an electron is involved, and therefore no positron is produced:

$$^1H + e^- + {}^1H \ \rightarrow \ {}^2D + \nu_e. \tag{19.35}$$

Both reactions produce a deuterium nucleus. This nucleus now can react again via the reaction (19.33) to form a ^{3}He nucleus.

At this point, there are three possible follow-up reactions, which occur with very different frequencies, see also Fig. 19.6. In the proton-proton reaction I (ppI), two ^{3}He fuse to form a ^{4}He nucleus, with a lifetime of about 10^6 years, i.e., the average time until a ^{3}He nucleus participates in the reaction,

$$^3He + {}^3He \ \rightarrow \ {}^4He + {}^1H + {}^1H. \tag{19.36}$$

This process dominates in the Sun. The second possibility (ppII) is a three-step reaction chain. A ^{4}He nucleus serves as a catalyst in the formation of another ^{4}He nucleus:

$$^3He + {}^4He \ \rightarrow \ {}^7Be + \gamma, \tag{19.37a}$$

$$^7Be + e^- \ \rightarrow \ {}^7Li + \nu_e, \tag{19.37b}$$

$$^7Li + {}^1H \ \rightarrow \ {}^4He + {}^4He. \tag{19.37c}$$

This reaction chain takes place about 6 times less frequently in the Sun than the ppI chain. Alternatively, the ^{7}Be nucleus formed in the first substep (19.37a) can fuse

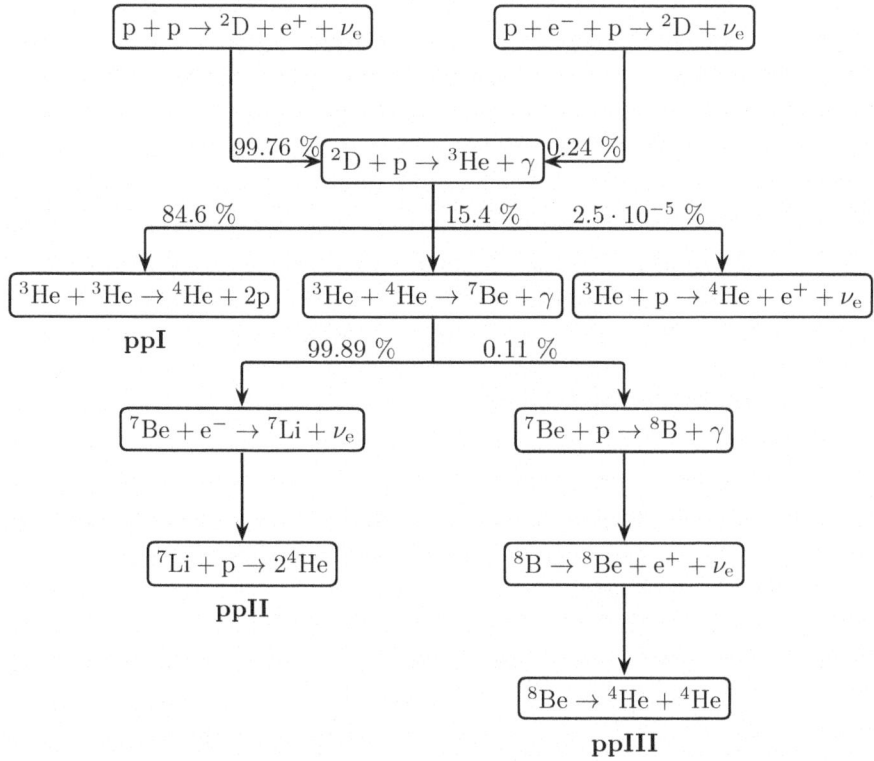

Fig. 19.6 The different subtypes of the pp reaction chain (ppI–ppIII) with their relative fractions in the Sun. The values are taken from Haxton et al. [10]. The reaction of a ^{3}He nucleus with a proton to form a ^{4}He has such a small fraction that it is not counted as part of the actual pp chain

with another proton to form ^{8}B. This leads to the ppIII reaction chain

$$^3\text{He} + {}^4\text{He} \rightarrow {}^7\text{Be} + \gamma, \tag{19.38a}$$

$$^7\text{Be} + {}^1\text{H} \rightarrow {}^8\text{B} + \gamma, \tag{19.38b}$$

$$^8\text{B} \rightarrow {}^8\text{Be} + e^+ + \nu_e, \tag{19.38c}$$

$$^8\text{Be} \rightarrow {}^4\text{He} + {}^4\text{He}. \tag{19.38d}$$

The nuclear charge of beryllium is 4, the Gamow energy is very high for this reaction, compared to the other reactions. Therefore, the ppIII reaction chain contributes only with a very small fraction, but it becomes important at higher temperatures. The resulting ^{8}B nucleus is very unstable and decays, according to (19.38c), via inverse beta decay into ^{8}Be. In the last substep (19.38d) the ^{8}Be nucleus decays into two ^{4}He nuclei with an average lifetime of $6.7 \cdot 10^{-17}$ s. If the mass of the ^{8}Be nucleus were only a fraction $1:10^{-5}$ smaller, and thus the nucleus more strongly bound, this decay would not be possible, with far-reaching consequences, since then heavier elements could have been formed in stars during the hydrogen fusion phase, but also in the early universe after the big bang, and today's element abundance would look completely different. This circumstance is called the *beryllium barrier*. Since ^{8}Be is unstable, only elements up to lithium can be formed in stars, and also could be formed in the early universe. Only at the end of a star's lifetime, when the hydrogen supply in the center slowly runs out, the fusion of heavy elements up to iron can happen, or even beyond iron, e.g., in supernova explosions, or mergers of two neutron stars, see Sect. 19.6.

19.4.2 Bethe-Weizsäcker Cycle

Besides the reaction chain just discussed, there is another important cycle, which includes heavier elements as catalysts. Because of the beryllium barrier, only elements up to and including lithium can be formed in stars, therefore these heavier elements must already be present to initiate the Bethe-Weizsäcker cycle. This is possible if the corresponding star contains supernova remnants of a previously exploded star. Therefore, this reaction chain cannot have taken place in the first stars in the universe, which consisted only of H and He. Furthermore, reactions of ^{4}He with ^{1}H cannot take place since there exists no nuclide with mass number $A = 5$ with a sufficiently long lifetime. For example, ^{5}He decays by neutron emission again into ^{4}He within $8 \cdot 10^{-22}$ s, and ^{5}Li within $3 \cdot 10^{-22}$ s by emission of a proton also into ^{4}He. The hydrogen isotope ^{5}H decays even faster.

The reactions of protons with deuterium, lithium, beryllium and boron all take place very fast, and therefore consume the reactants in a short time. For this reason, these elements are relatively rare both in the Sun and on the Earth. Carbon, on the other hand, is a relatively abundant element, and accounts for around 1% of newly formed stars. The reason is the existence of a cycle in which carbon acts as a catalyst for the fusion of protons into helium:

$$^{12}\text{C} + {}^1\text{H} \rightarrow {}^{13}\text{N}, \tag{19.39a}$$

$$^{13}\text{N} \rightarrow {}^{13}\text{C} + e^+ + \nu_e, \tag{19.39b}$$

$$^{13}\text{C} + {}^1\text{H} \rightarrow {}^{14}\text{N} + \gamma, \tag{19.39c}$$

$$^{14}\text{N} + {}^1\text{H} \rightarrow {}^{15}\text{O} + \gamma, \tag{19.39d}$$

$$^{15}\text{O} \rightarrow {}^{15}\text{N} + e^+ + \nu_e, \tag{19.39e}$$

$$^{15}\text{N} + {}^1\text{H} \rightarrow {}^{12}\text{C} + {}^4\text{He}. \tag{19.39f}$$

This reaction chain is called the *Bethe-Weizsäcker cycle*, or the CNO cycle, because of the elements involved in the reaction chain. All these nuclides are periodically

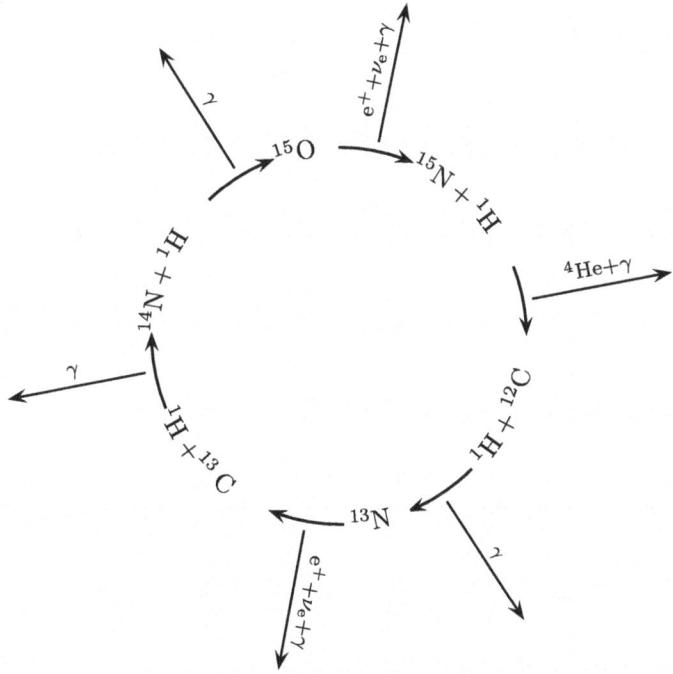

Fig. 19.7 The Bethe-Weizsäcker cycle. Four protons are combined into a ^{4}He nucleus using carbon, nitrogen, and oxygen as catalysts. Because of the high atomic numbers of the nuclides involved, this process takes place efficiently only at higher temperatures than the pp chain

created and annihilated in the reaction chain. However, the reaction rates of the individual reactions are very different. But once an equilibrium situation has been reached, the particle densities of the heavy nuclides remain constant. In fusion reactions of nitrogen and carbon nuclei with a proton, the Gamow energy is much higher than in the reactions of the pp chain. In the Sun, this cycle therefore contributes only insignificantly to the energy production.

The temperature dependence of the pp reactions and the CNO cycle are very different, for example, $R_{pp} \sim T^4$ and $R_{CNO} \sim T^{16-20}$ [17]. If the temperature increases only slightly, the number of CNO reactions therefore can increase strongly. In the course of the further evolution of the Sun, the temperature in the center will also rise, and the contribution of the CNO cycle to the energy production will increase to around 20% [4]. Figure 19.7 summarizes the reactions of the CNO cycle. The periodic sequence of the reactions can be clearly seen. More massive stars reach higher temperatures in their centers. Due to its stronger temperature dependence, the contribution of the CNO cycle quickly becomes dominant with increasing mass of the star, compared to the pp reactions. The process of hydrogen fusion therefore differs fundamentally in low-mass and in high-mass stars.

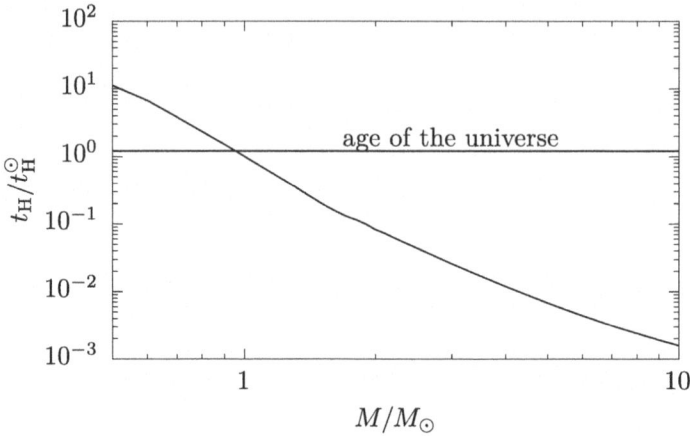

Fig. 19.8 Duration of the hydrogen burning phase as a function of the masses of the stars in comparison with the value of $t_{\rm H}^{\odot} \approx 1.15 \cdot 10^{10}$ y for the Sun, for models of stars with the same chemical composition as the Sun. For stars with $M \lesssim 0.9\,M_{\odot}$, this period is already longer than the age of the universe of about $13.8 \cdot 10^{9}$ y [16]. More massive stars live through this phase during much shorter periods. The data in this figure is taken from Salaris and Cassisi [20]

19.4.3 Duration of the Hydrogen Burning Phase

The strong temperature dependence of the reactions occurring during hydrogen fusion leads to different durations of the hydrogen burning phase for stars with different masses. Figure 19.8 shows the duration $t_{\rm H}$ of the hydrogen burning phase normalized with respect to the duration $t_{\rm H}^{\odot}$ of the Sun. Even for stars that are only 10% less massive than the Sun, the hydrogen burning phase would already last longer than the age of the universe, of about 13.8 billion years. By contrast, a star with $M = 10\,M_{\odot}$ goes through this phase in approximately 18 million years.

19.5   Nuclear Fusion After Hydrogen Burning

When a star runs out of hydrogen in its core, a sequence of further fusion reactions takes place, depending on the mass, which we will now discuss briefly.

19.5.1 Helium Burning

Due to the ongoing fusion of hydrogen, the fraction of helium in the center of a star continues to rise, and eventually a stellar core made of helium is formed. There the density and temperature increase until helium nuclides can fuse to form carbon. As with the onset of hydrogen fusion, the prerequisite for this to happen is that the temperature in the core of the star becomes high enough, before the electron degeneracy pressure renders further contraction, and thus temperature increase, impossible. Since for the fusion of helium higher temperatures are necessary than with hydrogen, the minimum mass also is higher, and amounts to about $M \approx 0.5\ M_\odot$. Therefore, less massive stars cannot fuse helium. On the other hand, Fig. 19.8 shows that less massive stars in the universe can remain in the hydrogen burning phase for a very long time. It is only when a star loses matter during its evolution, for example to a companion, that this limitation is relevant. However, even with stars in the mass range of the Sun, the electron degeneracy pressure already plays an important role in the transition to the helium burning phase. When nuclear fusion begins in the partially degenerate helium core of the star, and the temperature starts to rise, this does not increase the Fermi pressure, as it is only density-dependent. With increasing temperature also the reaction rate increases, and thus again the temperature. This chain reaction leads to a sequence of short, extreme increases of the luminosity in the center of the star, known as *helium flashes*. In every flash, the temperature in a shell around the core rises to such an extent that electron degeneracy again plays no role, and at the end of every flash phase the degeneracy is completely removed. The flashes do not change the luminosity of the star observable from the outside, but this phase has a major effect on the evolution of the star. In Sect. 20.4 we shall see that white dwarf stars—the remnants of low- or intermediate-mass stars—can explode on account of a similar process.

The fusion of helium is also known as the triple-alpha process, and takes place in two steps:

$$^4\mathrm{He} + {}^4\mathrm{He} \rightarrow {}^8\mathrm{Be}, \tag{19.40a}$$

$$^8\mathrm{Be} + {}^4\mathrm{He} \rightarrow {}^{12}\mathrm{C}^*, \tag{19.40b}$$

$$^{12}\mathrm{C}^* \rightarrow {}^{12}\mathrm{C} + 2\gamma. \tag{19.40c}$$

Here, Eq. (19.40a), the fusion of two helium nuclei to beryllium, is the exact reversal of the last step of the ppIII chain in (19.38d). This step is endothermic, and requires an energy of roughly 92 keV. Since the $^8\mathrm{Be}$ nucleus decays within $6.7 \cdot 10^{-17}$ s, the subsequent reaction (19.40b) must take place almost simultaneously. Effectively, three helium nuclei fuse simultaneously to form carbon [19]. The carbon nuclide is produced in an excited state and usually decays immediately. It is only in a few cases that it makes a transition to the ground state by emitting gamma radiation, like in (19.40c).

In addition to the triple-alpha process, other reactions take place in the helium burning phase, reactions, in which the carbon nuclide formed fuses with another helium nucleus:

$$^{12}\mathrm{C} + {}^4\mathrm{He} \rightarrow {}^{16}\mathrm{O} + \gamma, \tag{19.41a}$$

$$^{16}\mathrm{O} + {}^4\mathrm{He} \rightarrow {}^{20}\mathrm{Ne} + \gamma, \tag{19.41b}$$

$$^{16}\mathrm{Ne} + {}^4\mathrm{He} \rightarrow {}^{24}\mathrm{Mg} + \gamma, \tag{19.41c}$$

$$^{16}\mathrm{Mg} + {}^4\mathrm{He} \rightarrow {}^{28}\mathrm{Si} + \gamma. \tag{19.41d}$$

The reaction (19.41a) is of great importance since it changes the ratio of carbon to oxygen in the star, which in turn has a major effect on the further evolution of the star. At the same time, determining the S factor of this reaction is relatively difficult [19].

19.5.2 Later Phases of Fusion

When elements with greater nuclear charges are to be formed in fusion processes, then the corresponding star must be more massive in order to generate the temperatures in its interior necessary for the fusion of these higher elements. Following the helium burning phase, the fusion of carbon sets in when the star can generate the necessary temperature of $T \approx 5 \cdot 10^8$ K. The star must possess a mass of about 8 $M_\odot$ for this fusion process to happen. In this case, two carbon nuclei merge to form a highly excited ^{24}Mg nucleus, which subsequently decays into different lighter nuclides:

$$^{12}\text{C} + {}^{12}\text{C} \rightarrow {}^{24}\text{Mg} \rightarrow \begin{cases} ^{23}\text{Mg} + \text{n}, \\ ^{20}\text{Ne} + {}^4\text{He}, \\ ^{23}\text{Na} + \text{p}. \end{cases} \tag{19.42}$$

At even higher temperatures, $T \approx 10^9$ K, in stars with $M \gtrsim 10\ M_\odot$, the neon nucleus that is formed can be split by gamma radiation,

$$^{20}\text{Ne} + \gamma \rightarrow {}^{16}\text{O} + {}^4\text{He}, \tag{19.43}$$

a process which is referred to as *photo disintegration*. The helium nucleus, and also free neutrons, then can react with other nuclides:

$$^{20}\text{Ne} + {}^4\text{He} \rightarrow {}^{24}\text{Mg} + \gamma, \tag{19.44a}$$

$$^{20}\text{Ne} + \text{n} \rightarrow {}^{21}\text{Ne} + \gamma, \tag{19.44b}$$

$$^{21}\text{Ne} + {}^4\text{He} \rightarrow {}^{24}\text{Mg} + \text{n}. \tag{19.44c}$$

The carbon burning phase is followed by phases of oxygen and silicon burning. Oxygen nuclei first fuse to form heavier nuclides via

$$^{16}\text{O} + {}^{16}\text{O} \rightarrow {}^{32}\text{S} \rightarrow \begin{cases} ^{31}\text{S} + \text{n}, \\ ^{31}\text{P} + \text{p}, \\ ^{30}\text{P} + {}^2\text{D}, \\ ^{28}\text{Si} + {}^4\text{He}. \end{cases} \tag{19.45}$$

After the fusion of two silicon nuclei in the reaction

$$^{28}\text{Si} + {}^{28}\text{Si} \rightarrow {}^{56}\text{Ni} + \gamma, \tag{19.46}$$

and the possible photo disintegrations

$$^{28}\text{Si} + \gamma \rightarrow {}^{27}\text{Al} + {}^{1}\text{H}, \tag{19.47a}$$

$$^{28}\text{Si} + \gamma \rightarrow {}^{24}\text{Mg} + {}^{4}\text{He}, \tag{19.47b}$$

or the β^+ decays

$$^{56}\text{Ni} \rightarrow {}^{56}\text{Co} + e^+ + \nu_e, \tag{19.48a}$$

$$^{56}\text{Co} \rightarrow {}^{56}\text{Fe} + e^+ + \nu_e \tag{19.48b}$$

the tightly bound ^{56}Fe nucleus is formed in the star, one of the most stable nuclei, as we have seen in Fig. 19.1. At this point, no further energy can be released by fusion. In Chap. 21 we shall discuss the processes that can happen then. In general, the later burning phases release much less energy than the fusion of hydrogen. The duration of these fusion phases therefore is much shorter than the period of hydrogen burning. Table 19.1 lists the duration of the individual burning phases for stars with different masses. For example, a star with about the mass of the Sun still fuses helium for another 100 million years, after the H burning phase, which lasts approximately 11 billion years. In the helium burning phase, the star expands, and its radius increases by roughly a factor of 10, or more.

19.6   Formation of Heavy Elements

We have seen that fusion processes in stars can produce elements only up to iron and nickel. Higher elements cannot be formed by fusion since no energy is released. Moreover, the Gamow energy (19.18) of the corresponding reactions would be very large, and thus the Coulomb barrier very high, and the tunneling probability very small.

Yet elements with very much higher nuclear charge numbers do exist in nature, in non-negligible amounts, therefore these must be formed by other processes. This happens predominantly by capture of free neutrons. Since neutrons are uncharged, they do not have to tunnel through a Coulomb barrier, and therefore can also react with heavy nuclei.

However, when an isotope $^A_Z X$ captures a neutron, only a heavier isotope of the same element is produced in the reaction

$$^A_Z X + \text{n} \rightarrow {}^{A+1}_Z X + \gamma. \tag{19.49}$$

Table 19.1 Overview of the different fusion phases for stars with initial masses $1M_\odot$–$75M_\odot$. In all cases, the table lists the temperature in the burning phase, the mass at the beginning of the burning phase (which may be smaller than the initial mass due to mass losses), and the duration of the individual burning phases. The model marked with an asterisk * is a very low-metal star with only 0.01% the abundance of heavier elements present in the Sun. The data in the table is taken from Woosley et al. [21]

M_{init} [$M_\odot$]	T [K]	M [$M_\odot$]	τ	T [K]	M [$M_\odot$]	τ
	H burning phase			He burning phase		
1	$1.57 \cdot 10^7$	1.00	$1.10 \cdot 10^{10}$ y	$1.25 \cdot 10^8$	0.71	$1.10 \cdot 10^8$ y
13	$3.44 \cdot 10^7$	12.9	$1.35 \cdot 10^7$ y	$1.72 \cdot 10^8$	12.4	$2.67 \cdot 10^6$ y
25	$3.81 \cdot 10^7$	24.5	$6.70 \cdot 10^6$ y	$1.96 \cdot 10^8$	19.6	$8.39 \cdot 10^5$ y
75	$4.26 \cdot 10^7$	67.3	$3.16 \cdot 10^7$ y	$2.10 \cdot 10^8$	16.1	$4.78 \cdot 10^5$ y
75*	$7.60 \cdot 10^7$	75.0	$3.44 \cdot 10^7$ y	$2.25 \cdot 10^8$	74.4	$3.32 \cdot 10^5$ y
	C burning phase			Ne burning phase		
13	$8.15 \cdot 10^8$	11.4	$2.82 \cdot 10^3$ y	$1.69 \cdot 10^9$	11.4	0.341 y
25	$8.41 \cdot 10^8$	12.5	$5.22 \cdot 10^2$ y	$1.57 \cdot 10^9$	12.5	0.891 y
75	$8.68 \cdot 10^8$	6.37	$1.07 \cdot 10^3$ y	$1.62 \cdot 10^9$	6.36	0.569 y
75*	$10.4 \cdot 10^8$	74.4	$2.7 \cdot 10^1$ y	$1.57 \cdot 10^9$	74	0.026 y
	O burning phase			Si burning phase		
13	$1.89 \cdot 10^9$	11.4	4.77 y	$3.28 \cdot 10^9$	11.4	17.8 d
25	$2.09 \cdot 10^9$	12.5	0.402 y	$3.65 \cdot 10^9$	12.5	0.733 d
75	$2.04 \cdot 10^9$	6.36	0.908 y	$3.55 \cdot 10^9$	6.36	2.09 d
75*	$2.39 \cdot 10^9$	74	0.010 y	$3.82 \cdot 10^9$	74	0.209 d

The resulting isotope $^{A+1}_{Z}X$ is frequently created in a highly excited state, and emits gamma radiation by making a transition to the ground state. But no new element has been formed so far. Yet isotopes with a very high excess of neutrons, i.e., with many more neutrons than protons, are unstable with respect to β^- decay:

$$^{A}_{Z}X \ \rightarrow \ ^{A}_{Z+1}Y + e^- + \bar{\nu}_e. \tag{19.50}$$

If such a radioactive isotope is created by neutron capture, an isotope of a heavier element is produced.

These capture processes are categorized into two subcategories.

19.6.1 s Process

If the average time between neutron capture reactions is much longer than the average decay time for the β^- decay, i.e., $\tau_n \gg \tau_{\beta^-}$, this is referred to as an *s process* (s for slow). In this case, isotopes which are β^- unstable do not have enough time to capture more neutrons before they decay. Figure 19.9 shows a possible reaction chain for the *s process* starting from the ^{56}Fe isotope. The iron isotopes ^{57}Fe and ^{58}Fe are also stable, and therefore can capture more neutrons.

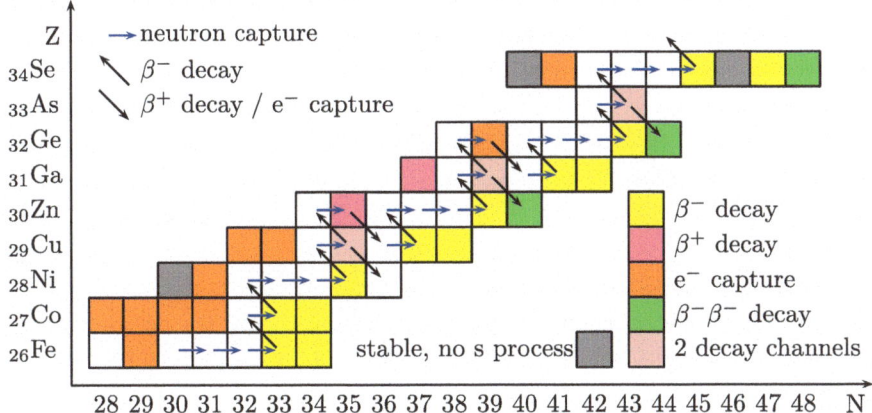

Fig. 19.9 The *s process* starting from the ^{56}Fe isotope. The isotopes ^{58}Ni and ^{74}Se are screened with respect to the *s process* by the stable isotopes ^{58}Fe and ^{74}Ge, respectively

The isotope ^{59}Fe, however, is unstable and decays into ^{59}Co, which in turn is stable. When this isotope then captures a neutron, the unstable nucleus ^{60}Co is formed, which subsequently decays into ^{60}Ni. In this way, heavier and heavier isotopes can be produced. However, no isotopes can be created in this way for which a stable isobar exists, i.e., a stable isotope of a lighter element with the same mass number. In Fig. 19.9 this applies to ^{58}Ni and ^{74}Se. Similarly, this process does not lead to stable isotopes for which there exists a lighter β^- unstable isotope, since before the β^- decay can happen, the isotope decays into an isobar with a nuclear charge higher by 1. This case applies to ^{80}Se. Such isotopes are called *screened* with respect to the *s process*.

For completeness, Fig. 19.9 also shows isotopes with other decay channels. Isotopes with a proton excess decay via the β^+ decay

$$^A_Z X \ \rightarrow \ ^A_{Z-1}Y + e^+ + \bar{\nu}_e \tag{19.51}$$

or by electron capture

$$^A_Z X + e^- \ \rightarrow \ ^A_{Z-1}Y + \nu_e. \tag{19.52}$$

For some neutron-rich elements, simple beta decay is energetically forbidden because the next-highest isobar has a lower binding energy. Such elements can exhibit double beta decay, in which they decay into an isobar with a nuclear charge number greater by two. This reaction, however, is very improbable, and these isotopes have very long half-lives, in the range of 10^{20} y, and therefore can be considered stable with respect to the *s process*.

A side note: Double beta decay is particularly interesting because it is hoped that one can detect the so-called neutrinoless double beta decay, i.e., instead of the reaction

$$_Z^A X \;\rightarrow\; _{Z+2}^A Z + 2e^- + 2\bar{\nu}_e \tag{19.53}$$

the reaction

$$_Z^A X \;\rightarrow\; _{Z+2}^A Z + 2e^-, \tag{19.54}$$

which is only possible if neutrinos are their own antiparticles. The detection of the neutrinoless double beta decay would be evidence for physics beyond the standard model.

19.6.2 r Process

When the time between neutron capture reactions is much smaller than the average time for the β^- decay, i.e., $\tau_n \ll \tau_{\beta^-}$, this is referred to as an *r process* (*r* for rapid).

In this case, β^- unstable isotopes can capture more neutrons before they decay, and isotopes are created with an even higher neutron excess, which then decay to isotopes with higher nuclear charge numbers, when the neutron flux drops.

Figure 19.10 provides a sketch for this case. It can be seen that isotopes with stable isobars of elements, with lower nuclear charge numbers, also are screened with respect to the *r process*, in the figure, e.g., ^{86}Sr.

Which of these two extremes occurs, *s* or *r process*, ultimately depends on the number of free neutrons, which determines the reaction rate according to (19.29).

Fig. 19.10 The *r process*. The legend is analogous to Fig. 19.9. Isotopes marked in gray are shielded by other stable or very long-lived isotopes. The isotope marked in brown, $_{37}^{87}$Rb, is unstable with respect to β^- decay, but with a very long half-life of $4.81 \cdot 10^{10}$ y, and therefore can screen $_{38}^{87}$Sr

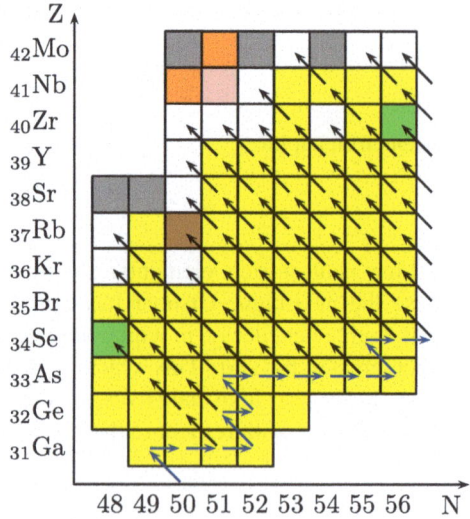

The *s process* takes place in stars in their later burning phases. Free neutrons are then produced, for example, in the reactions we have discussed, (19.42), (19.44c), and (19.45), but above all also in other reactions that we have not discussed, see, e.g., Salaris and Cassisi [18]. However, free neutrons are themselves unstable, and decay with a half-life of about 10 minutes:

$$n \rightarrow p + e^- + \bar{\nu}_e. \tag{19.55}$$

The *r process* probably takes place in supernovae, which we shall discuss briefly in Chaps. 20 and 21. Elements freshly synthesized by the *r process* could be detected in the spectra of the kilonova AT2017gfo formed by the merger of two neutron stars, as we have already discussed in Sect. 15.5.4.

19.7 Neutrino Oscillations

The electron neutrinos released in the pp reactions can be detected in extensive experiments on the Earth, primarily the Sudbury Neutrino Observatory [12] in Canada, and the Super-Kamiokande detector [11] in Japan. It turned out in these experiments that only roughly a third of the expected neutrino flux arrives at the Earth. The reason is the effect of the so-called *neutrino oscillations*.

In the standard model of particle physics, the three neutrino "flavors" ν_e, ν_μ, and ν_τ are assumed massless, although in many approaches of an extension of the standard model, small neutrino masses different from zero arise. In processes of the weak interaction, such as in the fusion reactions discussed above, the neutrinos are created in one of the flavor eigenstates $\{\nu_e, \nu_\mu, \nu_\tau\}$. However, their propagation through space takes place in eigenstates with definite mass, which do not coincide with the flavor eigenstates. During its flight to the Earth, an electron neutrino produced in the Sun oscillates back and forth between the flavor eigenstates, and can be detected also as a muon or tau neutrino. Neutrino oscillations are essentially determined by the three mixing angles θ_{12}, θ_{13}, and θ_{23}, and the corresponding squared mass differences. The exact measurement of these quantities therefore allows for investigations of physics beyond the standard model, on the one hand, and for a better understanding of the fusion processes in stars, on the other. Haxton et al. [10] provides a current review of this topic.

The principal investigators of the Canadian and Japanese research teams, A. B. McDonald[10] and T. Kajita,[11] were awarded the Nobel Prize in physics in 2015 [13].

---

[10] Arthur B. McDonald, ⋆1943, Canadian physicist.

[11] Takaaki Kajita, ⋆1959, Japanese physicist.

References

1. Adelberger, E.G., et al.: Solar fusion cross sections. II. The *pp* chain and CNO cycles. Rev. Mod. Phys. **83**, 195–245 (2011)
2. Antognini, A., et al.: Proton structure from the measurement of 2S-2P transition frequencies of muonic hydrogen. Science **339**(6118), 417–420 (2013)
3. Audi, G., Wapstra, A.H., Thibault, C.: The Ame2003 atomic mass evaluation: (II). Tables, graphs and references. Nucl. Phys. A **729**, 337–676 (2003)
4. Bahcall, J.N., Pinsonneault, M.H., Basu, S.: Solar models: current epoch and time dependences, neutrinos, and helioseismological properties. Astrophys. J. **555**(2), 990 (2001)
5. Bahcall, J.N., Serenelli, A.M., Basu, S.: New solar opacities, abundances, helioseismology, and neutrino fluxes. Astrophys. J. Lett. **621**(1), L85 (2005)
6. Bethe, H.A.: Energy production in stars. Phys. Rev. **55**(5), 434–456 (1939)
7. Burbidge, E.M., Burbidge, G.R., Fowler, W.A., Hoyle, F.: Synthesis of the elements in stars. Rev. Mod. Phys. **29**, 547–650 (1957)
8. Dunkel, J.: Relativ heiß (Relatively hot). Phys. J. **10**, 49–53 (2011)
9. Gamow, G.: Zur Quantentheorie des Atomkernes. (On the quantum theory of the atomic nucleus) Z. Phys. **51**(3–4), 204–212 (1928)
10. Haxton, W.C., Hamish Robertson, R.G., Serenelli, A.M.: Solar neutrinos: status and prospects. Annu. Rev. Astron. Astrophys. **51**(1), 21–61 (2013)
11. Homepage of Super-Kamiokande: https://www-sk.icrr.u-tokyo.ac.jp/en/sk/
12. Homepage of the Sudbury Neutrino Observatory: https://falcon.phy.queensu.ca/SNO/
13. https://www.nobelprize.org/prizes/physics/2015/press-release/
14. Luhman, K.L.: The formation and early evolution of low-mass stars and brown dwarfs. Annu. Rev. Astron. Astrophys. **50**(1), 65–106 (2012)
15. Mohr, P.J., Tiesinga, E., Newell, D.B., Taylor, B.N.: Codata Internationally Recommended [sic!] 2022 Values of the Fundamental Physical Constants. https://physics.nist.gov/constants
16. Planck Collaboration: Planck 2013 results. I. Overview of products and scientific results. Astron. Astrophys. **571**, A1 (2014)
17. Ryan, S.G., Norton, A.J.: Stellar Evolution and Nucleosynthesis. Cambridge University Press, Cambridge (2010)
18. Salaris, M., Cassisi, S.: The advanced evolutionary phases. In: Equation of State of the Stellar Matter, S., pp. 187–237. Wiley, Chichester (2006)
19. Salaris, M., Cassisi, S.: The helium burning phase. In: Equation of State of the Stellar Matter, S., pp. 161–186. Wiley, Chichester (2006)
20. Salaris, M., Cassisi, S.: The hydrogen burning phase. In: Equation of State of the Stellar Matter, S., pp. 117–159. Wiley, Chichester (2006)
21. Woosley, S.E., Heger, A., Weaver, T.A.: The evolution and explosion of massive stars. Rev. Mod. Phys. **74**, 1015–1071 (2002)

White Dwarfs

20

Contents

White dwarfs are the final stage in the evolution of low- or intermediate-mass stars. Since many more low-mass stars exist than high-mass stars, most stars, probably more than 97%, will end up as a white dwarf [1], including our Sun.

After the hydrogen burning phase, stars bloat to radii of 100 million kilometers, or bigger, and become *red giants*. An example of this class of stars is Betelgeuse in the constellation Orion with a radius of approximately 530 million kilometers [7] or about $764 R_\odot$, see also Table 1.1. Hydrogen continues to be fused in the outer layers, while helium and carbon are fused further inside at ever higher temperatures. The outer shell is only weakly gravitationally bound to the core. If the core gets hot enough, it can ionize the surrounding gas. A planetary nebula is formed. Figure 20.1 shows an image of a planetary nebula, the Ring Nebula in Lyra. Because low-mass stars cannot go through the entire fusion chain up to silicon burning, at the end of the fusion chain their core consists of the products of the last fusion reactions, mainly helium, carbon or oxygen, depending on the mass of the star. The planetary nebula itself becomes invisible relatively quickly as the ionized gas recombines. What then remains is the highly compact core, a white dwarf has been formed, which is stabilized by the electron degeneracy pressure.

© The Editor(s) (if applicable) and The Author(s), under exclusive license to
Springer-Verlag GmbH, DE, part of Springer Nature 2026
S. Boblest et al., *Special and General Relativity*,
https://doi.org/10.1007/978-3-662-71332-7_20

Fig. 20.1 The Ring Nebula in Lyra is a planetary nebula. In its center sits a young white dwarf. ©NASA, ESA, and the Hubble Heritage (STScI/AURA)—ESA/Hubble Collaboration

The first indications of white dwarfs emerged around the year 1914. At the time, *Russell*[1] discovered the star 40 Eridani B, which had a very high effective surface temperature, and at the same time a very low luminosity [20].

The Stefan-Boltzmann law $L = 4\pi R^2 \sigma T_{\text{eff}}^4$ in Eq. (18.18) links the effective temperature and the luminosity of a star to its radius. With a high effective temperature and simultaneously a low luminosity, the radius of the star must be very small. This is why the name "white dwarf" was introduced for this class of stars. In a white dwarf, no more fusion processes occur, as they do in main sequence stars and their intermediate stages, and it cannot gain energy through further contraction like protostars. It can therefore only continue to shine due to its high temperature, i.e., white dwarfs only have a thermal energy reservoir, and slowly cool down over time.

Between 1915 and 1925, first quantitative estimates [12] could be obtained for the radius and density of the white dwarf Sirius B, the companion of the brightest star in the night sky in the northern hemisphere, see Table 1.5. These estimates still

---

[1] Henry Norris Russell, 1877–1957, US-American astronomer. The Hertzsprung-Russel diagram is named after him, see Chap. 22.

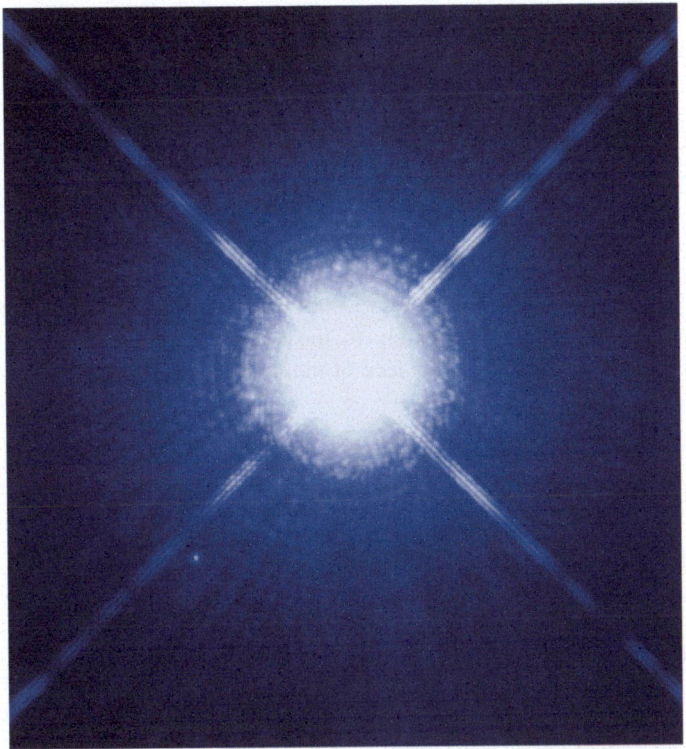

Fig. 20.2 Hubble Space Telescope image of the bright Sirius A and its faint, tiny companion, the white dwarf Sirius B. ©NASA, ESA, H. Bond and E. Nelan (Space Telescope Science Institute, Baltimore, Md.); M. Barstow and M. Burleigh (University of Leicester, UK); and J.B. Holberg (University of Arizona)

had large errors, as it later turned out. Figure 20.2 shows an image of Sirius A and B taken by the Hubble Space Telescope. Current values for this star are $M_{\text{SiriusB}} \approx 0.98\, M_{\odot}$ and $T_{\text{Sirius B}} \approx 25{,}000\,\text{K}$ [17]. The mass can be determined very precisely because Sirius B is part of a binary system. For an isolated white dwarf, this is much more difficult. The values for T, M, and the Stefan-Boltzmann law can then be used to determine the radius.

Another possibility to determine the radius for a given mass is via absorption lines in the spectrum of the white dwarf. These can be identified by comparing with known atomic transitions, although this is not trivial, see Sect. 20.5.2. From a comparison with the corresponding transition wavelengths in the laboratory, the redshift of the spectrum can be determined.

From (13.89), a distant observer sees the redshift

$$z = \frac{\Delta\lambda}{\lambda} = \frac{1}{\sqrt{1 - r_{\text{s}}/R}} - 1. \qquad (20.1)$$

Thus, the redshift is a function of the ratios r_s/R, or M/R. If the mass is known, the radius can be determined from the redshift. The current value for Sirius B is

$$R_{\text{Sirius B}} \approx 6000\,\text{km}. \tag{20.2}$$

Therefore we have the result

$$\rho_{\text{Sirius B}} \approx 2.2 \cdot 10^9\,\text{kg}\,\text{m}^{-3}. \tag{20.3}$$

This corresponds approximately to 15,000 times the density (18.21c) in the center of the Sun, and approximately to the value of the critical density $\rho_c = 1.95 \cdot 10^9\,\text{kg}\,\text{m}^{-3}$ from (18.58). A piece of matter from this star the size of a lump sugar has a mass of about 2 tons!

20.1 Qualitative Considerations

Shortly after these discoveries, it was already understood that at these extreme densities the degeneracy pressure of the electrons is responsible for the stabilization. We will use the equation of state of the degenerate electron gas from Sect. 18.4 to understand the properties of white dwarfs. However, we can already derive essential properties of white dwarfs using physically reasonable estimates. To do this, we start from the relation (18.15), $f(\rho_m) \sim p/(\rho_m c^2) \approx r_s/R$, and later compare with the polytropic equations of state for the non-relativistic and ultra-relativistic limit in (18.66). We shall omit all constant numerical factors, since they were also not considered in deriving (18.15). Then Eq. (18.66) becomes $f \sim \rho_m^{n/3}$, with $n = 2$ in the non-relativistic limit, and $n = 1$ in the ultra-relativistic case. We express the radius by the mass and the density, $R = (M/\rho_m)^{1/3}$, and obtain the relation $f(\rho_m) \sim M^{2/3}\rho_m^{1/3}$, because of $r_s \sim M$. We solve with respect to the mass, and get

$$M \sim f^{3/2}\rho_m^{-1/2}. \tag{20.4}$$

When we insert $f \sim \rho_m^{n/3}$ in (20.4), we find $M \sim \rho_m^{1/2}$ in the non-relativistic limit of low densities $\rho_m \ll \rho_c$, and $M \sim \text{const}$ in the ultra-relativistic case of very large densities $\rho_m \gg \rho_c$. Thus, in the ultra-relativistic case the mass is equal to M_c and independent of the density. For $\rho_m \to \rho_c$ the non-relativistic expression must tend to the ultra-relativistic expression. Therefore, we can write

$$M(\rho_m) \approx M_c \cdot \begin{cases} (\rho_m/\rho_c)^{1/2} & \text{for} \quad \rho_m \ll \rho_c, \\ 1 & \text{for} \quad \rho_m \gg \rho_c. \end{cases} \tag{20.5}$$

The maximum mass M_c of white dwarfs is called the *Chandrasekhar limit*, named after *Chandrasekhar*,[2] who derived the value of M_c quantitatively in the 1930s.

A very simple physical rationale for the existence of a maximum mass of white dwarfs, and at the same time an estimate of the mass, was presented by *Landau*[3] in 1932 [13]. Here we follow the line of argument given in Shapiro and Teukolsky [23], which is based on Landau's work, and consider a star with radius R, which contains N electrons. Again we shall omit all numerical factors. The particle density is $n \sim N/R^3$, and therefore the space available for every single particle in each of the three dimensions is $\Delta x \sim n^{-1/3}$. From *Heisenberg's uncertainty principle*[4]

$$\Delta x \Delta p \geq \frac{\hbar}{2} \tag{20.6}$$

then follows $p \sim \hbar n^{1/3}$. Thus, for ultra-relativistic electrons, the energy is $E_F \approx p_F c \sim \hbar c n^{1/3} \sim (\hbar c/R) N^{1/3}$.

The second contribution to the total energy is the gravitational binding energy E_G. It is given by the mass of the baryons which make up the star, i.e., $E_G \sim -G M m_{bar}/R$ per baryon. The number of baryons is proportional to the number of electrons, $M \sim N m_{bar}$, and thus $E_G \sim -G N m_{bar}^2/R$. Therefore, the total energy per baryon can be estimated by

$$E \approx \frac{\hbar c}{R} N^{1/3} - G \frac{N m_{bar}^2}{R}, \tag{20.7}$$

where for simplicity we choose $m_{bar} = m_p$ in the following. The crucial point is the different N-dependence of the Fermi energy and the gravitational energy. The contribution of the Fermi energy dominates for small N, and the total energy is positive. Since both terms in (20.7) have the same R-dependence, the star can lower its energy by expanding. For sufficiently wide expansion, the electrons become non-relativistic, i.e., the Fermi energy is $E_F \sim p_F^2 \sim \hbar^2 n^{2/3} \sim R^{-2}$. Then the Fermi energy decreases faster with R than the gravitational energy, and becomes negligible. The total energy then turns negative above a certain value of R, and tends to zero for $R \to \infty$. This implies the existence of an energy minimum in between, and thus the existence of a stable equilibrium.

On the other hand, for large particle numbers N, the total energy is always negative. Then $E \to -\infty$ for $R \to 0$, and there is no equilibrium: the gravitational

---

[2] Subrahmanyan Chandrasekhar, 1910–1995. US-American physicist. Nobel Prize in physics in 1983 for his work on stellar evolution.

[3] Lew Dawidowitsch Landau, 1908–1968, Soviet physicist, Nobel Prize in physics 1962.

[4] Werner Heisenberg, 1901–1976, German physicist, Nobel Prize in physics 1932.

Fig. 20.3 Illustration of the rationale for the existence of a maximum mass of white dwarfs according to Landau [13, 23]. The gravitational energy dominates at large radii, therefore the total energy is negative. For large particle numbers the energy diverges for $R \to 0$, i.e., the star collapses. In the case of low-mass stars with $M < M_c$, the energy becomes positive for small radii, therefore there must exist an energy minimum in between, i.e., an equilibrium. For orientation, the line $E = 0$ is plotted, which is used in estimating the maximum mass

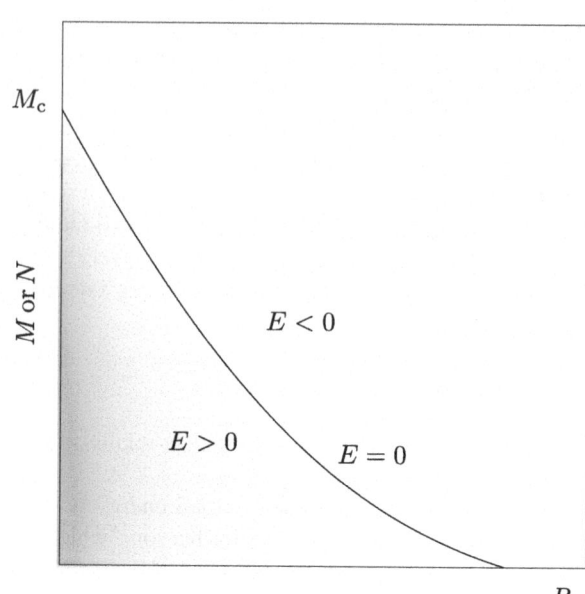

collapse sets in. In order to estimate the maximum particle number N possible, we set $E = 0$ in (20.7). This leads to

$$N_c = \left(\frac{\hbar c}{G m_p^2} \right)^{3/2} \approx 2.2 \cdot 10^{57}, \qquad (20.8)$$

and thus to the approximate (overestimated) maximum mass

$$M_c \sim N_c m_p \approx 1.9 \, M_\odot. \qquad (20.9)$$

In Fig. 20.3, the energy function $E(N, R)$ is sketched to illustrate the previous line of argument.

In Sect. 1.4 we pointed out the parallels between gravitation and electrostatics, and in (1.22) introduced the fine structure constant of gravitation $\alpha_G = G m_p^2 / (\hbar c)$. When we compare this expression with Eq. (20.8), we see that using this constant, one can represent the Chandrasekhar limit in a very compact way (apart from yet unknown numerical prefactors)

$$M_c \sim (\alpha_G)^{-3/2} \, m_p. \qquad (20.10)$$

Planck's quantum of action enters the definition of α_G, a quantity important in quantum mechanics. From this it is evident that Planck's quantum of action not only determines the microcosmos, but also the mass scale and the structure of degenerate stars. Of course, this must be so since the Fermi pressure is a quantum mechanical effect due to the Pauli principle. Stars made up of degenerate matter are macroscopic objects that are determined quantum-mechanically.

What is the size of a white dwarf? To answer this question we use again $R(\rho_m) = (M(\rho_m)/\rho_m)^{1/3}$, with $M(\rho_m)$ from (20.5). In the non-relativistic case then follows

$$R(\rho_m) = \left(\frac{M_c\,(\rho_m/\rho_c)^{1/2}}{\rho_m}\right)^{1/3} = \left(\frac{M_c}{\rho_c}\right)^{1/3} \cdot \left(\frac{\rho_c}{\rho_m}\right)^{1/6} = R_c \cdot \left(\frac{\rho_c}{\rho_m}\right)^{1/6}.$$

(20.11)

In the second step we inserted a "one" in the form ρ_c/ρ_c. By R_c we denote the critical radius, or *Chandrasekhar radius*.

From (18.55) for ρ_c and (20.10) for M_c we find

$$R_c \sim \alpha_G^{-1/2}\lambdabar_e.$$

(20.12)

Inserting the corresponding values yields

$$R_c \approx 5 \cdot 10^3 \text{ km}.$$

(20.13)

We therefore expect radii of white dwarfs in the range of a few thousand kilometers, which corresponds to the size of the small planets in our solar system, and also agrees nicely with the value of the radius of Sirius B in (20.2). With (20.5) and (20.11) we also find, in the non-relativistic limit,

$$M(\rho_m)R(\rho_m)^3 = M_c\left(\frac{\rho_m}{\rho_c}\right)^{1/2} R_c^3 \left(\frac{\rho_m}{\rho_c}\right)^{-1/2} = M_c R_c^3 = \text{const},$$

(20.14)

or the functional relation

$$M(R) \sim M_c\left(\frac{R_c}{R}\right)^3.$$

(20.15)

The radii of white dwarfs *drop* with increasing mass, see also Fig. 20.4. This is clear since to stabilize a higher mass a higher Fermi pressure is necessary, and therefore a stronger confinement of the electrons.

20.2 Numerical Solution of the Equation of State

In the quantitative evaluation we only expect numerical correction factors in front of the expressions derived before, in particular for M_c. In addition to pure numerical factors, we have also neglected the average baryon number Y_e per electron so far, which enters into the expression (18.55) for the critical mass density, and therefore also in the equation of state $p(\rho_m)$. Depending on the numerical value of the ratio

Fig. 20.4 Approximate mass function $M(R)$ in the non-relativistic and the ultra-relativistic approximation. For a constant mass, the density would diverge for $R \to 0$, $\rho_m \to \infty$, therefore the curve is unphysical in that range

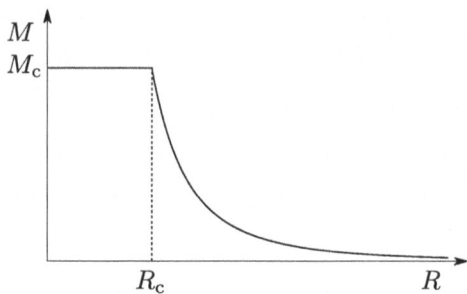

of baryons to electrons, a slightly different relation between pressure and density is obtained. Therefore, the value of Y_e must necessarily have an effect on the maximum mass of a white dwarf.

The starting point of a more detailed analysis of the structure of white dwarfs is the condition (18.7) for hydrostatic equilibrium. It links the pressure gradient at the radial coordinate r to the density at this position, and to the mass enclosed within a sphere with radius r. In addition, with (18.8) we have an expression for the gradient of the enclosed mass. We can combine the two relations:

$$\frac{dp}{dr} = -G\frac{\rho_m(r)M(r)}{r^2} \quad \text{and} \quad \frac{dM(r)}{dr} = \rho_m(r)4\pi r^2. \tag{20.16}$$

We rephrase the condition for hydrostatic equilibrium as

$$\frac{r^2}{\rho_m(r)}\frac{dp}{dr} = -GM(r). \tag{20.17}$$

We take another derivative of this expression with respect to r, and insert the expression for the gradient of the enclosed mass on the right-hand side. This leads to

$$\frac{1}{r^2}\frac{d}{dr}\left(\frac{r^2}{\rho_m(r)}\frac{dp}{dr}\right) = -4\pi G\rho_m(r). \tag{20.18}$$

Note that we have divided by r^2 on both sides of the equation. At this point, we can insert the equation of state $p(\rho_m)$ for the degenerate electron gas, which we had found in the parametrized forms (18.48) and (18.54). The resulting differential equation for ρ_m cannot be solved analytically, but must be integrated numerically. The integration starts from $r = 0$ with some density ρ_c in the center, and extends out to the surface of the star, i.e., until the condition $p = 0$ is fulfilled. In this way, the radius is determined, and by the functional behavior of the mass density also its mass. If this method is carried out for different initial densities, the mass-radius relation $M(R)$ for white dwarfs is obtained.

When we use the non-relativistic and ultra-relativistic expressions for $p(\rho_m)$ in (18.62), we find less complicated, but also only numerically solvable equations.

In the non-relativistic limit we find a more accurate form of (20.15) [23]

$$M(R) = 2.8709 \, (2Y_e)^5 \, \frac{\hbar^6}{G^3 m_{bar}^5 m_e^3} \cdot R^{-3}$$

$$= 0.699808 \, (2Y_e)^5 \, M_\odot \, \frac{R^{-3}}{(10^4 \, km)^{-3}}. \tag{20.19}$$

In the ultra-relativistic limit we obtain a constant expression for the mass, viz.

$$M = 0.774495 \, (2Y_e)^2 \, m_{bar} \left(\frac{\hbar c}{G m_{bar}^2} \right)^2$$

$$= 1.45549 \, (2Y_e)^2 \, M_\odot. \tag{20.20}$$

In both equations we have expressed the dependence on Y_e by the factor $2Y_e$. We have already seen that for many important nuclides such as helium, carbon or oxygen, $Y_e = 0.5$. Then this factor is equal to 1. Furthermore, for most nuclides we find at most $Y_e < 0.5$—except for hydrogen, which, however, is rare in white dwarfs because it was burnt in the core in earlier phases of the stars. We therefore obtain the more accurate value for the Chandrasekhar limit

$$M_c \simeq 1.46 \, M_\odot. \tag{20.21}$$

It must be emphasized that the specified limiting mass applies to the *white dwarf*, i.e., to the residual mass of a star that has entered its final stage. Since the star sheds its shell before this process, thereby losing a considerable part of its mass, the mass of the original star may well be significantly greater than $1.5M_\odot$, roughly up to $8M_\odot$ [24].

The limiting mass can also be somewhat larger, if the white dwarf rotates very fast, because then the centrifugal forces partially can compensate for gravity, or if it has a strong magnetic field, as the quantum mechanical states of electrons change in a magnetic field [3].

20.3 Corrections to the Equation of State

The accuracy of the mass-radius relation considered so far can be increased by taking into account further corrections. These corrections were worked out in great detail by *Salpeter*[5] [21], and then applied to models of stars [4]. He considered

---

[5] Edwin Ernest Salpeter, 1924–2008, US-American astrophysicist of Austrian descent.

various corrections to the Fermi energy. Taking into account these corrections, he arrived at this expression for the Fermi energy:

$$E_F = E_0 + E_{coul} + E_{TF} + E_{ex} + E_{cor}. \tag{20.22}$$

Here, E_0 is our previous expression for the Fermi energy, and the other terms are the corrections. The largest correction is the Coulomb interaction energy E_{coul}, which is due to the interaction of the electrons with the ions in the stellar material. The Thomas-Fermi energy E_{TF} takes into account the polarization of the electron distribution by the ions. Strictly speaking, it is a correction to the Coulomb interaction energy. The exchange energy E_{ex} accounts for the fact that electrons are indistinguishable particles, and the correlation energy E_{cor} represents the electron-electron interaction. A change in the expression for the Fermi energy also leads to a change in the electron degeneracy pressure. Figure 20.5 shows the ratio of the value p_S of Salpeter to the simpler result p_c of Chandrasekhar. A change of the equation of state also results in a changed mass-radius relation. Figure 20.6 shows the results of Salpeter and Hamada [4] for white dwarfs with different chemical compositions. In comparison with the Chandrasekhar curve, the Salpeter-Hamada functions bend back at large masses. The reason is a further "softening" of the equation of state. At very high densities, the lattice ions in the material of the white dwarf can ignite fusion reactions simply on account of their quantum mechanical zero point motions. These are referred to as *pycnonuclear reactions*.

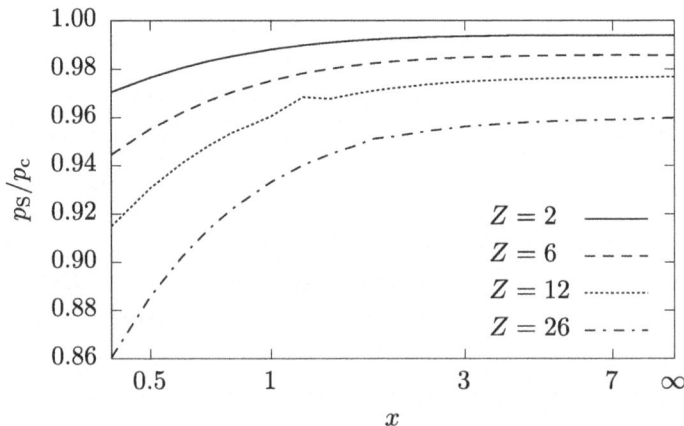

Fig. 20.5 The electron degeneracy pressure p_S with the corrections of Salpeter [21] normalized to the simpler result p_c of Chandrasekhar, for different nuclear charges, as a function of the relativity parameter $x = p_F/m_e c$ in Eq. (18.41). The corrections become bigger for increasing nuclear charge numbers, and smaller for highly relativistic electrons

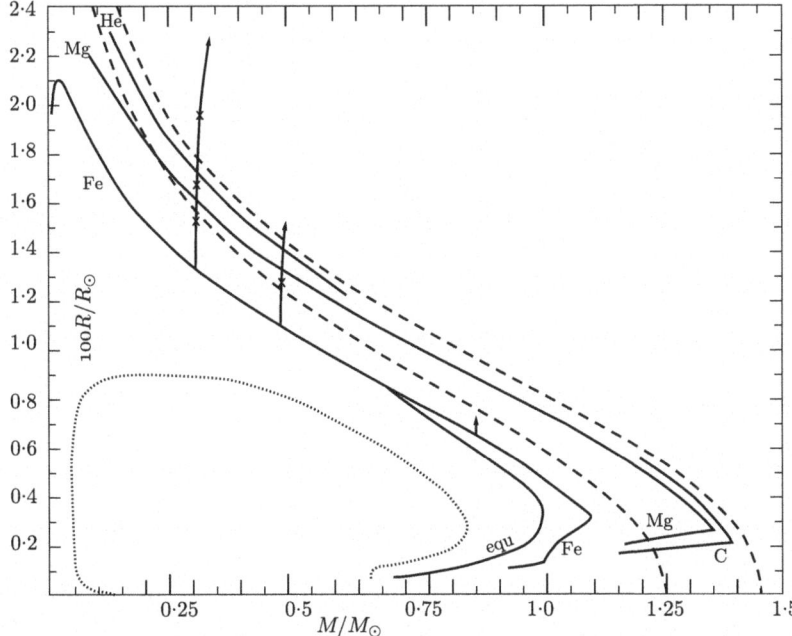

Fig. 20.6 Mass-radius relation for white dwarfs with different chemical compositions. The two *dashed curves* represent the results of the Chandrasekhar equation of state for $Y_e = 0.5$ (*top*) and $Y_e = 0.465$ (*bottom*), which corresponds, e.g., to the nuclide ^{56}Fe. The curve marked equ denotes the equilibrium composition at each density. From Hamada and Salpeter [4], © AAS. Reproduced with permission

But of even greater importance is the onset of the inverse β decay. When the Fermi energy of the electrons exceeds the mass difference of the nuclei (A, Z) and $(A, Z - 1)$, the reaction

$$\begin{matrix} A \\ Z \end{matrix}X + e^- \;\rightarrow\; \begin{matrix} A \\ Z-1 \end{matrix}Y + \nu_e \tag{20.23}$$

is allowed. Then the electrons involved in the reaction no longer contribute to the degeneracy pressure, and the maximum mass is further reduced.

This brief insight into further refinements of the equation of state of white dwarfs shall be sufficient in our context, even though we have not considered other important points, such as the effect of finite temperatures. However, the essential aspects remain unaffected by these further corrections. Readers interested in more details are referred to Althaus et al. [1], Koester and Chanmugam [12], and Shapiro and Teukolsky [23].

Figure 20.7 shows the distribution of masses of white dwarfs determined from data of the Sloan Digital Sky Survey (SDSS) [10]. One clearly recognizes an accumulation in the range 0.6–0.8 $M_\odot$. The lightest white dwarf known has

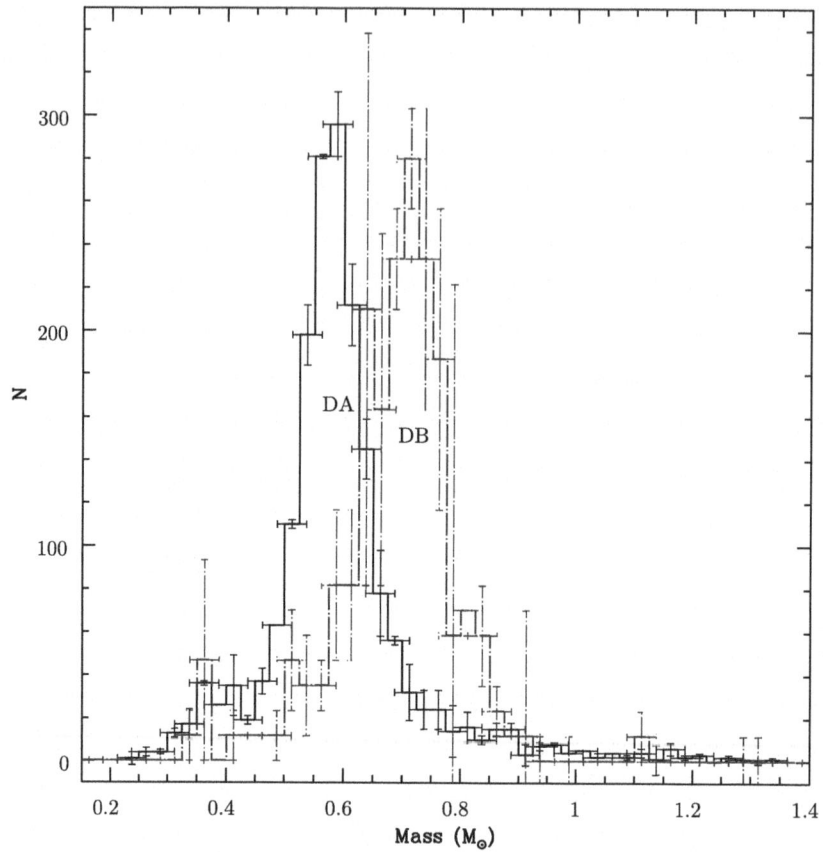

Fig. 20.7 Distribution of the masses of white dwarfs in the Sloan Digital Sky Survey (SDSS). The designations DA and DB stand for white dwarfs with hydrogen or helium lines in their spectra, respectively. From Kepler et al. [10], © 2007 by RAS. Reproduced with permission

a mass of $0.17\,M_\odot$ [11], the most massive white dwarf currently known is ZTF J1901+1458, the result of the merger of two smaller white dwarfs, with a mass of 1.327–$1.365\,M_\odot$. Being so massive, this star also has the *smallest* radius known for a white dwarf so far with $R = 2140^{+160}_{-230}\,$km [2]. As a matter of fact, all objects shown in the figure have masses below the Chandrasekhar limit.

In Fig. 20.8 the masses and radii of a few observed white dwarfs are compared with the results of the equations of state by Salpeter and Hamada [4]. One can state a good agreement with the theoretical results, within the relatively large error bars.

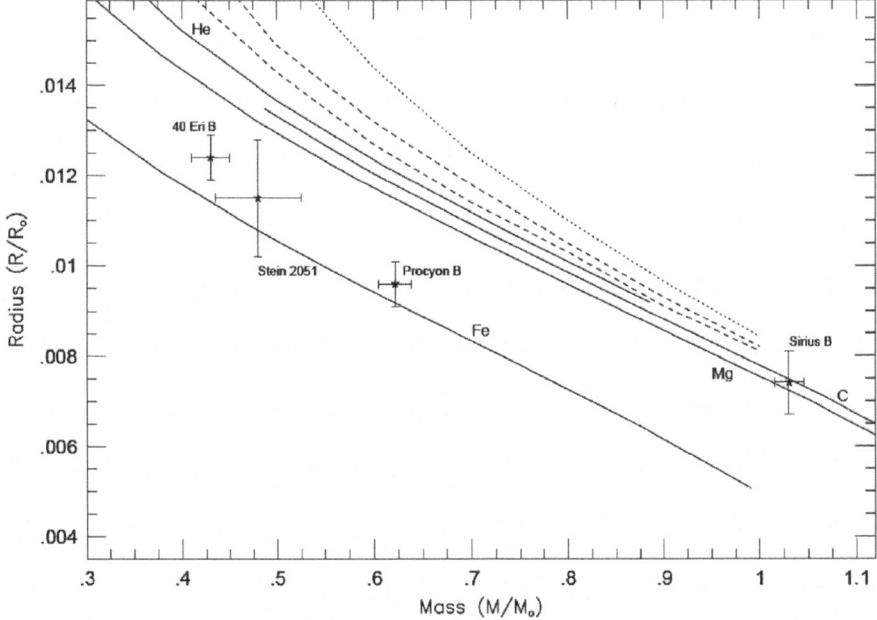

Fig. 20.8 Masses and radii of a few white dwarfs, determined from data in the Hipparcos catalog, in comparison with the results of the equations of state of Salpeter and Hamada [4,21] in Fig. 20.6. From Provencal et al. [18], © AAS. Reproduced with permission

20.4 Type Ia Supernovae

A *supernova* (SN) is the explosion of a star. The classification of supernovae dates back to the beginnings of supernova research, and classifies them according to the properties of their spectra. Supernovae of type I have no hydrogen lines in their spectra, whereas supernovae of type II do.

Supernovae of type I are subdivided further into three subtypes Ia, Ib, and Ic, where again properties of the spectrum are used for the classification. Type II supernovae are explosions of high-mass stars, which result in a neutron star or a black hole. We shall come back to this in Chap. 21. Type Ib and Ic probably undergo similar processes, in which a star explodes whose outer hydrogen shell is missing, or has been stripped by a companion.

By contrast, type Ia supernovae (SN Ia) are explosions of a white dwarf in a close binary system, which has accreted so much matter from a companion that an explosion is triggered. As a white dwarf consists of helium, carbon, or oxygen, it could in principle still initiate fusion processes, but does not reach the extremely high temperatures necessary for these processes because of the dominance of the electron degeneracy pressure. However, when a white dwarf withdraws matter from a companion until it exceeds the Chandrasekhar limit, it can no longer resist the gravitational collapse. During the collapse, the temperatures then become high enough that nuclear fusion processes can be ignited. As discussed in Sect. 19.5.1 in the context of helium flashes at the beginning of the helium burning phase of low-mass stars, the fusion processes that set in cannot stabilize the star, as the degeneracy pressure does not depend on temperature. The speed of fusion therefore increases in a self-amplifying process, until the star finally explodes, and is destroyed completely.

SN Ia are of particular importance in cosmology. Since the maximum mass of white dwarfs is well known, and the initial conditions are about the same in all cases, the brightness of each SN Ia should also be alike. Therefore, they can be used as "standard candles". In addition, as these explosions of white dwarfs are very bright, they can be observed at great distances. Therefore, they can serve to determine the distances to far distant galaxies. In Chap. 28 we shall discuss in more detail what can be learned from observing type Ia supernovae in distant galaxies.

On account of this possible application, supernovae Ia have received a great deal of attention in the last few decades. The exact details of the formation and the temporal evolution of these supernovae have not yet been completely understood. Extensive investigations have revealed that not every SN Ia has the same maximum brightness. On the contrary, there are considerable differences. However, there appears to exist a close relationship between the maximum brightness and the decrease in luminosity over time. Figure 20.9 shows several modeled curves together with observed values of a few SN Ia. With the help of these light curves,

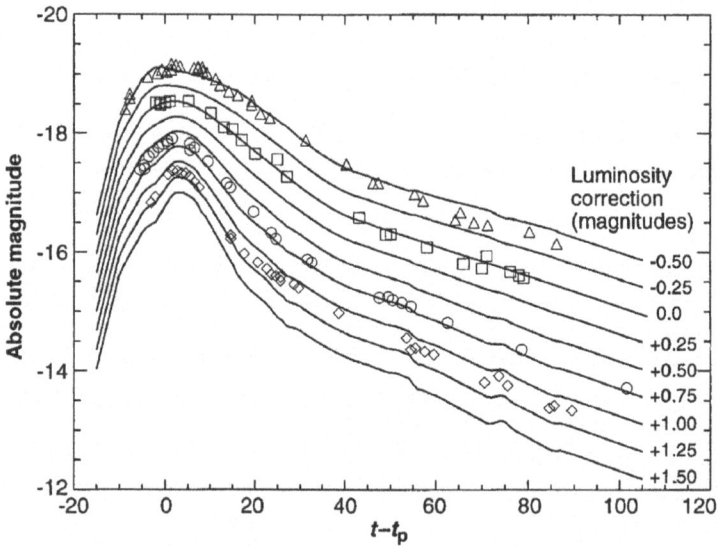

Fig. 20.9 Supernovae Ia show a characteristic correlation between their maximum brightness and the decay time of their luminosity. The figure shows empirical light curves for different type Ia supernovae. The *symbols* represent observations of different supernovae. From Nomoto et al. [15], © AAAS. Reproduced with permission

Fig. 20.10 Images of the supernova SN2011fe in the galaxy Messier 101, taken on August, 23, 24, and 25, 2011. From Nugent et al. [16], Reprinted by permission from Macmillan Publishers Ltd: © 2011

the absolute brightness can be determined with reasonable precision, but only from empirical data, and not on the basis of a detailed understanding of physical processes. For example, it is not yet clear whether an SN Ia is formed in a binary system of two white dwarfs, or whether a white dwarf has a main sequence star as its companion. The investigation of these relations is hampered by the fact that SN Ia are very rare events; in our galaxy such an event happens on the average once a year. By contrast, far distant SN Ia events can be observed quite frequently, simply because one can observe many far distant galaxies. However, to clarify the state of the initial system, observational data prior to the explosion is needed, and this would be possible only for SN Ia at small distances. For this reason, an active search for SN Ia is going on in nearby galaxies, for example in the project "Nearby Supernova Factory" [5]. Figure 20.10 shows the temporal evolution of the type Ia supernova SN2011fe. The explosion happened in the galaxy Messier 101, only about 22 million light years away. Because of this relatively small distance, this supernova was discovered very early, and its temporal evolution could be observed over 40 days [16].

20.5   Conservation Laws During Collapse

When a white dwarf is formed in the collapse of a star, one can assume that its angular momentum and the magnetic flux are approximately conserved. This has far-reaching consequences for the properties of white dwarfs.

20.5.1 Conservation of Angular Momentum

White dwarfs are much smaller than their progenitor stars. As the core of the star gets more compact and compact, the total angular momentum of the star remains conserved, since no external torque is exerted during this process. Therefore, we have for the angular momentum

$$L = \Theta\omega = aMR^2\omega \approx \text{const.} \qquad (20.24)$$

Here, Θ is the *moment of inertia*, which, apart from constant numerical prefactors, is given by the mass of the star multiplied by its radius squared. For a homogeneous solid sphere, e.g., one has $\Theta = 2MR^2/5$. Since there is no external torque acting on the star, one has $\dot{L} = 0$, therefore

$$\omega R^2 \sim \text{const} \quad \text{or} \quad P \sim R^2, \qquad (20.25)$$

because of $\omega \sim 1/P$, with the rotation period P. When we take the radius of the Sun as a representative value for a star in the hydrogen burning phase, and compare with the typical radius (20.13) of white dwarfs, we find $R_H/R_{wd} \approx R_\odot/R_{wd} \approx 10^2$. Therefore, in the transition to a white dwarf, the radius is reduced by a factor of roughly 100.

We consider the rotation period of the Sun as a typical rotation period of a star, $P_\odot \approx 28\,\text{d} \approx 2.5 \cdot 10^6\,\text{s}$, i.e., we assume rotation periods on the order of $P \approx 10^6$ to $10^7\,\text{s}$. Then we obtain a rotation period of a white dwarf which is shorter by a factor of 10^4

$$P_{wd} \approx 10^2\text{–}10^3\,\text{s}, \qquad (20.26)$$

in other words, in the range from minutes to hours. In fact, this estimate is rather too low, since we have not taken into account that the star has removed a considerable amount of its mass during the collapse of the core to a white dwarf, which reduces the angular momentum. The majority of observed values lie in the range from 20 minutes to a few days [8]. The fastest rotating white dwarf currently known has a rotation period of about 5.3 min [19], so we at least estimated the right orders of magnitude.

20.5.2 Conservation of Magnetic Flux

In the 1970s, detailed investigations of white dwarfs showed that some white dwarfs possess extremely strong magnetic fields, in the range $B \lesssim 7 \cdot 10^4\,\text{T}$ [9], therefore many times stronger than they can be realized in laboratories on the Earth.

An intuitive consideration can help us to understand how such strong magnetic fields can come about. The magnetic flux of a magnetic field B through an area A is defined as

$$\Phi = \int B \mathrm{d}A. \tag{20.27}$$

It is reasonable to assume that the plasma in the interior of a star is an excellent electric conductor, because of the many free charges present. Then it is possible to show that the magnetic flux through a surface comoving with the plasma is approximately conserved during the evolution of the star to a white dwarf.

Ohm's law[6]

$$j = \sigma(E + v \times B) \tag{20.28}$$

establishes a relationship between the electric and magnetic fields and the electric current density, where σ is the electric conductivity. For a highly conducting plasma we may assume $\sigma \to \infty$.

Then, for the current density to remain finite, $E = -v \times B$ must apply. Of course, in real physical situations this relation only applies approximately. Faraday's law of induction $\nabla \times E = -\dot{B}$ from (1.5a) leads to

$$\dot{B} = \nabla \times (v \times B), \quad \text{or} \quad 0 = \dot{B} - \nabla \times (v \times B). \tag{20.29}$$

We integrate this expression over small surface elements comoving with the plasma, which means that the charges defining the boundary of the total surface also comove with the plasma. Then

$$0 = \int \dot{B} \cdot \mathrm{d}A - \int \nabla \times (v \times B) \cdot \mathrm{d}A = \int \dot{B} \cdot \mathrm{d}A - \int (v \times B) \cdot \mathrm{d}s. \tag{20.30}$$

In the second step we have exploited Stoke's law, and $\mathrm{d}s$ is the line element on the boundary ∂A of A. Using the rule for the scalar triple product, $(v \times B) \cdot \mathrm{d}s = -B \cdot (v \times \mathrm{d}s)$, we can rephrase this in the form

$$0 = \int \dot{B} \cdot \mathrm{d}A + \int B \cdot (v \times \mathrm{d}s). \tag{20.31}$$

To understand the second term, we take a look at Fig. 20.11. In the time $\mathrm{d}t$, the line element $\mathrm{d}s$ sweeps over the area $\mathrm{d}v \times \mathrm{d}s \, \mathrm{d}t$, which means that we have

$$\frac{\mathrm{d}A}{\mathrm{d}t} = v \times \mathrm{d}s. \tag{20.32}$$

---

[6] Georg Simon Ohm, 1789–1854, German physicist.

Fig. 20.11 The change of
the oriented surface A during
the motion is given by the
term $v \times \mathrm{d}s\,\mathrm{d}t$

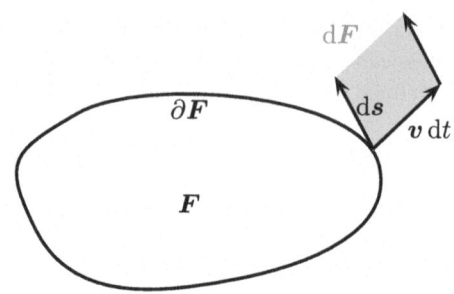

Thus, the second integral in (20.31) characterizes the change of the surface during
the course of the motion. Then, taking all together, we have

$$\int \dot{B} \cdot \mathrm{d}A + \int B \cdot (v \times \mathrm{d}s) = \frac{\mathrm{d}}{\mathrm{d}t} \int B \cdot \mathrm{d}A = \dot{\Phi} = 0. \qquad (20.33)$$

Therefore, the magnetic flux through the comoving surface is conserved. In a
plasma, as an ideal electric conductor, the magnetic field lines are "frozen" (frozen
magnetic flux). This means they directly participate in the motion. This result is
Alfvén's theorem.[7]

We now apply this result to the situation occurring in the course of the evolution
of the star to a white dwarf. Let B_H and R_H be the magnetic field and radius of the
star during the hydrogen burning phase, and B_wd and R_wd the magnetic field and
radius of the resulting white dwarf. Any arbitrary area in the star changes with the
square of the radius, $A \sim R^2$. This leads us to

$$B_\mathrm{H} A \propto B_\mathrm{H} R_\mathrm{H}^2 = \Phi_0 = B_\mathrm{wd} R_\mathrm{wd}^2, \qquad (20.34)$$

and thus to

$$B_\mathrm{wd} = B_\mathrm{H} \left(\frac{R_\mathrm{H}}{R_\mathrm{wd}} \right)^2 . \qquad (20.35)$$

We can use the magnetic field strength of the Sun as a representative value for the
field strength of a star during the hydrogen burning phase,

$$B_\mathrm{H} \sim 10^{-1} - 10^0\,\mathrm{T}. \qquad (20.36)$$

Furthermore, $R_\mathrm{H}/R_\mathrm{wd} \approx R_\odot/R_\mathrm{wd} \approx 100$, as before. From (20.35) we then obtain
for the magnetic field strength of a white dwarf

$$B_\mathrm{wd} \sim 10^3 - 10^4\,\mathrm{T}. \qquad (20.37)$$

---

[7] Hannes Olof Gösta Alfvén, 1908–1995, Swedish physicist, Nobel Prize in 1970.

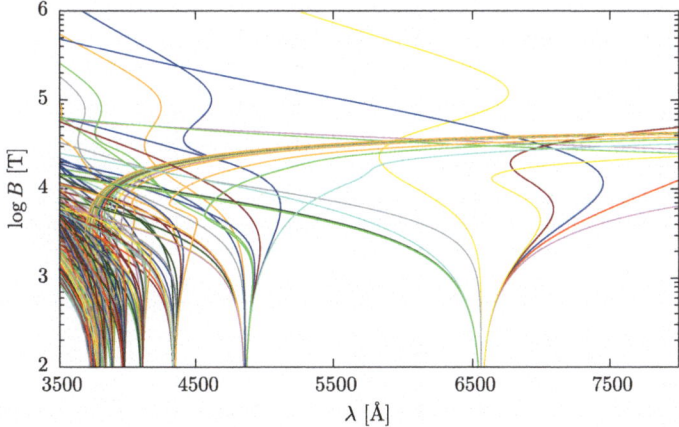

Fig. 20.12 Transition wavelengths of the Balmer series as a function of the magnetic field strength. The results are taken from Schimeczek and Wunner [22]. We thank Christoph Schimeczek for providing the data

This explains the observed strong magnetic fields. It is assumed that at least 10% of all white dwarfs possess magnetic fields with $B \geq 10^2$ T [6, 14]. These magnetic fields offer the opportunity to investigate the behavior of matter in magnetic fields that are much stronger than those which ever can be produced on the Earth.

Figure 20.12 shows results of calculations for the transition wavelengths of the Balmer series of the hydrogen atom as a function of the magnetic field strength [22]. We can see that the wavelength spectrum is very complicated in the range of field strengths of magnetic white dwarfs. Results as shown in Fig. 20.12 are used by astronomers to model the observed spectra of magnetic white dwarfs. This interplay of astrophysics and atomic physics allows for a better understanding of white dwarfs, but also of the behavior of matter in strong magnetic fields.

20.6 Exercises

20.1. Mass-radius relation for white dwarfs In this exercise we solve Eq. (20.18) numerically for a polytropic equation of state of the form (18.63), with the aim of deriving the mass-radius relation of white dwarfs. The calculation is rather extensive. A step-by-step guide and discussion can be found on the book's website.

References

1. Althaus, L.G., Córsico, A.H., Isern, J., García-Berro, E.: Evolutionary and pulsational properties of white dwarf stars. Astron. Astrophys. Rev. **18**(4), 471–566 (2010)
2. Caiazzo, I., et al.: A highly magnetized and rapidly rotating white dwarf as small as the Moon. Nature **595**, 39–42 (2021)

3. Das, U., Mukhopadhyay, B.: New mass limit for white dwarfs: super-Chandrasekhar type Ia supernova as a new standard candle. Phys. Rev. Lett. **110**, 071102 (2013)
4. Hamada, T., Salpeter, E.E.: Models for zero-temperature stars. Astrophys. J. **134**, 683–698 (1961)
5. Homepage of the Nearby Supernova Factory: https://snfactory.lbl.gov
6. Jordan, S., et al.: The fraction of DA white dwarfs with kilo-Gauss magnetic fields. Astron. Astrophys. **462**(3), 1097–1101 (2007)
7. Joyce, M., Leung, S.-C., Molnár, L., Ireland, M., Kobayashi, Ch., Nomoto, K.: Standing on the shoulders of giants: new mass and distance estimates for Betelgeuse through combined evolutionary, asteroseismic, and hydrodynamic simulations with MESA. Astrophys. J. **902**(1), 63 (2020)
8. Kawaler, S.D.: White dwarf rotation: observations and theory. In: Stellar Rotation Proceedings IAU Symposium, S. 215 (2003)
9. Kemp, J.C., Swedlund, J.B., Landstreet, J.D., Angel, J.R.P.: Discovery of circularly polarized light from a white dwarf. Astrophys. J. **161**, L77–L79 (1970)
10. Kepler, S.O., et al.: White dwarf mass distribution in the SDSS. Mon. Not. R. Astron. Soc. **375**, 1315–1324 (2007)
11. Kilic, M., Prieto, C.A., Brown, W.R., Koester, D.: The lowest mass white dwarf. Astrophys. J. **660**(2), 1451–1461 (2007)
12. Koester, D., Chanmugam, G.: Physics of white dwarf stars. Rep. Prog. Phys. **53**(7), 837 (1990)
13. Landau, L.D.: On the theory of stars. Phys. Z. Sowjet. **1**, 285 (1932)
14. Liebert, J., Bergeron, P., Holberg, J.B.: The true incidence of magnetism among field white dwarfs. Astronom. J. **125**(1), 348 (2003)
15. Nomoto, K., Iwamoto, K., Kishimoto, N.: Type Ia supernovae: their origin and possible applications in cosmology. Science **276**, 1378–1382 (1997)
16. Nugent, P.E., et al.: Supernova SN 2011fe from an exploding carbon–oxygen white dwarf star. Nature **480**, 344–347 (2011)
17. Press release on the observation of Sirius B with the Hubble space telescope. https://esahubble.org/images/heic0516a/
18. Provencal, J.L., Shipman, H.L., Høg, E., Thejll, P.: Testing the white dwarf mass-radius relation with hipparcos. Astrophys. J. **494**(2), 759 (1998)
19. Reding, J., et al.: An isolated White dwarf with 317 s rotation and magnetic emission. Astrophys. J. **894**, 19 (2020)
20. Russel, H.N.: Relations between the spectra and other characteristics of the stars. Pop. Astron. **22**, 275–294 (1914)
21. Salpeter, E.E.: Energy and pressure of a zero-temperature plasma. Astrophys. J. **134**, 669–682 (1961)
22. Schimeczek, C., Wunner, G.: Atomic data for the spectral analysis of magnetic DA white dwarfs in the SDSS. Astrophys. J. Suppl. Ser. **212**(2), 26 (2014). Data base: https:/doi.org/10.18419/darus-2118
23. Shapiro, S.L., Teukolsky, S.A.: Black Holes, White Dwarfs, and Neutron Stars: The Physics of Compact Objects. Wiley-Interscience, New York (1983)
24. Woosley, S.E., Heger, A., Weaver, T.A.: The evolution and explosion of massive stars. Rev. Mod. Phys. **74**, 1015–1071 (2002)

Neutron Stars

21

Contents

In Chap. 20 we saw that white dwarfs, the end products of the evolution of low- or intermediate-mass stars, are objects with extreme density that are stabilized by the degeneracy pressure of electrons. In this chapter we discuss the end product of more massive stars, namely neutron stars. Compared to white dwarfs, their properties are even more extreme and for their comprehensive description both quantum mechanics and special as well as general relativity are indispensable.

Supplementary Information The online version contains supplementary material available at https://doi.org/10.1007/978-3-662-71332-7_21.

21.1   Formation of Neutron Stars

Stars that enter the neon burning phase are so massive that they cannot end up as white dwarfs. At no point in their further evolution can the core of these stars be stabilized by the electron degeneracy pressure. Therefore, in their centers they go through all fusion phases up to iron. Fusion of lighter elements continues in the outer layers. At the end of the silicon burning phase the core consists mainly of iron, and the entire star has a layered structure as sketched in Fig. 21.1. In each layer, a certain element is dominant.

The iron core can no longer gain any energy through further fusion processes, and therefore cannot build up a radiation pressure that compensates for the gravitational pressure. As the electron degeneracy pressure is also unable to stabilize the core, the core moves out of hydrostatic equilibrium, and collapses. During the collapse, the Fermi energy of the electrons increases beyond the critical value

$$E_c \geq (m_n - m_p - m_e)c^2 = 782.33(38)\,\text{keV}, \tag{21.1}$$

and, as in the case of very massive white dwarfs, the energetic conditions are favorable for the inverse β decay

$$e^- + p \rightarrow n + \nu_e. \tag{21.2}$$

In white dwarfs, new nuclides are formed by the inverse β decay, which are stable with respect to further β decays. As the core continues to collapse, more and more neutrons are formed. Neutrons, too, are fermions and therefore can build up a degeneracy pressure. We shall see in the following section that the latter can stabilize the star when the core has collapsed to a radius of approximately $R_{ns} \approx 10\,\text{km}$. Thus, while in white dwarfs stabilization is achieved by the Fermi pressure of electrons, neutron stars are stabilized by the corresponding pressure of neutrons. We can again understand the process of core collapse in a simple picture by going

Fig. 21.1 Layered structure of a massive star before core collapse. The thickness of the individual layers is not drawn to scale

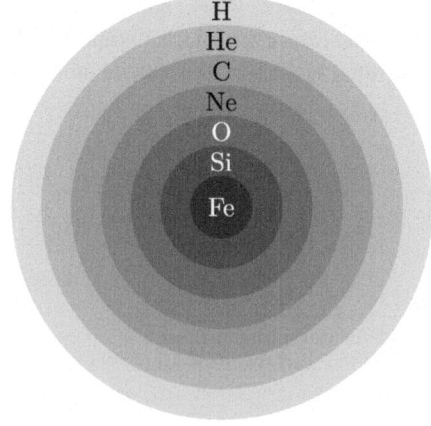

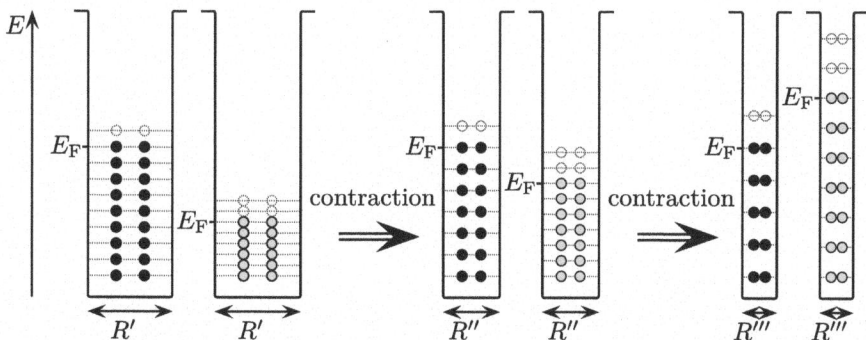

Fig. 21.2 Collapse to a neutron star: We refer to Fig. 18.4. When the density of the degenerate matter in the collapsing star exceeds a critical value, the inverse β decay sets in: neutrons (*gray*, in the right-hand wells) are formed from electrons (*black*, in the left-hand wells) and protons. Like electrons, the resulting neutrons are fermions and occupy previously empty energy levels (*white*) in their own potential well. Eventually, the Fermi pressure of the neutrons becomes so strong that it can stabilize the star

back to the potential-well model in Fig. 18.4, see Fig. 21.2. During the collapse, more and more electrons disappear while neutrons are formed. Therefore, the Fermi energy of the electrons no longer increases. The resulting neutrons occupy empty energy levels with higher and higher energies in their own potential well. When the Fermi energy of the neutrons becomes sufficiently large, the collapse stops.

The gravitational binding energy released during this collapse is enormous. If we consider the neutron star in a simplified picture as a homogeneous solid sphere with the binding energy $E_G = -3M^2 G/(5R)$ from (1.44), and also take into account that $R_{ns} \ll R_{Fe}$, assuming that the radius of the iron core after the collapse is much smaller than before, we can estimate the energy released as

$$\Delta E_m = \frac{3}{5} G M_{Fe}^2 \left(\frac{1}{R_{ns}} - \frac{1}{R_{Fe}} \right) \simeq \frac{3}{5} \frac{G M_{Fe}^2}{R_{ns}} \approx 3 \cdot 10^{46} \, \text{J}. \tag{21.3}$$

When we compare this value with the luminosity $L_\odot$ of the Sun in (1.58), we realize that this corresponds to the amount of energy radiated by the Sun in roughly $3 \cdot 10^{12}$ y!

Finally, in a series of very complicated physical processes the star explodes as a *supernova*. The exact sequence of these stellar explosions is still a topic of current research. But the fundamental process is most probably that the core collapses to an overdense state, and then expands again. The inward falling matter impacts the re-expanding core and is hurled outwards in a shock wave, see Fig. 21.3. A comprehensive discussion of the explosion of very massive stars can be found in Woosley et al. [31]. A very famous supernova remnant is the Crab Nebula in the constellation Taurus, see Fig. 21.4.

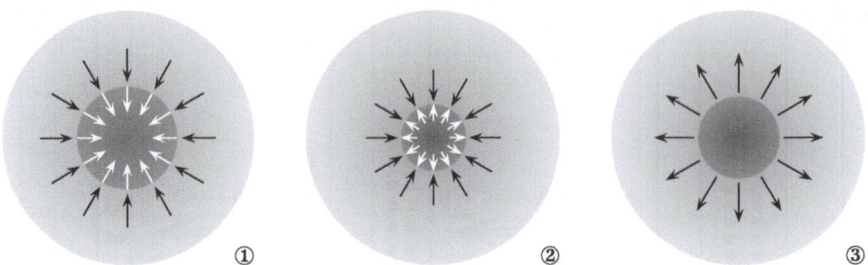

Fig. 21.3 Simplified sequence of a supernova explosion: The iron core exceeds the Chandrasekhar limit and collapses. ① The outer layers of the star fall inwards. ② The collapse of the core is stopped by the neutron degeneracy pressure, the over-dense core expands again slightly. The inward falling layers impact the re-expanding core. ③ A shock wave is triggered, the star explodes and only the now degenerate core remains

Fig. 21.4 The Crab Nebula in the constellation Taurus is the remnant of a supernova explosion in the year 1054, which could be observed from Earth even during the day. A neutron star resides in its center. Because the Crab Nebula can be assigned to an observed supernova, its age is precisely known. ©NASA, ESA, J. Hester and A. Loll (Arizona State University)

Only a small proportion of the energy (21.3) is released as electromagnetic radiation, the vast majority is presumably lost in the form of neutrinos. Nevertheless, the radiation component is large enough for a supernova to be as bright as a galaxy for a few days.

21.2   Radii of Neutron Stars

Why are neutron stars so much smaller than white dwarfs? The answer can be found in the definition (18.44) of the characteristic number density of electrons in degenerate matter. It corresponds to one electron in a cube with an edge length equal to the Compton wavelength of the electron, $\lambda_e = \hbar/(m_e c)$. In like manner, we can define a characteristic number density for neutrons,

$$n_{n,c} = \lambda_n^{-3}, \tag{21.4}$$

with the reduced Compton wavelength for neutrons

$$\lambda_n = \frac{\hbar}{m_n c}. \tag{21.5}$$

Then

$$\frac{n_{n,c}}{n_{e,c}} = \frac{\lambda_e^3}{\lambda_n^3} = \frac{m_n^3}{m_e^3}. \tag{21.6}$$

The mass ratio of neutron to electron is $m_n/m_e \simeq 1839$, and thus

$$n_{n,c} \approx 6 \cdot 10^9 n_{e,c}. \tag{21.7}$$

In contrast to white dwarfs, in neutron stars we need no longer distinguish between the baryons as responsible for the mass density and the electrons responsible for the degeneracy pressure, since now the neutrons determine both the mass density and the degeneracy pressure. From a number density bigger by a factor of $6 \cdot 10^9$ evidently follows a mass density bigger by a factor of $6 \cdot 10^9$, since $\rho_{n,c} = m_n n_{n,c}$, or

$$\rho_{n,c} = \frac{m_n}{\lambda_n^3} \cong 10^{18} - 10^{19} \, \text{kg m}^{-3}. \tag{21.8}$$

The electron mass does not enter the expression (20.10) for the Chandrasekhar limit. Therefore, for neutron stars we expect the same maximum mass, and the relation

$$M(\rho_m) \sim M_c \cdot \begin{cases} (\rho_m/\rho_{n,c})^{1/2} & \text{for} \quad \rho_m \ll \rho_{n,c}, \\ 1 & \text{for} \quad \rho_m \gg \rho_{n,c} \end{cases} \tag{21.9}$$

to hold at least approximately (compare the corresponding relation (20.5) for white dwarf stars). The radius of the neutron star then can be found from (20.12) when we replace $\lambdabar_e$ with $\lambdabar_n$:

$$R_{ns} \approx \alpha_G^{-1/2} \lambdabar_n \approx 10\,\text{km}. \tag{21.10}$$

Note that we have rounded up generously. In fact, we will soon see that we have not yet taken into account other very important effects on the structure of neutron stars, and that therefore the radii will be larger than in our simple estimate.

21.3 Rotational Periods and Magnetic Fields

In Sect. 20.5 we derived, starting from very simple considerations, that white dwarfs possess rotational periods shorter by a factor of $R_H^2/R_{wd}^2 \approx 10^4$ as compared to main sequence stars—in the range of hours, and also have magnetic fields stronger by the same factor. We can apply these considerations directly to neutron stars. Then we find possible rotational periods

$$P_{ns} \gtrsim 10^{-3}\,\text{s}. \tag{21.11}$$

This extremely short period means a rotational speed of $v = 2\pi R_{ns}/T_{ns} \approx 0.2c$ at the equator! On the other hand, the duration of the period must satisfy the condition that the centripetal acceleration $a_{zp} = R\omega^2$ does not exceed the gravitational attraction, because otherwise the star would break apart. The relation

$$\frac{GM_{ns}}{R_{ns}^2} = R_{ns}\omega_{max}^2 \tag{21.12}$$

leads to a minimum period of

$$P_{min} = \frac{2\pi}{\omega_{max}} = 2\pi \left(\frac{R_{ns}^3}{GM_{ns}}\right)^{1/2}. \tag{21.13}$$

For $M_{ns} = 1.5\,M_\odot$ and $R_{ns} = 10\,\text{km}$ we find a lower bound of $P \approx 0.4\,\text{ms}$. The fastest rotating neutron star known to date, PSR J1748-2446ad [8], has a period of roughly 1.4 ms, and is thus actually in the estimated range.

One has to realize the large amount of energy contained in such a rapidly rotating neutron star. The rotational energy is defined as $E_{rot} = \Theta\omega^2/2$, with the moment of inertia Θ. For simplicity, let us assume the neutron star as a homogeneous solid sphere with

$$\Theta_{\text{solid sphere}} = 2MR^2/5. \tag{21.14}$$

With the values $M_{ns} = 1.5\,M_\odot$ and $R_{ns} = 10\,\text{km}$ and the rotational frequency of PSR J1748-2446ad this leads to

$$E_{rot} \approx 10^{45}\,\text{J}, \tag{21.15}$$

i.e., only one order of magnitude smaller than the energy released in a supernova explosion.

In addition, we obtain an estimate of possible magnetic field strengths

$$B_{ns} \lesssim 10^9\text{--}10^{10}\,\text{T}. \tag{21.16}$$

This prediction, too, is confirmed by actual observations, as we shall in Sect. 21.5.

21.4 Mass-Radius Relation for Neutron Stars

With very basic considerations about the mass-radius relation of white dwarfs, we already achieved a relatively good agreement with observations. This, however, is not the case for neutron stars. In fact, these compact objects are much less well understood than white dwarfs because numerous physical phenomena become important that we have not considered so far.

The first important corrections result from general relativity. For a neutron star with $M \approx 1.5 M_\odot$ and $R \approx 10\,\text{km}$ the ratio of Schwarzschild radius and physical radius is

$$\frac{r_s}{R_{ns}} \approx \frac{1}{3}. \tag{21.17}$$

Therefore, we must expect very strong effects of general relativity. By contrast, for a white dwarf with

$$\frac{r_s}{R_{wd}} \approx 10^{-3} \tag{21.18}$$

these effects will be comparatively small. In a general relativistic treatment, Einstein's field equations have to be solved for a spherically symmetric mass distribution *inside* the mass. This will not yield the Schwarzschild metric in equation (13.16), which, as we know, describes the spacetime of a spherically symmetric mass distribution only outside the mass. The result of these considerations will be a modified equation for the hydrostatic equilibrium.

21.4.1 Hydrostatic Equilibrium in General-Relativistic Form

In Sect. 18.1 we derived the condition for the hydrostatic equilibrium of a star. We now investigate in which way this condition is changed when general-relativistic effects are taken into account. This calculation was carried out, already with a view on neutron stars, by *Tolman*,[1] *Oppenheimer*[2] and *Volkoff*[3] in 1939 [17, 23]. The derivation goes similar to that of the Schwarzschild metric in Chap. 13. We also begin with the form

$$
g_{\mu\nu} = \begin{pmatrix} -e^{2\Phi(r)}c^2 & 0 & 0 & 0 \\ 0 & e^{2\Psi(r)} & 0 & 0 \\ 0 & 0 & r^2 & 0 \\ 0 & 0 & 0 & r^2\sin^2(\vartheta) \end{pmatrix}
\tag{21.19}
$$

for a spherically symmetric static metric.

Since we are no longer outside the mass distribution, i.e., in vacuum, the energy-momentum tensor no longer vanishes. For simplicity, we assume that the stellar material can be described as an ideal liquid. This means, we neglect contributions of strains to the energy-momentum tensor. This approach leads to the form

$$
T^{\mu\nu} = (\rho_{\mathrm{m}}c^2 + p)u^\mu u^\nu + pg^{\mu\nu}.
\tag{21.20}
$$

Matter in the star is assumed to be at rest, i.e., all spatial components of the four-velocity are zero, and because of the choice of our coordinates we have $u^\mu = (e^{-\Phi(r)}/c, 0, 0, 0)$ and $u_\mu = (-ce^{\Phi(r)}, 0, 0, 0)$ in the rest frame of the matter. At this point we deviate from the calculation in Chap. 13 and work with Einstein's field equations in the form $G^\mu{}_\nu = \kappa T^\mu{}_\nu$. In doing so, we follow the convention in the literature. In particular, we follow closely the calculation in Oppenheimer and Volkoff [17].

The energy-momentum tensor is obtained in the very simple form

$$
T^\mu{}_\nu = \mathrm{diag}(-\rho_{\mathrm{m}}c^2, p, p, p).
\tag{21.21}
$$

However, in this case the calculation of the Einstein tensor from the Ricci tensor (13.6) belonging to the metric tensor (21.19), via the relation (12.39) $G_{\mu\nu} = R_{\mu\nu} - Rg_{\mu\nu}/2$, is very unwieldy because of the long mathematical expressions that turn up. Therefore, we only quote the result

$$
G^t{}_t = e^{-2\Psi}\left(-\frac{2\Psi_r}{r} + \frac{1}{r^2}\right) - \frac{1}{r^2},
\tag{21.22a}
$$

$$
G^r{}_r = e^{-2\Psi}\left(\frac{2\Phi_r}{r} + \frac{1}{r^2}\right) - \frac{1}{r^2},
\tag{21.22b}
$$

---

[1] Richard Chace Tolman, 1881–1948, US-American physicist.

[2] Julius Robert Oppenheimer, 1904–1967, US-American physicist. Well known in particular by his leading role in the Manhattan Project, and from the Born-Oppenheimer approximation.

[3] George Michael Volkoff, 1914–2000, Canadian physicist.

$$G^{\vartheta}{}_{\vartheta} = \frac{-\Psi_r + \Phi_r + \Phi_{rr}r - \Psi_r\Phi_r r + \Phi_r^2 r}{re^{2\Psi}}, \tag{21.22c}$$

$$G^{\varphi}{}_{\varphi} = G^{\vartheta}{}_{\vartheta}. \tag{21.22d}$$

To continue the calculation we only need the equations $G^t{}_t = -\kappa\rho_{\mathrm{m}}c^2$ and $G^r{}_r = \kappa p$. Later, additionally pressure and density must be interconnected again by an equation of state $p = p(\rho_{\mathrm{m}})$. This, however, is not yet sufficient. We still need a differential equation for the pressure gradient, which, after all, must produce the hydrostatic equilibrium. Fortunately, we are still left with one more condition, which will provide us with such an equation. We know that the divergence of the energy-momentum tensor must vanish, i.e., $T^{\mu\nu}{}_{;\nu} = 0$, with

$$T^{\mu\nu} = \mathrm{diag}(\rho_{\mathrm{m}}e^{-2\Phi}, pe^{-2\Psi}, p/r^2, p/(r^2\sin^2(\vartheta))). \tag{21.23}$$

We evaluate this equation for the radial component, and find, with the definition (11.84) of the divergence,

$$\begin{aligned}
T^{r\nu}{}_{;\nu} &= T^{r\nu}{}_{,\nu} + T^{\sigma\nu}\Gamma^r{}_{\sigma\nu} + T^{r\sigma}\Gamma^\nu{}_{\sigma\nu} \\
&= T^{rr}{}_{,r} + T^{tt}\Gamma^r{}_{tt} + T^{rr}\Gamma^r{}_{rr} + T^{\vartheta\vartheta}\Gamma^r{}_{\vartheta\vartheta} + T^{\varphi\varphi}\Gamma^r{}_{\varphi\varphi} + T^{rr}\Gamma^\nu{}_{r\nu}.
\end{aligned} \tag{21.24}$$

Using the relation

$$\frac{\mathrm{d}}{\mathrm{d}r}T^{rr} = \frac{\mathrm{d}}{\mathrm{d}r}\left(pe^{-2\Psi}\right) = \frac{\mathrm{d}p}{\mathrm{d}r}e^{-2\Psi} - 2p\Psi_r e^{-2\Psi} \tag{21.25}$$

and the components of the energy-momentum tensor from (21.23), and the Christoffel symbols (13.5), we can write this equation as

$$\frac{\mathrm{d}}{\mathrm{d}r}p = -\Phi_r\left(\rho_{\mathrm{m}}c^2 + p\right). \tag{21.26}$$

We can further strongly simplify the relatively complicated form of $G^t{}_t$ by the substitution

$$u(r) = \frac{1}{2}r\left(1 - e^{-2\Psi}\right), \quad \text{or} \quad e^{-2\Psi} = 1 - 2\frac{u}{r}. \tag{21.27}$$

Noting that

$$\frac{\mathrm{d}u}{\mathrm{d}r} = e^{-2\Psi}\left(-\frac{1}{2} + r\Psi_r\right) + \frac{1}{2} \tag{21.28}$$

we see that

$$G^t_{\ t} = -\frac{2}{r^2}\frac{du}{dr}. \tag{21.29}$$

Therefore, the equation $G^t_{\ t} = \kappa T^t_{\ t}$ turns into

$$\frac{du}{dr} = \frac{\kappa}{2}r^2\rho_\mathrm{m}c^2. \tag{21.30}$$

We can now replace Φ_r and $e^{-2\Psi}$ in $G^r_{\ r}$ using (21.26) and (21.27), and arrive at

$$\frac{dp}{dr} = -\left(\rho_\mathrm{m}c^2 + p\right)\frac{u + \frac{\kappa}{2}pr^3}{r(r - 2u)}. \tag{21.31}$$

This result is the *Tolman-Oppenheimer-Volkoff equation* (TOV equation). We can still reformulate this equation in order to compare with the non-relativistic result. For this purpose, we take a look again at (21.30). We do not know the precise form of the metric inside the mass distribution, since we do not know Ψ and Φ. But we do know that on the surface, i.e., with $r = R$, they must reproduce the Schwarzschild metric,

$$e^{-2\Psi(R)} = 1 - \frac{r_\mathrm{s}}{R} = 1 - 2\frac{u}{R}, \tag{21.32}$$

where we have used (21.27). From this relation, we can finally infer that for all r

$$u(r) = \frac{GM(r)}{c^2}. \tag{21.33}$$

With this result, and $\kappa = 8\pi G/c^4$ from (12.37), (21.30) becomes

$$\frac{dM(r)}{dr} = 4\pi r^2\rho_\mathrm{m}. \tag{21.34}$$

This is the same equation for the mass enclosed by the sphere with radius r as in the non-relativistic case in (18.8). We insert the definition (21.33) and $\kappa = 8\pi G/c^4$ in (21.31), and draw a factor $M(r)$ and a factor $\rho_\mathrm{m}(r)$ in front of the equation. We then obtain the form

$$\frac{dp(r)}{dr} = -G\frac{\rho_\mathrm{m}(r)M(r)}{r^2}\left(1 + \frac{p(r)}{\rho_\mathrm{m}(r)c^2}\right)\left(1 + \frac{4\pi r^3 p(r)}{M(r)c^2}\right)\left(1 - 2\frac{GM(r)}{c^2 r}\right)^{-1} \tag{21.35}$$

of the TOV equation. Here, the first factor corresponds to the Newtonian result (18.7), which is multiplied by two factors greater than one, and divided by one factor smaller than one. This means, the relativistic pressure gradient is bigger than the non-relativistic one.

In the derivation of the TOV equation we have assumed a nonrotating neutron star. For fast rotating neutron stars, further corrections turn up, which, however, we shall not go into.

21.4.2 Relativistic Effects in the Observation of Neutron Stars

Effects of general relativity are also relevant when observing neutron stars. We have seen that the radii of white dwarfs can be inferred from the redshift of their spectra. For white dwarfs the redshift is very small, in the range of $z \approx 10^{-4}$. By contrast, for neutron stars we find from (20.1), with $r_s/R_{ns} \approx 1/3$, the big redshift

$$z_{ns} \approx 0.22. \tag{21.36}$$

This leads to an apparent change in the temperature of the star of the form

$$T_{eff}^\infty = T_{eff}\sqrt{1 - r_s/R_{ns}}. \tag{21.37}$$

Here the effective temperature T_{eff} is defined in such a way that a black body with this temperature would have the same luminosity as the star, see Sect. 18.2.

Moreover, the deflection of light results in an enlarged apparent radius

$$R^\infty = R(1 + z_{ns}), \tag{21.38}$$

which implies that more than half of the surface of the star can be observed, see Fig. 21.5. The radius of a neutron star *decreases* when its mass increases (just as it did for white dwarfs). But with increasing mass the redshift also increases. Thus, for every mass of a neutron star there exists a minimally possible observable value of its radius [18], independent of the actual radius.

21.4.3 Neutron Star Models for the Equation of State

The transition from (18.7) to (21.35) does complicate the numerical solution of equation (20.18), but poses no serious problem. A much more difficult problem arises from the fact that neutrons are baryons, and are therefore subject to the strong interaction of nucleons. The density in a neutron star corresponds to that in an atomic nucleus, or is even higher. The influence of the strong interaction is therefore crucial, but at the same time its theoretical treatment within the context of quantum chromodynamics is very difficult. For this reason the equation of state $p(\rho_m)$ of neutron matter is not exactly known. However, neutron star models do exist, which

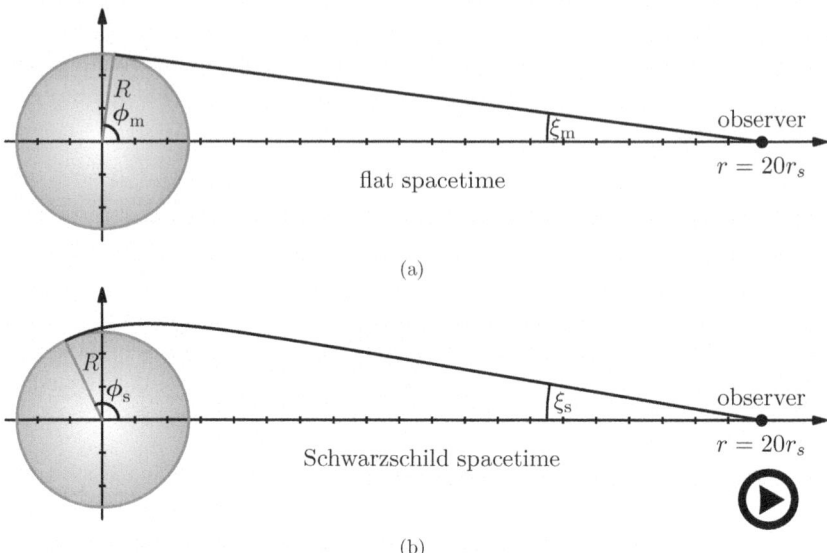

Fig. 21.5 Apparent size of a neutron star with actual radius $R = 2.65r_s$ as seen from an observer sitting at $r_{obs} = 20r_s$. In flat spacetime (**a**), the viewing angle is $2\xi \approx 15.23°$, and the observer sees less than half of the surface, $\phi_m \approx 82.39°$. In the Schwarzschild spacetime (**b**), the viewing angle is $2\xi \approx 18.84°$, and the observer sees more than half of the surface, $\phi_s \approx 115.74°$ (▶ https://doi.org/10.1007/000-hw5)

only differ in what assumptions are made and which mathematical methods are used in the theoretical calculation of the equation of state [12].

The models can be roughly divided into three groups, see Fig. 21.6.

Traditional Neutron Stars

Traditional neutron stars consist of hadronic matter. At the surface, density and pressure go to zero. Most of the neutron stars known to date probably belong to this category. As with white dwarfs, their radii decrease with increasing mass.

Exotic Matter Stars

As a modification of the first group, models have been proposed that predict exotic states of matter in the center of the neutron star, including the occurrence of various baryon resonances (Σ, Λ, Ξ, Δ), among them hyperons containing strange quarks (Σ, Λ, Ξ), or even quark matter, consisting of up, down and strange quarks, and dibaryon particles (H) [27].

Strange-Quark Matter Stars

The last group are neutron stars which, apart from a possible small crust, consist completely of strange-quark matter (SQM). The radius of these stars increases with mass. Should such stars exist, the density at the surface could possibly be higher than

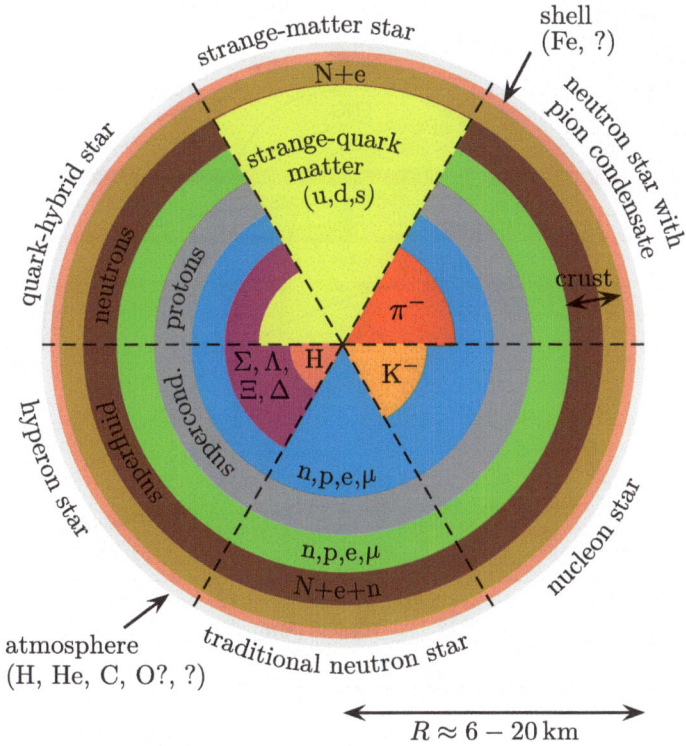

Fig. 21.6 Structure of different types of neutron stars. Among other quantities, the resulting radii vary strongly from model to model. Uncertainties are indicated by question marks. Adapted from [27]

in an atomic nucleus. However, recent observations seem to disprove their existence, as will be shown in Fig. 21.7.

A major problem in the exact determination of the equation of state also stems from the fact that certainly atomic nuclei can be studied in detail in the laboratory, i.e., in principle one has access to matter in the density range of neutron stars, even if these probably may have a density up to ten times that of atomic nuclei. However, atomic nuclei consist of roughly equal proportions of protons and neutrons, but matter in neutron stars only has a proton content of a few per cent. The effects of this difference on the properties of neutron star matter are not well understood. For the reasons mentioned, the maximum mass of neutron stars is also only known with a great uncertainty. Calculations in the framework of general relativity limit the maximum mass to $M \leq 3\,M_\odot$. The most massive neutron star known, PSR J0952-0607 [19] has a mass of $M = 2.35 \pm 0.17\,M_\odot$ thus in any case clearly above the Chandrasekhar limit. The abundance of physically extreme phenomena makes neutron stars a subject of very diverse research. Questions from most different subfields of physics come into play. Plasma physics, nuclear and particle physics and

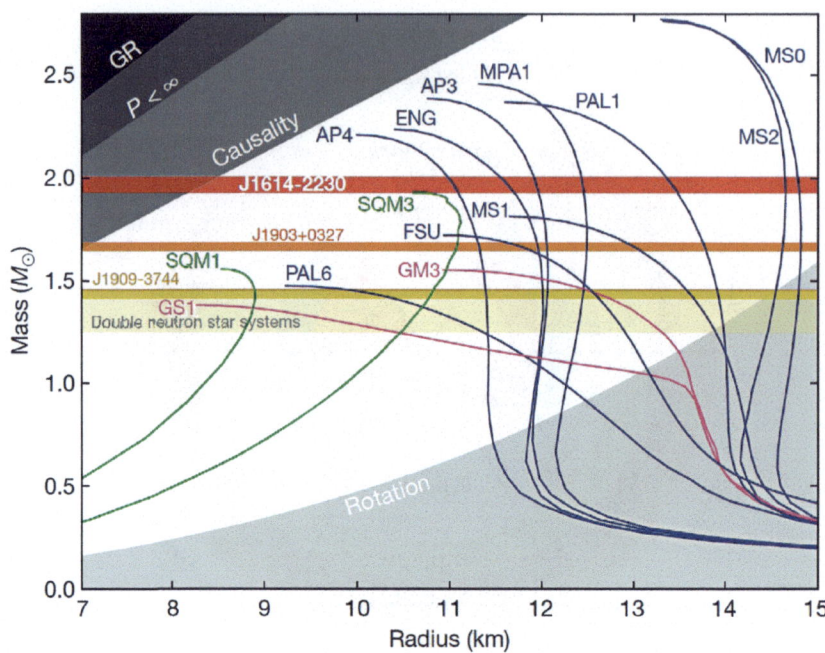

Fig. 21.7 Mass-radius relations for different neutron star models. Blue lines represent nucleon models, pink lines models with exotic matter, and green lines strange-quark-star models. The regions shaded in gray are excluded on the basis of other considerations. Lower bounds on the radius can be found from different considerations. These are: General relativity: the general relativistic requirement $R > r_s$; $P < \infty$: the requirement of a finite pressure leads to the stronger constraint $R > (9/8)r_s$; causality: the requirement of a sound velocity smaller than c leads to the even more stringent constraint $R > 1.45 r_s$. The region denoted by "rotation" yields an upper bound on the radius for given mass, as the maximum rotational frequency possible is limited by comparing with the rotational frequency of the previously mentioned star PSR J1748-2446ad [8]. Further details concerning these considerations can be found, e.g., in Lattimer and Prakash [13]. The discovery of the $2M_\odot$-star J1614-2230 [5] (*red bar*), the most massive neutron star known in 2010, already completely rules out a few models, in which such large masses are impossible. The modern record holder PSR J0952-0607 [19] with $M = 2.35 \pm 0.17\,M_\odot$ puts even stronger constraints on possible neutron star models. From Demorest et al. [5]. Reprinted by permission from Macmillan Publishers Ltd: © 2010

atomic physics altogether are important to the description of the star. To understand the structure of the outer crust, consisting partly of very exotic nuclides, very accurate knowledge of the binding energies of these nuclei is required, which must be determined in experiments [30]. The structure of the layers deeper inside, i.e., the outer and inner core, is investigated using particle physics models [27]. Assuming a certain equation of state, a mass-radius relation can be determined for the individual neutron star models. Figure 21.7 shows the results of such calculations.

To further narrow down the equations of state, it would be of great importance to precisely determine the mass and, in particular, the radius of a neutron star. The

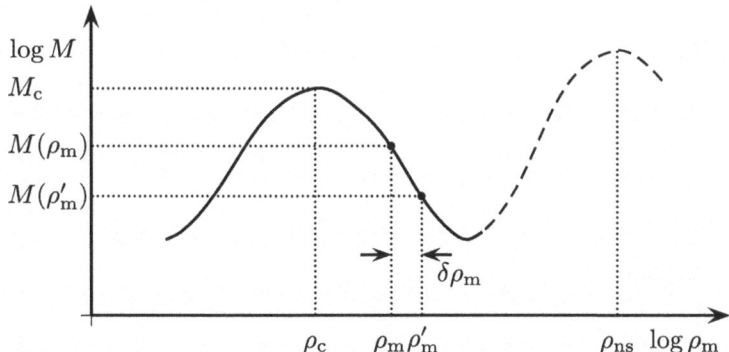

Fig. 21.8 No stable stars exist in the range of densities between white dwarfs and neutron stars, because the equilibrium mass $M(\rho_m)$ decreases with increasing density. If the density increases in this range because of a small perturbation, then only the smaller mass $M(\rho'_m)$ can be stabilized at the new density $\rho'_m = \rho_m + \delta\rho_m$, and the star collapses. If the density decreases, a bigger mass can be stabilized, the star then further reduces its density

observation of the thermal emission of neutron stars is, however, more difficult than that of white dwarfs, as they emit radiation mainly in the range of X-rays, and are much smaller than white dwarfs. The red bar in Fig. 21.7 characterizes the mass $M = 2 M_\odot$ of the neutron star J1614-2230, whose radius, though, is not known. J1614-2230 was the most massive neutron star at the time of publication of Demorest et al. [5]. The mere existence of such a massive star already rules out some neutron star models. The even more massive star PSR J0952-0607 puts even stronger constraints on possible neutron star models.

21.4.4 Mass Densities in Between White Dwarfs and Neutron Stars

In the large range of values for the density between white dwarfs and neutron stars, in which the equilibrium mass becomes smaller with increasing density, i.e., for densities $10^{11}\,\mathrm{kg\,m^{-3}} < \rho_m < 10^{16}\,\mathrm{kg\,m^{-3}}$, stable stars cannot exist. This can be understood by looking at how such a star would react to a small perturbation. With a small increase in density in a certain region of the star, the associated stable mass decreases, which leads to a further reduction in the size of the star and a further increase in density: the star will collapse. If, on the other hand, the density decreases, an even larger mass is stable, and therefore the star would continue to reduce its density, i.e., it would continue to expand. This behavior is illustrated in Fig. 21.8.

21.4.5 Mass Densities Beyond Neutron Stars

If the stellar mass is so large that even the Fermi pressure of the neutrons can no longer stabilize the star, there is, according to current knowledge of physics, no process that could stop the collapse. As a hypothetical intermediate step quark stars are still being discussed, which partly consist of a quark-gluon plasma. However, if they were to exist, it would be very difficult to distinguish them from ordinary neutron stars. The star then continues to collapse, and forms a Schwarzschild *black hole*, as have we discussed in Sect. 13.3, or, due to the conservation of angular momentum, the star ends as a Kerr black hole.

21.5   Pulsars

Most of the neutron stars that are known today very regularly emit ever recurring signals in the range of radio wavelengths. The first of these "radio pulsars" was discovered at Cambridge University in 1967 by *Bell*[4] in the research group of *Hewish*,[5] and today is designated PSR B1919+21 (PSR for pulsar, and 1919+21 for its celestial coordinates). The extreme regularity of the signals posed a serious puzzle to the researchers. When a source on Earth had been ruled out as the cause of the regular pulses, even a message from extraterrestrial beings was briefly considered as a possible explanation. The first pulsar therefore was designated LGM1 for "Little Green Men 1". A short time later, Gold correctly suggested a rotating neutron star as the source of the signals [7].

We already know that neutron stars with very short rotation periods exist, in the range of one second and significantly less, and that they also have extremely strong magnetic fields, in the range 10^8 T and above. Their rotational energy also provides a huge reservoir of energy. When we combine these pieces of information, we can gain a good qualitative understanding of the properties of pulsars. To do this we assume, for simplicity, a pure dipole structure for the magnetic field. If the axis of rotation of the neutron star does not coincide with the magnetic field axis but is tilted against it by an angle θ, as shown in the sketch in Fig. 21.9, then this leads to the emission of electromagnetic radiation with the power [21]

$$\dot{E}_{\text{rad}} = \frac{2}{3c^3} \frac{\mu_0}{4\pi} m^2 \sin^2(\theta) \, \omega^4. \tag{21.39}$$

Here m designates the *dipole moment* of the star, and μ_0 is the magnetic field constant from (1.7). This energy comes from the rotational energy of the pulsar, i.e.,

$$\dot{E}_{\text{rad}} = -\dot{E}_{\text{rot}} = -\Theta \omega \dot{\omega}, \tag{21.40}$$

---

[4] Jocelyn Bell Burnell, ★1943, British radio astronomer.

[5] Antony Hewish, 1924–2021, British radio astronomer, Nobel Prize 1974 together with Martin Ryle.

Fig. 21.9 When the rotational axis and the magnetic field axis of a neutron star are tilted with respect to each other by an angle θ, then the star can emit pulsating electromagnetic radiation originating at the polar caps

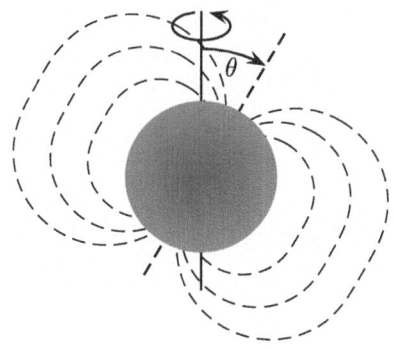

where Θ is the unknown *moment of inertia* of the pulsar. Since $\dot{E}_{\text{rad}} > 0$, one must have $\dot{E}_{\text{rot}} < 0$, and thus $\dot{\omega} < 0$, which means that the rotational frequency *decreases*. Therefore, the magnetic dipole moment of the star can be estimated from the decrease of the rotational frequency. However, this only results in a lower bound on m, since only the expression $m \sin(\theta)$ can be estimated, while θ is unknown.

If the dipole moment is known, we can estimate the magnetic field strength on the surface of the neutron star using the formula for the field strength of a magnetic dipole

$$B = \frac{\mu_0 m}{4\pi R^3} \left(1 + 3 \cos^2(\vartheta) \right)^{1/3} . \tag{21.41}$$

Here ϑ is the angular distance from the magnetic pole. The dependence on ϑ results in a decrease in the magnetic field strength from the pole to the equator by roughly 37%. In continuing our estimate, we do not consider this angular dependence any further, as we have other, presumably greater, uncertainties in the radius R, the angle θ and the moment of inertia.

When we combine Eqs. (21.39) and (21.40), we obtain

$$\Theta \omega \dot{\omega} = -\frac{2}{3c^3} \frac{\mu_0}{4\pi} m^2 \sin^2(\theta) \, \omega^4 , \tag{21.42}$$

i.e., we have the relation

$$\dot{\omega} = -A\omega^3 \quad \text{with} \quad A = \frac{2}{3c^3} \frac{\mu_0}{4\pi} \frac{m^2 \sin^2(\theta)}{\Theta} , \tag{21.43}$$

from which we can obtain an expression for m. We then have

$$B = \frac{\mu_0}{4\pi R^3} \left(-\frac{3}{2} c^3 \frac{4\pi}{\mu_0} \Theta \frac{\dot{\omega}}{\omega^3} \right)^{1/2} \frac{1}{\sin(\theta)} . \tag{21.44}$$

Because of the dependence on $1/\sin(\theta)$, this estimate only yields a lower bound on the magnetic field strength.

We can also estimate the age of the pulsar from the decrease of the rotational frequency.

We got back to Eq. (21.43), separate variables and integrate over time from $t = 0$ to the age τ of the pulsar:

$$\int\limits_{\omega_0}^{\omega} \frac{d\omega'}{\omega'^3} = -A \int\limits_0^{\tau} dt. \tag{21.45}$$

Here ω_0 is the angular frequency of the pulsar at the time of its formation. We solve with respect to the age of the pulsar and obtain

$$\tau = \frac{1}{2A}(\omega^{-2} - \omega_0^{-2}) = \frac{1}{2}\frac{\omega}{\dot{\omega}}\left[\left(\frac{\omega}{\omega_0}\right)^2 - 1\right]. \tag{21.46}$$

In the second step we have exploited the relation $A = -\dot{\omega}/\omega^3$ from (21.43). Of course the initial frequency ω_0 is unknown. However, an upper bound on the age can be obtained by the reasonable assumption that the rotational frequency at the time of the pulsar's formation was much bigger than it is today, i.e., $\omega/\omega_0 \approx 0$. This leads to

$$\tau \leq -\frac{1}{2}\frac{\omega}{\dot{\omega}}, \tag{21.47}$$

where, again, we point out that $\dot{\omega} < 0$.

We can also write B and τ as functions of the period P and its rate of change $\dot{P}$, instead of as functions of ω and $\dot{\omega}$. With $\omega = 2\pi/P$ we have $\dot{\omega} = -2\pi\dot{P}P^{-2}$, and therefore $\dot{\omega}/\omega = -\dot{P}/P$ and $\dot{\omega}/\omega^3 = -\dot{P}P/(4\pi^2)$.

Finally, we again approximate the moment of inertia with the expression for a homogeneous solid sphere, $\Theta = (2/5)MR^2$, from Eq. (21.14). For our purposes here, this approximation is sufficient. However, obtaining accurate values for the moment of inertia of neutron stars is a subject of ongoing research, because it would allow to learn more about the equation of state of neutron star matter, see Sect. 21.5.2 and Exercise 21.3. We insert $\Theta = (2/5)MR^2$ in (21.44), and find the two relations

$$\tau \leq \frac{1}{2}\frac{\dot{P}}{P} \tag{21.48}$$

and

$$B \simeq \frac{1}{R^2}\left(\frac{3}{80\pi^3}\mu_0 Mc^3\right)^{1/2}\frac{1}{\sin(\theta)}\left(P\dot{P}\right)^{1/2}. \tag{21.49}$$

Therefore we can derive estimates for the age and the magnetic field strength of a pulsar from its observed period and its observed period change. Through long-term observations of individual pulsars, periods, in particular, can be determined with very high accuracy, in the range of 10 significant digits. However, there is a restriction, especially for pulsars in binary systems. For example, if a pulsar forms a binary system with a main sequence star, such as Hercules X-1, which we will discuss in the next section, it is possible that it pulls off material from its companion, and thus also angular momentum, which means that its period can drop sharply. This is why the fastest rotating pulsars are often much older than given by the estimate (21.48). In the literature, this is aptly referred to as "recycled pulsars".

Figure 21.10 shows a doubly logarithmic presentation of pulsars with known periods and known period changes. In the logarithmic representation, the two relations (21.48) and (21.49) lead to straight lines of constant age and constant magnetic field, because of

$$\log(2\tau) = \log(\dot{P}P^{-1}) = \log(\dot{P}) - \log(P) \tag{21.50}$$

and

$$\log(\text{const} \cdot B^2) = \log(\dot{P}P) = \log(\dot{P}) + \log(P). \tag{21.51}$$

Therefore pulsars having the same age lie on straight lines with positive slope, while pulsars having the same magnetic field strength lie on straight lines with negative slope. In the figure, the pulsars listed are further subdivided into different types. We can see that many pulsars have been discovered in binary systems (red triangles). In Sect. 21.5.2 we shall see that pulsars in binary systems allow highly accurate tests of general relativity. Some relatively young pulsars, including the Crab Pulsar, still have a supernova remnant (orange dots). The group of magnetars (cyan blue triangles) feature particularly extreme properties. Neutron stars in this group possess very high magnetic fields with $B \gtrsim 10^{10}$ T. Another very unusual and interesting group for research are the "Magnificent Seven" (green diamonds). The members of this group, of which only 7 are known to date, possess an almost purely thermal emission spectrum [26]. In Fig. 21.10 6 of these 7 objects are included.

The precise measurement of the pulsar periods has revealed another phenomenon: From time to time many pulsars exhibit an abrupt increase of their rotational frequency, by a factor of about $1 + 10^{-6}$. The precise cause of these *pulsar glitches* is not known, but they must be due to some change of the internal structure of the neutron star ("star quakes"). The investigation of these glitches should provide further information about the internal structure of neutron stars. This is also the reason why this phenomenon is studied in great detail, see e.g. [6].

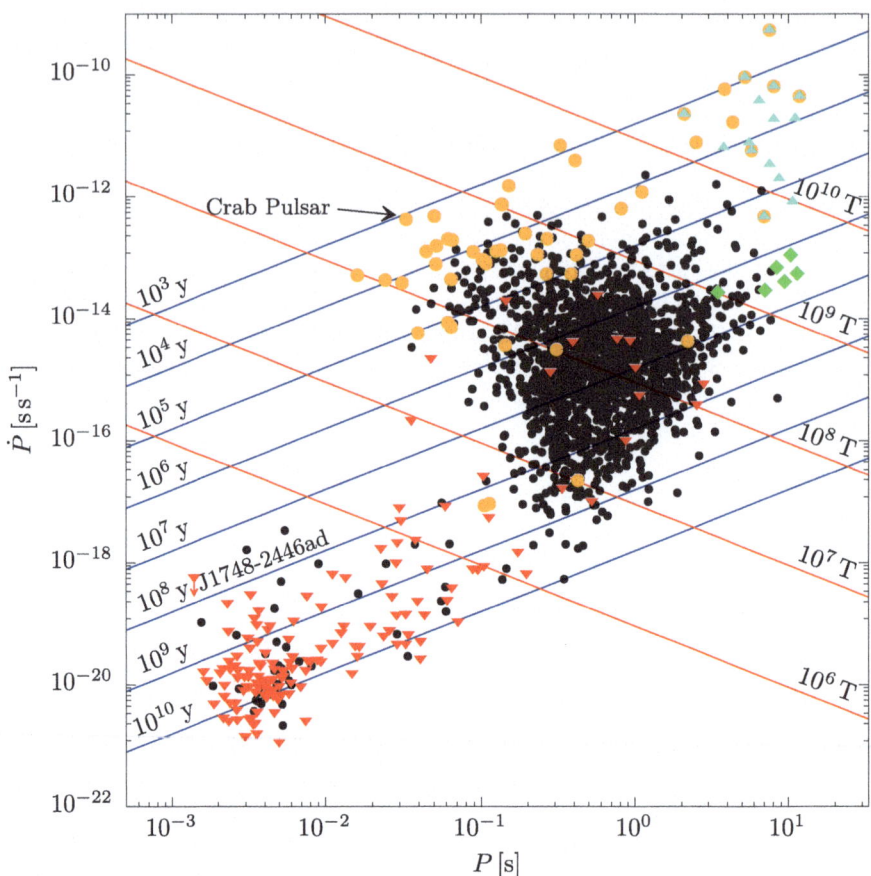

Fig. 21.10 $P\text{-}\dot{P}$ diagram of all pulsars for which both quantities are known, together with different deceleration ages according to (21.48) (*blue lines*), and magnetic field strengths according to (21.49) (*red lines*). *Red triangles* indicate pulsars in binary systems, *cyan blue triangles* represent magnetars. The *green diamonds* represent those 6 of the "Magnificent Seven" for which P and $\dot{P}$ are known [26]. *Orange circles* are pulsars with a known associated supernova remnant. The Crab Pulsar is an important representative of this group. The fastest rotating pulsar PSR J1748-2446ad [8] is shown for comparison. Only an upper bound on $\dot{P}$ is known for this pulsar, which is indicated by the arrow. This figure is based on data from the *ATNF pulsar catalog* [1, 16]

21.5.1 The X-ray Pulsar Hercules X-1

To illustrate the wealth of information that can be gathered from detailed observations of pulsar systems, we take a closer look at one specific system, namely the X-ray pulsar Hercules X-1. It is part of a binary system and orbits a normal star. The interplay between these two celestial bodies leads to a variety of interesting properties.

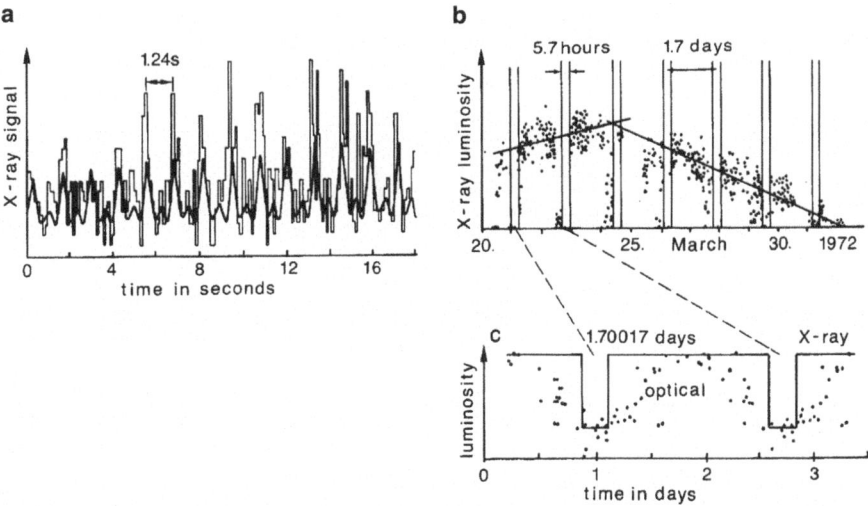

Fig. 21.11 Observations of the Uhuru satellite of the HZ Her system. (**a**) Short-time measurements of the X-ray signal show a periodicity of $t = 1.24$ s. (**b**) Long-time measurements additionally show a breakdown of the X-ray signal for 5.7 hours every 1.7 days (*top*). During the same time the luminosity of the companion star decreases. From [20]; with the kind permission of © Springer-Verlag Berlin Heidelberg 1994. All Rights Reserved

In the 1970s, the X-ray satellite Uhuru[6] discovered an X-ray source near the already known normal star HZ Her in the constellation Hercules. An X-ray signal was measured whose intensity changed with a period of 1.24 s, see Fig. 21.11a. In addition, long-term observations showed that the intensity of the X-ray radiation dropped to zero for 5.7 hours every 1.7 days. It had been known before that the luminosity of the star in the optical range varied with the same period. The decrease in X-ray intensity was accompanied by a decrease in intensity in the optical range (Fig. 21.11b).

The interpretation of the measurements was that HZ Her has a neutron star as its companion, which rotates with a period of 1.24 s and orbits the star in 1.7 days (Fig. 21.12). Because of its strong gravitation, the neutron star pulls off matter from the star, and an accretion disk is formed

The strong magnetic field of the neutron star transports the ionized matter from the disk along the magnetic field lines to its poles. In this way about 10^{11} tons of matter fall onto the polar caps of the neutron star every second, reaching a free-fall velocity of about 40% of the speed of light (Fig. 21.13), as we had already estimated in Sect. 1.4.4. The impact of the ionized matter on the surface of the neutron

---

[6] Uhuru was the first satellite launched specifically for the purpose of X-ray astronomy. The satellite's name, "Uhuru", is the Swahili word for "freedom". It was named in recognition of the hospitality of Kenya from where it was launched.

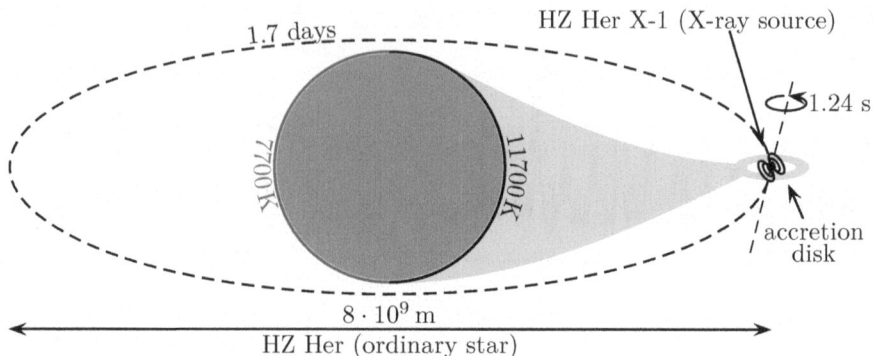

Fig. 21.12 The Hercules system consists of a normal star orbited by a neutron star which in turn rotates with a period of 1.24 s. The intense X-ray radiation from the neutron star heats up the side of the normal star facing it. Therefore, the temperature and the luminosity vary. The neutron star pulls off matter from the normal star, and an accretion disk is formed

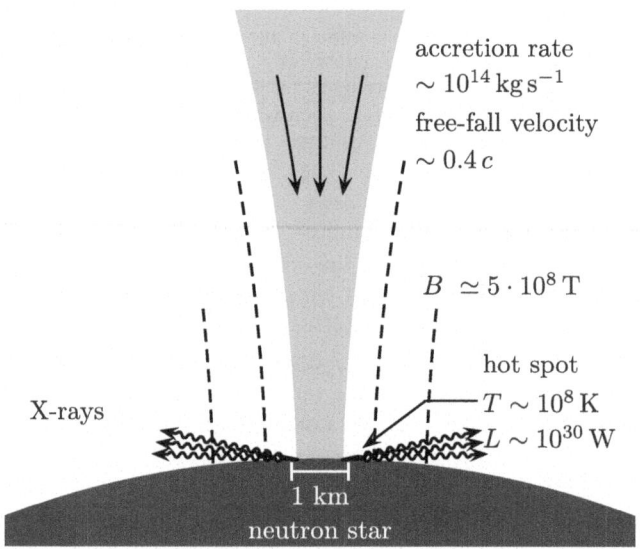

Fig. 21.13 Ionized matter from the accretion disk falls along the magnetic field lines on the poles of the neutron star. It reaches a free-fall velocity in the range of 0.4c. The impact of the charged particles on the surface of the neutron star generates X-ray bremsstrahlung

star generates X-ray bremsstrahlung ("hot spot"). The power radiated amounts to roughly 10^{30} W, which corresponds to approximately 2000 times the luminosity of the Sun.

The X-rays heat the normal star on one side, causing its temperature and luminosity to increase. If the pulsar is located behind the star, the X-ray signal is lost, and then the colder, less luminous side of the normal star is facing the Earth.

In addition, absorption lines at 54 and 108 keV were detected in the X-ray spectrum of the neutron star. Figure 21.14 shows the results of the corresponding measurements. To explain these absorption lines three possibilities were considered: atomic transitions, nuclear transitions, and magnetic transitions between Landau levels. A possible atomic transition would be a strongly gravitationally shifted Lyman-α emission of 77 times ionized hydrogen-like platinum. However, the existence of a sufficient amount of platinum on the neutron star to explain the intensity of the feature seems completely absurd. A nuclear transition with the suitable energy would be possible, e.g., in americium 241, but again this origin is not very plausible. The most convincing explanation is cyclotron transitions, i.e., transitions of electrons between different Landau levels, the quantum mechanical eigenstates of electrons in a strong magnetic field [24]. If this is the correct explanation, one can directly measure the magnetic field strength on the neutron star from the energy of the transition with the *cyclotron frequency*

$$\omega = \frac{eB}{m_{\mathrm{e}}} \tag{21.52}$$

using the relation $E = \hbar\omega$. From the energy difference 54 keV follows a magnetic field strength of

$$B = \frac{E}{\hbar} \cdot \frac{m_{\mathrm{e}}}{e} \simeq 5 \cdot 10^8 \, \mathrm{T}. \tag{21.53}$$

This was the first direct measurement of such a strong magnetic field in a neutron star.

The absorption line at 108 keV can then be explained as the second harmonic of the transition.

21.5.2 Precision Tests of General Relativity in Pulsar Binary Systems

Pulsars in binary systems, in particular in systems with another neutron star or a white dwarf, are ideal candidates for testing the predictions of general relativity, and for comparing these predictions with alternative gravitational theories. Since the periods of pulsars can be measured with extreme precision, pulsars provide high-precision clocks in the gravitational field of their binary companions. Through long-term observations of up to several decades, the orbital parameters then can be measured very accurately, and compared with the results of calculations.

In this section we will discuss which physical information can be gleaned from the study of pulsar binary systems. Specifically, we shall look at two systems. The most famous, and first detected, binary system consisting of a pulsar and another neutron star is PSR B1913+16. The name of the system indicates its position in the sky: The latest published position [28] of the system has a right ascension of

Fig. 21.14 X-ray spectrum of Her X-1: Absorption dips can be seen at 54 and 108 keV. The interpretation as cyclotron transitions allows determining the corresponding magnetic field strengths. From Trümper et al. [25], © AAS. Reproduced with permission

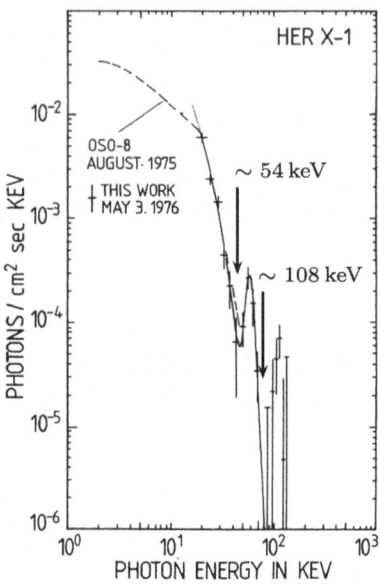

$$\delta = 16°06'27''.3868(5)$$

and a declination of

$$\alpha = 19^{h}15^{m}27^{s}.99942(3).$$

We have explained the meaning of these quantities in Sect. 1.5.7.

PSR B1913+16 was discovered in the 1970s by *Hulse*[7] and *Taylor*[8] using the Arecibo radio telescope in Puerto Rico [9], and has been observed continuously ever since. In 1993, the two researchers were awarded the Nobel Prize in physics for their investigations of PSR B1913+16. The system consists of two neutron stars circling each other on nearly elliptical orbits, with an orbital diameter of about 700,000 km and a period of about 7.75 h, i.e., roughly one working day. One of the two neutron stars is a radio pulsar which is orientated in such a way that its pulses can be received on Earth, with a period of about $P = 60$ ms, see Fig. 21.15. In the literature this binary system of two neutron star is somewhat misleadingly referred to as a double pulsar, even though pulses are received only from one of the two neutron stars. The second system is PSR J0737-3039. It is of particular significance since it is the only true double pulsar known to date, i.e., the pulses of both neutron stars can be received on Earth.

---

[7] Russel Alan Hulse, ★1950, US-American physicist, Nobel Prize 1993.

[8] Joseph Hooton Taylor Jr., ★1941, US-American physicist, Nobel Prize 1993.

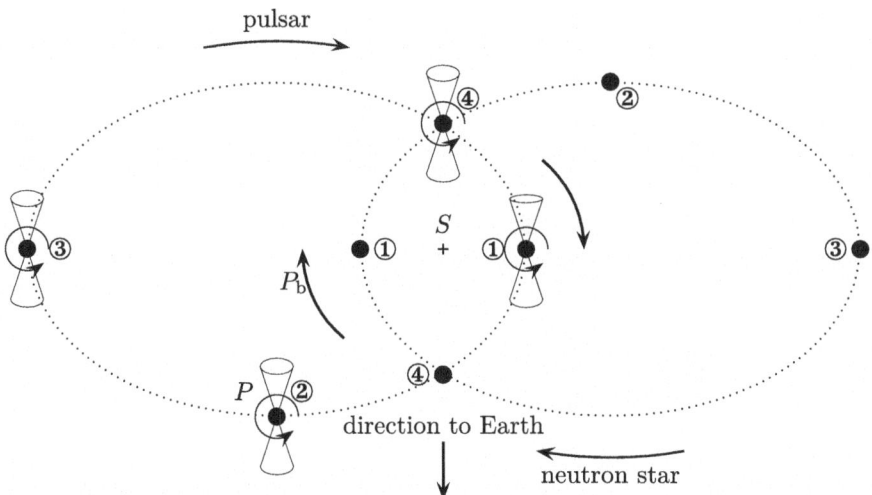

Fig. 21.15 Illustration of the physical situation in pulsar binary systems. The two neutron stars orbit each other with some period P_b; for PSR B1913+16, e.g., one has $T = 7.75$ h. At least one of the two neutron stars is a pulsar, with a rotational period P, and must be orientated in such a way that its signals can be received on Earth. For PSR B1913+16 one has $P \approx 60$ ms. The numbers in the figure correspond to different constellations of the system, in which different pronounced physical effects become important, see the discussion in the text

In classical mechanics, a Kepler orbit can be described by 5 quantities, viz. the *Keplerian parameters*, referred to a given plane. We use the tangential plane to the celestial sphere as a plane of reference. It is also called the celestial plane, and lies perpendicular to the viewing direction towards the binary star system. The orbital plane in which the system moves is tilted by an angle i with respect to the celestial plane. Then the Keplerian parameters are the projected semi-axis $a \sin i$, the eccentricity e of the elliptical orbit, the angle Ω between some referential direction and the intersection of the orbital trajectory with the celestial sphere, the orbital period P_b, the angular distance ω of the periastron relative to the celestial plane, and the point in time when the object goes through the periastron during each orbit. The situation in three dimensions is illustrated in Fig. 21.16.

The deviation of an orbit from the predictions of classical mechanics can be characterized by a few more parameters, the *post-Keplerian parameters* (PK). We have already discussed one effect of general relativity on the orbital motion in a gravitational potential, namely the motion of the perihelion for Mercury in Sect. 13.4.2. In relation to a star in a binary star system, the point of closest approach to the focus is called the *periastron*, instead of perihelion, and correspondingly the point of greatest distance is called *apastron*, but the motion of these points is analogous to that in the solar system. Since the angle to the periastron is denoted by ω, the symbol $\dot{\omega}$ is used for this PK.

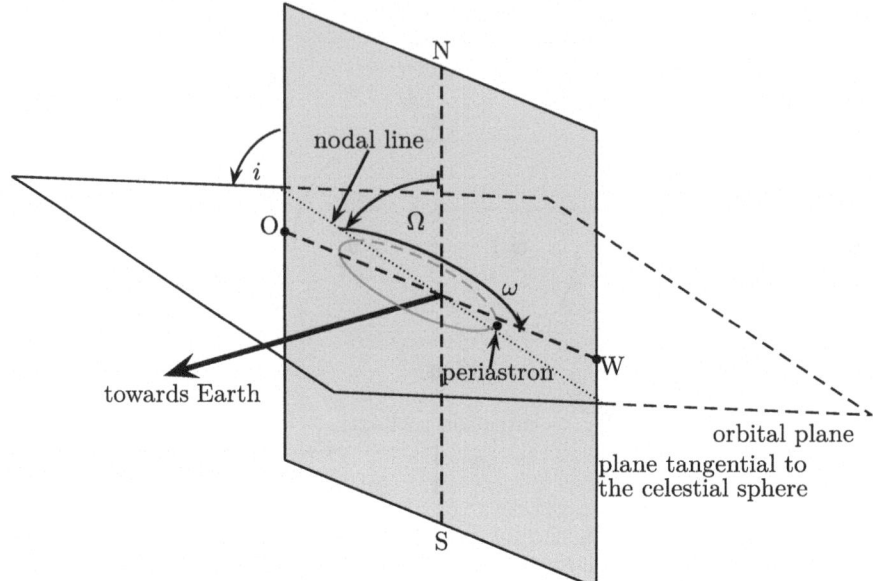

Fig. 21.16 Sketch for the orbital parameters of binary systems. The observation of pulsar binary systems from the Earth yields information only on the projected semi-axis $a_\perp$

In PSR B1913+16, with an orbital period of approximately 8 hours and a major semi-axis in the range of 700,000 km, velocities of the pulsar relative to the system's center of gravity occur on the order $v \approx 100\text{--}500\,\mathrm{km\,s^{-1}}$, where, by Kepler's second law, the velocity is very different at different points on the orbit due to the large orbital eccentricity. These high velocities lead to a relatively strong Doppler effect according to (8.13). The frequency of the signals increases when the two stars are in the *periastron* (situation ① in Fig. 21.15, the pulsar moves towards Earth), and decreases in the *apastron* (situation ③, the pulsar moves away from Earth). However, these modifications are effects purely of special relativity, and therefore do not provide any information about gravity.

Another modulation of the signals which is independent of gravitation results from the changing distances of the pulsar on its orbit from Earth. The signals arrive earlier on Earth when the pulsar is closer to Earth. By contrast, the signal arrives late, when it is at its furthest point from Earth (situation ④). For PSR B1913+16 this geometric effect sums up to about 2 s.

In addition to the longitudinal Doppler effect, the transverse Doppler effect also occurs, i.e., a frequency change also occurs in the situation ② due to the relative longitudinal velocity of the pulsar. Together with the gravitational redshift caused by the companion, this effect is combined in the PK γ_E. Furthermore, a Shapiro time-delay occurs in the situation ④, like in the distance measurements to Venus in Sect. 13.4.4. The Shapiro time-delay depends on the mass of the companion, but also on the angle i, since it determines the maximum distance at which the

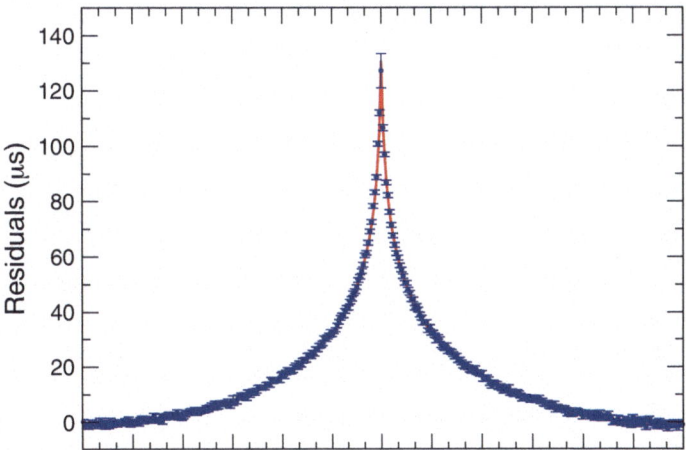

Fig. 21.17 Measurements of the Shapiro time-delay in the double pulsar system PSR J0737-3039, in comparison with calculated values [11]. ©APS. Reused under the terms of the Creative Commons Attribution 4.0 License

pulsar is positioned exactly behind the companion. For $i = 0°$ the effect of Shapiro time-delay would be weakest, for $i = 90°$ the effect would be strongest. The time-delay is described by two parameters, s (shape), i.e., the shape of the time-delay, and r (range), i.e., the size of the time-delay. Figure 21.17 shows the results of measurements of the Shapiro time-delay in the genuine double pulsar system PSR J0737-3039 [11], that we will discuss below. Also compare with the theoretically determined curve in Fig. 13.13.

The two masses rotating around each other radiate gravitational waves and therefore lose energy, which implies that the orbital period changes. This change $\dot{P}_b$ provides another PK. In addition, there occurs a geodetic precession of the rotational axis of the pulsar by relativistic spin-orbit coupling, i.e., the angular momentum associated with the rotation of the pulsar and its orbital angular momentum couple to each other, similar to the coupling of the electron spin to its orbital angular momentum in atoms.

Therefore, the direction of the rotational axis of the pulsar is not fixed in space but depends on time. The rate of the change of this angle is the PK $\Omega_{\text{SO,pu}}$. Within the framework of general relativity quantitative expressions can be derived for the PKs. The most important ones are [10]

$$\dot{\omega} = 3T_\odot^{2/3} \left(\frac{P_b}{2\pi} \right)^{-5/3} \frac{(m_{\text{pu}} + m_{\text{ns}})^{2/3}}{1 - e^2}, \tag{21.54a}$$

$$\gamma_E = T_\odot^{2/3} \left(\frac{P_b}{2\pi} \right)^{1/3} e \frac{m_{\text{ns}}(m_{\text{pu}} + 2m_{\text{ns}})}{(m_{\text{pu}} + m_{\text{ns}})^{4/3}}, \tag{21.54b}$$

$$r = T_\odot m_{\text{ns}}, \tag{21.54c}$$

Table 21.1 Properties of the double pulsar PSR B1913+16 [29]

	Symbol	Value
Projected major semi-axis	$a_\perp$ [km]	701,901.0(9)
Rotational period	P [s]	0.0590300032180(6)
Change of the period	$\dot{P}$ [s s^{-1}]	8.628(4) · 10^{-18}
Orbital eccentricity	e	0.6171334(5)
Orbital period	P_b [s]	27906.979586(4)
Mass of the pulsar	m_p	1.4398(2) $M_\odot$
Mass of the companion	m_{ns}	1.3886(2) $M_\odot$
Post-Keplerian parameters		
Change of the orbital period	$\dot{P}_b$ [s s^{-1}]	$-$ 2.423(1) · 10^{-12}
Average motion of the periastron	$\langle\dot{\omega}\rangle$	4.226598(5)° y^{-1}
Gravitational redshift and Doppler effect	γ_E [ms]	4.2919(8)

$$s = \sin i = T_\odot^{-1/3} \left(\frac{P_b}{2\pi}\right)^{-2/3} x_{pu} \frac{(m_{pu} + m_{ns})^{2/3}}{m_{ns}}, \tag{21.54d}$$

$$\dot{P}_b = -\frac{192}{5} T_\odot^{5/3} \left(\frac{P_b}{2\pi}\right)^{-5/3} f(e) \frac{m_{pu} m_{ns}}{(m_{pu} + m_{ns})^{1/3}}, \tag{21.54e}$$

$$\Omega_{SO,pu} = T_\odot^{2/3} \left(\frac{P_b}{2\pi}\right)^{-5/3} \frac{1}{1 - e^2} \frac{m_{ns}(4m_{pu} + 3m_{ns})}{2(m_{pu} + m_{ns})^{4/3}}, \tag{21.54f}$$

with

$$f(e) = \frac{1 + (73/24)e^2 + (37/96)e^4}{(1 - e^2)^{7/2}} \tag{21.54g}$$

and

$$T_\odot = \frac{GM_\odot}{c^3} = 4.925490947 \text{ μs.} \tag{21.54h}$$

In the Eqs. (21.54), the masses of the stars have to be inserted in units of the solar mass, and x_{pu} is the projected major semi-axis of the pulsar's orbit. It can be seen that the expressions in (21.54) are always functions of the two (yet unknown) stellar masses, of the orbital eccentricity, and of the period P_b.

It follows that, once two PKs have been determined, the masses of the two stars can be determined. Every other measured value of a PK then provides a test of general relativity! Table 21.1 lists the parameters of the system PSR B1913+16, including the measured PKs.

The great significance of PSR B1913+16 results from the observation of $\dot{P}_b$ for almost half a century now. A decrease in the orbital period means that the two stars come closer together, and the system is losing energy. The agreement between

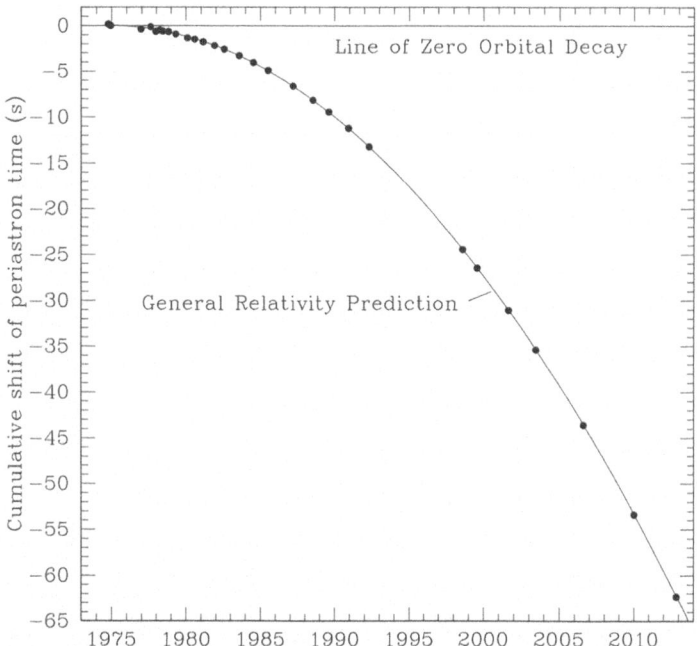

Fig. 21.18 Change in the orbital period in the system PSR B1913+16. What is shown is how much earlier the system goes through its periastron, compared to Newtonian mechanics (line at top). The measured points lie exactly on the theoretical curve calculated using (21.54e), which was derived under the assumption that the system radiates gravitational waves. Note that, in 2013, the time difference to the Newtonian prediction was already about 63 s. From Weisberg and Yuang [28], © AAS. Reproduced with permission

theoretical predictions and the actual observations is excellent, see Fig. 21.18. These measurements represented the first indirect proof of the existence of gravitational waves! Due to the loss of energy, the two stars approach each other by about 3.1 mm per orbit, i.e., 3.5 m per year, and will merge in about 300 million years. Note that we computed a decrease of the orbital radius of 6.3 m for a similar example system in exercise 15.1. The lower value for PSR B1913+16 results from the strong eccentricity of the system's orbit.

In the years after the discovery of the binary system PSR B1913+16 several more systems were found in which a pulsar forms a binary system with a neutron star or a white dwarf. In the meantime around 10 systems are known which consist of two neutron stars [14].

The genuine double pulsar PSR J0737-3039 is a unique system, and therefore we will discuss it in a little more detail. In this binary system of two neutron stars, in 2003 one of the stars could be identified as a pulsar with a pulse frequency of about 22 ms [3]. The great significance of this system became evident, when it was shortly after detected that the second star of the system also is a pulsar [15]. It has a longer rotational period of 2.8 s. Thus, in this system one has two high-precision clocks, moving in their mutual gravitational potential. The two stars orbit each other in only

Table 21.2 Properties of the double pulsar system PSR J0737-3039 [10]

	Symbol	Value
Projected major semi-axis pulsar A	$a_{\perp A}$	424,126.8(3) km
Projected major semi-axis pulsar B	$a_{\perp B}$	454,419(480) km
Rotational period pulsar A	P_A	0.022699378599624(2) s
Rotational period pulsar B	P_B	2.77346077007(8) s
Change of the period pulsar A	$\dot{P}_A$	$1.75993(6) \cdot 10^{-18}$ s s^{-1}
Change of the period pulsar B	$\dot{P}_B$	$8.92(8) \cdot 10^{-16}$ s s^{-1}
Orbital eccentricity	e	0.0877775(9)
Orbital period	P_b	8834.53500(5) s
Change of the orbital period	$\dot{P}_b$	$-2.423(1) \cdot 10^{-12}$ s s^{-1}
Mass pulsar A	m_A	1.3381(7) $M_\odot$
Mass pulsar B	m_B	1.2489(7) $M_\odot$

Table 21.3 Values of the post-Keplerian parameters for the double pulsar system PSR J0737-3039 in comparison with the general relativistic predictions as of 2008 [10]

	Observation	ART	Ratio
$\langle\dot{\omega}\rangle$ [°y^{-1}]	16.89947(68)	–	–
$\dot{P}_b$ [s s^{-1}]	1.252(17)	1.24787(13)	1.003(14)
γ_E [ms]	0.3856(26)	0.38418(22)	1.0036(68)
s	0.99974^{+16}_{-39}	0.99987^{+13}_{-48}	0.99987(50)
r [µs]	6.21(33)	6.153(26)	1.009(55)
$\Omega_{SO,B}$ [°y^{-1}]	$4.77^{+0.66}_{-0.65}$	5.0734(7)	0.94(13)

147 minutes, an orbital period much shorter than in PSR B1913+16. Therefore, even more pronounced relativistic effects can be expected. In Table 21.2 the properties of this system are summarized. Due to the pulsar property of both stars, a short time after the discovery all 6 PKs in (21.54) could already be determined, which allowed for 4 independent tests of general relativity, see Table 21.3.

According to (21.54), every single PK defines a curve in the (m_A, m_B) plane. Within the error tolerance, all these curves must intersect in one point. Otherwise, general relativity is disproved as the correct theory of gravitation.

Figure 21.19 demonstrates clearly the impressive agreement between general relativity and the observations, while some alternative theories of gravitation can be excluded. The data used in Fig. 21.19 have been collected in observations up to 2019 [11]. Pulsar B, due to relativistic spin-orbit coupling has not been observable since March 2008, and it is unclear when it will again return to our view. For Pulsar A, however, the additional observations in the years since 2009 greatly reduced the uncertainties for its parameters, see Table 21.4. The greatly increased accuracy of the PKs makes the inclusion of higher-order corrections necessary for the first time, in particular for $\dot{P}_b$ and s. Also, a first estimate for the moment of inertia

$$\Theta_A \approx 1.15 - 1.48 \, [10^{45} \, \text{g cm}^{-2}] \tag{21.55}$$

of Pulsar A could be obtained from constraints on the radius of the pulsar and theoretical considerations on the relation between radius and moment of inertia [11].

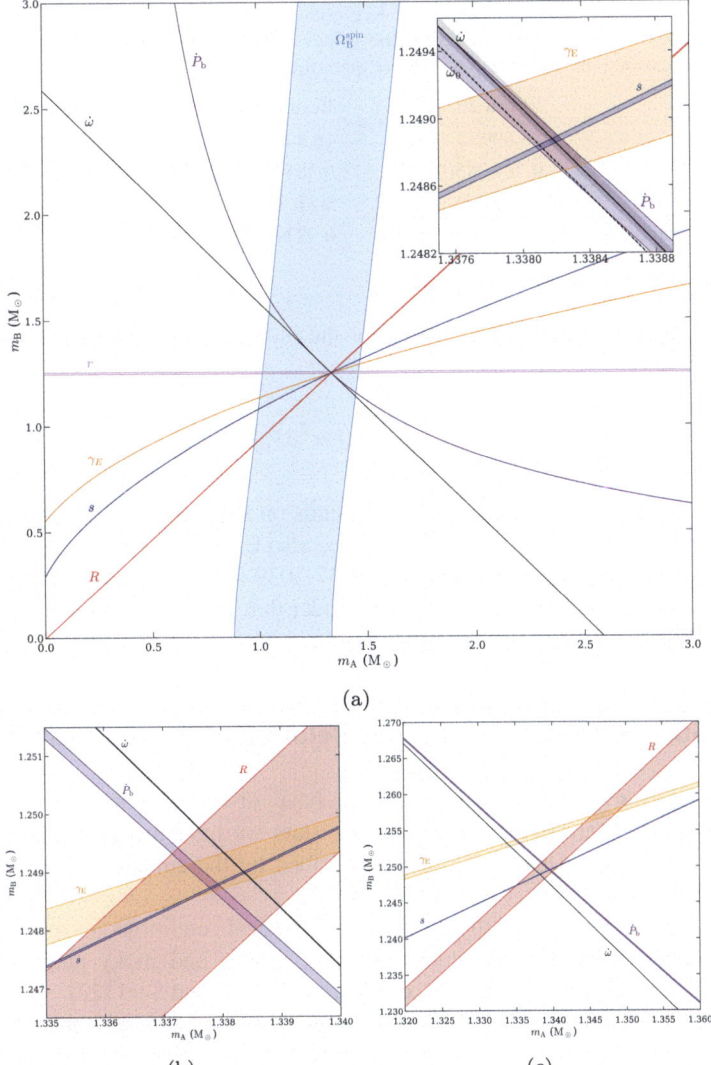

Fig. 21.19 (a) Evaluation of the post-Keplerian parameters for the double pulsar system PSR J0737-3039 [11] in a mass-mass diagram. Within the uncertainty, the curves for the individual PKs calculated in the framework of general relativity all intersect at one point. This confirms general relativity as the correct theory of gravitation. The additional parameter R is the ratio of the major semi-axes of the orbits of the two pulsars, it can be determined only in a genuine double pulsar system [10]. The post-Keplerian parameters can also be used to test alternative theories of gravity or extensions of general relativity. (b) Shows the mass-mass diagram for one version of Damour-Esposito-Farèse gravity [4] and (c) the same diagram for Bekenstein's TeVeS [2]. Both theories are excluded by the measurements. ©APS. Reused under the terms of the Creative Commons Attribution 4.0 License

Table 21.4 Updated values for some post-Keplerian parameters for the double pulsar system PSR J0737-3039 as of 2021 compared to the values of 2009. The ongoing observations have drastically decreased the uncertainties of these quantities

	Result 2021	Result 2009	Error reduction factor
$\langle \dot{\omega} \rangle$ [°y^{-1}]	16.899323(13)	16.89947(68)	52.3
γ_E [ms]	0.384 045(94)	0.3856(26)	27.7
r [μs]	6.162(21)	6.21(33)	15.7
Orbital eccentricity e	0.087777023(61)	0.0877775(9)	14.8

The expression in classical mechanics for the moment of inertia of a solid sphere with radius R is

$$\Theta = \int \rho(r) r^2 \mathrm{d}V. \tag{21.56}$$

It is therefore apparent that Θ strongly depends on the equation of state of neutron star matter due to its dependence on $\rho(r)$, see also Exercise 21.3.

The observations of the system PSR J0737-3039 as well as other binary neutron star systems continue. We can therefore expect that our understanding of neutron stars will continue to increase in the years to come.

21.6 Supplement: Mass-Radius Relation of Moons and Planets

With white dwarfs and neutron stars we have discussed objects made up of degenerate matter, and with main sequence stars we have discussed gas spheres, which, to good approximation, can be described by the equation of state for the ideal gas.

In this section, we will briefly consider the physical conditions for much smaller masses, namely objects in the mass range of moons and rocky planets. In our presentation, we closely follow the explanations in Sexl and Sexl [22].

The equation of state (18.48) for degenerate matter predicts that the pressure is zero when the density is zero. But we all know from experience that the density of cold matter is finite, even if the pressure is zero, i.e., $\rho_m(p = 0) = \rho_p \neq 0$. The reason is that the density of normal matter depends on its chemical composition. Take, for example, the density of hydrogen atoms located at a distance of the Bohr radius $a_B = \lambdabar_e / \alpha$

$$\rho_p = \frac{m_p}{a_B^3} \approx 8000 \, \mathrm{kg} \, \mathrm{m}^{-3} = 8 \, \mathrm{g} \, \mathrm{cm}^{-3}. \tag{21.57}$$

This value is roughly typical of planets and moons. For the Earth, e.g., we obtain with the values from Table 1.3 $\rho_{\oplus} = 5.5 \, \mathrm{g} \, \mathrm{cm}^{-3}$. The stability of these objects is guaranteed by the atomic structure, i.e., by the electromagnetic interaction, and not determined by some Fermi pressure. Within certain limits, the atomic structure is independent of pressure, and to good approximation also of temperature. Yet, if the mass becomes too big, atomic structure will break down, and these objects will also collapse until the Fermi pressure of the electrons stabilizes them again. Thus, below some critical pressure $p < p_p$, and therefore also $\rho_m = \rho_p$, one will have the mass-radius relation of moons and planets. It is nothing else but the relation well known from everyday life

$$mass = density \cdot volume,$$

or

$$M \cong \rho_p R^3 \quad \text{or} \quad M \sim R^3. \tag{21.58}$$

But for $p > p_p$ the equation of state must turn into that for non-relativistic degenerate matter in (18.62),

$$p(\rho_m) = \frac{\varepsilon_{e,c}}{3\pi^2} \cdot \frac{1}{5} \left(\frac{\rho_m}{\rho_c} \right)^{5/3} \tag{21.59}$$

and we will recover the mass-radius relation $M \sim R^{-3}$ for white dwarfs with degenerate matter.

By comparing with white dwarfs we can estimate the maximum mass and the maximum radius of planets. To do this we set $\rho_m = \rho_p$ in the qualitative mass-density relation (20.5), i.e.

$$M_p \approx M_c \cdot \left(\rho_p / \rho_c \right)^{1/2} \approx 6 \cdot 10^{27} \, \text{kg}. \tag{21.60}$$

At the same time, M_p yields an approximate lower bound on the mass of a white dwarf. Below this mass, the density is too low to produce a degenerate electron gas. As already mentioned, the lightest known white dwarf has a mass of about $M = 0.17 \, M_\odot \approx 3.4 \cdot 10^{29} \, \text{kg}$, and therefore is much more massive. However, the lower limit we have determined is also only based on the purely physical properties of white dwarfs, and neglects other considerations entering the possible formation of such a low-mass white dwarf. We already know from Sect. 19.4.3 that such a low-mass star cannot have lived through the main sequence stage, since the universe is too young for this to happen. Thus, white dwarfs can exist only in the narrow mass range $M_p < M_{wd} < M_c$ at all, with the actual mass range occurring in nature very likely being considerably smaller. Comparing with the mass of the Sun this yields

$$3 \cdot 10^{-3} M_\odot < M_{wd} < 1.5 \, M_\odot. \tag{21.61}$$

These are three orders of magnitude. The mass range of planets, on the other hand, is enormous, and lies in the range of 54 orders of magnitude, with the mass of the hydrogen atom constituting the (extreme) lower bound:

$$2 \cdot 10^{-27}\,\text{kg} < M_p < 6 \cdot 10^{27}\,\text{kg} = 3 \cdot 10^{-3} M_\odot. \tag{21.62}$$

This definition does of course not agree with the more stringent categorization of a planet as a nearly spherical object, which dominates its orbit, i.e., has cleared it of all other material (this stringent definition also has expelled Pluto from the family of planets). However, the present definition applies to minor planets, asteroids, and similar bodies.

To find an estimate for the maximum radius R_P we exploit the relation $\rho_p \simeq M_p/R_p^3$, i.e.,

$$R_p \approx \left(\frac{M_p}{\rho_p} \right)^{1/3} \approx 10^8\,\text{m}. \tag{21.63}$$

We had found a value of $R_c \approx 5 \cdot 10^6$ m for the Chandrasekhar radius in (20.13), it is therefore roughly an order of magnitude smaller.

We can get an idea of the accuracy of these estimates by comparing with Jupiter, the biggest planet in the solar system. Jupiter has a mass of $M_{\mathrm{2\!\!+}} = 1.899 \cdot 10^{27}$ kg and a radius of $R_{\mathrm{2\!\!+}} \simeq 7.1 \cdot 10^7$ m, see Table 1.1. One can see this is relatively close to the above limits.

A clear representation of these relationships is obtained by plotting the mass-radius relationships for planets and for white dwarfs on a logarithmic scale, see Fig. 21.20. Combining the equations of state discussed in Chap. 18 with the findings sketched in this section, we obtain the relation

$$f(\rho_m, T) = \frac{p}{\rho_m c^2} = \begin{cases} f(T_m), & \text{for normal stars, see (18.25)} \\ f(\rho), & \text{for degenerate stars, see (18.66)} \\ \sim p & \text{for planet-like objects.} \end{cases} \tag{21.64}$$

Of course, in our discussion we have made strong approximations in some places. For example, large gas planets such as Jupiter certainly behave very differently from small planets such as the Earth. In particular, there must be a transition region between these planets and small stars that are massive enough to ignite nuclear fusion processes.

21.7 Exercises

21.1. TOV equation for constant density In general, the TOV equation (21.35) cannot be solved analytically. However, under the strong assumption of a constant density this is possible.

Find this solution, and compare with the non-relativistic result that follows from (18.7).

Fig. 21.20 A schematic sketch of the mass-radius relations of white dwarfs and planets. White dwarfs can exist only in the mass range $M_p < M_{wd} < M_c$. For planets, $M \sim R^3$, while for white dwarfs $M \sim R^{-3}$

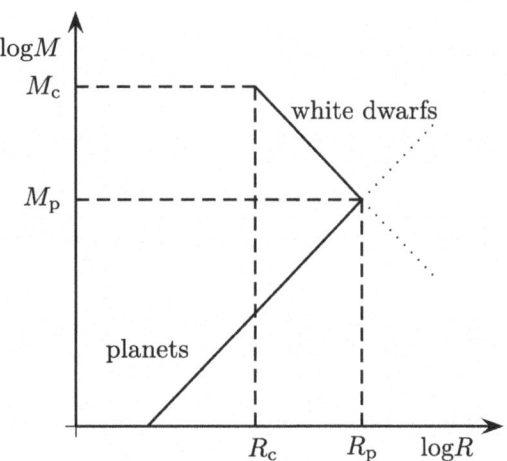

21.2. Millisecond pulsars We consider a millisecond pulsar with a period of $P = 1.558$ ms and a change in period of $\dot{P} = 1.051 \cdot 10^{-19}$ s s^{-1}.

(a) In music, the standard pitch a^1 has a frequency 440 Hz. Which tone on the piano comes closest to the frequency of the pulsar?
(b) Assuming a constant $\dot{P}$, how long does it take the pulsar to "sing" one semitone lower? (Note: 1 octave = 12 semitone steps = frequency doubling)

21.3. Moment of inertia and the equation of state The value for the moment of inertia of Pulsar A in the system PSR J0737-3039 in Eq. (21.55) was not measured independently, but obtained using estimates for the radius and theoretical relations between the mass and the moment of inertia.

For the purpose of this exercise we want to assume that we have independently measured the mass and the moment of inertia of a neutron star and derived in theoretical considerations that the density is given by $\rho(r) = \rho_c(1 - r/R)$. We used a similar approach in Chap. 18 for the pressure to get a basic understanding of the interior of stars, see Eq. (18.9). Show that we can then compute the radius and the density at the center of the neutron star.

References

1. ATNF Pulsar Katalog: https://www.atnf.csiro.au/people/pulsar/psrcat
2. Bekenstein, J.D.: Relativistic gravitation theory for the modified Newtonian dynamics paradigm. Phys. Rev. D **70**(8), 083509 (2004). https://link.aps.org/doi/10.1103/PhysRevD.70.083509
3. Burgay, M., et al.: An increased estimate of the merger rate of double neutron stars from observations of a highly relativistic system. Nature **426**(6966), 531–533 (2003)
4. Damour, Th., Esposito-Farèse, G.: Nonperturbative strong-field effects in tensor-scalar theories of gravitation. Phys. Rev. Lett. **70**(15), 2220–2223 (1993). https://link.aps.org/doi/10.1103/PhysRevLett.70.2220

5. Demorest, P.B., et al.: A two-solar-mass neutron star measured using Shapiro delay. Nature **467**, 1081–1084 (2010)
6. Espinoza, C.M., Lyne, A.G., Stappers, B.W., Kramer, M.: A study of 315 glitches in the rotation of 102 pulsars. Mon. Not. R. Astron. Soc. **414**(2), 1679–1704 (2011)
7. Gold, T.: Rotating neutron stars as the origin of the pulsating radio sources. Nature **218**, 731–732 (1968)
8. Hessels, J.W.T., et al.: A radio pulsar spinning at 716 Hz. Science **311**(5769), 1901–1904 (2006)
9. Hulse Jr., R.A., Taylor, J.H.: Discovery of a pulsar in a binary system. Astrophys. J. **195**, L51–L53 (1975)
10. Kramer, M., Wex, N.: The double pulsar system: a unique laboratory for gravity. Class. Quantum Gravity **26**(7), 073001 (2009)
11. Kramer, M., et al.: Strong-field gravity tests with the double pulsar. Phys. Rev. X **11**(4) (2021). 041050 https://link.aps.org/doi/10.1103/PhysRevX.11.041050
12. Lattimer, J.M., Prakash, M.: Neutron star structure and the equation of state. Astrophys. J. **550**, 426–442 (2001)
13. Lattimer, J.M., Prakash, M.: Neutron star observations: prognosis for equation of state constraints. Phys. Rep. **442**(1–6), 109–165 (2007)
14. Lorimer, D.R.: Binary and millisecond pulsars. Living Rev. Relativ. **11** (2008)
15. Lyne, G., et al.: A double-pulsar system: a rare laboratory for relativistic gravity and plasma physics. Science **303**, 1153–1157 (2004)
16. Manchester, R.N., Hobbs, G.B., Teoh, A., Hobbs, M.: The Australia telescope national facility pulsar catalogue. Astronom. J. **129**(4), 1993 (2005)
17. Oppenheimer, J.R., Volkoff, G.M.: On massive neutron cores. Phys. Rev. **55**, 374–381 (1939)
18. Potekhin, A.Y.: The physics of neutron stars. Physics-Uspekhi **53**(12), 1235 (2010)
19. Romani, R.W., et al.: PSR J0952-0607: the fastest and heaviest known galactic neutron star. https://dx.doi.org/10.3847/2041-8213/ac8007
20. Ruder, H., Wunner, G., Herold, H., Geyer, F.: Atoms in Strong Magnetic Fields. Springer, Berlin (1994)
21. Ryan, S.G., Norton, A.J.: Stellar Evolution and Nucleosynthesis. Cambrigde University Press, Cambridge (2010)
22. Sexl, R., Sexl, H.: Weiße Zwerge – Schwarze Löcher (White dwarfs – Black holes). Rowohlt Taschenbuch Verlag (1975)
23. Tolman, R.C.: Static solutions of Einstein's field equations for spheres of fluid. Phys. Rev. **55**, 364–373 (1939)
24. Trümper, J., et al.: Evidence for strong cyclotron emission in the hard X-ray spectrum of Her X-1. Ann. N Y Acad. Sci. **302**(1), 538–544 (1977)
25. Trümper, J., Pietsch, W., Reppin, C., Voges, W.: Evidence for strong cyclotron line emission in the hard X-ray spectrum of Hercules X-1. Astrophys. J. **219**, L105–L110 (1978)
26. Turolla, R.: Isolated Neutron Stars: The Challenge of Simplicity. In: Becker, W. (ed.) Neutron Stars and Pulsars. Astrophysics and Space Science Library, vol. 357, pp. 141–163. Springer, Berlin (2009)
27. Weber, F., Negreiros, R., Rosenfield, P.: Neutron star interiors and the equation of state of superdense matter. In: Becker, W. (Hrsg.) Neutron Stars and Pulsars. Astrophysics and Space Science Library, vol. 357, pp. 213–245. Springer, Berlin (2009)
28. Weisberg, J.M., Yuang, Y.: Relativistic measurements from timing the binary pulsar PSR B1913+16. Astrophys. J. **829**, 55 (2016)
29. Weisberg, J.M., Nice, D.J., Taylor Jr., J.H.: Timing measurements of the relativistic binary pulsar PSR B1913+16. Astrophys. J. **722**, 1030–1034 (2010)
30. Wolf, R.N., et al.: Plumbing neutron stars to new depths with the binding energy of the exotic nuclide ^{82}Zn. Phys. Rev. Lett. **110**, 041101 (2013)
31. Woosley, S.E., Heger, A., Weaver, T.A.: The evolution and explosion of massive stars. Rev. Mod. Phys. **74**, 1015–1071 (2002)

Classification of Stars

22

Contents

In the observation and study of stars it is helpful, like in other branches of science, to subdivide the objects of investigation into different types. On the Earth we obtain information about a star from its apparent magnitude and its spectrum. Simply speaking, cool stars appear reddish, and hot stars bluish. A distinction then can be made between different spectral classes of stars. As we have already seen in Sect. 1.5.3, the apparent magnitude of a star does not, of course, provide any information about the star's physical properties because the apparent magnitude depends on the distance. It therefore makes more sense to classify stars according to their absolute magnitude. However, this presupposes that the latter can be determined with sufficient accuracy, i.e., that the distance to the individual star is known.

22.1   Hertzsprung-Russell Diagram

If stars are plotted in one figure according to their absolute magnitude and spectral class, the outcome is the *Hertzsprung-Russell diagram*[1] (HRD).

---

[1] Ejnar Hertzsprung, 1873–1967, Danish astronomer. Introduced the concept of absolute magnitude and was the first to determine the distance of a Cepheid in the Small Magellanic Cloud.

© The Editor(s) (if applicable) and The Author(s), under exclusive license to
Springer-Verlag GmbH, DE, part of Springer Nature 2026
S. Boblest et al., *Special and General Relativity*,
https://doi.org/10.1007/978-3-662-71332-7_22

Fig. 22.1 In the Hertzsprung-Russell diagram, stars are sorted according to their absolute magnitude and spectral classes. This figure is an adaptation of a public domain original [5], with kind permission

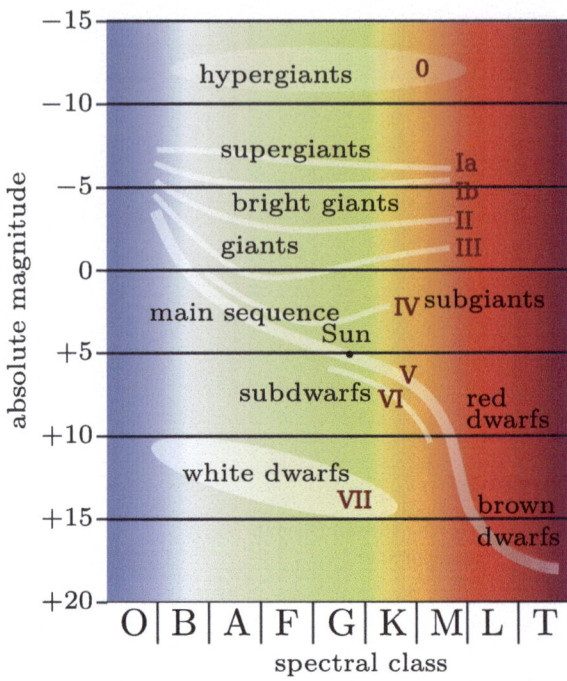

Figure 22.1 shows an example of such a diagram. The names of the spectral classes are of historical origin and have no real physical meaning. A mnemonic for these classes is

Oh, Be A Fine Girl Kiss Me!

Here the classes L and T are not included, they comprise the faint brown and red dwarfs, and were introduced only later. If a great number of known stars is plotted in this way, then one realizes in the diagram that most of the stars lie on a line from bottom right to top left, the so-called main sequence. Correspondingly, these stars are called *main sequence stars*. The smallest stars on the main sequence are red dwarfs. No hydrogen fusion can take place in objects of even lower mass. However, in this intermediate region between stars and planets lie the brown dwarfs, without hydrogen fusion, but possibly with deuterium fusion, which can already proceed at lower temperatures as we have seen in Sect. 19.4. Yet the energy production through fusion in brown dwarfs is only small, and the brightness of these objects is therefore very low. Below the main sequence lie the hot, but still very faint white dwarfs, above the main sequence lie the red giants. The Sun is a star of spectral type G and, as already mentioned, has an absolute magnitude of 4.7^m.

Besides plotting the absolute magnitude as a function of the spectral class, other diagrams can be plotted using other, but related quantities. For example, there exists a relationship between the spectral class and the effective temperature, which was introduced in (18.18). The absolute magnitude, on the other hand, is a measure

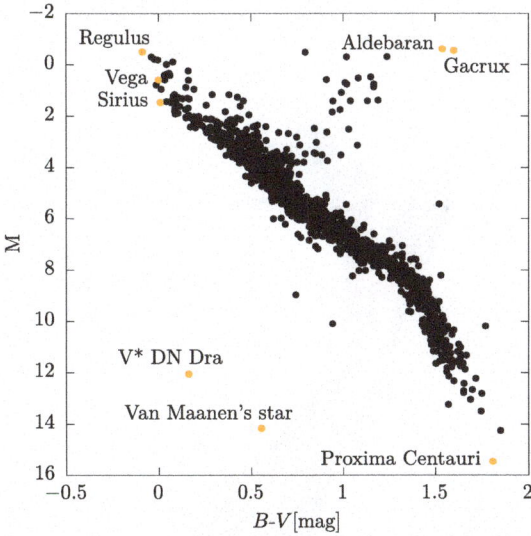

Fig. 22.2 HRD of stars with distances of up to 100 ly from the Earth. Only stars are shown for which the error in parallax is less than 5%, and the maximum error in the B-V magnitude is 0.025^m. Proxima Centauri is a red dwarf in the southern sky with the spectral class M. Sirius is the brightest star in the night sky, and part of the winter hexagon, Vega is the brightest star in the constellation Lyra. V* DN Dra and Van Maanen's star (named after its discoverer) are two white dwarfs. The data is taken from the Hipparcos catalog [3], retrieved via the *NASA HEASARC* [4]

of the luminosity of the stars, which in turn can be represented in units of the solar luminosity. Often also the B-V magnitude is used in place of the effective temperature. One compares the magnitude of the star at shorter (blue) wavelengths with that in the visual (green) range. A high luminosity in one range of wavelengths implies a small magnitude (remember: bright means small values of the magnitude). In hot stars the B magnitude is smaller than the V magnitude, correspondingly the B-V is small (negative). For cooler stars the situation is vice versa. As an example, Fig. 22.2 shows the HRD of all stars in the Hipparcos catalog [3] within a distance of 100 light years. All in all this includes about 1700 stars. The accumulation of the main sequence stars in the diagram on a curve from bottom right to top left is clearly visible.

From 1989 until 1993, the Hipparcos satellite has measured the stellar positions, parallaxes and proper motions of more than 118,000 stars with a previously unattained precision of around 0.003″ or 0.002″ per year, respectively. Its scientific successor was the Gaia satellite, launched in 2013, which could extend these observations to 1.8 billion stars. A refined HRD of four million stars within a distance of up to 5000 light years from the Earth, based on these measurements, is shown in Fig. 22.3.

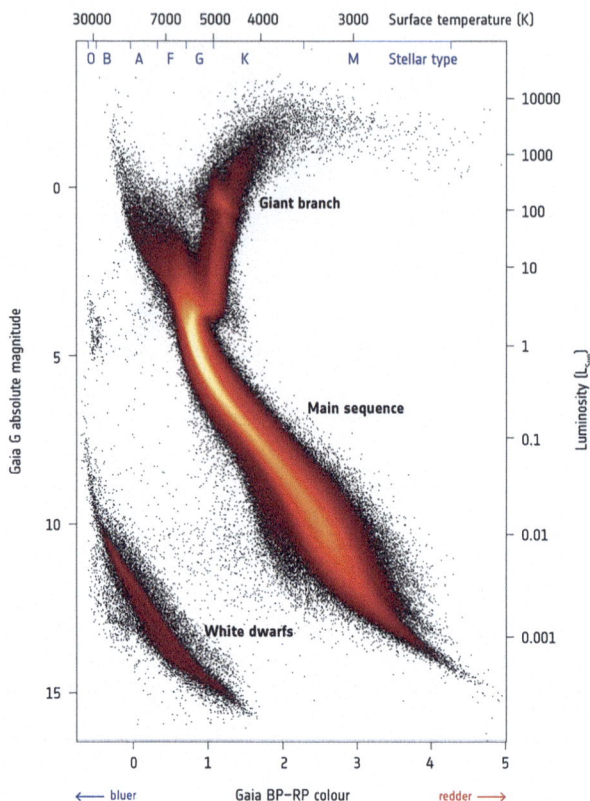

Fig. 22.3 The HRD obtained by the Gaia satellite of four million stars within a distance of 5000 light years away from the Earth. BP and RP denote the magnitudes measured by the blue and red photometer, respectively. ©ESA/Gaia/DPAC, CC BY-SA 3.0 IGO

22.2 Evolution of Stars

Since the properties of stars change over time (for example during the transition between different fusion phases), stars do not always remain in the same place in the HRD during their lifetime. Rather they pass through a certain path in the HRD, e.g., from the main sequence to the red giants, and from there to the white dwarfs. As an example, Fig. 22.4 shows the trajectories for three stars with masses $M = 0.6M_\odot$, $M = 1M_\odot$, and $M = 5M_\odot$, and a chemical composition similar to that of the Sun. The curves in Fig. 22.4 demonstrate again that low-mass stars have much longer lifetimes than mass-rich stars.

Finally, Fig. 22.5 summarizes the life cycle of stars once again. From a gas cloud, protostars form through contraction and fragmentation. The mass of the protostar determines whether it can ignite hydrogen fusion and become a main sequence star. At the end of the hydrogen burning phase the star expands into a red giant, and

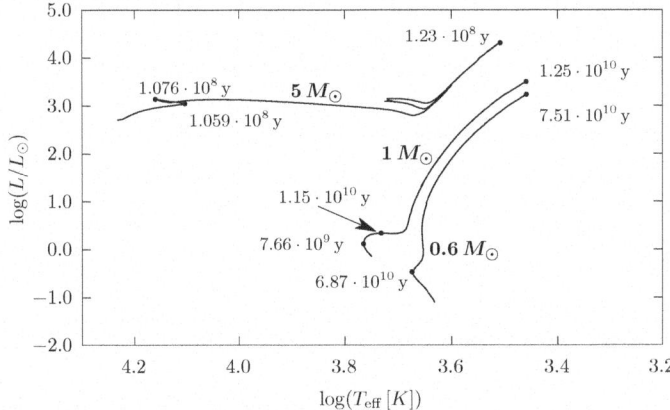

Fig. 22.4 Evolution of three stars in the HRD, with a chemical composition similar to that of the Sun, from the start of the hydrogen burning phase until the end of the helium burning phase. The stars have masses $M = 0.6\,M_\odot$ (*bottom*), $M = 1\,M_\odot$ (*middle*), and $M = 5\,M_\odot$ (*top*). For a few points on the trajectories, the time is given that has elapsed since the star arrived on the main sequence. One can recognize the extremely long lifetime of low-mass stars, and the relatively short lifetime of mass-rich stars, which we had already seen in Fig. 19.8. The data is taken from Bertelli et al. [1, 2]

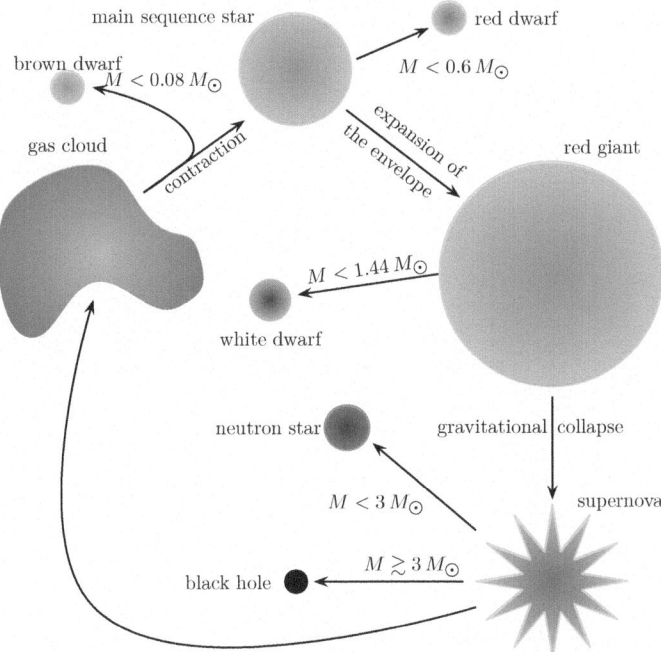

Fig. 22.5 Life cycle of stars, from their formation as protostars until their end as a white dwarf, a neutron star, or a black hole. The mass of the core remaining after the phase as a red giant decides on what end product is formed

finally ends up as a white dwarf, a neutron star, or a black hole, depending on its mass. The gas envelopes ejected during the collapse from a giant to a white dwarf or a neutron star are then available as building material for new stars.

In our discussion we have not distinguished between the properties of the different types of stars. Readers interested in further details on stellar evolution and the different types of stars are referred to Ryan and Norton [6].

References

1. Bertelli, G., Girardi, L., Marigo, P., Nasi, E.: Scaled solar tracks and isochrones in a large region of the Z-Y plane* - I. From the ZAMS to the TP-AGB end for 0.15-2.5 $M_\odot$ stars. Astron. Astrophys. **484**(3), 815–830 (2008)
2. Bertelli, G., Nasi, E., Girardi, L., Marigo, P.: Scaled solar tracks and isochrones in a large region of the Z-Y plane* - II. From 2.5 to 20M$_\odot$ stars. Astron. Astrophys. **508**(1), 355–369 (2009)
3. Homepage of the Hipparcos catalog: https://www.cosmos.esa.int/web/hipparcos/catalogues
4. Homepage of the *NASA High Energy Astrophysics Science Archive Research Center*. https://heasarc.gsfc.nasa.gov
5. Public domain figure: https://en.wikipedia.org/wiki/Hertzsprung-Russell_diagram
6. Ryan, S.G., Norton, A.J.: Stellar Evolution and Nucleosynthesis. Cambridge University Press, Cambridge (2010)

Introduction to Cosmology

23

Contents

Cosmology deals with the structure, origin and evolution of the cosmos as a whole. It attempts to understand how the universe has developed into the state we observe today, and what its future evolution will look like.

Mankind has always thought about these questions in one form or another. Even in ancient times there were models for our universe, and the philosophers in Greece, in particular, reflected on such questions. *Ptolemy*[1] developed a view of the world in which the Earth is at the center of the universe, and is orbited by the Moon, the Sun, and all the planets. His idea prevailed for centuries, even though for example *Aristarchus of Samos*[2] had already argued in favor of a heliocentric view of the world. It was only *Copernicus*[3] and Kepler, who rediscovered the heliocentric view of the world, with the Sun at the center of the planetary system.

---

[1] Claudius Ptolemy, around 100–160, Greek mathematician, astronomer, and philosopher.

[2] Aristarchus of Samos, around 310 BC–230 BC, Greek astronomer and mathematician.

[3] Nicolaus Copernicus, 1473–1543, Polish mathematician, astronomer, physician, and Catholic canon.

© The Editor(s) (if applicable) and The Author(s), under exclusive license to
Springer-Verlag GmbH, DE, part of Springer Nature 2026
S. Boblest et al., *Special and General Relativity*,
https://doi.org/10.1007/978-3-662-71332-7_23

23.1   Overview

Since the beginning of the twentieth century, ever more powerful telescopes, in combination with the newly developed theory of general relativity, made it possible to pursue cosmology as a natural science based on observations. Previously, it had been more of a philosophical discipline. Hubble's observations of galaxies had a particularly strong impact on the development of cosmology. He was able to prove that practically all galaxies are moving away from us, and the further away they are, the faster they move. We shall discuss his observations in more detail in Sect. 23.3.

General relativity was first applied in a cosmological context by Einstein in 1917 [2]. Gravity is the dominant interaction on cosmological length scales, therefore cosmology is based on general relativity, since, according to current knowledge, it is the theory that best describes gravity. In our treatment of general relativity we have seen that the field equations establish a relationship between the geometry of space and the distribution of matter, or, more generally, the distribution of energy in space. However, general relativity provides no information on the distribution of matter and energy in the universe. In Chap. 24, on the basis of symmetry considerations and physical arguments, we can restrict the set of conceivable energy distributions to a smaller subset. We shall examine this subset mathematically in more detail. We shall see that, even with this restriction, a great number of free parameters remains. It is worth noting that the corresponding considerations are already contained in Einstein's paper cited above [2].

General relativity therefore does not provide us with a single picture of our universe, but with a whole range of possible models. From Einstein's field equations we can derive simpler differential equations that describe these models using the symmetry considerations mentioned above. In the Chaps. 25 and 26 we derive these equations and discuss a variety of solutions. The final selection of the model that best describes our universe must be based on observational results. In Chap. 27 we shall discuss how observations of the universe can be compared with these models.

In the last two decades new findings have greatly expanded and changed our view of the universe. Observations of distant supernovae with the Hubble Space Telescope have led to the conclusion that, contrary to all expectations, our universe is expanding at an accelerated rate. Chapter 28 is dedicated to these observations.

The discovery of the microwave background radiation in the sixties provided strong evidence that the universe emerged from a phase of high temperature and high pressure, the Big Bang. The precise analysis of the microwave background, in particular with the satellites COBE, WMAP, and more recently *Planck*, has provided, and continues to provide, detailed information about the young universe, as we will see in Chap. 29.

Even the very first fractions of a second after the Big Bang are being considered today, and compared with observations. We will see that for the understanding of the evolution of the very young universe, but also the discussion of the possible future of our universe, further-reaching aspects have to be included, some of which go beyond general relativity. They require, in particular, taking into account effects of quantum gravity, for which, however, a comprehensive theory is not yet available. We discuss these ideas in Chap. 30.

We shall begin our discussion with a centuries-old problem that has puzzled numerous researchers over the course of time, and which will convey a first idea of the problems appearing in cosmology.

23.2 Olbers's Paradox

This paradox, or problem, named after *Heinrich Olbers*[4] can be summarized in the simple question: Why is it dark at night? Olbers thought about this question at the beginning of the nineteenth century, but before him the question was put forward by many other researchers, and it even can be dated back to Thomas Digges and Johannes Kepler in the sixteenth century.

At first glance, the answer to this question seems to be obvious: It is dark at night because the Sun is behind the Earth and its light cannot reach us. At the same time, the other stars are much too far away to give rise to any significant amount of light on the Earth compared to the light of the Sun. To realize alone that the answer to this question is far from obvious is therefore an achievement that deserves great respect in its own right.

A closer look reveals problems with this line of argument. Let us assume that our universe is infinitely large, and the density n of stars is constant, i.e., uniform on average. Let us also ignore the differences in the types of stars, and assume that all stars have the same luminosity L. The radiation flux $S(r)$ reaching us from a star at distance r is related to the distance via

$$S(r) = \frac{L}{4\pi r^2}. \tag{23.1}$$

We already know this relationship from Chap. 1, where in Eq. (1.59) we derived the solar constant from the solar luminosity in an analogous way. To calculate the total flux of radiation $\mathrm{d}S_{\mathrm{tot}}$ of all stars at a distance between r and $r + \mathrm{d}r$, we have to multiply the radiation flux of one of these stars by the number $\mathrm{d}N = n\,\mathrm{d}V$ of stars in the spherical shell between r and $r + \mathrm{d}r$. As the surface of the shell $S_{\mathrm{shell}} = 4\pi r^2$ grows quadratically with r, see Fig. 23.1, we obtain $\mathrm{d}V = 4\pi r^2\,\mathrm{d}r$, and thus

$$\mathrm{d}S_{\mathrm{tot}} \sim S(r)\mathrm{d}N(r) \sim \frac{n4\pi r^2}{4\pi r^2}\mathrm{d}r = n\,\mathrm{d}r. \tag{23.2}$$

The total flux of radiation received on the Earth from all stars in our hypothetical universe then amounts to

$$S_{\mathrm{tot}} \sim \int_0^\infty n\,\mathrm{d}r \to \infty. \tag{23.3}$$

---

[4] Heinrich Wilhelm Olbers, 1758–1840, German physician and astronomer.

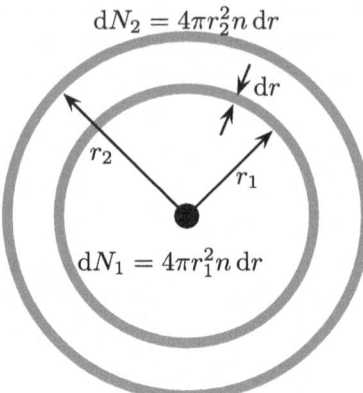

$$dN_2 = 4\pi r_2^2 n\, dr$$

$$dN_1 = 4\pi r_1^2 n\, dr$$

Fig. 23.1 If the number density of stars in the universe is constant then the number dN of stars in the volume between the spheres with radii r and $r + dr$ increases quadratically with r, whereas the radiation flux of every star decreases quadratically with r. In the figure one therefore has $dN_2/dN_1 = r_2^2/r_1^2$, but for two stars at r_2 and r_1 results a ratio $S_2/S_1 = r_1^2/r_2^2$ of the radiation flux arriving at the Earth. Thus, every spherical shell yields the same contribution to the total radiation flux S_{tot} arriving at the Earth, which therefore diverges

According to this simple reasoning, the night sky should therefore be infinitely bright. How can we explain this obvious discrepancy to the observed dark night sky?

A first step is to realize that we calculated as if stars were point-like objects. When we take into account that in reality stars are extended objects then our prediction changes considerably. With a probability that increases with the distance, very far distant stars will be hidden by stars nearer to us. No matter into what direction we look, after a certain distance our line of sight will end on the surface of some star. The stars still further away can no longer contribute to the radiation flux arriving at the Earth. An intuitive analog is the situation where an observer stands in the middle of a forest. No matter which way he looks, eventually a tree will be in the way and prevent him from looking out of the forest, at least as long as the forest is big enough. In an infinitely large forest, it does not matter how thinly distributed the trees are, or whether they are arranged in groups, just like stars are gathered in galaxies. Eventually, a tree stands in every direction. In Fig. 23.2 this situation is illustrated for a section of the sky.

In this case, however, we could still expect in every direction a night sky that is about as bright as the Sun. The distance to the stars is not crucial here, as we can easily see. Imagine yourself standing inside a sphere whose inner surface shines uniformly with a constant power. Then this sphere will appear to you with a certain brightness, regardless of its radius. The reasoning here is also that, for a sphere with say twice the size, only a quarter of the radiation power per surface area will reach us, but at the same time the surface area of the sphere is four times as large. Since the Sun only fills about $1/180,000$ of the sky, $180,000$ times its radiation power would

Fig. 23.2 In an infinitely large universe with extended stars, the surface of some star lies somewhere in any viewing direction, blocking the view to other stars. The situation is analogous to a person standing in a forest: in every direction a tree will eventually block the view to trees further away

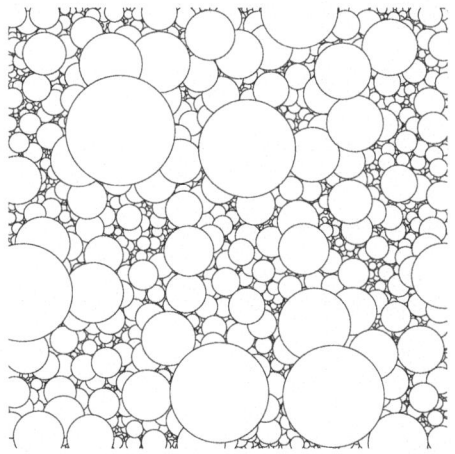

illuminate the Earth in this case, enough to vaporize all the oceans and also melt the Earth in a short time. Thus, simply taking into account the finite size of stars does not help us to resolve the paradox. Also, the assumption that, for example, interstellar dust, obscures our view of distant stars does not help us either. The radiation from the stars would heat up the dust until it reached the same temperature as the surface of the stars. Therefore, the dust would shine with the same brightness.

We therefore have to invoke additional assumptions in order to resolve the paradox. They all will ultimately have to account for this obvious fact: Since the sky is dark, there cannot stand a star in the sky in every direction that we can look.

If this is the case, then one of our preconditions must be wrong. It could be, for example, that the radiation power of stars decreases faster than the square of the distance. As we shall see, in certain cosmological models this is indeed possible due to the curvature of space. Or from a certain distance from the Earth there are no more stars. A solution to the paradox could therefore be a finite universe. Kepler in fact assumed this to be the case [6]. However, it is also possible that our universe is infinite, but has a finite age, so that the light from very distant stars did not yet have enough time to reach us. Olbers's paradox therefore already gives us an indication that our universe either is not infinite, or that our universe is not static, or not infinitely old. Figure 23.3 shows sketches that illustrate the different cases discussed above.

As already mentioned, Olbers's paradox has puzzled the minds of many researchers over the centuries, and a wide variety of solutions was proposed. The interested reader will find a brief historical and scientific overview in an article by Harrison [5]. If the reader would like to delve deeper into the subject, we recommend the book "Darkness at Night: A Riddle of the Universe" by the same author.

Fig. 23.3 Sketches for the explanation of Olbers's paradox. (**a**) In an infinitely extended universe with point-like, infinitely old stars, the sky would appear infinitely bright. (**b**) Due to the finite extent of the stars, the view of stars further back is blocked by stars further in front. (**c**) In a finite universe, in some directions there stands no star in the sky. (**d**) The same applies in a finitely old universe, where the light from far distant stars did not have enough time to reach the observer

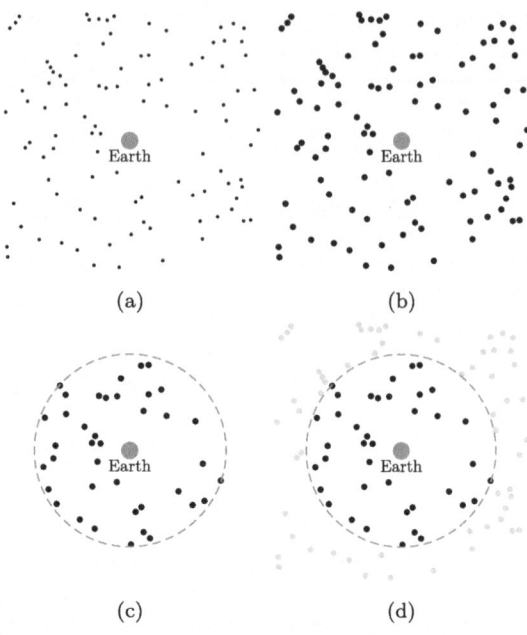

23.3   Edwin Hubble's Discoveries

Hubble's[5] discoveries in the 1920s and 1930s had a strong impact on the view of Science on the universe. He could answer one of the central questions of cosmology at the time: Does the universe only consist of the Milky Way, or are there also other galaxies? His most important discovery, however, was the observation that far distant galaxies are moving away from us. This laid the foundation for the idea of an expanding universe. It is worth noting that he was not the first to deal with this idea; as early as 1927, Lemaître had already thought about the possibility of an expanding universe.

For his work, Hubble used the *Hooker telescope* at Mt. Wilson observatory which had been commissioned in 1917. Its mirror had a diameter of 2.5 m. Until 1948 it remained the biggest telescope worldwide [7].

23.3.1 Discovery of Other Galaxies

When Hubble arrived at the Mt. Wilson Observatory in 1919, the prevailing opinion was that the universe consisted only of the Milky Way. Hubble succeeded in observing several spiral nebulae, of which it was unclear at the time whether they

---

[5] Edwin Hubble, 1889–1953, US-American astronomer.

were simply nebulae within the Milky Way, or were galaxies in their own right. In these nebulae Hubble was able to discover periodically pulsating variable stars, so-called *Cepheids*.

In 1908, *Leavitt*[6] had analyzed photo plates taken at different nights of the Large and Small Magellanic Cloud. There she could identify Cepheids, variable stars with periodically changing apparent magnitudes, and noticed a close relationship between the period P of the pulsations, ranging from 1 to 130 days, and the absolute magnitude M of the stars [3,9],

$$M = M_0 - f \log P. \tag{23.4}$$

Loosely speaking, the "heart beat" of brighter Cepheids is slower than that of less luminous Cepheids. The constant f can be easily determined from the measurement of several Cepheids. However, this relationship is only suitable for determining distances if the absolute magnitude M_0 of a reference Cepheid with period P_0 can also be determined. To do this, the distance to this star must be known. The first such measurement was carried out by Shapley [12].

The significance of Cepheids derives from the fact that they can be used as *standard candles*. This term refers to objects whose absolute magnitude can be derived from certain physical properties, without knowing their distance to us. The apparent magnitude can then be used to determine the distance to such an object. Therefore, Cepheids play an important role in astronomy even today. Another standard candle is the type Ia supernovae mentioned in Sect. 20.4, whose cosmological application we shall discuss in Chap. 28.

In the years 1922–1923, Hubble observed several such Cepheids, among others in the Andromeda Galaxy, and was able to prove that they were far too distant to be part of the Milky Way. Hubble presented his results at a conference of the American Astronomical Society in 1925. The results drastically changed the way astronomers had to think about the universe.

23.3.2 Distance-Dependent Red Shift

It was already in 1912 that *Slipher*[7] had begun to observe the shift of lines in the spectra of galaxies. He performed his first measurement, published in 1913, for the Andromeda Galaxy. Coincidentally, this is one of the few galaxies with a blueshifted spectrum, since the Andromeda Galaxy moves towards us [13]. In the years following, however, Slipher analyzed the spectra of many more galaxies, and discovered that most of them were redshifted.

In 1929 Hubble set out to find a possible relation between the redshift of spectra of galaxies and their distance from the Earth. While measuring the redshift

---

[6] Henrietta Swan Leavitt, 1868–1921, US-American astronomer.

[7] Vesto Slipher, 1875–1969, US-American astronomer.

of a galaxy is relatively easy, determining its distance is a much more difficult and error-prone task. One of the methods Hubble used to attempt this was based on the *P-M* relation just mentioned [8]. Hubble actually found a simple linear relationship between redshift and distance, namely that the greater the distance, the more redshifted were the spectra of the galaxies. For the objects that Hubble observed, the redshift was small, on the order of $z < 0.004 \ll 1$. If one assumes, as Hubble did, that the Doppler effect is the cause of the redshift, one can assign a speed to a certain redshift via the relation $z \approx \beta$, see (8.17), mentioned already in Sect. 8.1.3.

Hubble thus discovered that all galaxies, with a few exceptions that are not very far away, are moving away from us, and the further away they are, the faster they move. Between this *escape velocity* v and the distance d there holds the linear relation

$$v = H_0 d. \tag{23.5}$$

The proportionality factor H_0 is called the *Hubble constant*. From (23.5) one can immediately see that the Hubble constant must have the dimension s^{-1} (one per second). For historical and practical reasons, however, its value is usually given in the somewhat unusual, but of course equivalent, unit $km\,s^{-1}\,Mpc^{-1}$. Figure 23.4 shows the sketch from Hubble's original 1929 publication. The figure clearly exhibits the large scatter of the measured velocities compared to the expected velocity for the respective distance. This is due to the fact that the galaxies can

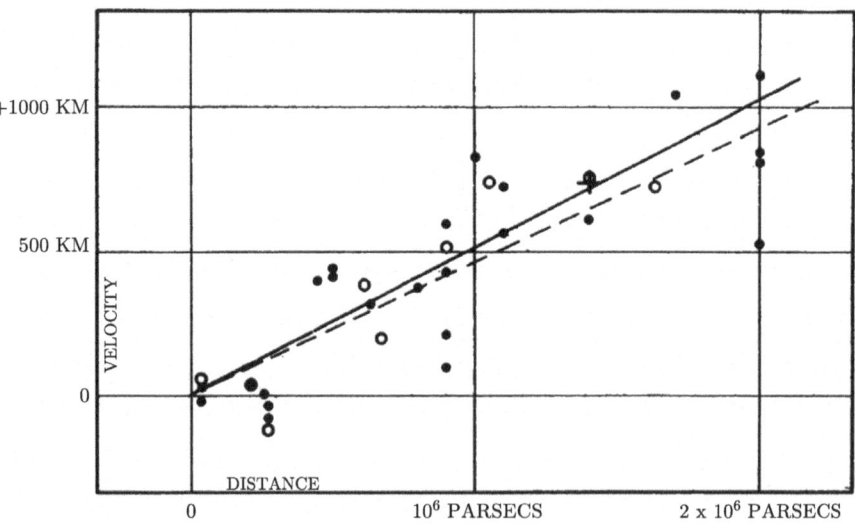

Fig. 23.4 Original presentation by Edwin Hubble on the distance-dependent escape velocity of galaxies. Later it turned out that the distances he had assumed were far too small. From Hubble [8], with kind permission

have random velocity components, such as the Andromeda Galaxy moving towards us.

These velocities, which are not caused by the expansion of the universe and differ from galaxy to galaxy, are called *peculiar velocities*. The peculiar velocity is therefore a velocity characteristic of each individual galaxy, and independent of the dynamics of the universe as a whole. Galaxies which show no peculiar motion, or those that are so far away that their peculiar velocity is negligible compared to the escape velocity, are called "comoving objects". The peculiar velocities can be used to define structures in the universe. For example, in 2014, *Tully*[8] and coworkers determined the size of the region around us in which the peculiar velocities point inwards [14]. They named this region of the universe *Laniakea*.[9] It has a diameter of about 160 Mpc and includes roughly 100,000 galaxies.

From his observational data, Hubble determined the value of H_0 to be about $500 \, \mathrm{km \, s^{-1} \, Mpc^{-1}}$. However, the distances he estimated to the observed galaxies were far too small, therefore his value was clearly too large. The most accurate value for H_0, which was determined using Hubble's method (distances measured using Cepheids) is $H_0 = 72 \pm 8 \, \mathrm{km \, s^{-1} \, Mpc^{-1}}$ [4]. The *Hubble Space Telescope* was used for these observations. Even more precise values result from observations of the cosmic microwave background radiation, namely from the 9-year data of the WMAP mission [1] and the measurements of the *Planck* satellite [10] with

$$H_{0,\mathrm{WMAP}} = 69.32 \pm 0.80 \, \mathrm{km \, s^{-1} \, Mpc^{-1}} \tag{23.6a}$$

$$H_{0,Planck} = 67.66 \pm 0.42 \, \mathrm{km \, s^{-1} \, Mpc^{-1}}, \tag{23.6b}$$

see Chap. 29.

The observations of the cosmic microwave background radiation look back into the *early* universe. Measurements of the Hubble constant in the *local* universe, on the other hand, have led to the much larger value of $74.22 \pm 1.82 \, \mathrm{km \, s^{-1} \, Mpc^{-1}}$ [11]. We shall return to this so-called Hubble controversy in Chap. 29.

For the small redshifts that Hubble observed, the interpretation as a Doppler shift was obvious. Yet we shall see that with today's instruments objects are observed for which $z \gg 1$ results. In this case, this would correspond to a velocity $v > c$. An escape velocity greater than the speed of light, however, does not contradict special relativity because we consider as the cause of the redshift the expansion of the space relative to which the respective galaxies are at rest. Furthermore, the definition of a velocity of two observers at different locations is trivially possible only in flat space, as we have already discussed in Sect. 11.2.7. We shall go into this topic in more detail later when we discuss the redshift quantitatively.

---

[8] Richard Brent Tully, 1943, US-American astronomer and cosmologist.

[9] The name "Laniakea" comes from the Hawaiian language and means "immeasurable sky", consisting of lani for "sky" and akea for "immeasurable/huge".

If all galaxies are moving away from each other at a speed proportional to their distance, obviously they must have been closer together in the past. If we assume that their speed has not changed, we can use the condition

$$vt = H_0 dt \stackrel{!}{=} d \tag{23.7}$$

to calculate how long it took them to reach their current distance from us. In this way we find

$$t = H_0^{-1}. \tag{23.8}$$

In other words: At a period of time $t = H_0^{-1}$ ago, which is called the *Hubble time*, all galaxies must have been in contact, provided that they have been moving away from each other at a constant speed ever since. Therefore, the time H_0^{-1} can be used as an estimate for the age of the universe. The metrics that we shall use to describe the universe (see Sect. 24.2), contain a time-dependent scaling factor $a(t)$ which determines the expansion of the universe. The distance between galaxies, for example, then is proportional to this scaling factor, $d \sim a(t)$, and the escape velocity is proportional to its time derivative, $v \sim a_{,t}(t) = \dot{a}(t)$. In analogy with the Hubble constant we can define the *Hubble parameter* as

$$H(t) = \frac{a_{,t}(t)}{a(t)}. \tag{23.9}$$

The Hubble constant then is equal to the value of the Hubble parameter today:

$$H_0 = H(t_0). \tag{23.10}$$

If we have obtained the explicit form of the function $a(t)$ in a certain model of the universe, then the tangent to the curve $a(t)$ at a point t_0 has the slope $a_{,t}(t_0)$. The value of H_0^{-1} would therefore be the age of the universe if it expanded linearly for all times, because then $a(t_0) = a_{,t}(t_0)t_0$. In the case of decelerated expansion, this value overestimates the age of the universe; in the case of accelerated expansion, both an overestimate and an underestimate are possible. Figure 23.5 shows these two cases.

With the value for the Hubble constant from (23.6b), we obtain an estimate for the age of our universe

$$H_0^{-1} = (1.445 \pm 0.012) \cdot 10^{10} \text{ y}. \tag{23.11}$$

To calculate this value we must convert H_0 into units of s^{-1}, or y^{-1}, using the relation $1 \text{ Mpc} = 3.2616 \cdot 10^6 \text{ ly}$. As we shall see later, (23.11) is a very good estimate.

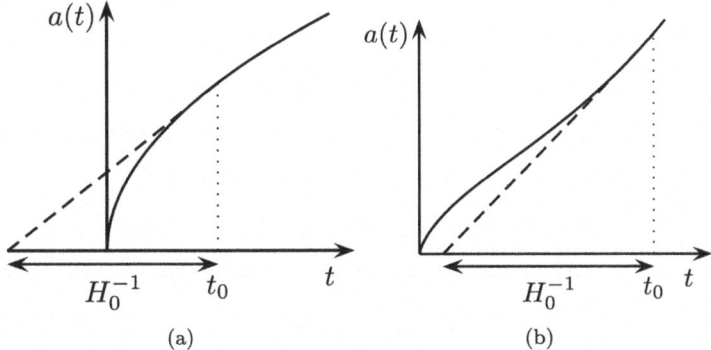

Fig. 23.5 The Hubble time $H_0^{-1} = a(t_0)/\dot{a}(t_0)$ can be used to estimate the age of the universe. (a) For decelerating expansion the estimated value of the age is too big. (b) For accelerating expansion the value can be both, overestimated and underestimated. Note that in both cases, a starts with an infinite slope at $a = 0$. See Sect. 25.4.1 for the discussion of why this is so

We can still derive another important quantity from (23.5). Obviously for

$$d = c/H_0 \tag{23.12}$$

the escape velocity at which a galaxy moves away from us becomes equal to the speed of light. This characteristic distance is called the *Hubble distance*. From (23.11) we see that the Hubble distance in our universe is

$$cH_0^{-1} = (1.445 \pm 0.012) \cdot 10^{10} \text{ ly.} \tag{23.13}$$

Since H_0^{-1} is an estimate for the age of the universe, the Hubble distance is also an estimate for the maximum distance that a ray of light has been able to travel in the universe up to now. No information can have reached us from regions further away than the Hubble distance. These regions lie beyond a horizon, similar to the regions $r < r_s$ in the Schwarzschild metric. This horizon is called the *particle horizon*.

References

1. Bennett, C.L., et al.: Nine-year Wilkinson microwave anisotropy probe (WMAP) observations: final maps and results. Astrophys. J. Suppl. Ser. **208**(2), 20 (2013)
2. Einstein, A.: Kosmologische Betrachtungen zur Allgemeinen Relativitätstheorie (Cosmological considerations on the general theory of relativity). Sitz. Preuß. Akad. Wiss. Berlin, 142–152 (1917)
3. Feast, M.W., Catchpole, R.M.: The cepheid period-luminosity zero-point from *Hipparcos* trigonometrical parallaxes. Mon. Not. R. Astron. Soc. **286**, L1–L5 (1997)
4. Freedman, W.L., et al.: Final results from the Hubble Space Telescope key project to measure the Hubble Constant. Astrophys. J. **553**, 47–72 (2001)
5. Harrison, E.R.: The dark night sky paradox. Am. J. Phys. **45**(2), 119–124 (1977)

6. Harrison, E.R.: Darkness at Night: A Riddle of the Universe. Harvard University Press, Cambridge (1987)
7. Homepage of the Mount Wilson Observatoriums: https://www.mtwilson.edu
8. Hubble, E.: A relation between distance and radial velocity among extra-galactic nebulae. Proc. Natl. Acad. Sci. **15**, 168–173 (1929)
9. Leavitt, H.S.: 1777 variables in the magellanic clouds. Ann. Harv. Coll. Observatory **60**, 87–108 (1908)
10. Planck Collaboration, Aghanim, N., et al.: *Planck* 2018 results. I. Overview and the cosmological legacy of *Planck*. Astron. Astrophys. **641**, A1 (2020)
11. Riess, A.G., et al.: Large Magellanic Cloud cepheid standards provide a 1% foundation for the determination of the Hubble constant and stronger evidence for physics beyond ΛCDM. Astrophys. J. **876**, 85 (2019)
12. Shapley, H.: Studies based on the colors and magnitudes in stellar clusters. VII. The distances, distribution in space and dimensions of 69 globular clusters. Astrophys. J. **48**, 154–181 (1918)
13. Slipher, V.M.: The radial velocity of the Andromeda Nebula. Lowell Obs. Bull. **58**, 56–57 (1913)
14. Tully, R.B., Courtois, H., Hoffman, Y., Pomarède, D.: The Laniakea supercluster of galaxies. Nature **513**(7516), 71–73 (2014)

The Cosmological Principle and Its Implication for the Metric of the Universe

24

Contents

We saw in Chap. 12 that the field equations of general relativity are non-linear, coupled differential equations, which can be solved analytically only in special cases. At the same time, we know that our universe is an extremely complicated entity. Our solar system consists of the Sun, planets and many smaller bodies. Together with numerous other stars, about whose systems we only have partial information, our Sun is part of the Milky Way. On length scales beyond the Milky Way, many galaxies form even larger structures of clusters and superclusters. It is therefore obvious that a mathematical description of the universe and its dynamics within the framework of general relativity is only possible if simplifying assumptions are made. These assumptions must of course be in accord with observations of the universe, or be based on them.

24.1   Homogeneity and Isotropy of the Universe

One answer to what such assumptions might look like was given by Einstein in 1917 with the *cosmological principle* in his article "Cosmological considerations on the general theory of relativity" [1], one of the first papers in which general relativity was applied to cosmological questions. However, in its basic statement the cosmological principle is much older, such ideas already existed in Ancient Greece, but Einstein was the first to use it in the context of general relativity. The cosmological principle can be formulated as follows:

> Space is homogeneous and isotropic, i.e., no point stands out from other points, and no direction stands out from other directions.

Formulated differently, but with the same basic message, we can also say:

> We on Earth are not in a special place in the universe. On large length scales, the universe looks the same for every observer, at any point.

This assumption is very fundamental, and will accompany us throughout our entire discussion of cosmology. It is therefore worth taking a closer look at it. First, let us realize that the cosmological principle obviously contradicts observations of the universe, both on Earth and on much larger length scales.

Our solar system is neither homogeneous nor isotropic: The Sun, by far the most massive and brightest body in the solar system, sets a preferred direction. The matter in the solar system is also in no way distributed homogeneously, with the Sun accounting for well over 99% of the total mass. In addition, the Earth is certainly a special point in our solar system; we could not survive on any other planet. But even if we look at even larger length scales, space is not homogeneous or isotropic: At night we can see the band of the Milky Way in the sky. The center of the Milky Way gives us again a preferred direction.

Obviously, we must therefore consider even larger length scales. Astronomers have discovered that galaxies form clusters and superclusters and therefore create inhomogeneous and anisotropic structures. In Sect. 23.3.2 we saw that around us, in a region of space with a diameter of about 160 Mpc, a coherent structure can be defined on the basis of the peculiar velocities. The typical length scale, from which the approximation inherent in the cosmological principle becomes valid, is thus about 10^8 or more light years. We can memorize as a benchmark $d \gtrsim 100$ Mpc, i.e., a very large length scale. Nevertheless, we will see that these scales are still significantly smaller than the extent of the visible universe.

The idea of "averaging away" entire galaxies or collections of galaxies is almost inevitably hard to accept at first. On the other hand, in this part of our discussion we are interested in the universe as a whole, and the extent of a galaxy is vanishingly small compared to the distances to the most distant objects we know today. In the following chapters the basis of our considerations is therefore always a picture of the universe in which it is homogeneously filled with matter and energy, although we still have to find out which forms of energy are conceivable.

Some readers may have noticed another point. The fact that all galaxies are moving away from us according to Hubble's law in Eq. (23.5) seems to contradict the cosmological principle, because one specific point stands out from all others, namely our position at the center of these movements, and, furthermore, a certain direction is preferred, namely the one towards us as the center of the movements. That this is not the case can be easily seen from Fig. 24.1.

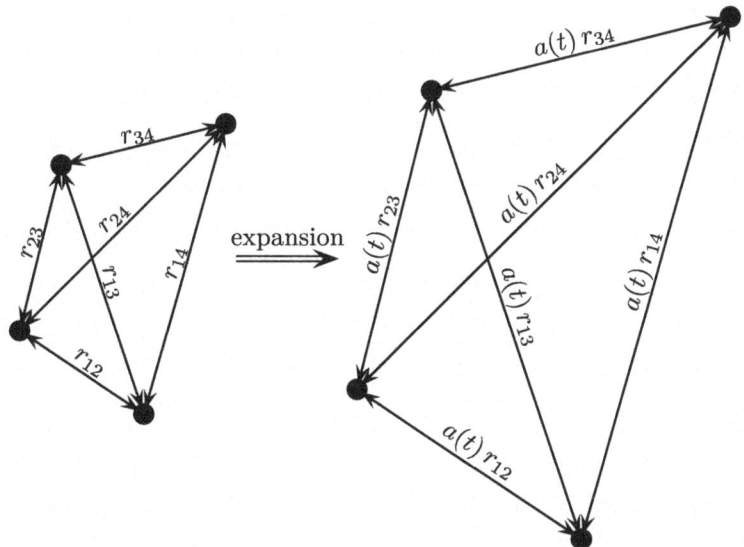

Fig. 24.1 Hubble's law does not contradict the cosmological principle. Just as we observe that all galaxies move away from us at a speed proportional to their distance from us, so all observers in other galaxies also observe the same. The expansion is homogeneous and isotropic

We consider 4 galaxies with relative distances r_{ij} in a universe in which the Hubble law applies. After a certain time t, these distances have increased by a common factor $a(t)$ to $a(t)r_{ij}$. All observers in all galaxies therefore see that all other galaxies are moving away from them at a speed proportional to their mutual distance. This simple reasoning shows that both the expansion observed by Hubble and the distance-dependent escape velocity are homogeneous and isotropic. The scaling factor $a(t)$ is the same in the entire universe and in every direction.

The situation can be illustrated by looking at an example in two dimensions. Let us imagine a swarm of ants. The ants in this swarm should crawl randomly and evenly distributed on the surface of a sphere, whose radius becomes bigger, i.e., its radius corresponds to $a = a(t)$.[1] As the radius of the sphere increases, the coordinate grid on the sphere is also "inflated", according to the temporal evolution of a. The ants move away from each other, but locally they are at rest on average as referred to the coordinate grid. Seen from each point on the sphere, all other points move "radially" away from it. The line of argument can also be applied to a sphere whose radius shrinks. In this case for each point on the sphere, all points move "radially" towards it.

---

[1] Alternatively, one could look at motion in a plane. Then $a(t)$ would be a scaling factor describing the stretching or contraction of the plane.

24.2   Friedmann-Lemaître-Robertson-Walker Metric

The most general form of a metric which is compatible with the assumptions introduced in the previous section is the *Friedmann-Lemaître-Robertson-Walker metric*[2,3,4] [3–6]. In the following we shall write FLRW metric for short. In the coordinates $(t, r, \vartheta, \varphi)$ its line element reads

$$ds^2 = -c^2dt^2 + a^2(t)\left[\frac{dr^2}{1 - qr^2} + r^2\left(d\vartheta^2 + \sin^2(\vartheta)d\varphi^2\right)\right] \text{ with } q = 0, \pm1.$$
(24.1)

The function $a(t)$ has the dimension length, while r, ϑ and φ are dimensionless. The parameter q is called the *curvature index*, it can assume the three values $0, \pm1$. We shall see that the FLRW metric describes spaces with constant curvature, where three cases must be distinguished according to the value of the curvature index: For $q = 0$ one obtains an Euclidean space, for $q = 1$ a spherical space, and for $q = -1$ a pseudospherical space. However, this should not give the impression that the constant curvature of space is a discrete quantity. The value of q only indicates the sign of a possible curvature, the actual size of the curvature is determined by the *curvature radius*. With our choice of coordinates, this corresponds to the scaling factor $a(t)$. The exact functional form of $a(t)$ cannot be derived from the requirements of homogeneity and isotropy. For this it will be necessary to solve the field equations of general relativity for physical models of the universe.

We shall now consider step by step which structure of the metric is compatible with the requirements of the cosmological principle of homogeneity and isotropy, and how the form (24.1) can be attained from these considerations.

24.2.1 Existence of a Universal Time

If no point and no direction stand out from any other in the universe, then this implies that there must exist a universal time coordinate t, which is independent of position, i.e., time passes in the same way at every single point in the universe. If this were not so, and time would pass say slower at some point, similar to the time in the Schwarzschild metric in Chap. 13, then the homogeneity of space would be broken, and the cosmological principle would be violated. Therefore, if a universal time is to exist, then space and time must decouple in the metric, i.e., the metric must have the general form

$$ds^2 = -c^2dt^2 + g_{ij}dx^idx^j = -c^2dt^2 + dl^2,$$
(24.2)

---

[2] Alexander Friedmann, 1888–1925, Russian cosmologist and mathematician.

[3] Georges Lemaître, 1894–1966, Belgian astrophysicist and Catholic priest. He was the first to propose a Big Bang Theory, which, however, initially was rejected by many scientists on the grounds that it was too closely related to the Christian doctrine of creation.

[4] Howard Percy Robertson, 1903–1961, US-American physicist and mathematician.

with the spatial distance dl^2. This does not imply, however, that the universe cannot evolve in the course of time. The only requirement is that the changes on average take place in the same way everywhere in the universe, in order to preserve homogeneity and isotropy. The more stringent requirement also of a "temporal homogeneity" would mean that the universe looks the same at all times, past, present, and future. This assumption of a *perfect cosmological principle*, so-called, is the foundation of the steady-state theory, which was popular in the 1950s and 1960s, see also Sect. 26.7.6.

24.2.2 Homogeneous and Isotropic Spaces

Now that we have established that our metric has to decompose in a g_{tt} component and g_{ij} components, we can turn to a discussion of the possible structure of the spatial part $dl^2 = g_{ij}dx^i dx^j$. The properties of the FLRW metric can be illustrated best in the two-dimensional case. Therefore, in this section we consider the metrics corresponding to the possible values $q = 0, \pm 1$ for spaces in two dimensions, using only the coordinates (r, φ). From (24.1) we obtain the line element in two dimensions

$$dl^2 = a^2(t) \left[\frac{dr^2}{1 - qr^2} + r^2 d\varphi^2 \right]. \tag{24.3}$$

It is obvious that only spaces with constant curvature fulfill the conditions of homogeneity and isotropy, in the two-dimensional case, for example, the plane with curvature zero, or the surface of a sphere with curvature $1/a$, where a denotes the radius of the sphere. On the plane and the surface of a sphere, no point is distinguished before any other points. Evidently, a position-dependent, non-constant curvature would destroy homogeneity.

Euclidean Plane
When we set $q = 0$ in (24.3), we obtain

$$dl^2 = a^2(t) \left[dr^2 + r^2 d\varphi^2 \right]. \tag{24.4}$$

One immediately recognizes the metric of the Euclidean plane in polar coordinates, with an additional scaling factor $a(t)$. The distances between points at rest on the plane (i.e., with constant r and φ) therefore can change over time.

Metric of the Surface of a Sphere
Next we consider the case $q = 1$, the metric of the surface of a sphere. We imagine the sphere embedded in a three-dimensional Euclidean space. This step should only aid to better understand this case. We derive the metric for the surface of the sphere in these coordinates, and show that we can bring it into the form of (24.3) with

$q = 1$. On no account should one conclude that the FLRW metric, too, is embedded in some higher-dimensional spacetime. We use the coordinates $\hat{x}_1$, $\hat{x}_2$ and $\hat{x}_3$ for describing the three-dimensional space, where the "hat" $\hat{}$ is to indicate that these are embedding coordinates. Let the radius of the sphere be a. The distance of two infinitesimally neighboring points on the surface of the sphere is

$$dl^2 = d\hat{x}_1^2 + d\hat{x}_2^2 + d\hat{x}_3^2, \tag{24.5}$$

with the constraint that the two points remain on the surface of the sphere, i.e.,

$$\hat{x}_1^2 + \hat{x}_2^2 + \hat{x}_3^2 = a^2 \quad \text{or} \quad \hat{x}_1 d\hat{x}_1 + \hat{x}_2 d\hat{x}_2 + \hat{x}_3 d\hat{x}_3 = 0. \tag{24.6}$$

Using this relation we can eliminate the dependent variable $\hat{x}_3$ and its differential $d\hat{x}_3$ from the line element, and obtain

$$\begin{aligned}
dl^2 &= d\hat{x}_1^2 + d\hat{x}_2^2 + \frac{(\hat{x}_1 d\hat{x}_1 + \hat{x}_2 d\hat{x}_2)^2}{a^2 - \hat{x}_1^2 - \hat{x}_2^2} \\
&= \frac{(a^2 - \hat{x}_2^2)d\hat{x}_1^2 + 2\hat{x}_1\hat{x}_2 d\hat{x}_1 d\hat{x}_2 + (a^2 - \hat{x}_1^2)d\hat{x}_2^2}{a^2 - \hat{x}_1^2 - \hat{x}_2^2}.
\end{aligned} \tag{24.7}$$

We now transform to spherical polar coordinates,

$$\hat{x}_1 = a\sin(\vartheta)\cos(\varphi), \quad \hat{x}_2 = a\sin(\vartheta)\sin(\varphi), \quad [\hat{x}_3 = a\cos(\vartheta)]. \tag{24.8}$$

The transformation equation for $\hat{x}_3$ is only given for the sake of completeness. An elementary calculation then yields

$$dl^2 = a^2\left(d\vartheta^2 + \sin^2(\vartheta)\,d\varphi^2\right). \tag{24.9}$$

This is the metric on the surface of a sphere with radius a and curvature $1/a$.

We carry out another transformation by the assignments

$$r = \sin(\vartheta), \quad dr = \cos(\vartheta)\,d\vartheta, \quad d\vartheta^2 = \frac{dr^2}{1 - r^2}, \tag{24.10}$$

and obtain

$$dl^2 = a^2(t)\left(\frac{dr^2}{1 - r^2} + r^2\,d\varphi^2\right), \tag{24.11}$$

with the constraint $r \in [0,1]$.

In this form we recognize the FLRW metric in the two-dimensional case for $q = 1$. Now the designation "spherical space" becomes evident by the link to the surface of a sphere. The coordinate lines correspond to those in polar coordinates,

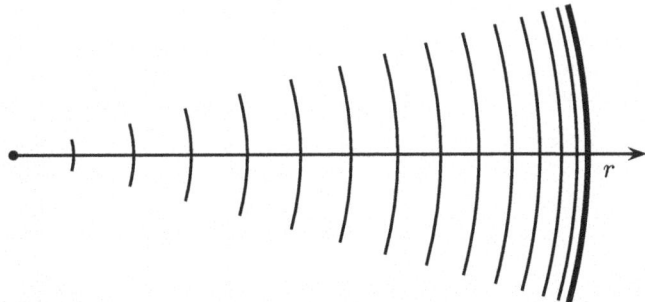

Fig. 24.2 Equidistant spatial distance lines in the metric of the surface of the sphere. Referred to the metric (24.11), the circular arcs all have the same spatial distance. However, one immediately recognizes that the radial coordinates are not equidistant. The circular arcs shown cover the range $r \in [0, 1]$

only the r length scale is extended by the factor $(1 - r^2)^{-1/2}$. The spatial distance between a point at $r = 0$ and another point at $r = r_1$ with arbitrary φ is therefore

$$l = \int_0^{r_1} \frac{\mathrm{d}r}{\sqrt{1 - r^2}} = \arcsin(r_1). \tag{24.12}$$

We see that the radial coordinate, like in the case of the Schwarzschild metric, does not have the meaning of an actual radial distance. This is particularly easy to see because we have transformed from the angular coordinate ϑ to the coordinate r. Figure 24.2 shows circular arcs for which the spatial interval changes by a constant value. The radial coordinate r, by contrast, does not change equidistantly. The maximum allowed value for r in the figure of course also is 1.

Metric of the Pseudosphere

In the case $q = -1$, unfortunately, we lose much of an intuitive view. If we could, we would like to embed a two-dimensional space with constant negative curvature in the three-dimensional Euclidean space $\mathbb{R}^3$. However, according to a theorem by *Hilbert*[5] this is impossible [2]. To bypass this problem we assume a metric different from the purely Euclidean one. We begin with a constraint on our embedding coordinates analogous to Eq. (24.6), namely

$$\hat{x}_1^2 + \hat{x}_2^2 - \hat{x}_3^2 = -a^2 \quad \text{or} \quad \hat{x}_1 \mathrm{d}\hat{x}_1 + \hat{x}_2 \mathrm{d}\hat{x}_2 - \hat{x}_3 \mathrm{d}\hat{x}_3 = 0. \tag{24.13}$$

This boundary condition characterizes a rotational hyperboloid of two sheets, also called the *pseudosphere*. Figure 24.3 shows such a rotational hyperboloid. Our

---

[5] David Hilbert, 1862–1943, German mathematician.

Fig. 24.3 Rotational
hyperboloid of two sheets
$\hat{x}_1^2 + \hat{x}_2^2 - \hat{x}_3^2 = -a^2$ together
with its embedding cones
$\hat{x}_3 = \pm\sqrt{\hat{x}_1^2 + \hat{x}_2^2}$. The two
poles N and S are at $\hat{x}_3 = \pm a$

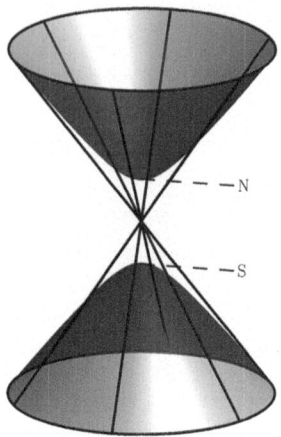

metric is no longer Euclidean but *pseudo-Euclidean*, and has the form

$$dl^2 = d\hat{x}_1^2 + d\hat{x}_2^2 - d\hat{x}_3^2. \tag{24.14}$$

Of course in Euclidean space the distance on the rotational hyperboloid still is $dl^2 = d\hat{x}_1^2 + d\hat{x}_2^2 + d\hat{x}_3^2$. It is only our special choice of the metric that leads to a constant negative curvature. In analogy with the coordinates on the surface of a sphere we introduce new coordinates via

$$\hat{x}_1 = a \sinh(\vartheta) \cos(\varphi), \quad \hat{x}_2 = a \sinh(\vartheta) \sin(\varphi), \quad [\hat{x}_3 = a \cosh(\vartheta)]. \tag{24.15}$$

We can again eliminate the $\hat{x}_3$ coordinate using the constraint (24.13). Like in the previous section we then arrive at the squared spatial distance

$$dl^2 = a^2 \left(d\vartheta^2 + \sinh^2(\vartheta) d\varphi^2 \right). \tag{24.16}$$

Instead of (24.10) we now use the transformation

$$r = \sinh(\vartheta), \quad dr = \cosh(\vartheta) \, d\vartheta, \quad d\vartheta^2 = \frac{dr^2}{1 + r^2} \tag{24.17}$$

and find

$$dl^2 = a^2(t) \left(\frac{dr^2}{1 + r^2} + r^2 d\varphi^2 \right). \tag{24.18}$$

Again we recognize the two-dimensional form of the FLRW metric, here for $q = -1$. The coordinates also correspond to polar coordinates, except for a compression of the r length scale by the factor $\left(1 + r^2\right)^{-1/2}$. In this case, the spatial

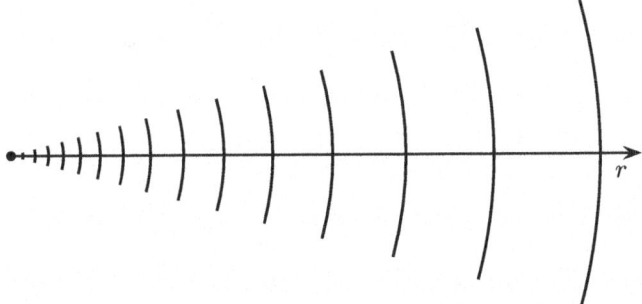

Fig. 24.4 Equidistant spatial distance lines in the metric of the pseudosphere. Referred to the metric (24.18), the circular arcs all have the same spatial distance. In this case the radial coordinate is not bounded

distance between the point at $r = 0$ and a second point $r = r_1$ with arbitrary φ is

$$l = \int_0^{r_1} \frac{dr}{\sqrt{1 + r^2}} = \text{arsinh}(r_1), \tag{24.19}$$

see also Fig. 24.4. It can be proved quite generally that the solutions with constant positive and negative curvature, which we have found in an intuitive way, follow from the requirement of homogeneity and isotropy.

24.2.3 FLRW Spacetime

To conclude we consider $ds^2 = -c^2\,dt^2 + dl^2$ in the four-dimensional space-time. The results of the previous sections can easily be transferred to the three-dimensional case with the spatial coordinates (r, ϑ, φ) or $(\chi, \vartheta, \varphi)$.

All three cases can then be combined in the following representations of the FLRW metric. On the one hand in the form

$$ds^2 = -c^2 dt^2 + a^2(t)\left[\frac{dr^2}{1 - qr^2} + r^2\left(d\vartheta^2 + \sin^2(\vartheta)d\varphi^2\right)\right] \text{ with } q = 0, \pm 1. \tag{24.20a}$$

The coordinate transformation $r = \chi,\ \sin(\chi),\ \sinh(\chi)$ leads to the second form

$$ds^2 = -c^2 dt^2 + a^2(t)\left[d\chi^2 + \left\{\begin{array}{c} \sin^2(\chi) \\ \chi^2 \\ \sinh^2(\chi) \end{array}\right\}\left(d\vartheta^2 + \sin^2(\vartheta)d\varphi^2\right)\right] \tag{24.20b}$$

with $q = 1$ (top), $q = 0$ (middle) and $q = -1$ (bottom). Finally, we can carry out a transformation to *conformal Euclidean coordinates* (see Exercise 24.1), and obtain the third form

$$ds^2 = -c^2 dt^2 + \frac{a^2(t)}{\left(1 + q\frac{\bar{r}^2}{4}\right)^2} \left(dx^2 + dy^2 + dz^2\right) \tag{24.20c}$$

with $\bar{r} = 2\tan(\chi/2)$ for $q = 1$ and $\bar{r} = \tanh(\chi/2)$ for $q = -1$.

The metrics in (24.20a) and (24.20b) are the starting point of our discussion of cosmology in the following chapters.

The essential task of cosmology now is to determine the value of the curvature q and the functional form of the scaling factor $a(t)$. As already mentioned, this can be done only on the basis of astronomical observations, but also requires theoretical input and further model assumptions in order to bring the observational results in agreement with the quantities which are searched for in the mathematical equations.

24.3   Exercises

24.1. FLRW metric in conformal Euclidean coordinates In this exercise, we show that the FLRW metric can be brought into the form (24.20c). For the sake of simplicity we restrict ourselves again to the two-dimensional case.

(a) For positive curvature use the transformation

$$\bar{r} = 2\tan(\vartheta/2) \tag{24.21}$$

and express the term $\sin^2(\vartheta)$ in (24.9) by $\bar{r}$ using the trigonometric relations

$$\sin^2(\vartheta) = 4\sin^2(\vartheta/2)\cos^2(\vartheta/2),$$

$$\sin^2(\vartheta/2) = \frac{\tan^2(\vartheta/2)}{1 + \tan^2(\vartheta/2)}, \tag{24.22}$$

$$\cos^2(\vartheta/2) = \frac{1}{1 + \tan^2(\vartheta/2)}.$$

Then set

$$\bar{x} = \bar{r}\cos(\varphi), \quad \bar{y} = \bar{r}\sin(\varphi), \quad \bar{r}^2 = \bar{x}^2 + \bar{y}^2, \tag{24.23}$$

to arrive at

$$dl^2 = \frac{a^2(t)}{\left(1 + \frac{\bar{r}^2}{4}\right)^2} \left(d\bar{x}^2 + d\bar{y}^2\right). \tag{24.24}$$

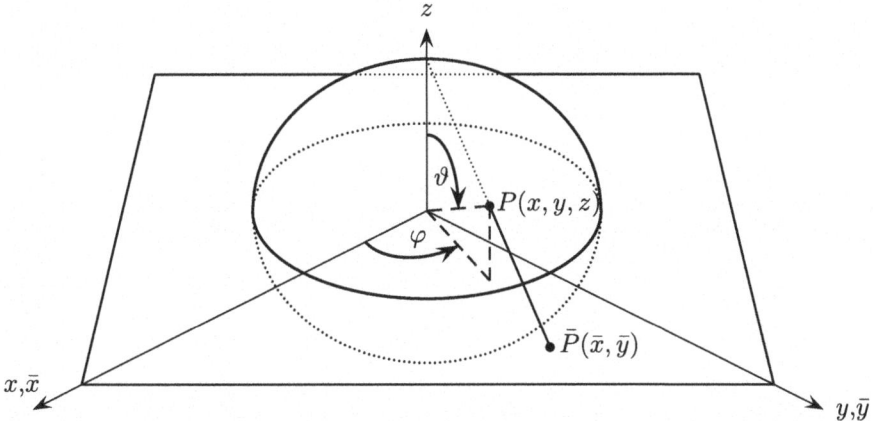

Fig. 24.5 Stereographic projection of the point P on the surface of a sphere onto the point $\bar{P}$ in the $\bar{x}\bar{y}$ plane

These conformal Euclidean coordinates correspond to a stereographic projection of the surface of the sphere onto the plane, see Fig. 24.5.

(b) For negative curvature proceed in an analogous way with

$$\bar{r} = 2\tanh(\vartheta/2) \tag{24.25}$$

to arrive at

$$dl^2 = \frac{a^2(t)}{\left(1 - \frac{\bar{r}^2}{4}\right)^2} \left(dx^2 + dy^2\right). \tag{24.26}$$

References

1. Einstein, A.: Kosmologische Betrachtungen zur Allgemeinen Relativitätstheorie (Cosmological considerations on the general theory of relativity). Sitz. Preuß. Akad. Wiss. Berlin, 142–152 (1917)
2. Hilbert, D.: Über Flächen von constanter Gaußscher Krümmung (On surfaces of constant Gaussian curvature). Trans. Am. Math. Soc. **2**, 87–99 (1901)
3. Robertson, H.P.: Kinematics and world-structure. Astrophys. J. **82**, 284–301 (1935)
4. Robertson, H.P.: Kinematics and world-structure 2. Astrophys. J. **83**, 187–201 (1936)
5. Robertson, H.P.: Kinematics and world-structure 3. Astrophys. J. **83**(4), 257–271 (1936)
6. Walker, A.G.: On Milne's theory of world-structure. Proc. Lond. Math. Soc. **42**, 90–127 (1937)

Field Equations for the FLRW Metric

25

Contents

In this chapter, we will explicitly set up the field equations from Chap. 12 for the FLRW metric, and solve them in a few cases. We start from the field equations $R_{\mu\nu} = \kappa T_{\mu\nu}^*$ in (12.40b), with $\Lambda = 0$ and $\kappa = 8\pi G/c^4$ from (12.37). We shall have to calculate the Ricci tensor $R_{\mu\nu}$ belonging to the FLRW metric, and find a physically reasonable form of the energy-momentum tensor $T_{\mu\nu}^* = T_{\mu\nu} - g_{\mu\nu} T/2$.

With these quantities, we can then write down the field equations, which will assume a relatively simple form due to the high symmetry of the FLRW metric.

25.1 Ricci Tensor of the FLRW Metric

To calculate the Ricci tensor of the FLRW metric, we recall the definition of the Christoffel symbols of the second kind

$$\Gamma^{\alpha}{}_{\mu\nu} = \frac{1}{2} g^{\alpha\sigma} \left(g_{\sigma\mu,\nu} + g_{\sigma\nu,\mu} - g_{\mu\nu,\sigma} \right) \tag{11.63}$$

and of the Ricci tensor

$$R_{\mu\nu} = \Gamma^{\alpha}{}_{\mu\nu,\alpha} - \Gamma^{\alpha}{}_{\mu\alpha,\nu} + \Gamma^{\alpha}{}_{\sigma\alpha}\Gamma^{\sigma}{}_{\mu\nu} - \Gamma^{\alpha}{}_{\sigma\nu}\Gamma^{\sigma}{}_{\mu\alpha}. \tag{11.99}$$

The non-vanishing metric components of the metric (24.20a) read

$$g_{tt} = -c^2, \quad g_{rr} = \frac{a(t)^2}{1 - qr^2}, \quad g_{\vartheta\vartheta} = a(t)^2 r^2, \quad g_{\varphi\varphi} = a(t)^2 r^2 \sin^2(\vartheta). \tag{25.1}$$

Due to the diagonal form of the metric, the corresponding components of the inverse simply become

$$g^{tt} = -\frac{1}{c^2}, \quad g^{rr} = \frac{1 - qr^2}{a(t)^2}, \quad g^{\vartheta\vartheta} = \frac{1}{a(t)^2 r^2}, \quad g^{\varphi\varphi} = \frac{1}{a(t)^2 r^2 \sin^2(\vartheta)}. \tag{25.2}$$

Inserting these quantities into the definition of the Christoffel symbols leads to the following non-vanishing expressions

$$\Gamma^{t}{}_{rr} = \frac{aa_{,t}}{c^2(1 - qr^2)}, \quad \Gamma^{t}{}_{\vartheta\vartheta} = \frac{r^2 aa_{,t}}{c^2}, \quad \Gamma^{t}{}_{\varphi\varphi} = \frac{aa_{,t} r^2 \sin^2(\vartheta)}{c^2}, \tag{25.3a}$$

$$\Gamma^{r}{}_{tr} = \frac{a_{,t}}{a}, \qquad \qquad \Gamma^{r}{}_{rr} = \frac{qr}{1 - qr^2}, \quad \Gamma^{r}{}_{\vartheta\vartheta} = -(1 - qr^2)r,$$

$$\Gamma^{r}{}_{\varphi\varphi} = -(1 - qr^2)r \sin^2(\vartheta), \tag{25.3b}$$

$$\Gamma^{\vartheta}{}_{t\vartheta} = \frac{a_{,t}}{a}, \qquad \qquad \Gamma^{\vartheta}{}_{r\vartheta} = \frac{1}{r}, \qquad \Gamma^{\vartheta}{}_{\varphi\varphi} = -\sin(\vartheta)\cos(\vartheta), \tag{25.3c}$$

$$\Gamma^{\varphi}{}_{t\varphi} = \frac{a_{,t}}{a}, \qquad \qquad \Gamma^{\varphi}{}_{r\varphi} = \frac{1}{r}, \qquad \Gamma^{\varphi}{}_{\vartheta\varphi} = \cot(\vartheta). \tag{25.3d}$$

The Ricci tensor therefore becomes

$$R_{tt} = -3\frac{a_{,tt}}{a}, \quad R_{rr} = \frac{\mathcal{A}}{c^2(1 - qr^2)}, \quad R_{\vartheta\vartheta} = \frac{r^2}{c^2}\mathcal{A}, \quad R_{\varphi\varphi} = \frac{r^2}{c^2}\sin^2(\vartheta)\mathcal{A}, \tag{25.4}$$

with

$$\mathcal{A} = aa_{,tt} + 2a_{,t}^2 + 2qc^2. \tag{25.5}$$

The simple structure of the Ricci tensor is of course not accidental. It reflects the assumption of a homogeneous and isotropic space, which we had imposed as a condition on the FLRW metric. Even though we do not need it for the following calculations, we evaluate the Ricci scalar. The result is

$$R = 6\frac{aa_{,tt} + a_{,t}^2 + qc^2}{a^2c^2}.$$ (25.6)

Through the scaling factor, it only depends on the t coordinate, and is independent of r, ϑ and φ, i.e., of position, as was to be expected for a metric with constant curvature.

25.2 Energy-Momentum Tensor of Matter

Our second task is to find an appropriate form of the energy-momentum tensor. Here, too, we must of course comply with the cosmological principle, and average over such large length scales that we can assume the universe to be homogeneous and isotropic.

As a matter of principle, all forms of energy contained in the universe are relevant as contributions to the energy-momentum tensor. We shall see later that various forms of energy are relevant in our universe. As an introduction, however, we will restrict ourselves to matter, and here only to its rest energy; we shall neglect the kinetic energy, i.e., we will only consider non-relativistic matter with

$$E_{\text{kin}} \ll mc^2.$$ (25.7)

This restriction is understandable from the current state of the universe; matter appears to be the dominant form of energy. As an additional contribution, we could consider so far only electromagnetic radiation, i.e., essentially the light of stars. However, it is evident that the energy emitted by stars in the form of electromagnetic radiation is much smaller than the energy density corresponding to the mass of these stars. In addition, matter has some properties that make its treatment very easy to handle. In particular, we will assume that the matter particles do not generate pressure. When we later discuss forms of energy with pressure, we must then extend the energy-momentum tensor by additional contributions. We can easily estimate how good the approximation $p = 0$ is. To do this, we go back to our considerations in Sect. 18.3, and again look at the ideal gas equation

$$pV = Nk_{\text{B}}T$$ (25.8)

in the form

$$p = \frac{\rho_{\text{m}}}{\langle m \rangle}k_{\text{B}}T,$$ (25.9)

with the average particle mass $\langle m \rangle$. For a non-relativistic gas the Maxwell distribution from Sect. 19.2.1 applies, and therefore we have the relation

$$3k_{\text{B}}T = \langle m \rangle \langle v^2 \rangle$$ (25.10)

for the average velocity squared $\langle v^2 \rangle$. We insert this relation in (25.9) and find

$$p = \frac{\langle v^2 \rangle}{3c^2} \varepsilon_{\mathrm{m}}. \tag{25.11}$$

Thus, pressure and the energy density

$$\varepsilon_{\mathrm{m}} = c^2 \rho_{\mathrm{m}}, \tag{25.12}$$

depend on each other linearly with the proportionality constant $\langle v^2 \rangle/(3c^2)$.

As an estimate, we can consider the situation in the interior of a main sequence star, which we discussed in Sects. 18.2 and 18.3. With $T \approx 1.5 \cdot 10^7$ K we had found $k_{\mathrm{B}} T \approx 1.3$ keV. At this temperature, for protons, e.g., with a rest energy of about 1 GeV, this yields

$$\frac{k_{\mathrm{B}} T}{m_{\mathrm{p}} c^2} \approx 1.4 \cdot 10^{-6} \ll 1. \tag{25.13}$$

The error we make by the approximation $p = 0$ is therefore on the order of $1{:}10^{-6}$ and completely insignificant, since all relevant observed quantities are known only with relative accuracies of a few per cent.

We now consider our universe to be uniformly filled with matter. Local density fluctuations are neglected. In addition, we will assume that there is no interaction between the particles constituting the matter. Our model then corresponds to a universe homogeneously filled with interaction-free dust.

In the co-moving coordinate system we use, which is given by the coordinates of the FLRW metric, the non-relativistic matter is always assumed to be locally at rest. This means, we neglect a possible disordered peculiar motion of the dust particles, which anyway is much slower than the speed of light, and is also supposed to be distributed in a disordered manner. Obviously such a choice is reasonable for an isotropic model. With another choice, the velocity direction of the matter would cause a non-equivalence of the different directions in space.

Then the energy tensor is quite similar to the one we considered in Sect. 21.4.1 for matter in the interior of a neutron star, namely

$$T^{\mu}{}_{\nu} = \mathrm{diag}(-\varepsilon_{\mathrm{m}}, 0, 0, 0), \tag{25.14a}$$

and

$$T^{\mu\nu} = \mathrm{diag}(\varepsilon_{\mathrm{m}}/c^2, 0, 0, 0), \tag{25.14b}$$

plus

$$T_{\mu\nu} = \mathrm{diag}(\varepsilon_{\mathrm{m}} c^2, 0, 0, 0). \tag{25.14c}$$

We have used the FLRW metrics (25.1) and (25.2) for lowering and raising indices. Furthermore, by complete contraction of the indices of $T_{\mu\nu}$ we obtain

$$T = g^{\mu\nu}T_{\mu\nu} = g^{tt}T_{tt} = -\varepsilon_{\mathrm{m}} \tag{25.15}$$

and thus with $T_{\mu\nu}^* = T_{\mu\nu} - (1/2)g_{\mu\nu}T$ the expression

$$T_{\mu\nu}^* = \frac{\varepsilon_{\mathrm{m}}}{2}\mathrm{diag}\left(c^2, \frac{a^2}{1-qr^2}, a^2r^2, a^2r^2\sin^2(\vartheta)\right). \tag{25.16}$$

25.3 Friedmann Equation for a Matter-Dominated Universe

Now that we have calculated the Ricci tensor of the FLRW metric and the energy-momentum tensor for our model matter universe, we can turn to the discussion of Einstein's field equations. In doing so, we still shall ignore the cosmological constant Λ from Chap. 12. Its influence on the type of possible solutions will be discussed in Chap. 26. The field equations read, with $R_{\mu\nu}$ from (25.4) and $T_{\mu\nu}^*$ from (25.16),

$$-3\frac{a_{,tt}}{a} = \frac{\kappa}{2}\varepsilon_m c^2, \tag{25.17a}$$

$$\frac{1}{c^2(1-qr^2)}\mathcal{A} = \frac{\kappa}{2}\varepsilon_m\frac{a^2}{1-qr^2}, \tag{25.17b}$$

$$\frac{r^2}{c^2}\mathcal{A} = \frac{\kappa}{2}\varepsilon_m a^2 r^2, \tag{25.17c}$$

$$\frac{r^2}{c^2}\sin^2(\vartheta)\mathcal{A} = \frac{\kappa}{2}\varepsilon_m a^2 r^2\sin^2(\vartheta), \tag{25.17d}$$

with $\mathcal{A} = aa_{,tt} + 2a_{,t}^2 + 2qc^2$ from (25.5). For the non-diagonal elements $\mu \neq \nu$ these equations are fulfilled in a trivial way. For $\mu = \nu$ one obtains an identical equation for all three cases $\mu = \{1, 2, 3\}$, viz. $\mathcal{A} = \kappa\varepsilon_m c^2 a^2/2$, or

$$aa_{,tt} + 2a_{,t}^2 + 2qc^2 = \frac{\kappa}{2}\varepsilon_m c^2 a^2. \tag{25.18}$$

It is no coincidence that all three spatial components yield the same equation, rather this is a consequence of the fact that the FLRW metric describes a homogeneous and isotropic space, and our matter model also is homogeneous and isotropic. For $\mu = \nu = 0$ we obtain

$$a_{,tt} = -\frac{\kappa}{6}\varepsilon_m c^2 a. \tag{25.19}$$

We insert (25.19) in (25.18) to eliminate $a_{,tt}$ from this equation. This leads to

$$a_{,t}^2 + qc^2 - \frac{\kappa}{3}\varepsilon_m c^2 a^2 = 0. \tag{25.20}$$

Furthermore, the energy-momentum tensor fulfills the continuity equation $T^{\mu\nu}{}_{;\nu} = 0$ from (12.18). This condition reads explicitly

$$T^{\mu\nu}{}_{,\nu} + T^{\omega\nu}\Gamma^\mu{}_{\omega\nu} + T^{\mu\omega}\Gamma^\nu{}_{\omega\nu} = 0. \tag{25.21}$$

In our case, where only T^{tt} is different from zero, $T^{tt} \neq 0$, we obtain a non-trivial equation only for $\mu = t$:

$$T^{tt}{}_{,t} + T^{tt}\Gamma^t{}_{tt} + T^{tt}\Gamma^\nu{}_{t\nu} = \varepsilon_{m,t} + 0 + \varepsilon_m \left(\Gamma^r{}_{tr} + \Gamma^\vartheta{}_{t\vartheta} + \Gamma^\varphi{}_{t\varphi}\right)$$
$$= \frac{1}{c^2}\left(\varepsilon_{m,t} + 3\varepsilon_m \frac{a_{,t}}{a}\right) = 0, \tag{25.22}$$

with $\Gamma^r{}_{tr} = \Gamma^\vartheta{}_{t\vartheta} = \Gamma^\varphi{}_{t\varphi} = a_{,t}/a$ from (25.3). Here the time derivative of the rest energy density ε_m appears, for which we can certainly assume that it is time-dependent when the scaling factor a changes over time. We can put the relationship (25.22) into a form that is very easy to interpret. For this purpose we multiply the last line by $a^3 c^2$. This leads to $\varepsilon_{m,t}a^3 + 3\varepsilon_m a^2 a_{,t} = 0$. In this expression we recognize the time derivative of $\varepsilon_m a^3$. We finally obtain

$$\left(\varepsilon_m a^3\right)_{,t} = 3\varepsilon_m a^2 a_{,t} + \varepsilon_{m,t}a^3 = 0 \quad \text{and thus} \quad \varepsilon_m a^3 = \text{const} = \varepsilon_{m0}a_0^3. \tag{25.23}$$

Here the index 0 designates the values of the quantities today. In the literature it is common to set today's scaling factor of the universe equal to one, therefore in what follows we also adopt this convention $a_0 = a(t_0) = 1$. In doing so we have to be careful with the physical units. If a_0 no longer appears explicitly in the equation we lose the unit of length. This will not cause any problems in our discussion, but readers who want to explicitly reproduce individual calculations should keep this in mind.

The expression $\varepsilon_m a^3/c^2$ has the dimension of a mass. When we consider a piece of volume subject to expansion then the total mass contained in this volume remains constant, according to (25.23). Thus, the conservation of mass is a consequence of the continuity equation of the energy-momentum tensor. From this we can infer how the mass density behaves as a function of the scaling factor. From (25.23) we immediately obtain

$$\varepsilon_m(t) = \varepsilon_{m0}\frac{a_0^3}{a(t)^3}. \tag{25.24}$$

In fact, this relation is also contained in (25.20). This does not come as a big surprise because we had already taken into account the requirement of vanishing divergence when we were setting up the field equations in Chap. 12. To see this explicitly we take the time derivative of (25.20). The result is

$$2a_{,t}a_{,tt} - \frac{\kappa c^2}{3}\left(2\varepsilon_{\mathrm{m}}aa_{,t} + \varepsilon_{\mathrm{m},t}a^2\right) = 0. \tag{25.25}$$

Again we insert $a_{,tt} = -\kappa\varepsilon_{\mathrm{m}}c^2a/6$ from (25.19) and multiply by $-3/\kappa$. This leads to

$$\frac{c^2}{a}\left(3\varepsilon_{\mathrm{m}}a^2a_{,t} + \varepsilon_{\mathrm{m},t}a^3\right) = 0, \tag{25.26}$$

where we also factored out the expression c^2/a. In the parentheses we recognize again the time derivative of $\varepsilon_{\mathrm{m}}a^3$. We insert the result (25.24) in (25.20), and then obtain the *Friedmann equation* [2, 3] (with $a_0 = 1$),

$$a_{,t}^2 + qc^2 - \frac{\kappa}{3}\frac{\varepsilon_{\mathrm{m}0}c^2}{a} = 0. \tag{25.27}$$

25.4   Qualitative Solution of the Friedmann Equation

Our ultimate goal is of course to solve the Friedmann equation (25.27). First, however, we will take a closer look at the individual terms in this equation, and interpret their meaning using a classical analogy.

25.4.1 "Energy Balance" in the Friedmann Equation

When we bring the curvature index q on the right-hand side of (25.27) we obtain

$$a_{,t}^2 - \frac{\kappa}{3}\frac{\varepsilon_{\mathrm{m}0}c^2}{a} = -qc^2. \tag{25.28}$$

We can interpret this equation as an analog to the equation $E_{\mathrm{kin}} + E_{\mathrm{pot}} = E_{\mathrm{tot}}$ of a particle with velocity $a_{,t}$ in a $(-1/a)$ potential. Understood in this way, the value of the curvature index q corresponds to the total energy of the particle. Depending on the value of q we obtain three types of solutions.

Negative Energy

For $q = 1$ the total energy is negative. This corresponds to a bound motion of the particle. In our case this means that the universe expands up to a maximum size

$$a_{\max} = \frac{\kappa}{3}\varepsilon_{m0}, \tag{25.29}$$

and then recollapses.

Zero Energy

For $q = 0$ the total energy is zero. This corresponds to the limit that the particle has exactly the energy needed to escape from the potential. In this universe, the expansion is becoming slower and slower until $a_{,t}$ goes to zero, when the scaling factor tends to infinity.

Positive Energy

For $q = -1$ the total energy is positive. This corresponds to an unbound particle. In this case, too, the universe expands for all time, the expansion velocity $a_{,t}$ is faster than in the case $q = 0$, and reaches the value $a_{,t}^\infty = c$ in the limit $t \to \infty$.

The three cases are illustrated in Fig. 25.1. We notice that the derivative $a_{,t}$ appears only quadratically in the Friedmann equation. Therefore, for every solution that expands over time we also obtain a corresponding solution that contracts over time. In other words, the Friedmann equation is invariant under the time reversal transformation $t \mapsto -t$. Another important point to note is that the potential goes to $-\infty$ for $a \to 0$. From this we can see that for $a = 0$ all solutions of the Friedmann equation must have infinite slope.

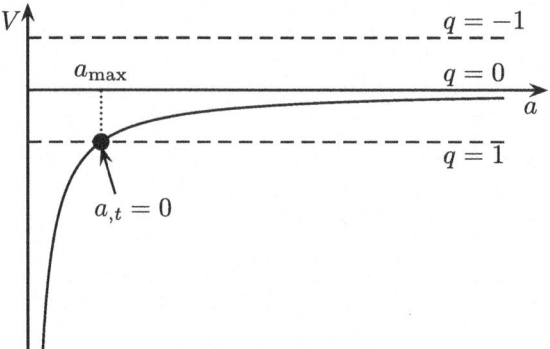

Fig. 25.1 Qualitative interpretation of (25.27). When one interprets this equation in terms of the energy balance of a particle in a $-1/a$ potential, then the case $q = 1$ corresponds to negative total energy. In this case the universe expands to some maximum value, and then must recollapse again. The cases $q = 0$, or $q = -1$ correspond to vanishing or positive total energy of the particle, respectively. Then the universe expands for all time

25.4.2 Newtonian Analogy to the Friedmann Equation

In Newtonian mechanics, we can derive an equation which has the same form as the Friedmann equation. We consider a homogeneous sphere with constant mass M, which either expands or contracts isotropically, i.e., in such a way that the spherical shape is preserved. The density within the sphere may be time-dependent but always assumes the same value everywhere in the sphere. Let us denote the time-dependent radius of the sphere by $a(t)$. The equation of motion for a particle on the surface of the sphere then reads

$$a_{,tt} = -G \frac{M}{a(t)^2}. \tag{25.30}$$

We multiply both sides of the equation by $a_{,t}$, and then integrate. This leads to

$$\frac{1}{2} a_{,t}^2 - G \frac{M}{a(t)} = E_{\text{tot}}. \tag{25.31}$$

This is the energy balance equation of the particle: The sum of the kinetic and the potential energy per mass of the particle is constant, and equal to the total energy E_{tot} (per mass), which is obtained as an integration constant. The solution for an expanding sphere corresponds to the solution of a contracting sphere for which the time is reversed. To demonstrate the equivalence to the Friedmann equation even more strikingly, we replace the mass with the initial radius a_0^3 by the relation

$$M = \frac{4}{3} \pi \frac{\varepsilon_{m0}}{c^2} a_0^3, \tag{25.32}$$

where we choose $a_0^3 = 1$. We multiply by 2, introduce κ, and obtain

$$a_{,t}^2 - \frac{\kappa}{3} \frac{\varepsilon_{m0} c^2}{a} = 2 E_{\text{tot}}. \tag{25.33}$$

This equation corresponds to (25.27), with the identification $qc^2 = -2E_{\text{tot}}$. Depending on the value of E_{tot}, this results again in the three cases, of an expansion to a maximum value of

$$a_{\max} = \frac{\kappa \, \varepsilon_{m0} c^2}{6 \, |E_{\text{tot}}|} \tag{25.34}$$

for $E_{\text{tot}} < 0$, or expansion to infinity with the expansion rate decreasing to zero for $E_{\text{tot}} = 0$, or infinite expansion with $a_{,t}(t \to \infty) = \sqrt{2E_{\text{tot}}}$ for $E_{\text{tot}} > 0$.

25.5   Scaling Factors for Matter-Dominated Universes

In this section we shall derive and discuss various solutions of the Friedmann equation (25.27). For our calculations we set

$$a_{\text{ref}} = \frac{\kappa}{6} \varepsilon_{m0} \tag{25.35}$$

and obtain the more compact form for our discussion

$$a_{,t}^2 = 2 \frac{a_{\text{ref}} c^2}{a} - q c^2. \tag{25.36}$$

The specific choice of the value of the quantity a_{ref} leads to a more transparent representation of the solutions.

25.5.1 Vanishing Curvature

For $q = 0$ we obtain

$$a_{,t} = \frac{da}{dt} = \pm c \sqrt{\frac{2a_{\text{ref}}}{a}}, \quad \text{or} \quad \sqrt{a}\, da = \pm \sqrt{2a_{\text{ref}}}\, c\, dt. \tag{25.37}$$

The two signs in this equation distinguish between universes expanding or collapsing over time. We would like to consider only expanding universes, and therefore restrict ourselves to the positive sign. The integration then leads to

$$\frac{2}{3} a^{3/2} = \sqrt{2a_{\text{ref}}} c (t - t_0), \tag{25.38}$$

with the integration constant t_0. We now make an exception in this section and the following sections, and choose the initial condition $a(0) = 0$, which implies $t_0 = 0$. Instead, with our usual convention $a(0) = 1$ we would have

$$t_0 = -\frac{1}{3c} \sqrt{\frac{2}{a_{\text{ref}}}}. \tag{25.39}$$

Solving with respect to a now leads to

$$a(t) = \left(\frac{9}{2} a_{\text{ref}} \right)^{1/3} c^{2/3} t^{2/3}. \tag{25.40}$$

The time derivative of the scaling factor is obtained as

$$a_{,t} = \frac{2}{3}\left(\frac{9}{2}a_{\text{ref}}\right)^{1/3} c^{2/3} t^{-1/3}. \tag{25.41}$$

Knowing a and $a_{,t}$ we can also calculate the Hubble parameter $H(t)$ defined in Eq. (23.9), and obtain

$$H(t) = \frac{a_{,t}}{a} = \frac{2}{3}\frac{1}{t}. \tag{25.42}$$

The distinction between the Hubble constant and the Hubble parameter is very important. The Hubble constant tells us at what rate the universe is expanding today, and is a quantity that can be determined by observations. The Hubble parameter, on the other hand, is characteristic of the respective model of the universe. From (25.41) and (25.42) we also see, that, for $t \to \infty$, $a_{,t} \to 0$ and $H \to 0$. Thus, this model exhibits a decelerated expansion, even though the scaling factor a grows beyond all bounds, exactly the way we had predicted it in our discussion in Sect. 25.4.1. This solution is called the *Einstein-de-Sitter universe*.[1] Einstein and de Sitter had proposed this model in a joint paper in 1932 [1].

25.5.2 Positive Curvature

In the case $q = 1$ we obtain

$$a_{,t}^2 = 2\frac{a_{\text{ref}}c^2}{a} - c^2, \quad \text{or} \quad \frac{da}{\sqrt{\frac{2a_{\text{ref}}}{a} - 1}} = \pm c\,dt. \tag{25.43}$$

The integral appearing in this expression can be evaluated analytically, but the resulting expressions are very complicated. Instead, we use the parametrization

$$a = a_{\text{ref}}(1 - \cos(\eta)) \quad \text{and} \quad da = a_{\text{ref}}\sin(\eta)\,d\eta. \tag{25.44}$$

From this follows

$$\pm c\,dt = a_{\text{ref}}\frac{\sqrt{1 - \cos(\eta)}}{\sqrt{1 + \cos(\eta)}}\sin(\eta)\,d\eta. \tag{25.45}$$

At this point we introduce another transformation by setting $x = \cos(\eta)$ and $dx = -\sin(\eta)\,d\eta$. From the relation

$$\int \frac{\sqrt{1 - x}}{\sqrt{1 + x}}\,dx = \sqrt{1 - x^2} + \arcsin(x), \tag{25.46}$$

---

[1] Willem de Sitter, 1872–1934, Dutch astronomer.

and using the identity $\arccos(x) = \pi/2 - \arcsin(x)$ for the back-transformation, we obtain the preliminary result

$$\pm (ct - ct_0) = -a_{\text{ref}} \left(-\arccos(x) + \sqrt{1 - x^2} \right). \tag{25.47}$$

Due to $\arccos(x) = \eta$, this form simply results in

$$\pm (ct - ct_0) = a_{\text{ref}} (\eta - \sin(\eta)). \tag{25.48}$$

The condition $a(0) = 0$ leads to an initial value for η of $\eta_0 = 0$ from (25.44). This leads again to $t_0 = 0$. In this way we have found a parametrized representation of the solution, viz.

$$a = a_{\text{ref}} (1 - \cos(\eta)),$$
$$ct = a_{\text{ref}} (\eta - \sin(\eta)). \tag{25.49}$$

Note that we have treated the case with the positive sign. Equation (25.49) is the representation of a *cycloid* with the radius a_{ref} of the rolling circle and the rolling angle η. As opposed to the case $q = 0$, this is a periodic solution, a first increases, reaches the maximum value

$$a(\eta = \pi) = 2a_{\text{ref}} = \frac{\kappa}{3}\varepsilon_{\text{m0}}, \tag{25.50}$$

at $ct = a_{\text{ref}}\pi$, the value that we already had found in our qualitative discussion in (25.29), and then decreases again down to zero.

The case $q = 1$ corresponds to a *closed* universe, which has a finite volume, extends to a maximum value, and then recollapses. In order to see that in the case $q = 1$ we have a finite volume, we use the FLRW metric in the form (24.20b). The volume is then given by

$$V = \int_0^{2\pi} \int_0^{\pi} \int_0^{\pi} \sin^2(\chi) \sin(\vartheta)\, d\chi\, d\vartheta\, d\varphi = 2\pi^2 a(t)^3. \tag{25.51}$$

Evidently the integral diverges for all other values of q, since then $\sinh(\chi)$ or χ stand in the place of $\sin(\chi)$, and one has $\chi \in [0, \infty]$. Quite generally a closed universe is understood to be a model of a universe with a finite volume. In Chap. 26 we shall extend the Friedmann equation, and encounter more general models. In those cases a closed universe need not necessarily recollapse again.

25.5.3 Negative Curvature

The calculation in the case of negative curvature $q = -1$ proceeds in the same way as for $q = 1$. One first obtains

$$a_{,t}^2 = 2\frac{a_{\text{ref}}c^2}{a} + c^2, \quad \text{or} \quad \frac{da}{\sqrt{2\frac{a_{\text{ref}}}{a} + 1}} = \pm c\, dt. \tag{25.52}$$

Setting

$$a = a_{\text{ref}}(\cosh(\eta) - 1) \quad \text{and} \quad da = a_{\text{ref}}\sinh(\eta)\, d\eta \tag{25.53}$$

leads to

$$ct - ct_0 = a_{\text{ref}}\left[\sqrt{x^2 - 1} - \text{arcosh}(x)\right]. \tag{25.54}$$

where $x = \cosh(\eta)$, $dx = \sinh(\eta)\, d\eta$. Here, too, the back-transformation is very easy because $\text{arcosh}(x) = \eta$, and one obtains

$$ct - ct_0 = a_{\text{ref}}(\sinh(\eta) - \eta). \tag{25.55}$$

The condition $a(0) = 0$ again implies the initial value $\eta_0 = 0$, and again one must have $t_0 = 0$. We obtain the parametrized representation of the solution

$$a = a_{\text{ref}}(\cosh(\eta) - 1),$$
$$ct = a_{\text{ref}}(\sinh(\eta) - \eta), \tag{25.56}$$

in complete analogy with the case $q = 1$. We can tell from (25.56) that for large values of η one has

$$\frac{ct}{a} \simeq \tanh(\eta), \quad \text{and thus} \quad \lim_{\eta \to \infty} \frac{ct}{a} = 1. \tag{25.57}$$

This means that for long times we have the limiting values

$$a(t) \to ct, \quad \text{or} \quad a_{,t}(t) \to c, \tag{25.58}$$

which we also had already found in our qualitative discussion.

In Fig. 25.2 the behavior of the three possible matter universes is shown. The important finding is that all universes exhibit a *Big Bang*, independent of the curvature, i.e., at some point in the past, with our choice of coordinates at the time $t = 0$, the scaling factor is equal to zero. The model with $q = 1$ also finally ends at $a = 0$, the *Big Crunch*.

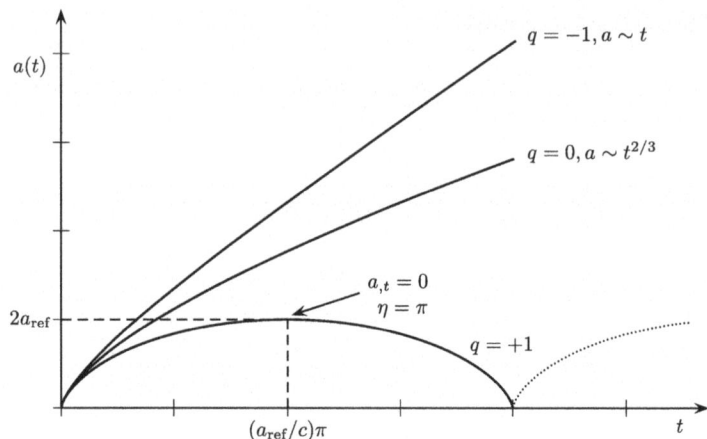

Fig. 25.2 Solutions of the Friedmann equation. For $q = 1$ the result is a finite universe with a finite volume, the functional form of the scaling factor can be represented by a cycloid. For $q = 0$ and $q = -1$ one obtains ever-expanding universes, where the expansion rate converges to zero for $q = 0$, and for $q = -1$ goes to c

References

1. Einstein, A., De Sitter, W.: On the relation between the expansion and the mean density of the universe. Proc. Natl. Acad. Sci. **18**(3), 213–214 (1932)
2. Friedmann, A.: Über die Krümmung des Raumes (On the curvature of space). Z. Physik **10**, 377–386 (1922)
3. Friedmann, A.: Über die Möglichkeit einer Welt mit konstanter negativer Krümmung des Raumes (On the possibility of a world with constant negative curvature of space). Z. Physik **21**, 326–332 (1924)

General Forms of Energy

26

Contents

Up to this point we have seen that Hubble's observations suggested an expanding universe. With our matter model universes we indeed found two solutions to the Friedmann equation that describe an eternally expanding, decelerating universe ($q = 0, -1$), and one solution ($q = 1$) describing a universe expanding at least for a certain period of time. This agreement between theory and observation seems to be very gratifying. However, observations in recent years have provided strong evidence for the fact that the expansion of the universe is accelerating. We cannot explain this within the framework of our present models. In Chap. 25 we only considered the contribution of matter to the energy density. This seemed reasonable because the contribution of the radiation energy density could easily be estimated as negligibly small, and we could not think of any other forms of energy. We will now take a closer look at this point, and shall see that we have to extend our previous analysis if we want to fully understand the evolution of the universe.

26.1   Reformulation of the Friedmann Equation

For a more general analysis, it is useful to reformulate the Friedmann equation.

We shall see that not all energy densities fulfill the condition (25.24), i.e., they do not decrease with the scaling factor as $1/a^3$. For this reason, the transformations from (25.20) to (25.27) cannot be used in general. We therefore go back to (25.20), and first look at models without curvature, i.e., we set $q = 0$. Now we will allow arbitrary contributions to the energy density, and start from a more general expression for ε. We then have

$$\frac{a_{,t}^2}{a^2} = \frac{\kappa}{3}c^2\varepsilon(t). \tag{26.1}$$

When we know, in some model, the value of a and $a_{,t}$, we can then calculate the value of the Hubble parameter at time $t_0 = t$,

$$H(t) = \frac{a_{,t}(t)}{a(t)}, \tag{26.2}$$

i.e., the value of the Hubble constant H_0 in this model universe. With this information we can write (26.1) in the form

$$H(t, q = 0)^2 = \frac{\kappa}{3}c^2\varepsilon(t). \tag{26.3}$$

Thus, in a flat universe there exists a proportionality between the energy density and the square of the Hubble parameter. However, we can also interpret this relationship in such a way that for a given Hubble parameter a *critical energy density* can be defined via

$$\varepsilon_c(t) = \frac{3}{\kappa c^2}H(t)^2. \tag{26.4}$$

We introduce the symbol

$$\varepsilon_{c0} = \frac{3}{\kappa c^2}H_0^2 \tag{26.5}$$

for this critical energy density. The density and the Hubble constant are quantities that can be measured, at least in principle. Equation (26.4) then tells us the following: If the value of the density is larger than the critical density, then $q = 1$ must apply, in order for the Friedmann equation to be fulfilled in the form

$$H(t)^2 = \frac{\kappa c^2}{3}\varepsilon(t) - \frac{qc^2}{a^2}. \tag{26.6}$$

In an analogous way, we must have $q = -1$ if $\varepsilon(t) < \varepsilon_c(t)$, and for $\varepsilon(t) = \varepsilon_c(t)$ we must have $q = 0$. Remember once again that q only determines the sign of the curvature, but not the radius of curvature, see Sect. 24.2. With the value from (23.6b) for the Hubble constant, $H_{0,Planck} = 67.66 \pm 0.42$ km s^{-1} Mpc^{-1}, we can calculate the critical density,

$$\varepsilon_{c0} = (4.82 \pm 0.06)\,\text{GeV m}^{-3}, \tag{26.7}$$

or an equivalent mass density

$$\rho_{c0} = (8.60 \pm 0.11) \cdot 10^{-27}\,\text{kg m}^{-3}. \tag{26.8}$$

When we compare with the rest energy of the proton, $m_p c^2 = 938.272$ MeV, we find that the critical density corresponds to about 5 hydrogen atoms per cubic meter.

Since the critical energy density is such a fundamental quantity, it is reasonable to measure actual energy densities in units of the respective critical density. We introduce the symbol

$$\Omega(t) = \frac{\varepsilon(t)}{\varepsilon_c(t)} \tag{26.9}$$

for this ratio. Introducing Ω into the Friedmann equation leads to the form

$$\Omega(t) - 1 = \frac{qc^2}{a(t)^2 H(t)^2}. \tag{26.10}$$

This equation is valid for all times, and therefore also at the time $t_0 = 0$. In this case, because of the convention $a_0 = 1$, we have the relation

$$q = \frac{H_0^2}{c^2}(\Omega_0 - 1), \tag{26.11}$$

Here we would like to point out once again the seemingly incorrect units due to the convention $a_0 = 1$. Still q is dimensionless, but we measure a in units of a_0. From (26.11) we can tell the value of q even more directly than by comparing the critical energy density and the actual energy density. The curvature index q cannot change its value during the evolution of the universe. Since H_0^2 is always positive, today's value $\Omega_0 \gtrless 1$ immediately implies $\Omega(t) \gtrless 1$ for all times. In particular, for $\Omega_0 = 1$, we must have $q = 0$, which then also must be valid for all times, past and

future. Since we did not restrict the derivation of (26.11) to a matter universe, this
equation does apply to all forms of energy.

In order to account for the possibility of several forms of energy contributing to
the total energy density, from now on we will write the energy density without an
index, when we mean the sum of all possible contributions:

$$\varepsilon = \sum_i \varepsilon_i. \tag{26.12}$$

We can also generalize the Ω parameter via

$$\Omega = \sum_i \Omega_i. \tag{26.13}$$

The individual terms Ω_i indicate how big their energy density is relative to the
critical density:

$$\Omega_i(t) = \frac{\varepsilon_i(t)}{\varepsilon_c(t)}. \tag{26.14}$$

It is assumed that the individual components do not interact with each other. It
is evident that solving the Friedmann equation will be much more difficult when
there are several energy density contributions, especially if they possess different
dependencies on a.

26.2   Generalized Energy-Momentum Tensor

If we wish to consider arbitrary forms of energy, we can no longer assume that the
pressure caused by these forms of energy is negligible. A non-vanishing pressure,
which of course may also be time-dependent, results in a contribution to the energy-
momentum tensor, which we therefore will have to generalize. To start with, we
consider the energy-momentum tensor of an ideal fluid, which has the form

$$T^{\mu\nu} = (\varepsilon + p)\, u^\mu u^\nu + p g^{\mu\nu}. \tag{26.15}$$

We know this form already from (21.20), where we also chose this same form for
the interior of a neutron star. Here, $u^\mu = (1/c, 0, 0, 0)$ and $u_\mu = (-c, 0, 0, 0)$
applies in the rest frame of the matter. Using the metric from (25.2), we find, as a
generalization of (25.14), the forms

$$T^\mu{}_\nu = \text{diag}(-\varepsilon, p, p, p), \tag{26.16a}$$

$$T^{\mu\nu} = \text{diag}\left(\varepsilon/c^2,\, p\,\frac{1-qr^2}{a^2},\, \frac{p}{a^2 r^2},\, \frac{p}{a^2 r^2 \sin^2(\vartheta)}\right) \tag{26.16b}$$

and

$$T_{\mu\nu} = \text{diag}\left(\varepsilon c^2, \, p\frac{a^2}{1-qr^2}, \, pa^2r^2, \, pa^2r^2\sin^2(\vartheta)\right), \tag{26.16c}$$

and by contraction

$$T = -\varepsilon + 3p. \tag{26.17}$$

Therefore, the generalization of the expression in (25.16) reads

$$T^*_{\mu\nu} = \frac{1}{2}\text{diag}\left(c^2(\varepsilon+3p), \, (\varepsilon-p)\frac{a^2}{1-qr^2}, \, (\varepsilon-p)a^2r^2, \, (\varepsilon-p)a^2r^2\sin^2(\vartheta)\right). \tag{26.18}$$

The additional diagonal entries now render the continuity equation $T^{\mu\nu}{}_{;\nu} = 0$, or written explicitly as in (25.22), more complicated. For $\mu = t$ three additional terms appear from the contribution

$$T^{\omega\nu}\Gamma^t{}_{\omega\nu} = T^{ii}\Gamma^t{}_{ii} = 3\frac{p}{c^2}\frac{a_{,t}}{a}, \tag{26.19}$$

where we have used again the Christoffel symbols from (25.3). Thus, we find the generalized equation

$$\varepsilon_{,t} + 3\frac{a_{,t}}{a}(\varepsilon+p) = 0. \tag{26.20}$$

For $\mu = \{r, \vartheta, \varphi\}$, one directly obtains $T^{\mu\nu}{}_{;\nu} = 0$. Analogous to the transition from (25.22) to (25.23), multiplying by a^3 brings Eq. (26.20) into the form

$$\left(\varepsilon a^3\right)_{,t} = -3a^2a_{,t}p. \tag{26.21}$$

We multiply by dt on both sides and obtain

$$d\left(\varepsilon a^3\right) = -p3a^2da. \tag{26.22}$$

On the left-hand side of this equation we can recognize a term of the form dE, with $E = \varepsilon V$, where, apart from constant prefactors, an arbitrary, sufficiently large volume in the universe is given by $V = a^3$. On the right-hand side we recognize the expression $dV = a^2da$. Therefore, we can write (26.22) formally as

$$dE = -pdV. \tag{26.23}$$

This equation has the form of the first law of thermodynamics

$$dE = dQ - pdV \qquad (26.24)$$

for processes with $dQ = 0$. As is well known, processes with no heat exchange, i.e., with $dQ = 0$, are called adiabatic. Thus, we see that the expansion of the universe, in this picture, is an adiabatic process. Because of

$$dS = \frac{dQ}{T}, \qquad (26.25)$$

this also implies $dS = 0$. Therefore, during the homogeneous and isotropic expansion of the universe, its entropy does not change. Due to the analogy to the first law of thermodynamics, and because (26.22) follows from the divergence-free nature of the energy-momentum tensor, we can also call this relationship the *law of conservation of energy in cosmology*.

With these results, we can now formulate the field equations $R_{\mu\nu} = \kappa T_{\mu\nu}^*$ in (25.17) for arbitrary forms of energy. It must be noted, however, that with $p(t)$ we have introduced another unknown variable.

26.3   Equations of State with Pressure

In our context, the equation of state is a mathematical relation that expresses the pressure p of a physical system as a function of the mass density ρ_m, or energy density, and possibly of temperature. We have encountered equations of state in Sects. 18.3 and 18.4 in the discussion of stars, white dwarfs, and neutron stars. For example, we used the ideal gas law as an equation of state for main sequence stars, and we also used it in order to justify that for non-relativistic matter the approximation $p \approx 0$ can be applied. Now we must find a general relation between the energy density in the universe and the pressure. Noting the relationship $\varepsilon = \rho c^2$ we can see that in cosmology we have a completely analogous problem as in the description of stellar matter. In cosmology, however, we can limit ourselves to equations of state which are much simpler than those we know, e.g., for the degenerate matter in white dwarfs and neutron stars. In fact, we only need to consider equations of state of the form

$$p = w\varepsilon. \qquad (26.26)$$

Here, w is a dimensionless, real-valued constant. Non-relativistic matter is described by this equation of state through the value $w = 0$. This simple form of the equation of state has the consequence that we can still solve the Friedmann equation analytically, at least in a few special cases.

26.4 Dark Matter

The matter that we encounter in everyday life is *baryonic*, which essentially means that it is composed of protons and neutrons. There are also electrons, which, due to their small mass, give only a small contribution to the rest energy density of matter.

However, there is strong evidence that a predominant portion of matter in the universe is non-baryonic. One of the first indications of the existence of this non-baryonic matter was furnished by *Zwicky*.[1] He found that the gravitational effect of visible matter in the Coma cluster, a galaxy cluster of about 1000 galaxies, is not sufficient to hold it together [10]. What he did was to measure the peculiar velocities of the galaxies. With an appropriate estimate for the mass of each galaxy, he could calculate the kinetic energy of the cluster. According to the virial theorem that we discussed in Sect. 17.1, the potential energy of the cluster is twice its kinetic energy. Zwicky found a potential energy which was ten times bigger than the gravitational potential energy produced by the masses of the visible galaxies.

Further indications emerged from the analysis of the orbital velocities of stars in spiral galaxies. In the 1960s, *Vera Rubin*[2] investigated the velocities of stars in the spiral galaxy NGC 3198. She found that, contrary to expectation, the velocity of stars in the outer region of the galaxy did not decrease but converged to a plateau when plotted as a function of the distance from the center of the galaxy. They moved much faster than was to be expected on account of the total visible mass. Also for other galaxies she inferred from the flatness of the rotation curves the presence of huge amounts of additional mass [9].

Finally, from the gravitational lensing effect discussed in Sect. 13.4.3 one can infer the total mass of the light-deflecting object. Such observations, too, indicate the existence of a very large fraction of non-baryonic matter, which should make up about 80% of the total mass.

This type of matter is called *dark matter*. The name refers to its property of only interacting gravitationally, and not with electromagnetic radiation. It would therefore be more correct to speak of *transparent matter*.

The nature of this invisible matter is still unknown. For a long time one possible candidate was neutrinos. As these move at relativistic velocities, they were referred to as *hot dark matter* (HDM). Favored today are still undiscovered elementary particles that are only subject to gravity and the weak interaction. This is known as *cold dark matter* (CDM). One candidate for such particles are supersymmetric partners of known elementary particles, such as those that are being searched for at the Large Hadron Collider.

---

[1] Fritz Zwicky, 1898–1974, Swiss physicist and astronomer, who worked in the United States for most of his life.

[2] Vera Cooper Rubin, 1928–2016, US-American astronomer.

26.5   Energy Density of Radiation

In today's universe, the energy density ε_r of radiation is very small. However, it
differs from the energy density of matter in one essential property. For the latter we
have the relationship $\varepsilon_m a^3 = \varepsilon_{m0} a_0^3$, which is ultimately a mathematical expression
of the conservation of mass. We can easily realize that this relationship does not
apply to the energy density of radiation. The number density of photons also
decreases with a^{-3} during expansion. At the same time, the expansion of space
also leads to a redshift of the photons, as Hubble found in his observations. We shall
derive the redshift in detail in Sect. 27.4, but a simple picture will suffice here, as it
already leads us to the correct result.

Let a source of light emit a light wave at a certain time t_e with a wavelength of λ_e.
When this light wave reaches us at a later time t_r, the scaling factor a has increased
by a factor $a(t_r)/a(t_e)$. Then the distance between the start point and the end point
of the light wave has increased by the same factor. This implies

$$\lambda_r = \lambda_e \frac{a(t_r)}{a(t_e)}. \tag{26.27}$$

Thus, the wavelength received by us changes, in comparison with the wavelength at
the time of emission, by the ratio of the values of the scaling factors then and now.
Because of the relation $\nu = c/\lambda$, we also have

$$\nu_r = \nu_e \frac{a(t_e)}{a(t_r)}. \tag{26.28}$$

Of course, the ray of light considered here is in no way special as compared to other
rays of light. Quite generally the change in frequency of electromagnetic radiation
is proportional to the inverse of the scaling factor. Because of $E = h\nu$, we also have
for every single photon

$$E \sim a^{-1}. \tag{26.29}$$

When we combine this result with the change in the number density proportional to
a^{-3}, we find for the energy density of radiation $\varepsilon_r a^4 = \varepsilon_{r0} a_0^4$, or

$$\varepsilon_r \sim a^{-4}. \tag{26.30}$$

Figure 26.1 illustrates the reason for this behavior. If the energy density of radiation
decreases faster during the expansion of the universe than the energy density of
matter, this means, conversely, that the ratio $\varepsilon_r/\varepsilon_m$ was higher in earlier times than
it is today. Thus, if we want to understand the overall evolution of the universe, we
cannot neglect radiation.

We still need the value of the w parameter for the energy density of radiation.
Calculating the radiation pressure of a photon gas is a task of statistical mechanics.

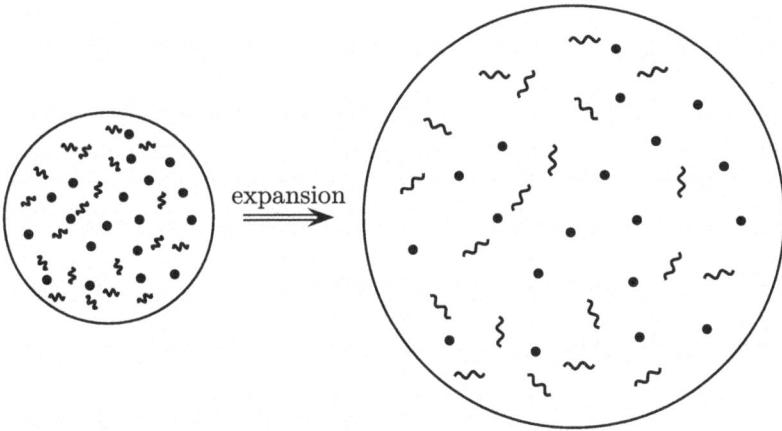

Fig. 26.1 As the universe expands, the energy density of matter decreases with $\varepsilon_m \sim a^{-3}$. Photons are also redshifted, therefore their energy density decreases with $\varepsilon_r \sim a^{-4}$

In order not to stray too far from the topic, we choose a slightly sloppy way, which leads to the correct result. For matter, we had derived from the ideal gas law that $p = \langle v^2 \rangle / (3c^2) \, \varepsilon_m$, see (25.11). We had neglected this contribution due to $v \ll c$. We obtain the correct result for a photon gas when we simply set $v = c$ in (25.11), since photons propagate at the speed of light. Thus, the value of the w parameter is

$$w_r = \frac{1}{3}. \tag{26.31}$$

26.6 Einstein's Cosmological Constant

Before setting up and analyzing the generalized Friedmann equation, we want to take into account, as a final extension, Einstein's *cosmological constant* Λ. We had already introduced Λ when we derived Einstein's field equations in Chap. 12, but did not consider it in the foregoing discussions. We shall see that we can interpret its contribution simply as an additional form of energy, though with very exotic properties. First we would like to briefly explain why Einstein originally introduced Λ in the first place.

In the discussion of matter-dominated universes, we had seen that for all values of q we obtain time-dependent scaling factors. But at the beginning of the twentieth century the notion of a static, eternally unchanging universe still prevailed. Therefore, Einstein was facing the problem that his theory did not at all produce the solution that he had expected. For this reason, in 1917 he proposed a modification of his field equations by adding a term $\Lambda g_{\mu\nu}$ [4]. We had already included this term in (12.40b), since it is compatible with all constraints on the field equations, except for the requirement that in the limit of the Minkowski spacetime the left-hand side of the field equations should vanish identically as $T_{\mu\nu}$. The Λ term does violate this requirement. However, if Λ were very small the difference would not be detectable in experiments.

Here we will give a short outline of Einstein's line of argument for introducing Λ, since we shall return to this line of argument later in a different context. Based on considerations of the behavior of the Newtonian gravitational potential for constant mass density at infinity, Einstein decided not to consider the Poisson equation

$$\Delta\phi = 4\pi G \rho_m \tag{26.32}$$

but the *Helmholtz equation*

$$\Delta\phi - \Lambda\phi = 4\pi G \rho_m, \tag{26.33}$$

with a free parameter Λ. The formal solution of the Poisson equation (26.32) is

$$\phi(r) = \int \frac{\rho_m(r')}{|r - r'|} d^3 r'. \tag{26.34}$$

This integral diverges for a spatially uniform mass density $\rho_m(r) = \rho_{m0}$. Therefore, it is not possible, in Newtonian mechanics, to have a static universe with constant density, which would obey the cosmological principle. In order to obtain a physically reasonable solution, it would be necessary to demand that the density is not constant everywhere, but goes to zero at infinity.

On the other hand, the formal solution of the Helmholtz equation reads

$$\phi(r) = \int \frac{e^{-\sqrt{\Lambda}|r-r'|}}{|r - r'|} \rho_m(r') d^3 r'. \tag{26.35}$$

Thus, the introduction of the cosmological constant leads to a screening term. Equation (26.35) formally describes a *Yukawa potential*,[3] which is of importance in particle physics, and in fact also has been proposed for gravitation time and again [2]. The crucial point is that this expression converges for a constant mass density. Contrary to the Poisson equation, the Helmholtz equation possesses a nonvanishing solution also for a spatially uniform density ρ_{m0}. In this case the solution is

$$\phi = -\frac{4\pi G}{\Lambda} \rho_{m0}. \tag{26.36}$$

Local variations of the matter density lead to corrections $\tilde{\phi}$ of this potential. However, for small values of Λ, $\tilde{\phi}$ would converge to a Newtonian gravitational

---

[3] Yukawa Hideki, 1907–1981, Japanese physicist, Nobel Prize in 1949.

potential with arbitrary accuracy. It is easy to recognize the point of this reasoning: With the correction term $\Lambda\phi$, a static universe with constant density everywhere is possible.

Einstein chose an approach via a modified Newtonian gravitation theory, which "in itself does not claim to be taken seriously" [4], but he did this only for the sake of illustration. In an analogous way, he then argued in favor of an additional term $\Lambda g_{\mu\nu}$ in his field equations. In fact, one can prove that the Helmholtz equation results as the non-relativistic limit of Einstein's field equation with cosmological constant. This means that one can find a Newtonian gravitation theory with cosmological constant, in the non-relativistic limit of a theory of general relativity with cosmological constant. When we supplement the non-relativistic approximation of the Einstein tensor (12.41) by the cosmological term, we have

$$\Delta\phi - \Lambda\phi = 4\pi G\rho_{\mathrm{m}} + \frac{1}{2}\Lambda c^2. \tag{26.37}$$

Thus, we obtain an equation of the form (26.33), in which the matter density is supplemented by a constant term $\Lambda c^2/2$.

In like manner, one proceeds in general relativity, and supplements the field equations by the cosmological constant to obtain

$$R_{\mu\nu} + \Lambda g_{\mu\nu} = \kappa T^*_{\mu\nu}, \tag{12.40b}$$

see Sect. 12.2.4.

We now wish to demonstrate that we can treat the additional Λ term as a contribution to the energy density. When we bring the term on the right-hand side, we have

$$R_{\mu\nu} = \kappa \left(T^*_{\mu\nu} - \frac{\Lambda}{\kappa} g_{\mu\nu} \right). \tag{26.38}$$

We now have to express the term $T^*_{\mu\nu} - (\Lambda/\kappa)g_{\mu\nu}$ as an energy-momentum tensor, in which the cosmological constant is included. This means that we have to analyze the components $(\Lambda/\kappa)g_{\mu\nu}$ in order to find the relationship between energy density and pressure. Written out explicitly, we have the following relations, where we omit the vanishing non-diagonal elements:

$$-\frac{\Lambda}{\kappa} = \frac{1}{2}(\varepsilon + 3p), \tag{26.39a}$$

$$\frac{\Lambda}{\kappa} \frac{a^2}{1 - qr^2} = \frac{1}{2}(\varepsilon - p)\frac{a^2}{1 - qr^2}, \tag{26.39b}$$

$$\frac{\Lambda}{\kappa}a^2 r^2 = \frac{1}{2}(\varepsilon - p)a^2 r^2, \tag{26.39c}$$

$$\frac{\Lambda}{\kappa}a^2 r^2 \sin^2(\vartheta) = \frac{1}{2}(\varepsilon - p)a^2 r^2 \sin^2(\vartheta). \tag{26.39d}$$

First, we see that the last three equations are equivalent. This is important, since otherwise we could not perform the assignment exactly without contradictions. The system of equations (26.39) then reduces to the two conditions

$$-2\frac{\Lambda}{\kappa} = \varepsilon + 3p, \qquad\qquad (26.40a)$$

$$2\frac{\Lambda}{\kappa} = \varepsilon - p. \qquad\qquad (26.40b)$$

Adding these equations, we immediately see that we have

$$p_\Lambda = -\varepsilon_\Lambda, \qquad\qquad (26.41)$$

and this directly implies

$$\varepsilon_\Lambda = \frac{\Lambda}{\kappa} = \frac{\Lambda c^4}{8\pi G}. \qquad\qquad (26.42)$$

In other words, the cosmological constant is associated with an energy density with negative pressure. This energy density fulfills (26.26) with $w = -1$. This very unusual property has the consequence that through ε_Λ the expansion of the universe can be accelerated.

Since Λ is merely a constant in the field equations, we can immediately tell from (26.42) the dependence of ε_Λ on the scaling factor, or rather its independence of the scaling factor, because ε_Λ just does not depend on a. The energy density belonging to the cosmological constant does not change during the expansion of the universe. While the universe keeps expanding, the energy densities of matter and radiation decrease as $\varepsilon_m \sim a^{-3}$, and $\varepsilon_r \sim a^{-4}$, respectively. Therefore, the relative importance of the cosmological constant is continuously increasing during the expansion of the universe. In addition, recalling the definition of the critical density $\varepsilon_c(t) = 3H(t)^2/(\kappa c^2)$ in (26.4), we can assign an Ω parameter to ε_Λ

$$\Omega_\Lambda(t) = \frac{\varepsilon_\Lambda}{\varepsilon_c(t)} = \frac{\Lambda c^2}{3H(t)^2}. \qquad\qquad (26.43)$$

After the discovery of the expansion of the universe by Hubble, the cosmological constant lost its justification, and it was generally assumed that its value is zero. However, in the 1990s observations of far distant supernovae suggested that the expansion rate of the universe is accelerating, and thus the cosmological constant may have a small, but non-vanishing value, or, alternatively a form of energy exists with similar properties, i.e., with $w \approx -1$. We shall discuss these observations in detail in Chap. 28, and also how one can infer from these observations the accelerated expansion of the universe. Thus, the cosmological constant plays again an important role in today's cosmology, even though in a completely different way from what Einstein had intended: Instead of allowing for a static solution of the field equations, the cosmological constant, or a form of energy behaving in a similar way, is responsible for the acceleration of the expansion.

After the discovery that Λ probably has a value different from zero, the question arose as to the origin of this energy density with its unusual properties, namely negative pressure and constancy during the expansion of the universe. It is fair to say that this problem is still completely unresolved. One idea is that ε_Λ originates from the *vacuum energy density*,

$$\varepsilon_\Lambda = \varepsilon_{\text{vac}}. \tag{26.44}$$

It is a result of quantum mechanics that an energy density different from zero can be assigned to the vacuum, which is produced by virtual particle-antiparticle pairs which continuously are created and annihilated. Estimates of the values ε_Λ resulting from these processes lie about 120 orders of magnitude above the actual value. We shall give a brief discussion of this topic in Chap. 30. At this point, we simply have to accept that evidently an energy contribution of this type exists, which drives the present accelerated expansion of the universe, and that this contribution must be taken into account when discussing possible models of the universe. To avail ourselves of a simple designation for ε_Λ, we shall use the word vacuum energy density in the following. The designation *dark energy* is also widespread in the literature, in analogy to dark matter.

26.7   Friedmann-Lemaître Equation

Inserting the new form of the energy-momentum tensor (26.18) into the field equations leads to the following set of equations

$$-3\frac{a_{,tt}}{a} = \frac{\kappa}{2}(\varepsilon + 3p)c^2, \tag{26.45a}$$

$$\frac{\mathcal{A}}{c^2(1 - qr^2)} = \frac{\kappa}{2}(\varepsilon - p)\frac{a^2}{1 - qr^2}, \tag{26.45b}$$

$$\frac{r^2}{c^2}\mathcal{A} = \frac{\kappa}{2}(\varepsilon - p)a^2r^2, \tag{26.45c}$$

$$\frac{r^2}{c^2}\sin^2(\vartheta)\mathcal{A} = \frac{\kappa}{2}(\varepsilon - p)a^2r^2\sin^2(\vartheta), \tag{26.45d}$$

with $\mathcal{A} = aa_{,tt} + 2a_{,t}^2 + 2qc^2$ from (25.5), as an extension of (25.17). Again, the equations for $\mu, \nu = \{1, 2, 3\}$ are identical, after all our assumption of a homogeneous and isotropic space has not changed. In Sect. 26.3 we have linked the pressure with a simple equation of state of the energy density, and have introduced the energy density of radiation and the vacuum energy density ε_Λ as important contributions, in addition to the energy density of matter. Therefore, we now can set out to explore which models for the universe result from this generalized form of the Friedmann equation.

26.7.1 Scaling-Factor Dependence of the Energy Densities

We are now in a position to derive the general relation between the energy density and the scaling factor. Inserting the equation of state (26.26) into the law of conservation of energy (26.22), and bringing the pressure term on the left-hand side of the equation, we obtain

$$a^3 \, d\varepsilon + 3a^2(1+w)\varepsilon \, da = 0. \tag{26.46}$$

Separation of variables leads to

$$\frac{d\varepsilon}{\varepsilon} = -3(1+w)\frac{da}{a}. \tag{26.47}$$

We integrate this equation and choose the integration constant in such a way that $\varepsilon(1) = \varepsilon_0$. We then obtain

$$\varepsilon(a) = \varepsilon_0 a^{-3(1+w)}. \tag{26.48}$$

Thus the law of conservation of energy, together with the equation of state, determines the functional dependence of each individual energy density on the scaling factor. It is important to note that Eq. (26.48) applies only for the individual forms of energy with a fixed w. As already discussed, in actual fact the total energy density in our universe will be the sum of different components, i.e., $\varepsilon = \sum \varepsilon_i$. Each individual contribution ε_i has its own parameter w_i, and fulfills (26.48) separately, provided the individual contributions do not interact with each other, which we will assume. As special cases, we obtain from (26.48) for $w = 0$, $w = 1/3$, and $w = -1$ our already well-known results $\varepsilon_m \sim a^{-3}$, $\varepsilon_r \sim a^{-4}$, and $\varepsilon_\Lambda \sim$ const.

26.7.2 Setting up the Equations

With the help of the equation of state it is now possible to eliminate the pressure and the energy density from $T^*_{\mu\nu}$ in (26.45). Let us write down the two equations once again:

$$a_{,tt} = -\frac{\kappa c^2}{6}(\varepsilon + 3p)a \tag{26.49a}$$

and

$$aa_{,tt} + 2a_{,t}^2 + 2qc^2 = \frac{\kappa c^2}{2}a^2(\varepsilon - p), \qquad (26.49b)$$

where we have inserted $\mathcal{A}$ from (25.5). We can see from (26.49a), that both the energy density and a positive pressure can decelerate the expansion, since they both yield a negative contribution to $a_{,tt}$. In fact, for matter-dominated universes we had found models with decelerated expansion.

At the same time, we can also see that, due to $p = w\varepsilon$, for $w < -1/3$ the second derivative of the scaling factor becomes positive. Forms of energy which fulfill this condition lead to an accelerated expansion. The vacuum energy density does fulfill this condition because of $w = -1$. If it is the dominant contribution to the total energy density of the universe, an accelerated expansion must be expected. When we insert the sums of the energy densities in (26.49), and use $p_i = w_i\varepsilon_i$, we obtain

$$a_{,tt} = -\frac{\kappa c^2}{6}a\sum_i \varepsilon_i(1 + 3w_i) \qquad (26.50a)$$

and

$$aa_{,tt} + 2a_{,t}^2 + 2qc^2 = \frac{\kappa c^2}{2}a^2\sum_i \varepsilon_i(1 - w_i). \qquad (26.50b)$$

When we insert (26.50a) in (26.50b), and bring the first term on the left-hand side on the right-hand side, we see that the w_i terms cancel in the sum. What remains is just

$$a_{,t}^2 + qc^2 - \frac{\kappa c^2}{3}a^2\sum_i \varepsilon_i = 0. \qquad (26.51)$$

This is the logical extension of (25.27), in which we had taken into account only the energy density of matter, $\varepsilon_m = \varepsilon_{m0}a^{-3}$. After all, we would have been able to directly write down this form. But it is only due to our careful considerations with the help of the equation of state and the law of conservation of energy that we are now in a position to replace the different contributions ε_i to the total energy density with functions of the scaling factor. In this way we end up with an ordinary differential equation for the scaling factor. The latter reads, using (26.48),

$$a_{,t}^2 + qc^2 - \frac{\kappa c^2}{3}a^2\sum_i \varepsilon_{i0}a^{-3(1+w_i)} = 0. \qquad (26.52)$$

This equation is referred to as the *Friedmann-Lemaître equation*. We draw the factor a^2 in front of the third term under the sum, and finally obtain

$$a_{,t}^2 - \frac{\kappa c^2}{3} \sum_i \varepsilon_{i0} a^{-(1+3w_i)} = -qc^2. \tag{26.53}$$

26.7.3 Qualitative Solution in an Effective-Potential Picture

As we did for the equations without the cosmological constant, we shall first carry out a qualitative discussion of the Eq. (26.52) with cosmological constant. To do so, we exploit again the fact that (26.53) can be interpreted as an energy balance equation. We restrict ourselves to models with $\varepsilon_r = 0$, since we can highlight the essential points already by just looking at models with matter and cosmological constant. Moreover, we use again the notation with Λ, in the place of ε_Λ, which follows from the relation $\varepsilon_\Lambda = \Lambda/\kappa$ between the vacuum energy density and the cosmological constant in (26.42). With this notation, Eq. (26.53) assumes the form

$$a_{,t}^2 - \frac{\kappa}{3} \frac{\varepsilon_{m0} c^2}{a} - \frac{\Lambda c^2}{3} a^2 = -qc^2. \tag{26.54}$$

Comparing with (25.28), we see that we have the additional term $-\Lambda c^2 a^2/3$ in the potential. As long as there appear contributions in the potential, apart from Λ, which diverge for $a \to 0$, we still have the finding that all these solutions will have an infinite slope at $a = 0$. In Fig. 26.2 the possible potential curves are shown for

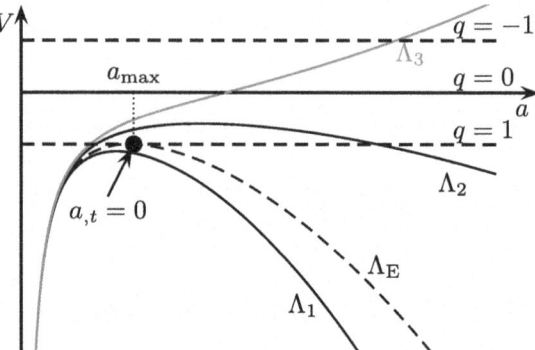

Fig. 26.2 Qualitative solution of the Friedmann-Lemaître equation (26.54). For $\Lambda < 0$ only recollapsing universes are possible (Λ_3). For $\Lambda > 0$ all universes expand for $q = 0$ and $q = -1$. For $q = 1$, there exist recollapsing solutions (Λ_2), forever expanding solutions (Λ_1), and, for Λ_E, a universe which expands up to a maximum value, or, for suitable initial conditions, a static universe

different values of Λ. The exact behavior of the solutions depends strongly on the value Λ, and, above all, its sign. The following types of solutions are possible:

Eternal Expansion
If Λ is large enough, the universe expands for all time, regardless of the value of q. This case corresponds to Λ_1.

Closed Universe for $q = 1$
If Λ is smaller, then for $q = 1$ there exists again a closed universe that reaches a maximum extension, and then collapses again (Λ_2).

Asymptotically Static Universe
Between these two cases there exists a value of Λ at which, for $q = 1$, the maximum of the potential energy corresponds exactly to the total energy. In this case, a asymptotically approaches the maximum value, with $a_{,t}$ going to zero (Λ_E). Thus, if one waits long enough, a static universe is reached. Alternatively, with the initial condition $a_{,t} = 0$, a completely static universe results, without a Big Bang, which is exactly the solution proposed by Einstein when introducing the cosmological constant. Evidently, the solution is unstable, since any deviation of Λ from the correct value, ever so small, causes the universe to either collapse or expand.

Recollapsing Universe Irrespective of Curvature
The case $\Lambda < 0$ corresponds to the bound motion of a particle for all three values of q. Every solution of this type first expands, and then must collapse again (Λ_3).

26.7.4 Formulation with Energy Density Fractions

In the following, we not only want to calculate solutions for the scaling factors from the Friedmann equation, but also later to compare our theoretical models with the results of actual astronomical observations. In essence, these will be the values of the Hubble constant H_0, and the Ω_0 parameters of the different forms of energy. For the direct comparison, we can bring (26.53) in a form that lends itself to practical applications. To do so, we first exploit (26.11) in order to express the curvature index q in terms of $\Omega_0 = \sum_i \Omega_{i0}$ and H_0. As an interim result, we then have

$$a_{,t}^2 + H_0^2(\Omega_0 - 1) - \frac{\kappa c^2}{3} \sum_i \varepsilon_{i0} a^{-(1+3w_i)} = 0. \tag{26.55}$$

We divide by H_0^2 and use the relation

$$\frac{\kappa c^2}{3H_0^2} = \frac{1}{\varepsilon_{c0}}, \tag{26.56}$$

which we obtain from (26.5), in order to replace the prefactor in front of the third term. The result is

$$\frac{a_{,t}^2}{H_0^2} + (\Omega_0 - 1) - \sum_i \frac{\varepsilon_{i0}}{\varepsilon_{c0}} a^{-(1+3w_i)} = 0. \tag{26.57}$$

The ratios $\varepsilon_{i0}/\varepsilon_{c0}$ correspond exactly to the values Ω_{i0} of the individual Ω_i parameters today. We make this replacement, shovel the second and third term onto the right-hand side, and multiply by H_0^2. We then get

$$a_{,t}^2 = H_0^2 \left[\sum_i \Omega_{i0} a^{-(1+3w_i)} + (1 - \Omega_0) \right]. \tag{26.58}$$

Finally, we take the root, and separate the variables. This leads to

$$\frac{da}{\sqrt{\sum_i \Omega_{i0} a^{-(1+3w_i)} + (1 - \Omega_0)}} = H_0 dt. \tag{26.59}$$

Now we have brought the Friedmann equation into a form in which the measurable quantities H_0 and Ω_{i0} appear. By integrating, we can obtain the formal solution of (26.59)

$$\int \frac{da}{\sqrt{\sum_i \Omega_{i0} a^{-(1+3w_i)} + (1 - \Omega_0)}} = H_0 t + \mathcal{C}. \tag{26.60}$$

We still have to choose the integration constant in such a way that $a_0 = 1$. Depending on which, and how many, contributions Ω_{i0} are taken into account in (26.60), the integration on the left-hand side is analytically possible, or not. In the general case, one has to resort to numerical integration. A good discussion of all sorts of possible forms of energy can be found in an article by Nemiroff [7]. One can see directly, that in a flat universe with just one additional contribution to the energy density, the scaling factor has the form

$$a(t) \sim t^{\frac{2}{3(1+w)}}. \tag{26.61}$$

In what follows we shall restrict ourselves to the three of types of energies which we are already familiar with, namely matter, radiation, and the cosmological constant. For this case

$$\Omega_0 = \Omega_{m0} + \Omega_{r0} + \Omega_{\Lambda 0}, \tag{26.62}$$

and we have, as a starting point for the following discussions, the relation

$$\int \frac{da}{\sqrt{\Omega_{\mathrm{m}0}a^{-1} + \Omega_{\mathrm{r}0}a^{-2} + \Omega_{\Lambda 0}a^2 + (1 - \Omega_0)}} = H_0 t + \mathcal{C}, \qquad (26.63)$$

where $w_m = 0$, $w_r = 1/3$ and $w_\Lambda = -1$.

Accurate values for the different Ω_0 parameters were determined by the WMAP [1] and the *Planck* [8] collaboration ($\Omega_{\mathrm{m}0}$ and $\Omega_{\Lambda 0}$) and by the COBE experiment ($\Omega_{\mathrm{r}0}$). The *Planck* values read

$$\Omega_{\Lambda 0} = 0.6889 \pm 0.0056,$$

$$\Omega_{\mathrm{m}0} = 0.3111 \pm 0.0056, \qquad (26.64)$$

$$\Omega_{\mathrm{r}0} = 8.4 \cdot 10^{-5},$$

where baryonic matter contributes only a fraction of

$$\Omega_{\mathrm{bary}} = 0.04897 \pm 0.00061 \qquad (26.65)$$

to the total energy density of matter. As already mentioned, the dominant fraction of the density of matter stems from dark matter, whose origin, however, is still completely unknown. In our universe, the vacuum energy density dominates today, whose origin also is unknown. Furthermore, it follows from these values that, within the error bars, $\Omega_0 = 1$ in our universe. This, however, is *not* equivalent to saying that really $q = 0$. This would be correct only for the exact equality $\Omega_0 \equiv 1$. Within the error bars of the measurements, the results presently are compatible with $q = 0$, but do not completely rule out the other two values $q = \pm 1$. We shall discuss the satellite missions COBE and WMAP just mentioned and their successor *Planck* in detail in Chap. 29.

After what has just been said, the current state of cosmology is somewhat strange: The current models describe the observational results very well, but at the same time, roughly 95% of the energy content of the universe is of an unknown origin. This situation is certainly unsatisfactory, and it demonstrates how many questions are still unanswered in modern physics.

Since we know the explicit dependence of the individual energy densities $\varepsilon_i(a)$ on the scaling factor, we can now use the numerical values from (26.64) to estimate at which values of the scaling factor two energy densities were of equal size, and determine what type of energy density dominated the evolution of the universe at a certain epoch.

From $\Omega_{\mathrm{r}}(a) = \varepsilon_{\mathrm{r}0}a^{-4}/\varepsilon_{\mathrm{c}}(a)$ and $\Omega_{\mathrm{m}}(a) = \varepsilon_{\mathrm{m}0}a^{-3}/\varepsilon_{\mathrm{c}}(a)$ we find

$$a_{\mathrm{rm}} = \frac{\Omega_{\mathrm{r}0}}{\Omega_{\mathrm{m}0}} \approx 2.7 \cdot 10^{-4}, \qquad (26.66)$$

evidently the critical densities in the denominators cancel each other. In like manner, we find with $\Omega_\Lambda(a) = \varepsilon_{\Lambda 0}/\varepsilon_c(a)$ that

$$a_{m\Lambda} = \left(\frac{\Omega_{m0}}{\Omega_{\Lambda 0}}\right)^{1/3} \approx 0.77. \tag{26.67}$$

Therefore, the universe was smaller by a factor of roughly 3000 than today when the energy densities of radiation and matter were of the same size. By contrast, the universe had already reached 77% of its current size when the vacuum energy density and the density of matter were of the same size. Using these two points in time we can roughly divide the history of the universe into 5 phases, where of course the transitions between the different phases proceeded smoothly:

1. The radiation-dominated phase for $\Omega_r \gg \Omega_m$.
2. The phase of radiation and matter for $\Omega_r \approx \Omega_m$.
3. The matter-dominated phase for $\Omega_m \gg \Omega_r$ and $\Omega_m \gg \Omega_\Lambda$.
4. The phase with contributions of comparable size from energy density of matter and vacuum energy density, i.e., $\Omega_m \approx \Omega_\Lambda$, the phase in which we are today.
5. The phase in which vacuum energy density dominates, i.e., for $\Omega_\Lambda \gg \Omega_m$.

Figure 26.3 shows these different phases with the corresponding values of the scaling factors, based on the values from (26.64). In the following we will solve the Friedmann equation separately for each of these phases, and discuss the properties of these models of the universe. (Of course, we are already familiar with the matter-dominated universe.)

In addition, it will also be worth looking at other models of the universe, even if they may be of lesser importance to our actual universe. These include models with non-vanishing curvature, which, as we have mentioned, cannot be completely ruled out by the observational measurements. Moreover, these models exhibit very interesting properties. Except for a very brief discussion at the end of the chapter, we shall only consider models with $\Lambda > 0$, i.e., with positive vacuum energy.

We begin our discussion by looking at the limiting cases $a \to 0$ and $a \to \infty$, i.e., the universes dominated either by radiation or Λ. In both cases we set $q = 0$, because these are mathematically easiest to handle. In all cases, we choose the initial condition $a(0) = 1$.

26.7.5 Flat Radiation Universe

When introducing the energy density of radiation we had already pointed out that it dominates the evolution of the universe for small scaling factors since it decreases faster with a than the energy density of matter. When we take only the contribution

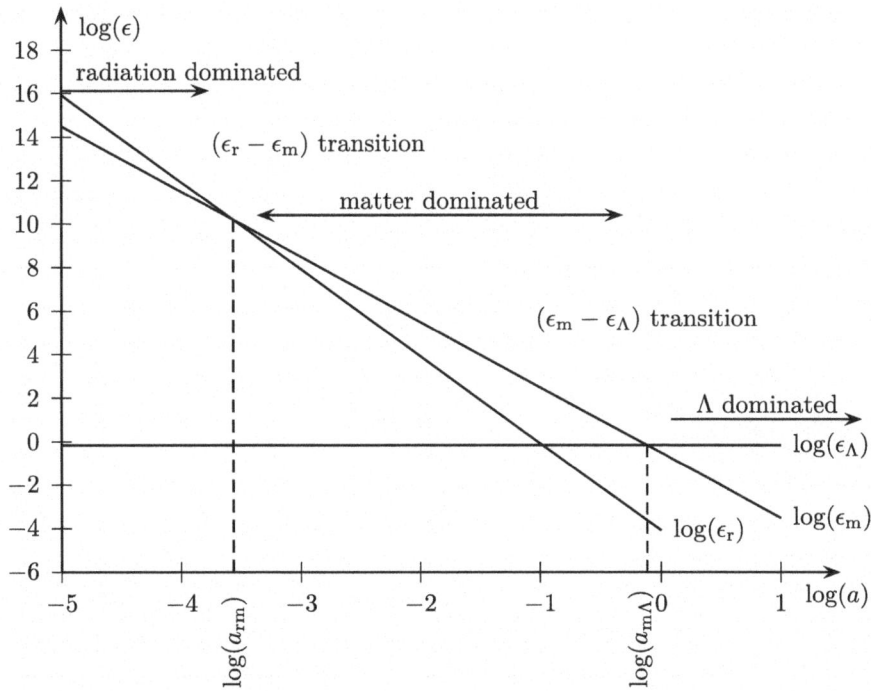

Fig. 26.3 Evolution of the radiation energy density, the matter energy density, and the vacuum energy density as a function of the scaling factor. In the diagram, today's status at $\log a = 0$ lies in the range of the $(\varepsilon_m - \varepsilon_\Lambda)$ transition

of radiation into account in (26.63), that is, when we consider a universe with $\Omega_{r0} = \Omega_0 = 1$, we obtain the equation

$$\int a\mathrm{d}a = H_0 t + \mathcal{C}, \tag{26.68}$$

which can be solved immediately:

$$a(t) = \sqrt{2H_0 t + 1}. \tag{26.69}$$

For

$$t_{\mathrm{BBr}} = -\frac{1}{2}H_0^{-1} \tag{26.70}$$

we have $a = 0$, therefore the age of the universe is given by t_{BBr} (BB = Big Bang). A flat universe that only contains radiation is of course described by (26.69) for all time.

26.7.6 Flat Λ Universe

In the limit of large scaling factors it is evident that the $\Omega_{\Lambda 0}$ term dominates in (26.63). For a universe with $\Omega_{\Lambda 0} = \Omega_0 = 1$ we have the result

$$\int \frac{da}{a} = H_0 t + C. \tag{26.71}$$

Here, too, the solution is simple, and we find

$$a(t) = e^{H_0 t}. \tag{26.72}$$

The metric associated with this behavior of $a(t)$ is called *de Sitter metric*. In this case, the Hubble parameter follows from (26.72) as

$$H(t) = \frac{a_{,t}(t)}{a(t)} = H_0. \tag{26.73}$$

Thus, in the de Sitter metric, the value of the Hubble parameter is equal to the Hubble constant for all time. The de Sitter metric describes an empty, flat universe in which only vacuum energy density is present. In this case, the value of the Hubble constant is directly related to the value of the cosmological constant. To see this, we exploit the fact that if Ω_0 is to be one, $\Omega_0 = \Omega_{\Lambda 0} = 1$, we then have $\varepsilon_{c0} = \varepsilon_\Lambda$. We combine (26.42) with (26.56) and find

$$H_0 = c\sqrt{\frac{\Lambda}{3}}. \tag{26.74}$$

In Sect. 24.2.1 we already briefly mentioned the perfect cosmological principle, and the *steady-state model* pertaining to it, according to which the universe is homogeneous not only with respect to space, but also does not change its essential properties over time (homogeneity in time). Obviously the de Sitter metric describes a model of the universe that features exactly this property. In particular, we also see that in the de Sitter metric, unlike in the other models we have come across so far, the scaling factor was never zero in the past. Thus, in the de Sitter model there is no Big Bang. In all other non-trivial models, $H(t)$ depends explicitly on time, and the perfect cosmological principle is therefore violated.

In the steady-model, proposed by *Bondi*[4] and *Gold*[5] [3] in 1948, the de Sitter metric is intended to describe a universe containing only matter. Since, unlike the vacuum energy density, the energy density of matter decreases during expansion, this is only possible if new matter is continuously created in order to keep the density

---

[4] Hermann Bondi, 1919–2005, British mathematician and cosmologist of Austrian descent.

[5] Thomas Gold, 1920–2004, US-American astrophysicist of Austrian descent.

constant. In a certain volume V the total mass contained is

$$M(t) = \rho_\mathrm{m}(t)V(t). \tag{26.75}$$

If the density of matter is to remain constant during expansion, matter must be created at a rate of

$$\dot{M}(t) = \rho_\mathrm{m0}\dot{V}(t). \tag{26.76}$$

With $V(t) \sim a(t)^3$, Eq. (26.72) implies $\dot{V}(t) = 3H_0V(t)$, and therefore

$$\dot{M}(t) = \rho_\mathrm{m0}3H_0V(t). \tag{26.77}$$

Using the critical mass density $\rho_\mathrm{c0} = (8.60 \pm 0.11) \cdot 10^{-27}\ \mathrm{kg\,m^{-3}}$ from (26.8), and $\Omega_\mathrm{m0} = 0.3111$ from (26.64), we obtain a mass density of our universe of

$$\rho_\mathrm{m0} \approx 2.68 \cdot 10^{-27}\ \mathrm{kg\,m^{-3}}. \tag{26.78}$$

This corresponds roughly to the mass of 2 protons per cubic meter. On the other hand, we have seen that the majority of matter in our universe is not baryonic in nature at all. This gives us the matter generation rate per volume

$$\frac{\dot{M}(t)}{V(t)} = 3\rho_\mathrm{m0}H_0 \approx 5.37 \cdot 10^{-37}\ \mathrm{kg\,m^{-3}\,y^{-1}}, \tag{26.79}$$

not quite one proton per cubic kilometer per year.

The discovery of the cosmic microwave background radiation, which we shall discuss in detail in Chap. 29, ultimately led to the realization that the steady-state model is an inappropriate description of our universe. However, since every infinitely expanding model of the universe asymptotically tends to the de Sitter metric for large scaling factors, the metric is important nonetheless. Furthermore, due to the very simple time dependence of the scaling factor, many theoretical calculations are possible analytically in the de Sitter spacetime, that are possible only numerically in other spacetimes. Therefore, the de Sitter metric is still considered as an example in many theoretical works.

26.7.7 Flat (Ω_{r0}-Ω_{m0}) Model

We can find an improved model for small scaling factors, when we include the energy density of matter as the second largest contribution in the limit $a \to 0$. Because of $\Omega_{r0} + \Omega_{m0} = \Omega_0 = 1$ we replace Ω_{m0} with $1 - \Omega_{r0}$, and find after a rather lengthy calculation

$$a_\mathrm{rm}(t) = B_\mathrm{rm}\left\{4\cos^2\left[\frac{1}{3}\mathrm{arctan2}\left(\sqrt{4 - f(t)^2}, f(t)\right)\right] - 1\right\}, \tag{26.80a}$$

with

$$f(t) = \frac{3}{2} \frac{(H_0 t + C_1)(\Omega_{r0} - 1)^2}{\Omega_{r0}^{3/2}}, \tag{26.80b}$$

and

$$C_1 = \frac{2}{3} \frac{(1 - 2B_{rm})\sqrt{B_{rm}(1 + B_{rm})}}{\sqrt{\Omega_{r0}}}, \tag{26.80c}$$

plus the abbreviation

$$B_{rm} = \frac{\Omega_{r0}}{1 - \Omega_{r0}}. \tag{26.80d}$$

In this definition we exploit the equality $\cos(ix) = \cosh(x)$. Therefore, the expression (26.80a) remains real valued also for $f(t) > 2$, when the argument of the arc tangent turns purely imaginary. The radiation-matter universe has a big bang at time

$$t_{BBrm} = \frac{2}{3} \frac{3\Omega_{r0} - 1 - 2\Omega_{r0}^{3/2}}{(1 - \Omega_{r0})^2} H_0^{-1}. \tag{26.81}$$

For $\Omega_{r0} = 0$, i.e., $\Omega_{m0} = 1$, this expression yields $t_{BBrm} = -2H_0^{-1}/3$ – the value for the matter universe –, and for $\Omega_{r0} \to 1$ one finds, using *L'Hospital's rule*,[6] $t_{BBrm} = -H_0^{-1}/2$, i.e., the age of the radiation-dominated universe. Therefore, a flat universe with energy contributions from both radiation and matter always has an age between $H_0^{-1}/2$ and $2H_0^{-1}/3$. In Fig. 26.4 the scaling factor of the $(\Omega_{r0}\text{-}\Omega_{m0})$ universe is shown together with its limiting models, the pure radiation universe from (26.69), and the pure matter universe.

26.7.8 Flat $(\Omega_{m0}\text{-}\Omega_{\Lambda0})$ Model

In complete analogy with the extension just discussed for small scaling factors, we obtain an improved model also for large scaling factors if the matter density is taken into account. In view of the values $\Omega_{\Lambda0} = 0.6889$ and $\Omega_{m0} = 0.3111$ from (26.64), it is evident that this model is particularly important for us, as it takes into account the main contributions to today's energy density. We proceed as in the previous

---

[6] Guillaume François Antoine, Marquis de L'Hospital, 1661–1704, French mathematician.

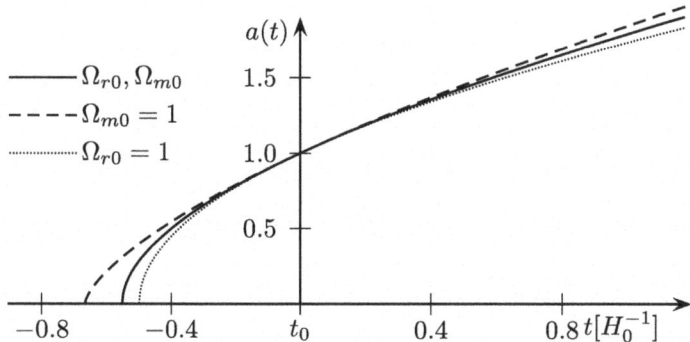

Fig. 26.4 Scaling factor $a(t)$ of the $(\Omega_{r0}\text{-}\Omega_{m0})$ universe with $\Omega_{r0} = \Omega_{m0} = 0.5$, in comparison with a matter-dominated and a radiation-dominated universe. In this model, the Big Bang happens at time $t_{\text{BBrm}}(0.5) \approx -0.552\, H_0^{-1}$, and thus between the two other limiting cases. For large scaling factors, the behavior approaches more and more that of the matter universe

section, insert $\Omega_{m0} = 1 - \Omega_{\Lambda 0}$ in all expressions (because of $\Omega_{m0} + \Omega_{\Lambda 0} = \Omega_0 = 1$), and find the scaling factor

$$a_{m\Lambda}(t) = \left[\frac{1}{\mathcal{B}_{m\Lambda}} \sinh^2\left(\frac{3}{2}\sqrt{\Omega_{\Lambda 0}}(H_0 t + C_1)\right)\right]^{1/3}, \tag{26.82a}$$

with

$$C_1 = \frac{2}{3}\frac{1}{\sqrt{\Omega_{\Lambda 0}}} \operatorname{arsinh}\left(\sqrt{\mathcal{B}_{m\Lambda}}\right) \tag{26.82b}$$

and

$$\mathcal{B}_{m\Lambda} = \frac{\Omega_{\Lambda 0}}{1 - \Omega_{\Lambda 0}}. \tag{26.82c}$$

An alternative form is obtained using the trigonometric relation $\sinh^2(x) = (\cosh(2x) - 1)/2$:

$$a_{m\Lambda}(t) = \left[\frac{1}{2\mathcal{B}_{m\Lambda}}\left(\cosh\left[3\sqrt{\Omega_{\Lambda 0}}(H_0 t + C_1)\right] - 1\right)\right]^{1/3}. \tag{26.82d}$$

Since $\sinh(0) = 0$, the age of the universe turns out to be

$$t_{\text{BBm}\Lambda} = -\frac{2}{3\sqrt{\Omega_{\Lambda 0}}} \operatorname{arsinh}\left(\sqrt{\mathcal{B}_{m\Lambda}}\right) H_0^{-1}. \tag{26.83}$$

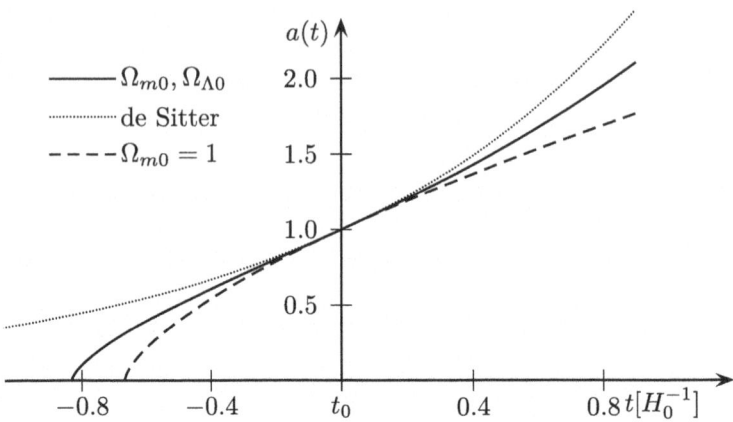

Fig. 26.5 Scaling factor $a(t)$ of the $(\Omega_{m0}\text{-}\Omega_{\Lambda 0})$ universe with $\Omega_{m0} = \Omega_{\Lambda 0} = 0.5$, in comparison with the matter-dominated and the de Sitter universe. In this model, the Big Bang is at $t_{\mathrm{BBm}\Lambda}(0.5) \approx -0.831\, H_0^{-1}$, and therefore this universe is older than the matter universe

For $\Omega_{\Lambda 0} = 0$ we obtain again the age of the matter universe with $t_{\mathrm{BBm}\Lambda} = -2H_0^{-1}/3$. For $\Omega_{\Lambda 0} \to 1$ this expression diverges. This is not surprising, because in this limiting case the model approaches the de Sitter universe, which has no Big Bang. However, we can firmly state that a flat universe with a positive cosmological constant and $\Omega_{m0} > 0$ always had a Big Bang a finite time ago.

We can also recognize that $a(t)$ exhibits the $t^{2/3}$ behavior for $t \to t_{\mathrm{BBm}\Lambda}$, i.e., $a \to 0$, when we insert the approximations $e^{\alpha t + y} \approx 1 + \alpha t$ and $e^{-(\alpha t + y)} \approx 1 - \alpha t$ in the hyperbolic sine, which is justified if t is close to $-y/\alpha$. Correspondingly, the de Sitter universe is obtained for $t \to \infty$, or $a \to \infty$, since in this case $e^{-(\alpha t + y)} \to 0$ and

$$\sinh^{2/3}\left(\frac{3}{2}\sqrt{\Omega_{\Lambda 0}}(H_0 t + C_1)\right) \approx e^{\sqrt{\Omega_{\Lambda 0}}(H_0 t + C_1)}. \tag{26.84}$$

In Fig. 26.5 the $(\Omega_{m0}\text{-}\Omega_{\Lambda 0})$ universe is shown together with its limiting cases.

The scaling factor in (26.82) describes a universe with an initially decelerated and later accelerated expansion. As we have already mentioned, it is exactly this type of evolution of the scaling factor that is assumed for the universe today on the basis of observations, and these observations are compatible with a flat universe.

When we calculate the time when the second derivative $a_{,tt}(t)$ of the scaling factor vanishes, we find

$$t_{\mathrm{b(m}\Lambda)} = \frac{2}{3\sqrt{\Omega_{\Lambda 0}}}\left[\operatorname{arcosh}\left(\frac{1}{2}\sqrt{6}\right) - \operatorname{arsinh}\left(\sqrt{\mathcal{B}_{\mathrm{m}\Lambda}}\right)\right]H_0^{-1}. \tag{26.85}$$

This is the turning point in time before which the expansion is decelerated, and after which the expansion is accelerated. Evaluating the scale factor at this time yields

Table 26.1 Values derived from the *Planck* results [8] for the $(\Omega_{m0}\text{-}\Omega_{\Lambda0})$ model. Negative times indicate a time span into the past

Characteristic times		In 10^9y	In H_0^{-1}
Beginning of accelerated expansion	t_b	−6.15	−0.4254
Age of the universe in the model	$t_{BBm\Lambda}$	−13.79	−0.9543
Estimated age of the universe	$-H_0^{-1}$	−14.45	−1.0

Scaling factor at $t = t_{b(m\Lambda)}$: $a(t_{b(m\Lambda)}) = 0.6089$

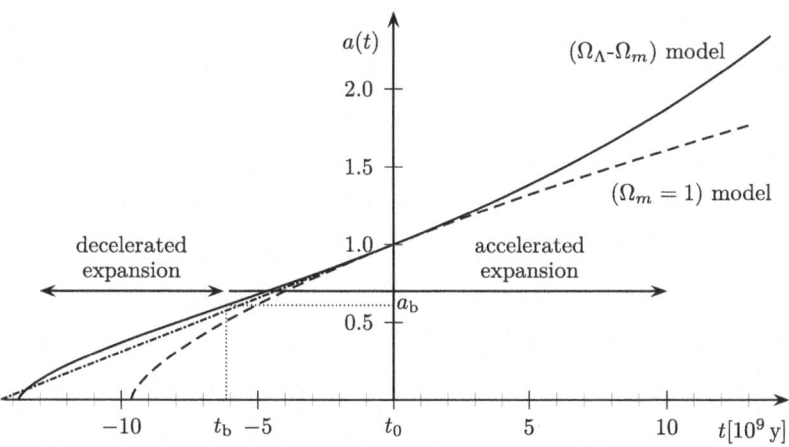

Fig. 26.6 Temporal evolution of the scaling factor $a(t)$ in a $(\Omega_m\text{-}\Omega_\Lambda)$ model adapted to our universe in units of billion of years. In this model, the Big Bang occurred at $t_{BBm\Lambda} \approx -0.9543\, H_0^{-1}$. For comparison, the *dashed* curve shows the $(\Omega_m = 1)$ model with a Big Bang at $t = -2H_0^{-1}/3$. The dashed-dotted straight line is the tangent to the curve at the present time, i.e., it is an estimate of the age of the universe using the Hubble time. At the time $t_b \approx -6.15\cdot10^9$ y, the transition occurred from decelerated to accelerated expansion

$$a(t_{b(m\Lambda)}) = \left[\frac{1}{2\mathcal{B}_{m\Lambda}}\left(\cosh\left[2\mathrm{arcosh}\left(\frac{1}{2}\sqrt{6}\right)\right] - 1\right)\right]^{1/3}$$

$$= \left(\frac{1}{2\mathcal{B}_{m\Lambda}}\right)^{1/3} \approx 0.7937\mathcal{B}_{m\Lambda}^{-1/3} \approx 0.6089. \tag{26.86}$$

Because of the small fraction of the energy density of radiation in (26.64), this model is an excellent approximation for the scaling factor of the actual universe for very large time periods into the past. It is only for very small values of the scaling factor, in time periods where the energy density of radiation is dominant, that the present model is not correct.

With the values of the Ω parameters in (26.64) and the Hubble constant in (23.6b), we can now adjust this model to our present universe according to the *Planck* results, where, of course, the energy contribution of radiation is neglected. We then obtain the values in Table 26.1. Figure 26.6 shows the temporal evolution of the scaling factor, assuming these values. As already mentioned, this is a very good approximation of the actual evolution of our universe over long periods of time. The deviation between $t_{BBm\Lambda}$ and the very crude approximation H_0^{-1} is less than 4%. In fact, the requirement $t_{BBm\Lambda} = -H_0^{-1}$ leads to the equation

$$\sqrt{\Omega_\Lambda} = \frac{2}{3} \operatorname{arsinh}\left(\sqrt{\frac{\Omega_\Lambda}{1 - \Omega_\Lambda}}\right), \tag{26.87}$$

with the numerical solution $\Omega_\Lambda \approx 0.737$, which is only a little larger than the *Planck* value.

26.7.9 Flat (Ω_{r0}-$\Omega_{\Lambda0}$) Model

This model is similar to the (Ω_{m0}-$\Omega_{\Lambda0}$) model. Actually, to us the model is far less important, since in our universe matter dominated at the time when the energy densities of radiation and the vacuum were of comparable size. However, this model has the mathematical allure that it can be treated analytically also with curvature, which is not possible for the (Ω_{m0}-$\Omega_{\Lambda0}$) model.

With $\Omega_{r0} + \Omega_{\Lambda0} = \Omega_0 = 1$ and $\mathcal{B}_{r\Lambda} = \Omega_{\Lambda0}/(1 - \Omega_{\Lambda0})$ we find the scaling factor

$$a_{r\Lambda}(t) = \left\{ \frac{1}{\sqrt{\mathcal{B}_{r\Lambda}}} \sinh\left[2\sqrt{\Omega_{\Lambda0}}\left(H_0 t + \frac{\operatorname{arsinh}\left(\sqrt{\mathcal{B}_{r\Lambda}}\right)}{2\sqrt{\Omega_\Lambda}} \right) \right] \right\}^{1/2}. \tag{26.88}$$

From this we can directly deduce the age of the universe

$$t_{BBr\Lambda} = -\frac{1}{2\sqrt{\Omega_\Lambda}} \operatorname{arsinh}\left(\sqrt{\mathcal{B}_{r\Lambda}}\right) H_0^{-1}. \tag{26.89}$$

In Fig. 26.7 this model is shown together with its limiting cases. One can recognize a behavior similar to that of the (Ω_{m0}-$\Omega_{\Lambda0}$) model, with the expansion rate initially being greater in the current case. In this model the change from decelerated to accelerated expansion takes place at

$$t_{b(r\Lambda)} = \frac{1}{4\sqrt{\Omega_{\Lambda0}}} \left[\operatorname{arcosh}(3) - 2\operatorname{arsinh}\left(\sqrt{\mathcal{B}_{r\Lambda}}\right) \right] H_0^{-1}. \tag{26.90}$$

For the value $\Omega_{\Lambda0} = 0.5$, used in the figure, one finds $t_{b(r\Lambda)} = 0$.

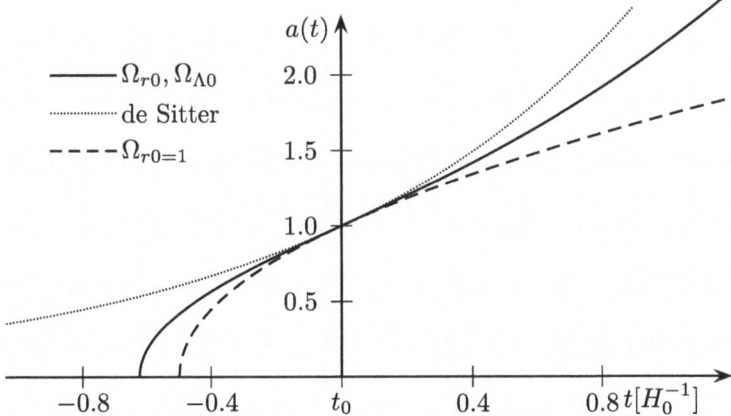

Fig. 26.7 Scaling factor $a(t)$ of the $(\Omega_{r0}\text{-}\Omega_{\Lambda0})$ universe with $\Omega_{r0} = \Omega_{\Lambda0} = 0.5$ in comparison with a radiation-dominated universe and the de Sitter universe. In this model, the Big Bang occurred at $t_{\mathrm{BBr}\Lambda}(0.5) \approx -0.623\,H_0^{-1}$, it is therefore older than the radiation-dominated universe. One can recognize the approach to the behavior of the de Sitter universe at large values of the scaling factor

26.7.10 Models with Curvature

Models with curvature result in the cases in which $\Omega_0 = \sum \Omega_i \neq 1$. Even though we shall later find good arguments in favor of the fact that the universe is flat, at least the universe observable to us, it is practically impossible to deduce a non-vanishing curvature from the observational results, since all quantities would have to be known exactly. We would therefore like to discuss in this section, how the previous models change when a curvature term is added. We recall once again that because of (26.11) the relations $\Omega_0 > 1 \Leftrightarrow q = 1$ and $\Omega_0 < 1 \Leftrightarrow q = -1$ apply.

However, for the model that is of most interest to us, the $(\Omega_{m0}\text{-}\Omega_{\Lambda0}\text{-}q)$ universe, no scaling factor can be found in analytical form. We will therefore have to restrict ourselves to the discussion of the other models, but will still be able to deduce many qualitative properties from the similar properties of the $(\Omega_{r0}\text{-}\Omega_{\Lambda0}\text{-}q)$ model.

Radiation and Curvature
For $q = 1$ we find the function

$$a_{rq\mathrm{pos}}(t) = \frac{1}{\sqrt{\Omega_{r0} - 1}}\sqrt{\Omega_{r0} - [(1 - \Omega_{r0})H_0t + 1]^2}. \tag{26.91}$$

In this model there exist two points in time, just as in the analogous case of the matter universe, when the scaling factor becomes zero. On the one hand at the time of the Big Bang,

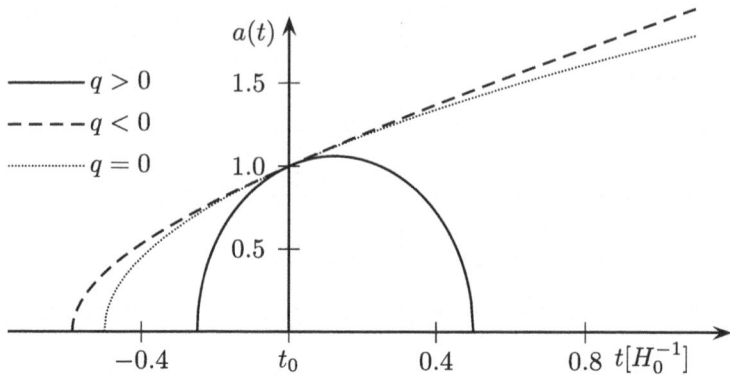

Fig. 26.8 Temporal evolution of the scaling factor $a(t)$ in a radiation-dominated universe with curvature. The model with positive curvature, in this case with $\Omega_{r0} = 9$, does recollapse. With the value chosen for Ω_{r0}, the collapse occurs at $H_0^{-1}/2$, as can be easily seen from (26.93). The model with negative curvature, in this case with $\Omega_{r0} = 0.5$, expands faster than the flat radiation-dominated universe

$$t_{\mathrm{BBr}q\mathrm{pos}} = \frac{1 - \sqrt{\Omega_{r0}}}{\Omega_{r0} - 1} H_0^{-1}, \tag{26.92}$$

on the other hand at the time of the "Big Crunch",

$$t_{\mathrm{BCr}q\mathrm{pos}} = \frac{1 + \sqrt{\Omega_{r0}}}{\Omega_{r0} - 1} H_0^{-1}. \tag{26.93}$$

For $q = -1$ the result is

$$a_{rq\mathrm{neg}}(t) = \frac{1}{\sqrt{1 - \Omega_{r0}}} \sqrt{[(1 - \Omega_{r0})H_0 t + 1]^2 - \Omega_{r0}}, \tag{26.94}$$

with the Big Bang taking place also at

$$t_{\mathrm{BBr}q\mathrm{neg}} = \frac{1 - \sqrt{\Omega_{r0}}}{\Omega_{r0} - 1} H_0^{-1}. \tag{26.95}$$

The two models are shown in Fig. 26.8, together with a flat radiation-dominated universe. The very high value $\Omega_{r0} = 9$ for the model with positive curvature was chosen only for the sake of comparison with the model that includes matter and curvature.

Matter and Curvature

Using these models, we had made ourselves familiar with the Friedmann equation, therefore they are already known. Here they are listed again for the sake of completeness. Furthermore, we present the results in the notation used in the foregoing sections. For example, comparing (25.36) with (26.63), we see that in our notation

$$a_{\text{ref}} = \pm \frac{1}{2} \frac{\Omega_{m0}}{\Omega_{m0} - 1} \quad \text{for} \quad q = \pm 1. \tag{26.96}$$

For $q = 0$ we then obtain

$$a_{\text{m}}(t) = \left(\frac{3}{2} H_0 t + 1 \right)^{2/3}, \tag{26.97}$$

with the Big Bang at $-2H_0^{-1}/3$. For the cases represented in parametrized form we can always find an angle η_0 from the condition $a(\eta_0) = 0$, and from the condition $t(\eta_0) = 0$ then the correct constant for t.

For $q = 1$ (25.49) becomes

$$t(\eta) = \frac{1}{2} \frac{\Omega_{m0}}{(\Omega_{m0} - 1)^{3/2}} H_0^{-1} \left[\eta - \sin(\eta) + \frac{2\sqrt{\Omega_{m0} - 1}}{\Omega_{m0}} - \arccos\left(\frac{2 - \Omega_{m0}}{\Omega_{m0}} \right) \right], \tag{26.98a}$$

$$a_{mq\text{pos}}(\eta) = \frac{1}{2} \frac{\Omega_{m0}}{\Omega_{m0} - 1} (1 - \cos(\eta)). \tag{26.98b}$$

For $q = -1$ (25.56) turns into

$$t(\eta) = \frac{1}{2} \frac{\Omega_{m0}}{(1 - \Omega_{m0})^{3/2}} H_0^{-1} \left[\sinh(\eta) - \eta + \operatorname{arcosh}\left(\frac{2 - \Omega_{m0}}{\Omega_{m0}} \right) - \frac{2\sqrt{1 - \Omega_{m0}}}{\Omega_{m0}} \right], \tag{26.99a}$$

$$a_{mq\text{neg}}(\eta) = \frac{1}{2} \frac{\Omega_{m0}}{1 - \Omega_{m0}} (\cosh(\eta) - 1). \tag{26.99b}$$

These scaling factors are shown in Fig. 26.9.

Cosmological Constant and Curvature

For this case it is worth taking another look at the potential picture, which changes considerably if only the vacuum energy density is present. Instead of the potential curve in Fig. 26.2 for positive Λ, with a potential maximum for a certain value of a, we now have a purely harmonic potential $V = -\Lambda c^2 a^2/3$. On the one hand, it is no longer true that for $a \to 0$ the rate of change $a_{,t}$ goes to infinity, as the potential now also vanishes in this limit. On the other hand, this leads to the fact that for $q = 1$ there exists a minimally permissible value for a, namely a value with

$$qc^2 = \frac{\Lambda c^2}{3} a_{\min}. \tag{26.100}$$

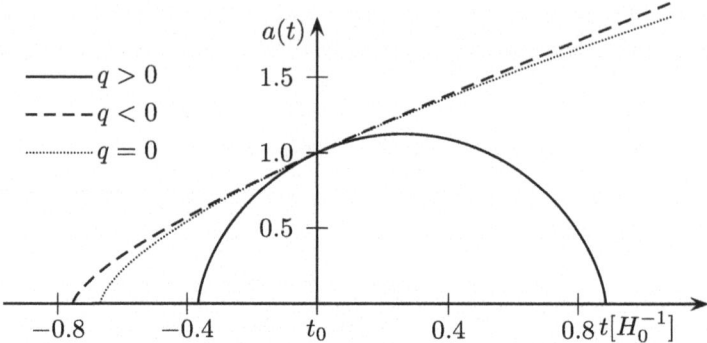

Fig. 26.9 Evolution of the scaling factor $a(t)$ in a matter universe with curvature. The model with positive curvature, here with $\Omega_{m0} = 9$, recollapses, the model with negative curvature, here with $\Omega_{m0} = 0.5$, expands faster than the flat model

Fig. 26.10 Qualitative picture for the Λ universe with curvature. For $\Lambda > 0$, without other contributions to the energy density, the potential is an inverted parabola, therefore for $q = 1$ forbidden values of the scaling factor $a < a_{\min}$ exist

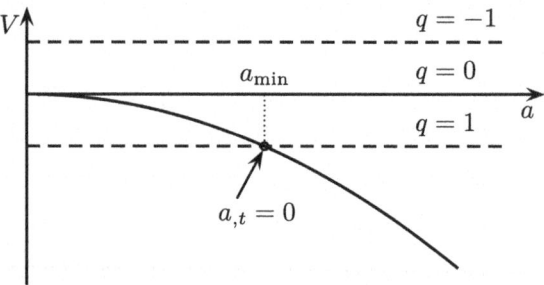

see Fig. 26.10. With the relations for Λ, q, and ε_{c0} from (26.5), (26.11), and (26.42) this leads to

$$a_{\min} = \sqrt{\frac{\Omega_{\Lambda 0} - 1}{\Omega_{\Lambda 0}}}, \qquad (26.101)$$

where here of course $\Omega_{\Lambda 0} > 1$ applies. The Friedmann equation then yields, independent of the sign of q, the scaling factor

$$a_{\Lambda q}(t) = \cosh(H_0\sqrt{\Omega_{\Lambda 0}}t) + \frac{1}{\sqrt{\Omega_{\Lambda 0}}} \sinh(H_0\sqrt{\Omega_{\Lambda 0}}t). \qquad (26.102)$$

In the case $q = -1$, the model has a Big Bang at

$$t_{\mathrm{BB}\Lambda q\mathrm{neg}} = -\frac{\mathrm{arcosh}\left(\frac{1}{\sqrt{1-\Omega_{\Lambda 0}}}\right)}{H_0\sqrt{\Omega_{\Lambda 0}}}. \qquad (26.103)$$

Figure 26.11 shows the scaling factors of the three models.

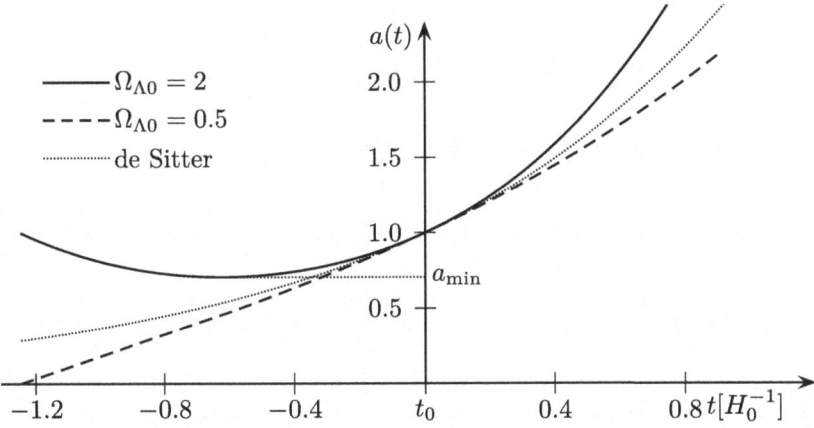

Fig. 26.11 Temporal evolution of the scaling factor in the Λ universe with curvature. Contrary to the de Sitter universe, the model with negative curvature, shown for $\Omega_{\Lambda 0} = 0.5$, features a Big Bang. By contrast, for positive curvature, shown for $\Omega_{\Lambda 0} = 2$, the scaling factor cannot fall below the value $a_{\min}$ from (26.101)

Radiation, Cosmological Constant, and Curvature

The (Ω_{r0}-$\Omega_{\Lambda 0}$) model with curvature is the most complex model that we will study analytically. This is the only case where we have contributions from two energy densities and can choose their sizes freely. In our previous models with two energy contributions, Ω_{i0} and Ω_{j0}, we always had $\Omega_{i0} + \Omega_{j0} = 1$. With this new freedom we find the scaling factor

$$
a_{r\Lambda q}(t) = \frac{1}{2\sqrt{\Omega_{\Lambda 0}}} \left\{ \left[(1 + \sqrt{\Omega_{\Lambda 0}})^2 - \Omega_{r0} \right] e^{2\sqrt{\Omega_{\Lambda 0}} H_0 t} \right.
$$
$$
\left. + \left[(1 - \sqrt{\Omega_{\Lambda 0}})^2 - \Omega_{r0} \right] e^{-2\sqrt{\Omega_{\Lambda 0}} H_0 t} - 2\Omega_q \right\}^{1/2},
$$
$$
(26.104)
$$

with the abbreviation

$$
\Omega_q = 1 - \Omega_{r0} - \Omega_{\Lambda 0}. \qquad (26.105)
$$

Big Bang and Big Crunch happen at the times

$$
t_{\text{BBr}\Lambda q} = \frac{1}{2\sqrt{\Omega_{\Lambda 0}} H_0} \ln \left(\frac{\Omega_q + 2\sqrt{\Omega_{\Lambda 0} \Omega_{r0}}}{(1 + \sqrt{\Omega_{\Lambda 0}})^2 - \Omega_{r0}} \right) \qquad (26.106)
$$

and

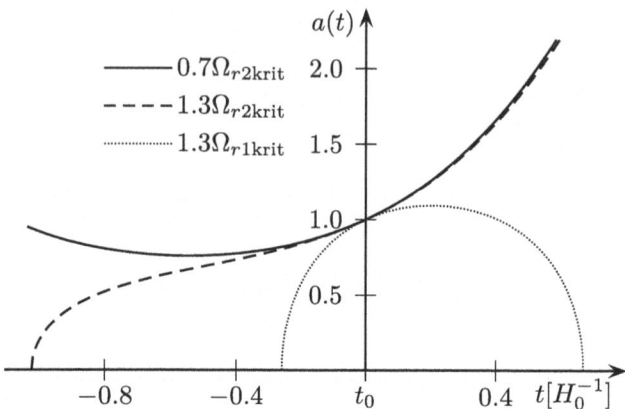

Fig. 26.12 Scaling factor in a model of the universe with radiation, vacuum energy density, and curvature for $\Omega_{\Lambda 0} = 3$. Depending on the values of the parameters Ω_{r0} and $\Omega_{\Lambda 0}$, the models do have a Big Bang, or can recollapse

$$t_{\mathrm{BCr\Lambda q}} = \frac{1}{2\sqrt{\Omega_{\Lambda 0}}H_0} \ln\left(\frac{\Omega_q - 2\sqrt{\Omega_{\Lambda 0}\Omega_{r0}}}{(1 + \sqrt{\Omega_{\Lambda 0}})^2 - \Omega_{r0}} \right), \qquad (26.107)$$

respectively. Depending on the values of the individual energy contributions these times can be real or complex valued. For example, in both expressions the denominator of the logarithm vanishes at

$$\Omega_{\mathrm{r1crit}} = (1 + \sqrt{\Omega_{\Lambda 0}})^2, \qquad (26.108)$$

and the numerator of the logarithm in (26.106) vanishes for

$$\Omega_{\mathrm{r2crit}} = (1 - \sqrt{\Omega_{\Lambda 0}})^2. \qquad (26.109)$$

Thus, models can be found with a Big Bang and a Big Crunch, with just a Big Bang, and also models with a minimum value of a. Figure 26.12 shows some examples.

26.7.11 Other Models

Of course there exist other models, in addition to the models discussed so far. A subgroup is made up by empty universes which contain only vacuum energy density. An example is the de Sitter universe (see Table 26.2). Apart from the latter, all other empty universes will be of minor importance in our further discussions. Yet some of these models played an important role historically. For example, early on the

Table 26.2 Scaling factors of models empty universes together with their designations

Λ	q	Scaling factor $a(t)$	Name
0	0	$a_0 \in \mathbb{R}$	Empty static universe
0	-1	$-ct + 1$	Milne model [6]
Positive	0	$\exp(H_0 t)$	de Sitter universe
Positive	1	$\cosh(H_0 t)$	–
Positive	-1	$\sinh(H_0 t)$	–
Negative	-1	$\sin(H_0 t)$	Anti-de Sitter-universe

Milne model[7] was proposed as a very simple cosmological model [6]. The de Sitter metric, in particular, nowadays plays an important role again, but beyond cosmology within the framework of the AdS/CFT correspondence, where a five-dimensional variant of this spacetime is being studied. Details about this interesting topic on an understandable level can be found in Zwiebach [11].

But even with these models all possible solutions of the Friedmann equation are far from being exhausted. A comprehensive article is devoted alone to classifying all two-component universes with different fractions, and arbitrary w parameters in the equation of state, plus additional vacuum energy density [5].

26.8   Exercises

26.1. Big Rip models In the models discussed so far we have encountered two possibilities for the future evolution of the universe. On the one hand models in which the universe eventually collapses again (*Big Crunch*), and models in which the universe expands forever, on the other. This case is also referred to as *Big Chill*.

Prove that in models which possess an energy contribution with $w < -1$ the scaling factor becomes infinitely large in *finite* time. This case is referred to as *Big Rip*.

26.2. Event horizon of the de Sitter metric We consider a universe equipped with the de Sitter metric, with the scaling factor $a(t) = e^{H_0 t}$ from (26.72). An observer emits a light signal at $r = r_B$ at time $t = t_B$. Calculate the time at which a second observer located in the origin, $r = 0$, receives the signal. Does he receive all signals?

26.3. Details for the model with radiation, cosmological constant, and curvature In this exercise we consider in some more detail a model with radiation, cosmological constant, and curvature. Which types of solutions show up for $\Omega_{r0} = \Omega_{r1crit}$ or $\Omega_{r0} = \Omega_{r2crit}$, and how do their properties depend on the value $\Omega_{\Lambda0}$?

---

[7] Edward Arthur Milne, 1896–1951, British astrophysicist and mathematician.

References

1. Bennett, C.L., et al.: Nine-year Wilkinson microwave anisotropy probe (WMAP) observations: final maps and results. Astrophys. J. Suppl. Ser. **208**(2), 20 (2013)
2. Berezhiani, Z., Nesti, F., Pilo, L., Rossi, N.: Gravity modification with Yukawa-type potential: dark matter and mirror gravity. J. High Energy Phys. **2009**(07), 083 (2009)
3. Bondi, H., Gold., T.: The steady-state theory of the expanding universe. Month. Not. R. Astron. Soc. **108**, 252–270 (1948)
4. Einstein, A.: Kosmologische Betrachtungen zur Allgemeinen Relativitätstheorie (Cosmological considerations on the general theory of relativity). Sitz. Preuß. Akad. Wiss. Berlin, 142–152 (1917)
5. Ha, T., et al.: Classification of the FRW universe with a cosmological constant and a perfect fluid of the equation of state $p = w\varrho$. Gen. Relativ. Gravit. **44**, 1433–1458 (2012)
6. Milne, E.A.: World-structure and the expansion of the universe. Z. Astrophys. **6**, 1–35 (1933)
7. Nemiroff, R.J., Patla, B.: Adventures in Friedmann cosmology: a detailed expansion of the cosmological Friedmann equations. Am. J. Phys. **76**(3), 265–276 (2008)
8. Planck Collaboration: Planck 2018 results - VI. Cosmological parameters. Astron. Astrophys. **641** A6 (2020). https://doi.org/10.1051/0004-6361/201833910
9. Rubin, V.C., Ford Jr., W.K.: Rotation of the Andromeda nebula from a spectroscopic survey of emission regions. Astrophys. J. **159**, 379–403 (1970)
10. Zwicky, F.: Die Rotverschiebung von extragalaktischen Nebeln (The redshift of extragalactic nebulae). Helv. Phys. Acta **6**, 110–127 (1933)
11. Zwiebach, B.: A First Course in String Theory. Cambridge University Press, Cambridge (2009)

Theoretical Prerequisites for Cosmological Observations

27

Contents

In the last chapter, we saw how differently the evolution of the various models of the universe can go on, depending on the contribution of the individual energy densities. In fact, many very different models are provided by the Friedmann equation. To find out which model best describes our universe, we need to determine from observations the value of the Hubble constant H_0, as well as the values of the individual Ω_{i0} parameters. However, these quantities cannot be measured directly, but must be extracted from other quantities. For example, if you wish to determine the Hubble constant by observing distant galaxies, you are facing the problem that, although you can measure the apparent magnitudes and the spectra of those galaxies, their distances cannot be determined directly.

In this chapter we would like to derive mathematical relations between the quantities that can actually be measured, and the relevant theoretical quantities describing a universe. In this way it will be possible to deduce the dynamics of our universe from the results of the observations.

© The Editor(s) (if applicable) and The Author(s), under exclusive license to
Springer-Verlag GmbH, DE, part of Springer Nature 2026
S. Boblest et al., *Special and General Relativity*,
https://doi.org/10.1007/978-3-662-71332-7_27

27.1   Linearization of the Scaling Factor

One of the goals of cosmology is to determine the scaling factor $a(t)$. However, extrapolating the complete behavior of the function $a(t)$ from measured values without incorporating further model assumptions is very hard. On the other hand, we can easily determine the behavior of the scaling factor for short periods of time around t_0, provided we know the value of the Hubble constant. It is simply $a(t) \approx 1 + H_0(t - t_0)$, which follows directly from the definition of H_0. A major step forward would be to measure deviations from this linear behavior. To describe this quantitatively we expand $a(t)$ in a Taylor series up to the second order around t_0:

$$
\begin{aligned}
a(t) &= 1 + H_0(t - t_0) + \frac{a_{,tt}(t_0)}{2}(t - t_0)^2 + \mathcal{O}\left((t - t_0)^3\right) \\
&= 1 + H_0(t - t_0) - \frac{1}{2}bH_0^2(t_0)(t - t_0)^2 + \mathcal{O}\left((t - t_0)^3\right).
\end{aligned}
\tag{27.1}
$$

Here we have introduced a new quantity

$$
b = -\frac{a_{,tt}(t_0)}{a_{,t}^2(t_0)} = -\frac{a_{,tt}(t_0)}{H_0^2}.
\tag{27.2}
$$

In cosmology, it is referred to as the *deceleration parameter*. The definition of b with a minus sign leads to $b > 0$, if $a_{,tt} < 0$, as was thought to be the case for our universe for a long time because, due to the limited observation possibilities, a matter-dominated model was taken for granted.

By contrast, the $(\Omega_{m0}$-$\Omega_{\Lambda 0})$ model of the universe assumed by us has $b < 0$. We can see this most straightforwardly when we express the deceleration parameter in terms of the Ω_{i0} parameters. A small transformation of the form of $a_{,tt}$ from (26.50a), using the definition of ε_{c0} in (26.5) and the definition of $\Omega_i(t)$ from (26.14), yields

$$
b = \frac{1}{2}\sum_i \Omega_{i0}(1 + 3w_i).
\tag{27.3}
$$

Inserting the values from (26.64) leads to

$$
b = \Omega_{r0} + \frac{1}{2}\Omega_{m0} - \Omega_{\Lambda 0} \approx -0.533.
\tag{27.4}
$$

In Table 27.1 the deceleration parameters of models discussed in the previous chapter are listed.

Table 27.1 Deceleration parameter of different models of the universe. We consider only models in which the Ω parameters add up to 1, and all contributions are the same size

Model	b
Radiation	1
Matter	1/2
de Sitter	−1
Ω_{r0}-Ω_{m0}	3/4
Ω_{m0}-$\Omega_{\Lambda0}$	−1/4
Ω_{r0}-$\Omega_{\Lambda0}$	0

27.2 Light Travel Distance

We have seen in Sect. 23.3 that Hubble was way off the mark in his estimate of the value of the Hubble constant because he extremely underestimated the distances to the galaxies he observed. Even today, it is still a very difficult task for astronomers and cosmologists to accurately determine the distance to a distant object. Apart from these methodological difficulties, we also have another problem. It is not at all easy to *define* what we actually mean by "distance" when we are dealing with the vast scales which occur in cosmology. The only information we have about galaxies is the light that reaches us from them. A possible measure would therefore be the distance travelled by photons emitted by a distant galaxy and detected on Earth. This is the definition of the *light travel distance*

$$d_{\mathrm{L}} = c(t_{\mathrm{d}} - t_{\mathrm{e}}). \tag{27.5}$$

However, during the time it takes for light to travel from emission at t_{e} to detection at t_{d}, the scaling factor of the universe is changing continuously. One specific signal takes the time $t_{\mathrm{d}} - t_{\mathrm{e}}$ to reach us. Another signal that starts later has a longer light travel time in the expanding universe, see Fig. 27.1. It is therefore evident that every distance definition in an expanding, or even collapsing, universe must have a time dependence.

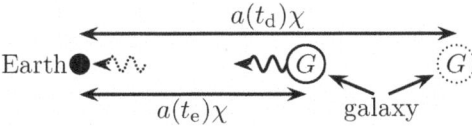

Fig. 27.1 The definition of distances on cosmological scales is complicated by the time-dependent scaling factor: While a ray of light is traveling from a galaxy to us, the universe is expanding continuously; therefore, the distance at the time of detection is not simply the light travel time multiplied by the speed of light

Fig. 27.2 A photon starts at $\vartheta_e = 0$ on the surface of a sphere at time $t = t_e$ (indicated by the short arrow), and travels to the point at $\vartheta = \vartheta_d$. When the photon arrives at time t_d, the distance to its point of origin is $d = a(t_d)\vartheta_d$, and not $c(t_d - t_e)$

27.3 Proper Distance

When we look at a distant galaxy then the finite speed of light implies that its light will have taken millions (or even billions) of years to have reached us. If we could construct a cosmological-sized ruler between the galaxy and the Earth then we would measure a quantity known as the *proper distance* d_{PD} to the galaxy. In other words, it is the separation between two distant objects at a fixed time t, that is, at a constant scaling factor $a(t)$.

To illustrate the concept of proper distance, we consider as a simple example the situation on a sphere that is being inflated. For example, if we emit a photon at $\vartheta_e = 0$ at time t_e to an object at polar angle ϑ_d, which arrives there at time t_d, the distance of the photon to its point of origin is

$$d_d = a(t_d)\vartheta_d, \tag{27.6}$$

see Fig. 27.2. However, this distance is of course not identical with the travel time of the ray of light multiplied by the speed of light.

An analogous relationship holds if we consider a comoving galaxy at the coordinates $(\chi_n, \vartheta_n, \varphi_n)$, where we use the FLRW metric in the form (24.20b), which we write down once again:

$$ds^2 = -c^2 dt^2 + a^2(t)\left[d\chi^2 + \left\{\begin{array}{c} \sin^2(\chi) \\ \chi^2 \\ \sinh^2(\chi) \end{array}\right\} \left(d\vartheta^2 + \sin^2(\vartheta)\,d\varphi^2\right)\right] \begin{cases} \text{for} & q = 1 \\ \text{for} & q = 0 \\ \text{for} & q = -1. \end{cases}$$
$$\tag{24.20b}$$

We shall assume that we are located at the coordinate $\chi = \vartheta = \varphi = 0$. Then the proper distance of a comoving galaxy with coordinates $(\chi, \vartheta, \varphi)$ is given by

$$d_{PD} = a(t)\chi. \tag{27.7}$$

Because of our convention $a(t_0) = 1$, the proper distance of each galaxy today coincides with its coordinate χ:

$$d_{PD}(t_0) = \chi. \tag{27.8}$$

The great advantage of the proper distance is its very simple definition. The proper distance is also what one would intuitively understand to be the distance of a galaxy today, contrary for instance to the light travel distance. In fact, the proper distance is the most widely used measure in cosmology, and when cosmologists use the term "distance" without further qualification they usually mean the proper distance. However, its great disadvantage is that it cannot be measured directly, since this would require determining instantaneously the distance of a far away galaxy by some cosmological ruler. A purely hypothetical possibility would be to position many (strictly speaking infinitely many) observers between us and the galaxy, and to sum up their momentary distances from each other at a given time.

Since the proper distance is defined as the product of the scaling factor and the coordinate χ, which is constant for comoving objects, we can write the Hubble law in a new form. Obviously the time derivative of d_{PD} is

$$\dot{d}_{PD} = \dot{a}(t)\chi, \tag{27.9}$$

and therefore

$$\dot{d}_{PD}(t) = H(t)d_{PD}(t). \tag{27.10}$$

Thus the change in time of the proper distance is given by the escape velocity, as it was determined by Hubble from using the redshift of galaxies, see (23.5). Here, however, the temporal change of the proper distance is not given by some velocity, but by the change of the scaling factor, since the coordinate χ of the comoving galaxy is constant, and therefore locally the galaxy is at rest.

To summarize: The proper distance provides a reasonable definition of distance, but we cannot measure it directly. The light travel time, too, of a detected signal cannot be determined directly. But what can be measured is the redshift of the signal. Using the redshift, we are able to introduce meaningful measurable definitions of distances.

27.4 Cosmological Redshift

We have already briefly explained that the expansion of space causes rays of light to be redshifted. In this section we wish to study the underlying relations once again in more detail. In contrast to the distance of a galaxy, the redshift of a galaxy can be determined very precisely by an accurate analysis of its spectrum and a comparison with laboratory data. The redshift is therefore one of the most important sources of information when observing distant objects.

27.4.1 Interpretation as a Doppler Effect

As discussed in Sect. 23.3, Hubble interpreted the redshifts he observed as Doppler shifts caused by the motion of galaxies away from us. According to (8.17), for $\beta \ll 1$ the frequency shift due to the longitudinal Doppler effect is given by

$$\omega_e = \omega_d \frac{\sqrt{1+\beta}}{\sqrt{1-\beta}} \approx \omega_d(1+\beta). \tag{27.11}$$

Here the velocity is defined with a positive sign if the galaxy is moving away from us. Only the longitudinal Doppler effect is important at this point since the comoving galaxy does not move perpendicularly to the line of sight, apart from a possible peculiar velocity. Then

$$\omega_d = \frac{\omega_e}{1+\beta} < \omega_e \tag{27.12}$$

and

$$z = \beta = \frac{v}{c}. \tag{27.13}$$

The redshift parameter therefore gives the apparent escape velocity in units of the speed of light. However, this line of argument is only correct for nearby galaxies with $\dot{d} \ll c$. Because of $v = \dot{d} = cz$ we also have

$$H_0 = c\frac{z}{d}. \tag{27.14}$$

According to this calculation, the escape velocity would become equal to the speed of light for $z = 1$. When we ask, in this picture, how far away a galaxy must be to have a redshift of $z = 1$, we arrive again at the Hubble distance cH_0^{-1} from (23.13).

The interpretation of redshift as an escape velocity is problematic for $v \geq c$. In the following section, we will therefore treat the redshift more generally, and resolve this problem.

27.4.2 Interpretation as an Effect of the Expansion of Space

In this section we shall establish a relation between the redshift and the change of the scaling factor during the time the light is traveling towards us. We shall see that there exists the relation $\lambda_d = \lambda_e a(t_d)/a(t_e)$ between the wavelengths at emission and detection, when the light was emitted at time t_e and detected at the later time t_d. In Sect. 26.5 we have already qualitatively explained in a few sentences why this relation should apply. Now we will present a quantitative derivation using the FLRW metric.

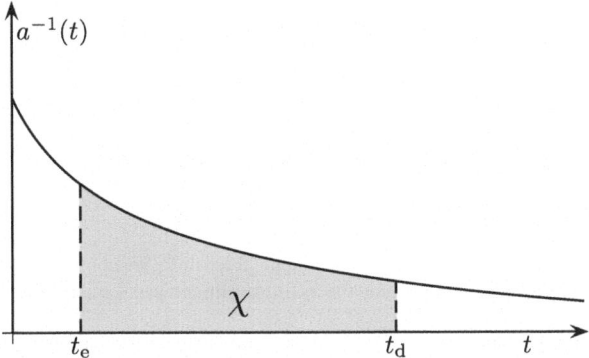

Fig. 27.3 Calculation of the cosmological redshift: The radial coordinate χ of a galaxy that we observe is proportional to the area below the curve $a^{-1}(t)$ during the complete time of flight

To do so we consider a ray of light that is propagating to us from a distant galaxy. For light, we have $ds = 0$, and therefore $-c^2 dt^2 + a^2(t)d\chi^2 = 0$, or

$$c \, dt = \pm a(t)d\chi. \tag{27.15}$$

The integration of this equation from the point in time of emission until the time of detection yields

$$\chi = c \int_{t_e}^{t_d} \frac{dt}{a(t)}. \tag{27.16}$$

For the trivial case $a(t) = a = $ const we would obtain $\chi = (c/a)(t_d - t_e)$, and thus the light travel distance $d_L = a\chi$.

In each case, the value of χ is proportional to the area below the curve $a^{-1}(t)$ as shown in Fig. 27.3. The crucial point to note is that the coordinate χ of the galaxy is independent of time, since it is constant in the comoving frame of reference. We now consider exactly one single wave train of the electromagnetic wave emitted by the galaxy. The head of the wave train starts at time t_e and reaches us at time t_d. The tail of the wave train starts at time $t_e + \Delta t_e$ and reaches us at time $t_d + \Delta t_d$ see Fig. 27.4. Due to the expansion, we know that $\Delta t_e \neq \Delta t_d$.

Since χ does not change, we must have

$$\frac{\chi}{c} = \int_{t_e}^{t_d} \frac{dt}{a(t)} = \int_{t_e + \Delta t_e}^{t_d + \Delta t_d} \frac{dt}{a(t)}, \tag{27.17}$$

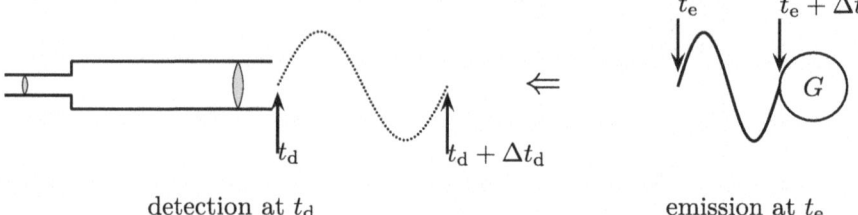

detection at t_d emission at t_e

Fig. 27.4 Sketch for the derivation of the cosmological redshift. We consider the head and the tail of a wave train of light which starts at time $t = t_e$, is traveling towards us, and arrives at time $t = t_d$

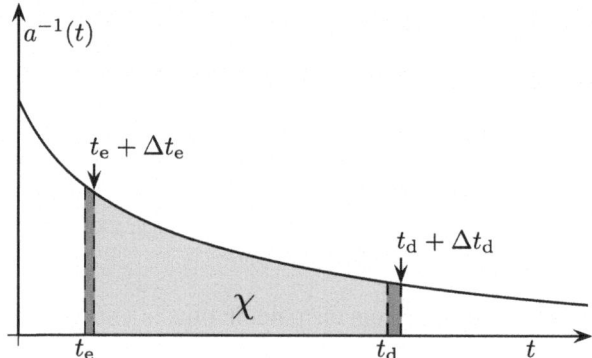

Fig. 27.5 Calculation of the cosmological redshift: The two integrals in (27.17) differ only by the small contributions shown in dark gray. The light-gray area under the curve is contained in both integrals. In the integration from t_e to t_d the dark-gray area on the left must be included, while in the integration from $t_e + \Delta t_e$ to $t_d + \Delta t_d$ it is the dark-gray area on the right

where for typical frequencies of light ν_0 we roughly have

$$\Delta t_e = \frac{1}{\nu_e} \sim 10^{-10} \text{ s.} \tag{27.18}$$

Over this short time span the expansion can be completely neglected, and a considered as constant. The two integrals in (27.17) differ only by small contributions, as illustrated by the dark-gray areas under the curve in Fig. 27.5. The first integral has an additional contribution of $\Delta t_e / a(t_e)$, the second an additional contribution of $\Delta t_d / a(t_d)$. The immediate consequence therefore is

$$\frac{\Delta t_e}{a(t_e)} = \frac{\Delta t_d}{a(t_d)}, \quad \text{with} \quad \Delta t_d = \frac{1}{\nu_d}, \tag{27.19}$$

i.e., we have

$$a(t_e)v_e = a(t_d)v_d, \quad \text{or} \quad \frac{\omega_e}{\omega_d} = \frac{a(t_d)}{a(t_e)}. \tag{27.20}$$

This yields for the redshift parameter

$$z = \frac{a(t_d)}{a(t_e)} - 1. \tag{27.21}$$

Since we are only interested in light that reaches us today, i.e., at $t_d = t_0$, we have $a(t_d) = a(t_0) = 1$ and therefore

$$z = \frac{1}{a(t_e)} - 1. \tag{27.22}$$

Conversely, we can determine from the redshift of a light signal how large the scaling factor was when this light was emitted:

$$a(t_e) = \frac{1}{1+z}. \tag{27.23}$$

Because the energy of the photons is also inversely proportional to the scaling factor according to Sect. 26.5, we additionally have the relation

$$E(t_0) = \frac{E(t_e)}{1+z} \tag{27.24}$$

for the energy $E(t_0)$ of a photon that was emitted with the energy $E(t_e)$.

The larger the value of z is, the further do we look back into the past, when the universe was smaller. However, this statement is only true if our universe has always been expanding. In the previous chapter we have found models of the universe in which an expansion phase was preceded by a contraction phase. But there is no evidence that this was the case with our universe.

It is a great advantage that the redshift depends only on the ratio of the scaling factors at emission and at detection, but not on the temporal evolution in the time in between. This is also the reason why we can infer directly from the observed redshift the value of the scaling factor of the universe at the time of emission. Would this be otherwise, and the redshift would depend on the temporal behavior of the scaling factor, the observed redshift would also contain model-dependent information that would be much more difficult to interpret.

In his 1929 measurements, Hubble looked at galaxies with redshifts up to about $z \leq 0.004$ [1]. The *Hubble Space Telescope* has made it possible to observe type Ia supernovae with much bigger redshifts, in the range up to $z \lesssim 1.7$. From (27.23) it follows that the universe was only about 37% of its current extension when the oldest of these supernovae occurred. As we will see, these observations allowed astronomers to draw conclusions regarding the expansion dynamics of the universe. The cosmic microwave background, which we shall discuss in Chap. 29, even has a redshift of $z \gtrsim 1000$. Analysis of the microwave background will provide information on the universe at a time when it had only about 0.1% of its current extension.

In principle, the redshift is relatively easy to determine from the spectrum of an object. To do so, one only has to identify specific wavelength windows of the spectrum, and compare the wavelengths with the values in the laboratory. Of course, there are various sources of error in such measurements. For example, the proper motion of our Milky Way in the local galaxy cluster must be taken into account. This motion superimposes a Doppler effect on the cosmological redshift. The same applies to galactic rotation, which results in a systematic redshift or blueshift with $v \approx 215 \, \mathrm{km \, s^{-1}}$, depending on the direction of observation. In particular, the peculiar velocities of the observed galaxies are problematic, especially at small redshifts. This can be seen very clearly from the large scatter of Hubble's own measurements in Fig. 23.4. The problem of the peculiar motion of galaxies can be solved by observing many galaxies, and then using statistical averaging. This procedure assumes that the peculiar velocities of the individual galaxies are distributed randomly. This assumption can again be justified by the cosmological principle.

27.4.3 Redshift in the Flat $(\Omega_{m0}\text{-}\Omega_{\Lambda 0})$ Model

In order to get an impression of the behavior of the redshift in a specific model of the universe, we consider the function $z(t)$ for the $(\Omega_{m0}\text{-}\Omega_{\Lambda 0})$ model, with the scaling factor from (26.82d). We can immediately write down this function, since this is an analytically solvable model,

$$z_{m\Lambda}(t) = \left[\frac{1 - \Omega_{\Lambda 0}}{2\Omega_{\Lambda 0}} \left(\cosh\left[3\sqrt{\Omega_{\Lambda 0}}(H_0 t + C_1) \right] - 1 \right) \right]^{-1/3} - 1. \qquad (27.25)$$

Figure 27.6 shows this curve in double-logarithmic representation. The redshift has a small value even for relatively long light travel times, as long as the signals were emitted at a sufficiently long time after the Big Bang. For example, a ray of light that left a galaxy just as the transition from decelerated to accelerated expansion was taking place, i.e., over 6 billion years ago, only has a redshift of about 0.64, because the scaling factor at that time was already about 61% of today's value, see (26.86). It is only for extremely old signals that the redshift increases strongly. The most distant known galaxy has a redshift of about 8.68 [3]. The cosmic microwave background, which was formed about 375,000 years after the Big Bang, finally has a redshift of $z \approx 1090$.

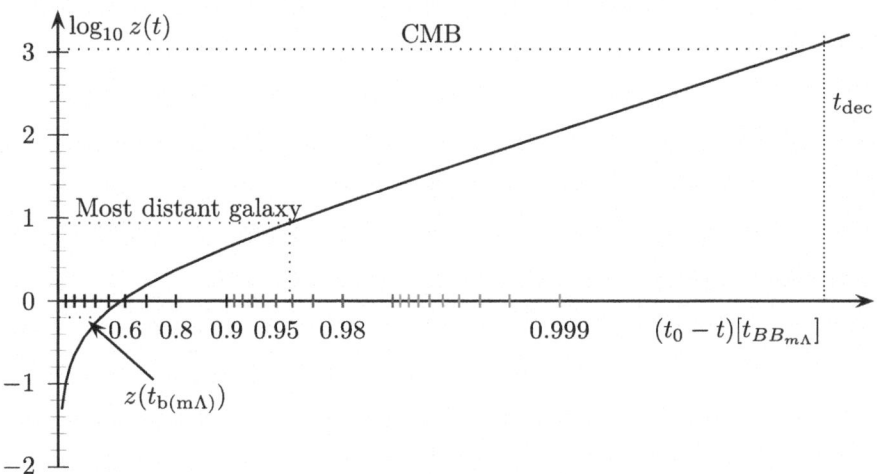

Fig. 27.6 Redshift $z_{m\Lambda}(t)$ in the $(\Omega_{m0}\text{-}\Omega_{\Lambda0})$ model in logarithmic representation going back to the time of the Big Bang $t_{BB_{m\Lambda}}$. The figure shows the values of the redshift at the time of the change from decelerated to accelerated expansion at $z(t_{b(m\Lambda)}) \approx 0.64$, with $t_{b(m\Lambda)}$ from (26.85), the galaxy with the highest observed redshift with $z \approx 8.68$, and the redshift of the CMB with $z \approx 1090$. The time of decoupling t_{dec} was about 375,000 years after the Big Bang, see Table 29.1

27.5 Measuring Distances in Cosmology

In Sect. 1.5.6 we looked at the problem of measuring distances on *astronomical length scales*, i.e., distances of objects within our solar system, and even to objects in nearby galaxies. Here we will discuss methods and problems that arise when astronomers try to determine the distances of even more distant objects. Of course, there is a smooth transition from measuring distances of nearby objects to more distant objects, but we will now focus on problems in measuring distances which we can only tackle using the findings that we have learned about in the framework of cosmology.

As already discussed, the method of parallaxes is not suitable for very distant objects, since the very small angles that appear cannot be resolved. However, it is still possible to determine distances using *standard candles*, i.e., objects whose absolute magnitude is known. In Sect. 1.5.6 we have seen that we can determine the distance of an object by measuring its radiation flux S per surface area on the Earth, if its luminosity L is known via

$$d = \left(\frac{L}{4\pi S}\right)^{1/2}. \tag{27.26}$$

The rationale behind this formula was that the entire radiation power must go through a sphere with radius d and a surface area of $4\pi d^2$. This simple relation is no longer valid, and we have to introduce two corrections.

Fig. 27.7 The radiation flux from a galaxy arriving on Earth on one square meter depends on the total radiation power of the galaxy going through the surface of a sphere with radius $r = d_{\text{LD}}$. However, due to the possibly curved geometry of spacetime and the redshift, the relationships here are more complicated than for a nearby star, as previously shown in Fig. 1.5

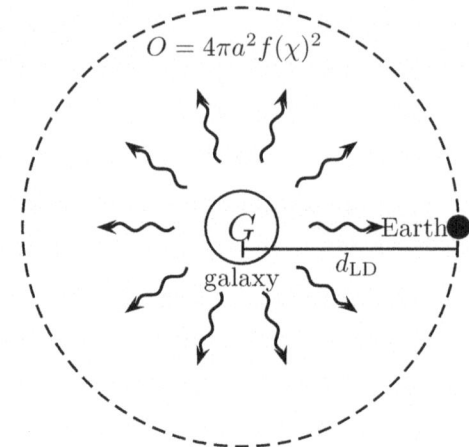

On the one hand, in curved spaces the surface of a sphere with radius $d = a\chi$ generally is no longer given by its Euclidean value $S_{\text{Euc}} = 4\pi a^2 \chi^2$, since the generalized solid angle element in (24.20b) is defined by

$$d\Omega^2 = a^2 f^2(\chi)(d\vartheta^2 + \sin^2(\vartheta)\,d\varphi^2) \quad \text{with} \quad f^2 = \begin{cases} \sin^2(\chi) & \text{for} \quad q = 1 \\ \chi^2 & \text{for} \quad q = 0 \\ \sinh^2(\chi) & \text{for} \quad q = -1 \end{cases}$$

(27.27)

As we observe the galaxies at the present time t_0, we can set $a = 1$ in what follows. This yields for the surface of a sphere

$$S_{\text{sphere}} = \begin{cases} 4\pi \sin^2(\chi) < S_{\text{Euc}} & \text{for} \quad q = 1 \\ 4\pi \chi^2 = S_{\text{Euc}} & \text{for} \quad q = 0 \\ 4\pi \sinh^2(\chi) > S_{\text{Euc}} & \text{for} \quad q = -1. \end{cases}$$

(27.28)

Depending on the curvature, the surface is smaller or larger than the Euclidean value, or equal to it, see Fig. 27.7. Accordingly, the incoming radiation flux for identical luminosity is smaller or greater than that in Euclidean space, or equal to it.

On the other hand, we have to take into account the change in energy of the photons according to (27.24), which contributes an additional factor $(1 + z)^{-1}$. Finally, it is not only the energy of the photons that changes but also the number density of photons: If, in a naive picture, we imagine a ray of light as a sequence of photons with a certain temporal or spatial distance, it becomes evident that these temporal or spatial distances change as $a(t)^{-1}$, since the total photon density decreases with $a(t)^{-3}$. This adds another factor $(1 + z)^{-1}$. Therefore, the final expression for the radiation flux is

$$S = \frac{L}{4\pi f^2(\chi)} \frac{1}{(1 + z)^2}. \tag{27.29}$$

We can define a *luminosity distance* from this expression

$$d_{LD} = f(\chi)(1 + z). \tag{27.30}$$

Because of the Taylor expansions $\sin(\chi) \approx \chi + \mathcal{O}(\chi^3)$ and $\sinh(\chi) \approx \chi + \mathcal{O}(\chi^3)$, we have in each case $f(\chi) \approx \chi$ for all objects with $\chi \ll 1$, and because of $d_{PD}(t_0) = \chi$ then

$$d_{LD}(t_0) = d_{PD}(t_0)(1 + z). \tag{27.31}$$

Thus, the luminosity distance overestimates the proper distance by a factor of $1 + z$.

Another possibility of determining distances follows from the relation between the angular diameter of the object and its distance. The necessary mathematical relations are essentially the same as for the method of parallaxes, discussed in Sect. 1.5.6, the only difference being that one does not determine the change of the angle of observation over one year—for large distances it is immeasurably small— but one determines the opening angle which an object of known size λ subtends in the sky.

In Euclidean space, for small angles $\Delta\vartheta$ one has a distance of

$$d = \frac{\lambda}{\Delta\vartheta}. \tag{27.32}$$

As expected, in curved spaces the situation is slightly more complicated. Let us consider again an object at the coordinates $(\chi_n, \vartheta_n, \varphi_n)$, but now we take into account its finite extension. This means, the coordinates just given are the position of one end of the object, the coordinates $(\chi_n, \vartheta_n + \Delta\vartheta, \varphi_n)$ give the position of the other end. Now let two rays of light be emitted simultaneously in our direction from both ends of the object. Both rays of light propagate radially towards us, i.e., we always have only $d\chi \neq 0$. That this must be so follows from homogeneity and isotropy, for symmetry reasons the ϑ and φ coordinates cannot change during the flight. For small angles $\Delta\vartheta$ the extension of the object seen from our point of view is

$$\lambda = a(t_e) f(\chi) \Delta\vartheta. \tag{27.33}$$

The correction to the Euclidean expression is again given by the general expression $f(\chi)$ instead of just χ. In addition, we must take into account the value of the scaling factor $a(t_e)$ at the time of emission. This leads again to another factor $(1 + z)^{-1}$.

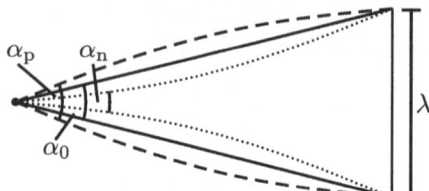

Fig. 27.8 Depending on the curvature of space, a structure in the sky of known extension λ appears differently in size. By measuring the viewing angle which an object of known size subtends in the sky, it is therefore possible to decide whether the space is flat or curved

Therefore, the *angular diameter distance* is

$$d_A = \frac{f(\chi)}{1+z}. \tag{27.34}$$

For $\chi \ll 1$ we get

$$d_A = \frac{d(t_0)}{1+z} = d(t_e). \tag{27.35}$$

The angular diameter distance makes it possible to draw conclusions as to the curvature of space. When one observes a distant object of known size λ, which subtends a certain angle in the sky, then one can compare with the angle that is expected theoretically. There are three possibilities:

1. The angle appears to be larger. This is evidence for a positive curvature of space.
2. The angle appears as expected. This is evidence for a flat space.
3. The angle appears to be smaller. This is an indication of a negatively curved space.

The three situations are illustrated in Fig. 27.8. This method can be applied to structures in the cosmic microwave background, which are produced by oscillatory phenomena with a specific wavelength λ. The latter can be derived independently using theoretical considerations, see Chap. 29. For this very reason we have used the symbol λ of the wavelength for the extension of the structure.

27.6   Luminosity-Redshift Relation

Our aim now would be to find a relation between the luminosity of an object and its redshift. To do so we would have to evaluate, for a model of the universe with given scaling factor function $a(t)$, Eq. (27.16) for χ, and then insert it in Eq. (27.29) for the luminosity, with the functions $f(\chi)$ from (27.27), and noting of course the definition of z as a function of $a(t)$. For arbitrary models of the universe and arbitrarily large values of χ, this can be done only numerically. However, an approximate solution can be found in the limit $\chi \ll 1$. We have seen that in this limit the function $f(\chi)$ becomes model-independent.

So our aim now is, for $\chi \ll 1$, to find an approximate expression for χ in such a way that only directly measurable parameters are included in our expression for S. By construction, this expression will be valid only for small redshifts, i.e., small χ, or short light travel times $t_0 - t_e$. Therefore, we can use the expansion of the scaling factor from (27.1), and insert it in the expression for χ in (27.17):

$$
\chi \approx c \int_{t_e}^{t_0} \frac{dt}{1 + H_0(t - t_0) - \frac{1}{2}bH_0^2(t - t_0)^2}
$$
(27.36)
$$
\approx c \left[(t_0 - t_e) + \frac{H_0}{2}(t_0 - t_e)^2 \right].
$$

The deceleration parameter would contribute only in third order of $t_0 - t_e$, and therefore has disappeared from this expansion. The redshift is obtained from the definition of z in (27.22) using the second-order expansion of the scaling parameter

$$
z \approx \frac{1}{1 + H_0(t_e - t_0) - \frac{1}{2}bH_0^2(t_e - t_0)^2} - 1.
$$
(27.37)

Resolved with respect to the denominator, (27.37) is a quadratic equation for $t_e - t_0$, with the solution

$$
t_e - t_0 \approx \frac{z}{H_0} \left[\left(1 + \frac{b}{2} \right) z - 1 \right].
$$
(27.38)

For the square we have

$$
(t_e - t_0)^2 = \frac{z^2}{H_0^2} + \mathcal{O}(z^3).
$$
(27.39)

Inserting (27.38) and (27.39) in (27.36) leads to

$$
\chi \approx \frac{cz}{H_0} \left[1 - \frac{(1 + b)z}{2} \right].
$$
(27.40)

When we insert this in the formula for the radiation flux, together with $a(t_e) = (1 + z)^{-1}$, we obtain

$$
S \approx \frac{L}{4\pi} \frac{H_0^2}{c^2} \frac{1}{(1 + z)^2} \frac{1}{z^2} \left[1 - \frac{(1 + b)z}{2} \right]^{-2}.
$$
(27.41)

We exploit once again the constraint $z \ll 1$, and make the approximations

$$(1 + z)^{-2} \approx 1 - 2z, \tag{27.42}$$

$$\left(1 - \frac{1 + b}{2} z\right)^{-2} \approx 1 + (1 + b)z \tag{27.43}$$

and

$$(1 - 2z)(1 + (1 + b)z) \approx 1 + (b - 1)z. \tag{27.44}$$

Thus, the final result is

$$S = \frac{L H_0^2}{4\pi c^2} \frac{1 - (1 - b)z}{z^2}. \tag{27.45}$$

Astronomers usually work with the apparent magnitude of an object, instead of its radiation flux. To arrive at an expression for the apparent magnitude, we have to reformulate (27.45). First we multiply by $4\pi c^2 / (L H_0^2)$, then take the common logarithm on both sides, and multiply by -1. We then have on the right-hand side

$$
\begin{aligned}
-\log_{10}\left(\frac{1 - (1 - b)z}{z^2}\right) &= \log_{10}(z^2) - \log_{10}(1 - (1 - b)z) \\
&\approx 2\log_{10}(z) + (1 - b)\log_{10}(e)z.
\end{aligned}
\tag{27.46}
$$

Note that we have Taylor-expanded in the second step. We can assume that $h = -(1 - b)z \ll 1$, because of $z \ll 1$. Then we can expand the logarithm around 1, i.e., the point with $h = 0$, and obtain

$$\log_{10}(1 + h) = \log_{10}(1) + \frac{d}{dh}\log_{10}(1 + h)|_0 h. \tag{27.47}$$

Note that the derivative is given by

$$\frac{d}{dh}\log_{10}(1 + h) = \frac{\log_{10}(e)}{1 + h}. \tag{27.48}$$

In accordance with the definition of the difference of classes of magnitude in (1.69), we still have to multiply by 2.5. In total, we then arrive at

$$-2.5\log_{10}\left(\frac{S}{\frac{L H_0^2}{4\pi c^2}}\right) = 5\log_{10}(z) + 2.5\log_{10}(e)(1 - b)z. \tag{27.49}$$

We now expand in the bracket on the left-hand side the numerator and the denominator by the square of a (yet unspecified) referential radius R_{ref} and split the logarithm into two terms,

$$-2.5 \log_{10}\left(\frac{S}{\frac{LH_0^2}{4\pi c^2}}\frac{R_{\text{ref}}^2}{R_{\text{ref}}^2}\right) = -2.5 \log_{10}\left(\frac{S}{\frac{L}{4\pi R_{\text{ref}}^2}}\right) + 2.5 \log_{10}\left(\frac{H_0^2 R_{\text{ref}}^2}{c^2}\right).$$

(27.50)

Note that the arguments of all logarithms are dimensionless. When we choose the referential radius $R_{\text{ref}} = 10$ pc, then the first term on the right-hand side is already the difference of the apparent and absolute magnitude, $m - M$. With this value of the referential radius, the second term on the right-hand side becomes

$$5 \log_{10}\left(\frac{H_0 R_{\text{ref}}}{c}\right) = 5 \log_{10}\left(\frac{H_0\, 10\,\text{pc}}{c}\frac{1\,\text{pc}}{1\,\text{pc}}\right) = 5 \log_{10}\left(\frac{H_0\, 1\,\text{pc}}{c}\right) + 5.$$

(27.51)

Note that again the arguments of the logarithms are dimensionless. Thus, the left-hand side of (27.49) becomes

$$-2.5 \log_{10}\left(\frac{S}{\frac{LH_0^2}{4\pi c^2}}\right) = m - M + 5 \log_{10}\left(\frac{H_0\, 1\,\text{pc}}{c}\right) + 5.$$

(27.52)

When we bring together the left-hand and right-hand sides of (27.49) we end up with the relation

$$m = M - 5\left(1 + \log_{10}\frac{H_0\, 1\,\text{pc}}{c}\right) + 5\log_{10}(z) + 1.086(1 - b)z + \mathcal{O}(z^2),$$

(27.53)

with $2.5 \log_{10}(e) \approx 1.086$. Thus, we have found an expression that links the observed apparent magnitude m of an object to its absolute magnitude M and its redshift z. The dynamics of the universe enters the relationship (27.53) via the Hubble constant H_0 and the deceleration parameter b. If m and z are measured simultaneously, conclusions can therefore be drawn regarding the expansion history of the universe, provided that the absolute magnitude of the observed object is known.

27.7 Corrections to the Luminosity-Redshift Relation

The expression $S(z)$ just derived for the luminosity-redshift relation is still too simplified with regard to two points, and requires modification by correction terms. Here we can only justify the reasons of these corrections, the discussion of their exact forms lies beyond the scope of this chapter.

Equation (27.53) refers to the apparent magnitude integrated over all wavelengths. However, observation instruments usually measure through a single filter or bandpass of wavelengths. Therefore, observers only measure a fraction of the total spectrum, redshifted into the frame of the observer, and this changes the value of the apparent magnitude. Model assumptions must therefore be made when determining this correction to the apparent magnitude. This is referred to as the K correction,[1] and an additional term $K(z)$ enters in (27.53).

Furthermore, (27.53) is built on the implicit assumption that the luminosity of the observed object does not change over time. This is not correct, because stars do change their luminosity during their lifetime. But then the luminosity of a galaxy also changes over time. The greater the value of z for an object, the earlier in the past we observe this object, and correspondingly the difference in luminosity can be large due to the evolution of the object. This difference enters (27.53) in the form of an evolution term $E(z)$. The calculation of this term requires models of galaxy evolution, and also models of stellar evolution.

With the corrections just discussed one obtains the formula for the relation between apparent magnitude and redshift

$$
\begin{aligned}
m = & M - 5\left(1 + \log_{10}\frac{H_0 1\,\text{pc}}{c}\right) + 5\log_{10}(z) + 1.086\,(1 - b)z \\
& - K(z) - E(z) + \mathcal{O}(z^2).
\end{aligned}
\tag{27.54}
$$

The comparison with observations shows that without these corrections no agreement with the measurements can be achieved. The important point for us to note is that by the two terms K and E now model-dependent assumptions enter in (27.54), which makes the evaluation of observational results even more difficult. Further details about these corrections can be found in the paper by Yoshii and Takahara [2].

References

1. Hubble, E.: A relation between distance and radial velocity among extra-galactic nebulae. Proc. Natl. Acad. Sci. **15**, 168–173 (1929)
2. Yoshii, Y., Takahara, F: Galactic evolution and cosmology. Astrophys. J. **326**, 1–18 (1988)
3. Zitrin, A., et al.: Lyman-alpha emission from a luminous $z = 8.68$ galaxy: implications for galaxies as tracers of cosmic reionization, Astrophys. J. Lett. **810**, L12 (2015)

---

[1] The designation was first introduced by the German astronomer Wirtz in 1918 as "Konstante k" correction when studying the redshift of spiral nebulae.

SN Ia as Standard Candles for the Young Universe

28

Contents

Using the luminosity-redshift relation derived in Chap. 27, two groups studied distant supernovae in the 1990s. One was the *Supernova Cosmology Project* group [3], and the other the *High-z Supernova Search Team* [2]. Their observational results provided strong evidence for the fact that we live in a universe that is expanding at an accelerated rate. The principal investigators of the two groups, Perlmutter,[1] and Schmidt[2] and Riess[3] were awarded the Nobel Prize in Physics for their discovery in 2011.

In this chapter we will present the results of this fascinating discovery, and discuss the methods involved in it. To do so we first have to understand how one can infer the dynamics of the universe from the luminosity-redshift relation. We have already pointed out that, to apply this relation, objects of known absolute magnitude are needed, in other words, standard candles. For cosmological studies it is desirable to look back deep into the past of the universe. This means one has to observe very distant objects. Therefore, these objects must also have a very high absolute luminosity. The two groups used type Ia supernovae (SN Ia) for this purpose.

---

[1] Saul Perlmutter, ⋆1959, US-American astrophysicist.

[2] Brian Paul Schmidt, ⋆1967, US-American/Australian astronomer.

[3] Adam Guy Riess, ⋆1969, US-American astrophysicist.

28.1   Determining the Dynamics of the Universe

It is evident, and we have said it several times, that the brightness of an object is a measure of its distance from us. For a given absolute magnitude M, an object will appear the darker, the further away it is from us. When we plot the apparent magnitude of objects with known absolute magnitude as a function of $\log_{10}(z)$, we obtain a straight line for small redshifts, according to (27.53), or (27.54). From the intercept with the y-axis we can determine the value of H_0. This was done in the *Calan/Tololo Supernova Survey*, in which nearby SN Ia events with $z \lesssim 0.1$ could be detected with terrestrial telescopes. For larger values of z, deviations from the straight-line form will show up, from which we can determine the deceleration parameter, and thus decide whether the universe is expanding at a decelerated or an accelerated rate.

We can illustrate very nicely what we can learn from the luminosity-redshift relation. Let us assume we have observed a cosmic object, and found that its spectrum has a redshift of, e.g., $z = 1$. From (27.22) we then see that the ratio of the scaling factor today, to the value of the scaling factor at the time when the light was emitted, is equal to 0.5, independent of the temporal evolution of the scaling factor in between.

However, the light travel time, and thus the distance of the observed object from us, will depend on the type of the temporal evolution of the universe. This is illustrated in Fig. 28.1 for fixed $a(t_e)$ at time of emission, i.e., for fixed redshift z_e. In an accelerating expanding universe, the light with this redshift has been emitted at an earlier time $t_{e,a}$ than in an expanding universe with constant acceleration $(t_{e,c})$ or in an expanding universe with decelerating expansion $(t_{e,d})$. Therefore, the observed object will appear darker in an accelerating expanding universe than in a

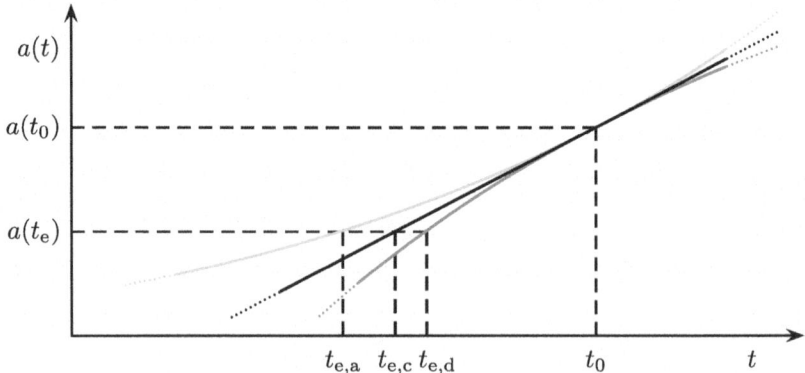

Fig. 28.1 Relation between the luminosity and the redshift of a cosmic object. The redshift of the detected spectrum depends only on the value of the scaling parameter $a(t_e)$ at the time of emission, $z_e = \frac{a(t_0)}{a(t_e)} - 1$. But the luminosity, and thus the apparent magnitude of the observed object, depends on the light travel time, which is different in each model

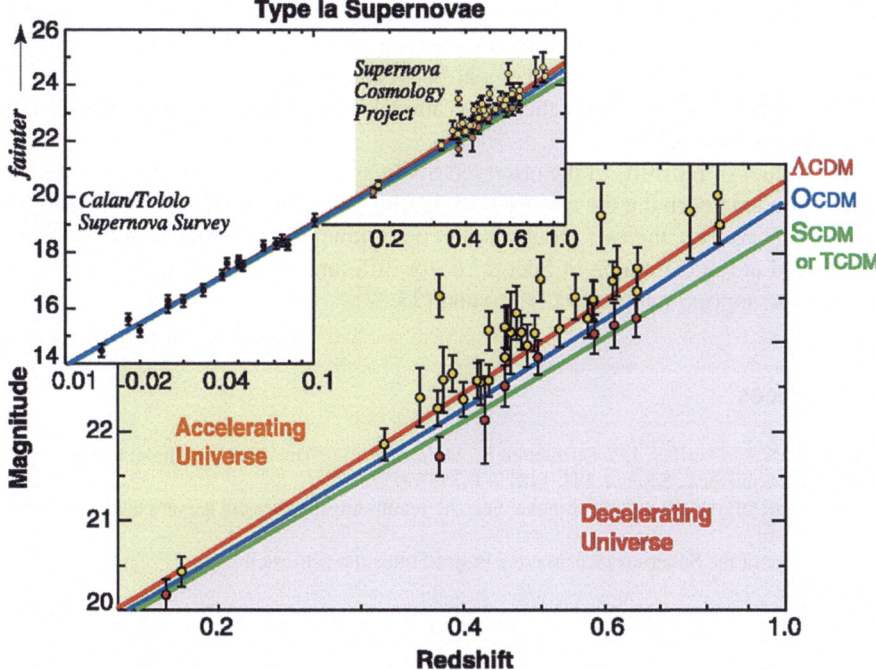

Fig. 28.2 Results of the measurements of the luminosity-redshift relations for supernovae Ia. The results point to an accelerated expansion of the universe. The apparent magnitude is plotted on the abscissa. From Bahcall et al. [1], © AAAS. Reproduced with permission

universe with constant acceleration, while it will appear brighter in a universe with decelerated expansion.

28.2   Observational Results for SN Ia

Figure 28.2 shows results of measurements of the luminosity-redshift relations for SN Ia measured by the Supernova Cosmology Project. These measurements were performed using the Hubble Space Telescope. The advantage of this telescope in orbit was that no turbulences blurred the images taken, and therefore also very distant supernovae in other galaxies, with z up to ≈ 1.7, could be detected. In the figure, the results of the measurements are compared with theoretical curves of different models of the universe. The acronym CDM stands again for "Cold Dark Matter", i.e., a universe in which the dark matter originates from as yet undiscovered elementary particles. The individual models are the flat S_{CDM} model with $\Omega_{m0} = 1$ and $\Omega_{\Lambda 0} = 0$, i.e., the classical matter-dominated universe from (26.97). The model Λ_{CDM} is the $(\Omega_{m0}\text{-}\Omega_{\Lambda 0})$ model, discussed in detail in Sect. 26.7.8, with the specific values $\Omega_{m0} = 1/3$ and $\Omega_{\Lambda 0} = 2/3$. These values differ slightly from the currently accepted values, but they could not yet be determined so precisely at that time. O_{CDM} is the model with negative curvature in (26.98), with the specific choice $\Omega_{m0} = 1/3$. The best agreement between the observational results and the theoretical curves is found for the Λ_{CDM} model, even though the error bars of the observations still are large.

The values of redshifts of the observed SN Ia are so high that the approximations that we made in deriving the approximate luminosity-redshift relation are no longer justified. Therefore, the two groups had to determine the m-z relation numerically, as outlined at the beginning of Sect. 27.6, for different models of the universe, i.e., without the approximations in (27.36) and (27.37).

References

1. Bahcall, N.A., Ostriker, J.P., Perlmutter, S., Steinhardt, P.J.: The cosmic triangle: revealing the state of the universe. Science **284**, 1481–1488 (1999)
2. Homepage of the High-z Supernova Search Team: https://www.cfa.harvard.edu/supernova/HighZ.html
3. Homepage of the Supernova Cosmology Project: https://supernova.lbl.gov

Cosmic Microwave Background Radiation

29

Contents

Hubble's astronomical observations and Lemaître's theoretical considerations suggested that our universe is expanding, therefore it was reasonable to assume that it had emerged from a state of very small extent and very high density. As early as the 1940s, Gamow speculated that the electromagnetic radiation from this period should still be detectable today, and predicted a temperature of around 5 K [11]. Together with *Alpher*[1] [3] he was working on a theory of the formation of elements, which eventually was to become known as Alpher-Bethe-Gamow theory[2] [4]. The work

---

[1] Ralph Asher Alpher, 1921–2007, US-American cosmologist.

[2] Hans Bethe was not involved in the formulation of this theory, and was only jokingly added as an author by Gamow in order to obtain author initials corresponding to the first letters α, β, γ of the Greek alphabet.

© The Editor(s) (if applicable) and The Author(s), under exclusive license to
Springer-Verlag GmbH, DE, part of Springer Nature 2026
S. Boblest et al., *Special and General Relativity*,
https://doi.org/10.1007/978-3-662-71332-7_29

Fig. 29.1 The Holmdel horn antenna with which Penzias and Wilson discovered the cosmic microwave background radiation. It was designated a *National Historic Landmark* of the United States in 1989 because of its association with the research work of the two radio astronomers. ©NASA GRiN GPN-2003-00013

by Alpher together with *Herman*[3] [1, 2] also should be mentioned in this context. However, the prediction of a radiation background remained relatively unnoticed at that time.

The detection of the cosmic microwave background (CMB) only succeeded in 1964 through *Penzias*[4] and *Wilson*[5] [22, 37].

The two worked with a radio antenna at Bell Laboratories in New Jersey, USA, see Fig. 29.1. In their observations, they noticed a background noise with a temperature $T = 3.5$ K at a wavelength of $\lambda = 7.35$ cm, which they could not explain, and also could not get rid of. In Princeton at the same time *Dicke*[6] and *Wilkinson*[7] together with others were preparing the search for the CMB. The two groups heard from each other, exchanged their findings, and simultaneously published two articles back-to-back in *Astrophysical Journal Letters* [9,23]. Penzias and Wilson were awarded the Nobel Prize in physics for their discovery in 1978.

29.1 Spectrum of the CMB

After the discovery of the CMB, several earthbound or balloon-borne experiments were launched to measure the CMB more precisely. However, only small sections of the sky, and not the whole sky, could be observed. In 1989, the *Cosmic Background Explorer* (COBE) was launched, a satellite that was to perform an all-sky survey of the CMB for several years. Figure 29.2 shows the spectrum obtained from the COBE data. Apart from tiny deviations, the CMB exhibits a highly precise, homogeneous, and isotropic Planck spectrum with the temperature 2.725 K. As we have already

---

[3] Robert Herman, 1914–1997, US-American physicist.

[4] Arno Allan Penzias, 1933–2014, US-American physicist of German descent. At the age of six, he was able to emigrate to England on a "children's transport". Nobel Prize 1978.

[5] Robert Woodrow Wilson, ∗1936, US-American astronomer, Nobel Prize 1978.

[6] Robert Henry Dicke, 1916–1997, US-American physicist.

[7] David Todd Wilkinson, 1935–2002, US-American cosmologist.

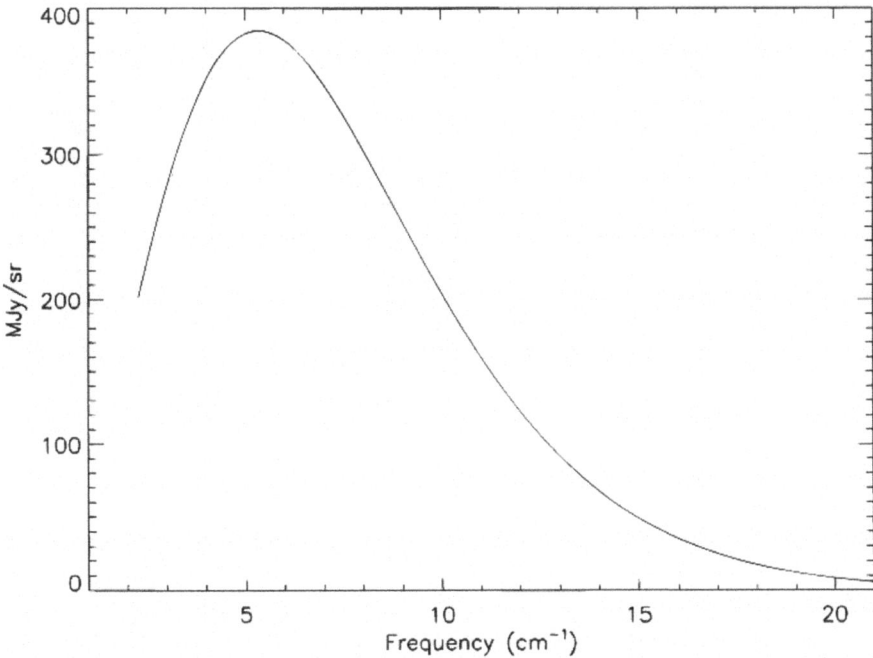

Fig. 29.2 Spectrum of the cosmic microwave background radiation measured by COBE. The deviations from a blackbody spectrum are smaller than the width of the solid line in the diagram. From Fixsen et al. [10], © AAS. Reproduced with permission

discussed, the CMB therefore "imposes" a definite coordinate system on us, namely the one in which the temperature of the CMB has the same value in all directions. This is not true when observed from Earth, because we are moving relative to the CMB, and a Doppler shift occurs as a result, see Fig. 29.3. Our own motion is composed of several relative motions. Explicitly, these are the motion of the Earth around the Sun, the rotation of the Sun around the center of the Milky Way, the motion of the Milky Way relative to the center of gravity of the local group, and finally the motion of the local group itself, which is moving towards the Virgo cluster at about $630\,\mathrm{km\,s^{-1}}$.

In the rest of this section, we will try to understand why we can expect such a spectrum for the CMB. We will then take a closer look at the small deviations from the perfect blackbody spectrum. As we will see, these are of extraordinary importance for understanding the evolution of the universe. In order to understand the spectrum of the CMB, we need to acquaint ourselves with the structure of the early universe. We must deal with that period in the evolution of the universe when the energy densities of radiation and matter were comparably high. However, at that period not only the radiation energy density was higher, but also the average energy per photon, because, as we have already discussed, the photon energy scales as a^{-1}. Because no stars existed at this stage of the universe's evolution which could produce heavier elements, the baryonic part of the energy density consisted

Fig. 29.3 The CMB can be
used to define a global inertial
frame of reference. An
observer is at rest relative to
this frame of reference if the
CMB is isotropic. Otherwise,
areas in the direction of his
motion appear blueshifted,
and areas against the direction
of motion appear redshifted

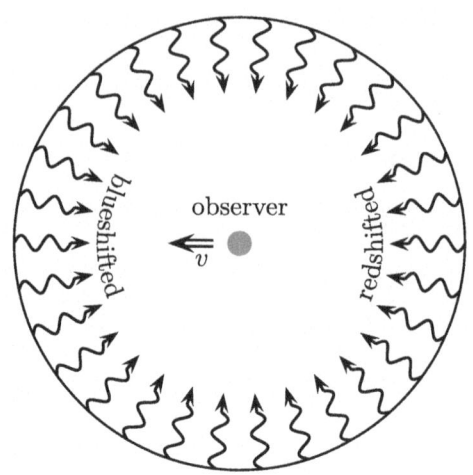

mainly of hydrogen and helium. When the average photon energy becomes of the order of magnitude of the Rydberg energy, 13.6 eV, these photons ionize hydrogen, and neutral hydrogen atoms can form only for a very short time. What one has is a plasma consisting of photons, protons, electrons, and a few helium nuclei, which is referred to as *photon baryon fluid* (PBF) ("fluid" since as in a fluid the individual constituents interact with each other). Because of the frequent interactions (scatterings) of photons with electrons the universe is opaque to light. At the same time, by virtue of this interaction the electrons and photons are in thermal equilibrium with each other, and also with the protons since these, in turn, interact with the electrons. Therefore, we can assign the same temperature T to the total system of photons, electrons, and protons. A body with a fixed temperature emits electromagnetic radiation that exhibits a Planck spectrum. The Planck radiation law reads

$$dN_\nu(t) = \frac{8\pi \nu^2}{c^3} V(t) \frac{d\nu}{\exp\left(\frac{h\nu}{k_B T(t)}\right) - 1}. \tag{29.1}$$

Here $dN_\nu(t)$ denotes the number of photons in the frequency interval $[\nu, \nu + d\nu]$ in a referential volume $V(t)$. When the universe expands, the scaling factor grows and thus the average energy of photons decreases. For the Planck spectrum, the average photon energy is obtained as the total energy divided by the total number of emitted photons. The energy of the $dN_\nu(t)$ photons in the referential volume is given by

$$dE_\nu(t) = h\nu dN_\nu(t). \tag{29.2}$$

The average photon energy is

$$\langle E_{\text{Ph}} \rangle = \int_0^\infty dE_\nu(t) \Big/ \int_0^\infty dN_\nu(t). \tag{29.3}$$

When we insert (29.1) in (29.3), we can draw a factor $8\pi V(t)/c^3$ in front of the integral in the numerator and in the denominator and cancel down. We obtain

$$\langle E_{\mathrm{Ph}} \rangle = \frac{\int_0^\infty h\nu^3 \left[\exp\left(\frac{h\nu}{k_{\mathrm{B}} T(t)}\right) - 1 \right]^{-1} \mathrm{d}\nu}{\int_0^\infty \nu^2 \left[\exp\left(\frac{h\nu}{k_{\mathrm{B}} T(t)}\right) - 1 \right]^{-1} \mathrm{d}\nu} = \frac{\pi^4}{30\,\zeta(3)} k_{\mathrm{B}} T(t) \approx 2.70\, k_{\mathrm{B}} T.$$

(29.4)

Here ζ denotes the *Riemann zeta function* with $\zeta(3) \approx 1.202$. Evidently the temperature is directly proportional to the average photon energy. Since we already know that for photons we have $E \sim a^{-1}$, we can conclude that the same applies for the temperature, $T \sim a^{-1}$. However, with this result we have not yet proved that the original Planck spectrum retains its form during the expansion of the universe, even though possibly with a different temperature. Let us consider the referential volume V. We assume that the photons contained in the volume fulfill Eq. (29.1) at time t. The same group of photons has the frequency $\nu' = a(t)/a(t')\nu$ at the later time t' due to the cosmological redshift, and correspondingly $\mathrm{d}\nu' = a(t)/a(t')\,\mathrm{d}\nu$. In the meantime, the volume has changed to $V(t') = V(t)[a(t')/a(t)]^3$. These relations imply

$$V(t)\nu^2 \mathrm{d}\nu = V(t')\nu'^2 \mathrm{d}\nu'. \tag{29.5}$$

The number of photons must have remained constant, i.e., $\mathrm{d}N'_{\nu'}(t') = \mathrm{d}N_\nu(t)$, since an effective current of photons into the volume or out of it would contradict the homogeneity and isotropy of the universe. This is of course only true when we choose a volume on the scale $V \gtrsim (100\ \mathrm{Mpc})^3$. Inserting these relations in (29.1) yields

$$\mathrm{d}N'_{\nu'}(t') = \frac{8\pi}{c^3} \nu'^2 V' \frac{\mathrm{d}\nu'}{\exp\left(\frac{h\nu'}{k_{\mathrm{B}} T'}\right) - 1} \tag{29.6}$$

with the modified temperature $T' = T(a/a')$. As claimed, the Planck form of the spectrum is retained, only the temperature changes proportional to $a^{-1}(t)$.

How low must the temperature drop for radiation and matter to decouple? An obvious approach would be to set the average photon energy equal to the ionization energy of the hydrogen atom, $E_{\mathrm{Ry}} = 13.6\ \mathrm{eV}$,

$$2.70\, k_{\mathrm{B}} T = E_{\mathrm{Ry}}. \tag{29.7}$$

This yields the rough estimate

$$T = \frac{E_{\mathrm{Ry}}}{2.70\, k_{\mathrm{B}}} \approx 58{,}000\ \mathrm{K}. \tag{29.8}$$

However, the actual temperature at decoupling is much lower. This is because, first, a small fraction of photons in the Planck spectrum has a much higher energy than the average photon, and, second, the number density of photons is about $2 \cdot 10^9$ times greater than the number density of baryons. So even if only very few photons have sufficient energy to ionize hydrogen, compared with the total number of photons, this is still sufficient if their number is comparable to that of the baryons.

In Ryden [30] interested readers will find a more comprehensive, easily comprehensible calculation using the methods of statistical mechanics, as well as a much more comprehensive discussion of the physics of recombination. The detailed calculation eventually leads to the decoupling temperature

$$T_{\text{dec}} \approx 3000 \text{ K,} \tag{29.9}$$

which corresponds to an average photon energy of about 0.7 eV. Since we know the relation between the temperature and the scaling factor, i.e., the redshift, we can easily calculate how big the universe was at the time of decoupling. The result is

$$a_{\text{dec}} = \frac{T_{\text{CMB}}}{T_{\text{dec}}} \approx 10^{-3}, \tag{29.10}$$

and therefore

$$z_{\text{dec}} \approx 10^3. \tag{29.11}$$

The cosmic microwave background radiation thus originates from an epoch when the universe had only 0.1% of its present extent, and the photons of the CMB are redshifted by a factor of about 10^3. The value of the WMAP group reads [5]

$$z_{\text{dec}} = 1090.97^{+0.85}_{-0.86}. \tag{29.12}$$

At this point, the photon energy was so low that hydrogen could no longer be ionized. Neutral hydrogen could form from protons and electrons. The PBF split into neutral hydrogen gas and photons. These two components were no longer coupled to each other, and the universe became transparent to radiation. CMB photons that reach us today have been traveling to us since the last scattering event with an electron. In the literature this is therefore referred to as the *horizon of last scattering*, see Fig. 29.4.

29.2   Anisotropies in the CMB

The high homogeneity and isotropy of the CMB cannot surprise us according to our previous considerations, otherwise the cosmological principle would be violated. We will see later that we run into difficulties in explaining this high degree of homogeneity, but for now we will take it for granted.

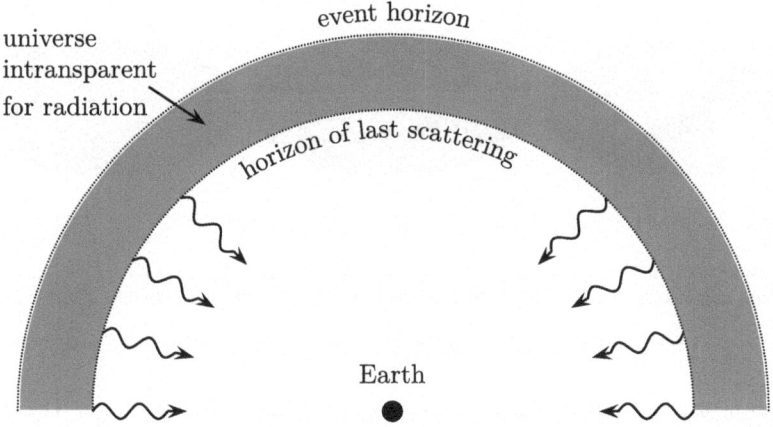

Fig. 29.4 Illustration of the horizon of last scattering. Photons of the CMB are on the way to us ever since they were scattered the last time in the PBF. We cannot look further back into the past with light since before the last scattering the universe was opaque to electromagnetic radiation

We know that the cosmological principle does not apply on small scales $d \lesssim 100$ Mpc in today's universe, because we see structures on such scales, specifically galaxies and galaxy clusters. Small inhomogeneities and anisotropies must have already been present in the very young universe in order that these structures could develop. To put it somewhat sloppily, we expect the cosmological principle to be fulfilled very well, but not perfectly exact. When applied to the CMB, this means that we can expect very high homogeneity and isotropy, but should still be able to find tiny deviations.

29.2.1 Detection of Anisotropies in the CMB

When the COBE satellite was launched in 1989, the most important goal of the mission was to detect such anisotropies in the CMB. COBE measured the CMB with an angular resolution of $7°$ across the entire sky. Figure 29.5 shows the results of these measurements. Measuring the anisotropies is not at all easy. First, as expected, they are very small, COBE detected relative deviations in temperature of

$$\left\langle \left(\frac{\Delta T}{T} \right)^2 \right\rangle^{1/2} \approx 10^{-5}. \tag{29.13}$$

Thus, fluctuations of roughly 30 μK are superimposed upon the mean temperature $T = 2.725$ K of the CMB. Second, massive interference signals result from the Doppler shift due to the relative motion of the Earth (Fig. 29.5a) and due to radiation sources within the Milky Way (Fig. 29.5b). These must first be factored out before

a   b   c

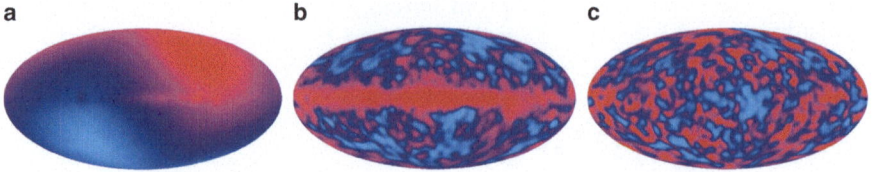

Fig. 29.5 Anisotropies of the cosmic microwave background radiation measured by the COBE satellite. Large perturbations are superimposed on the actual anisotropies. (**a**) Dipole anisotropy due the motion of the Earth relative to the CMB. (**b**) Perturbations due to sources in the Milky Way. (**c**) Actual anisotropies of the CMB after subtracting the perturbations. ©NASA-COBE Science Team [13]

the complete structure of the anisotropies in Fig. 29.5c can be seen. The actual detection of anisotropies in the CMB was a major scientific success. In 2006, the principal investigators *Mather*[8] and *Smoot*[9] were awarded the Nobel Prize in Physics for their discoveries made with COBE.

The anisotropies in the CMB are the precursors of today's structures in the universe. Understanding them better will therefore teach us a lot about the evolution of the universe as a whole, and the galaxies and galaxy clusters contained therein. As already said, the COBE satellite had an angular resolution of 7° and was therefore able to detect the anisotropies, but could not reveal any details that would have allowed a more precise analysis. For this reason, a successor to COBE was launched in 2001, the *Wilkinson Microwave Anisotropy Probe* (WMAP) [15], named after Wilkinson, whom we have already met at the beginning of this chapter, and who died shortly after the probe was launched. Compared to COBE, WMAP had a significantly improved resolution of 13′. Finally, in 2009, ESA launched the *Planck* satellite for a two-year mission in orbit [14]. With an angular resolution of down to 5′, it was able to measure the anisotropies with still higher accuracy.

Figure 29.6 shows a map of the CMB anisotropies as obtained from WMAP data, and Fig. 29.7 shows the corresponding sky map provided by the *Planck* satellite. Obviously, in that figure still much finer structures of the anisotropies can be resolved.

Another highlight of the *Planck* mission should also be pointed out: The *Planck* satellite could create the first complete sky map of polarized dust emission in the Milky Way at sub-millimeter-wavelengths [35]. The map provides new insights into the structure of the Galactic magnetic field and the properties of dust, and allows for the first statistical characterization of the Galactic foreground regarding CMB polarization.

---

[8] John Cromwell Mather, ⋆1946, US-American astrophysicist.

[9] George Fitzgerald Smoot III, 1945–2025, US-American astrophysicist.

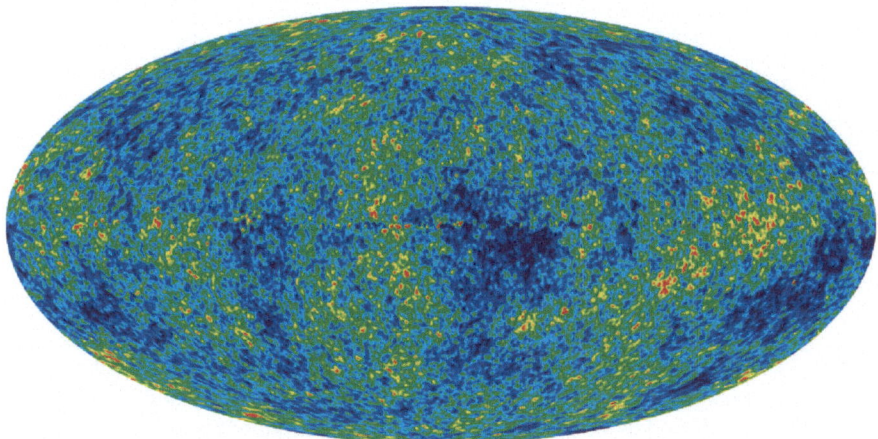

Fig. 29.6 WMAP map of the anisotropies of the CMB. The resolution is better by a factor 30 compared to COBE. ©NASA/WMAP Science Team [15]

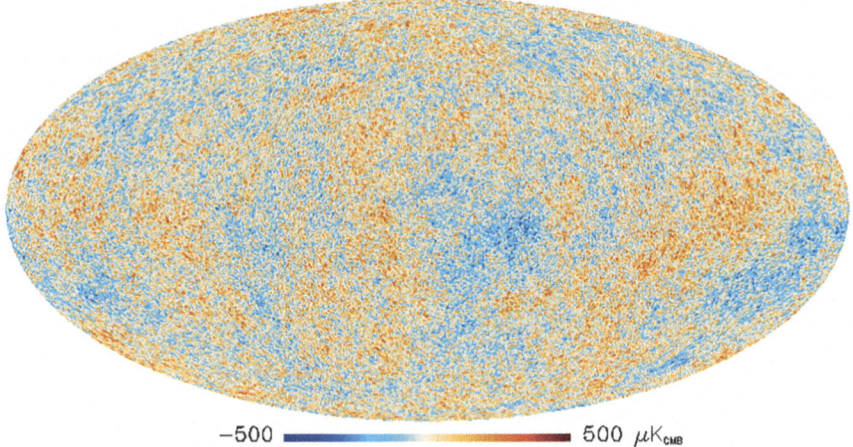

Fig. 29.7 Map of the anisotropies of the CMB as extracted from the still higher resolved data of the *Planck* satellite. Credit: Planck collaboration [25], reproduced with permission ©ESO

29.2.2 Statistical Analysis of the Anisotropies

Representations of the temperature distribution as shown in Figs. 29.6 and 29.7 are really impressive, but mathematical tools must be used for scientific evaluations of this data. The anisotropies of the CMB are analyzed using statistical methods. We will discuss the underlying mathematical theory in order to convey an impression of how scientific knowledge can be extracted from the CMB anisotropies. However, this is a wide and extremely complex subject area that we can only present in a simplified format.

Fig. 29.8 Illustration of the
Gaussian distribution of the
temperature anisotropies.
When we determine the
temperature of the spectrum
at many points in the sky, we
obtain a Gaussian distribution
of the temperature deviations
around the mean $\langle T \rangle$

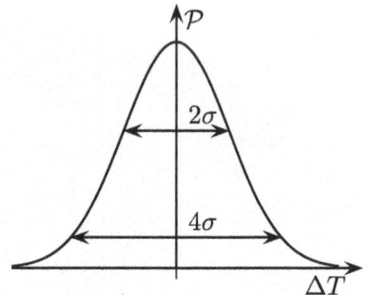

The basis for the description of the temperature fluctuations of the cosmic microwave background radiation is the theory of Gaussian random fields. The currently favored models of the universe with an inflationary phase, see Chap. 30, predict that the temperature fluctuations of the CMB obey a Gaussian distribution around the mean temperature. Although deviations from this form are being sought for, so far none have been found.

When we determine the temperatures that belong to the spectrum of a small section of the sky, we obtain a value of $T = \langle T \rangle + \Delta T$, with a deviation ΔT from the mean value. If we do this for many points in the sky, we essentially obtain a Gaussian distribution as shown in Fig. 29.8. In an interval $\Delta T \in [-\sigma, \sigma]$ lie 68% of the measured values, and in an interval $\Delta T \in [-2\sigma, 2\sigma]$ correspondingly 95%. We first consider these relations somewhat more abstract, and will later return to the implications on the analysis of the CMB. Our starting point is a random variable U with a continuous, Gaussian distributed range of values. If we want to know with what probability the variable U assumes a value in the interval $[u, u+du]$, we obtain the probability density function

$$\mathcal{P}(u) = \frac{1}{\sqrt{2\pi\sigma^2}} \exp\left[-\frac{1}{2}(u-\mu)^2/\sigma^2\right]. \tag{29.14}$$

Here μ denotes the mean value of the distribution, in our case $\langle T \rangle$, and σ is the *standard deviation*. In this way the statistical properties of the random variable U are completely characterized.

In the next step, we generalize our analysis to p random variables, each of which should be individually Gaussian distributed. This is necessary because we want to consider not just one point in the sky, but all points. If we only specify the mean and variance for each of the variables, the statistical properties are not yet completely characterized for this case. What we still need is information on the *correlations*, or the *covariance*, of the variables. The correlation results from a normalization of the covariance, and indicates a statistical pairwise dependence of the random variables

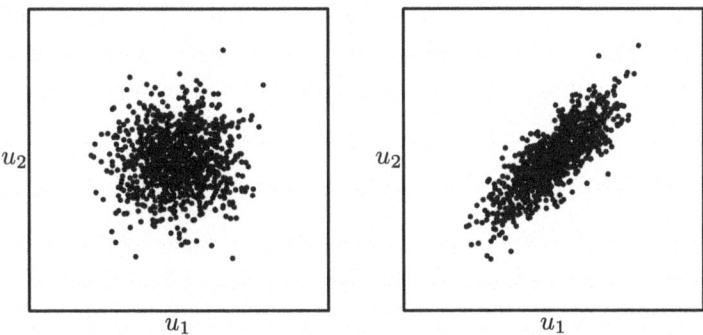

Fig. 29.9 Multivariate Gaussian distributions of two random variables u_1 and u_2. Left: No correlations between u_1 and u_2. Right: Strong positive correlations between the two variables. In both cases, the limiting densities are pure Gaussian distributions. Limiting densities are the distributions of the values u_1, or u_2, obtained when averaging over the other variable

on each other.[10] The statistical interdependencies of the variables are summarized in the symmetrical covariance matrix Σ:

$$\Sigma = \begin{pmatrix} \sigma_1^2 & \sigma_{12}^2 & \cdots \\ \sigma_{12}^2 & \sigma_2^2 & \cdots \\ \vdots & \vdots & \ddots \end{pmatrix}. \tag{29.15}$$

Roughly speaking, a positive correlation between U_1 and U_2 means that if U_1 assumes a value above the mean value, U_2 also tends to assume a value above its own mean value, and negative correlation means exactly the opposite. The multivariate Gaussian distribution, i.e., the probability $\mathcal{P}(u)$ that the p-component vector u assumes a value in the interval $[u, u + du]$, is given by

$$\mathcal{P}(u) = \frac{1}{(2\pi)^{p/2}\sqrt{\det \Sigma}} \exp\left[-\frac{1}{2}(u - \mu)^{\mathrm{T}} \Sigma^{-1}(u - \mu)\right]. \tag{29.16}$$

The correlation between the random variables has a strong effect on the distribution of the results. Figure 29.9 compares samples for the uncorrelated and the strongly positively correlated case for two random variables. In the uncorrelated case, $\sigma = \mathrm{diag}\left(\sigma_1^2, \sigma_2^2\right)$ obviously applies. Yet in any case the individual variables are normally distributed, i.e., they obey a probability density function as in (29.14).

---

[10] It should be pointed out that correlated variables need *not* be linked causally! For example, in hot summers consumption of drinking water and the number of people with circulatory problems both increase. But nevertheless high water consumption does not cause circulatory problems, or vice versa. In this case, the two variables are indirectly linked via another variable, the high temperature, but it is also possible that there is no causal relationship at all.

Now we move on to a two-dimensional continuum of random variables, i.e., each point x on the plane or the surface of a sphere is a separate random variable. This means that the vector of mean values turns into a function and, analogously, the covariance matrix turns into the covariance function:

$$\mu \mapsto \mu(x) \quad \text{and} \quad \Sigma \mapsto \Sigma(x_1, x_2) = \text{cov}(x_1, x_2). \tag{29.17}$$

In the case of the cosmic microwave background radiation, we can assume homogeneity of the statistical properties, or isotropy on the surface of a sphere. This follows directly from the cosmological principle. Then the mean value cannot depend on the individual positions, i.e.,

$$\mu(x) = \mu = \langle T \rangle, \tag{29.18}$$

and the covariance function may only depend on the *distance* between the points x_1 and x_2:

$$\text{cov}(x_1, x_2) = \text{cov}(|x_1 - x_2|). \tag{29.19}$$

On the surface of a sphere, the distance corresponds to an angle difference ϑ_{12}. The function $\text{cov}(\vartheta_{12})$ that depends on this intermediate angle is called *two-point correlation function*.

After the rather general theory in the foregoing paragraphs, we will now return to a more specific examination of the CMB. We can write the temperature deviation ΔT from the mean value $\langle T \rangle$ as a function of the viewing direction $\widehat{n}$, i.e., $\Delta T = \Delta T(\widehat{n})$. In the following, however, we consider the dimensionless function $\Theta(\widehat{n}) = \Delta T(\widehat{n}) / \langle T \rangle$ instead. For the investigation of the anisotropies the exact form of the function Θ is not crucial, because the explicit form of the anisotropies is only one possible realization of a Gaussian distribution, or in other words: The exact form of the anisotropies that we see could also be different, and yet lead to the same model of the universe. A simple example of a different distribution would be a shift by a specific angle in a specific direction. The resulting distribution of anisotropies would of course lead to the same model of the universe.

But what matters are the statistical properties of Θ, which are manifested in correlations. Our task therefore is to determine the two-point correlation function of the temperature fluctuations. To this end, we multiply the functional values of Θ at two points in the sky in the directions $\widehat{n}$ and $\widehat{n}'$, which are separated from each other by a given intermediate angle θ. Both vectors are assumed to be normalized. Then we have $\cos(\theta) = \widehat{n} \cdot \widehat{n}'$. Subsequently, we average the result over all points in the sky which have the same angular distance θ. We write this in the form

$$C(\theta) = \langle \Theta(\widehat{n}) \Theta(\widehat{n}') \rangle, \quad \text{with} \quad \cos(\theta) = \widehat{n} \cdot \widehat{n}'. \tag{29.20}$$

Here $C(\theta)$ denotes the value of the two-point correlation function for regions separated by an angle θ. For this calculation, we can also express Θ as a function of

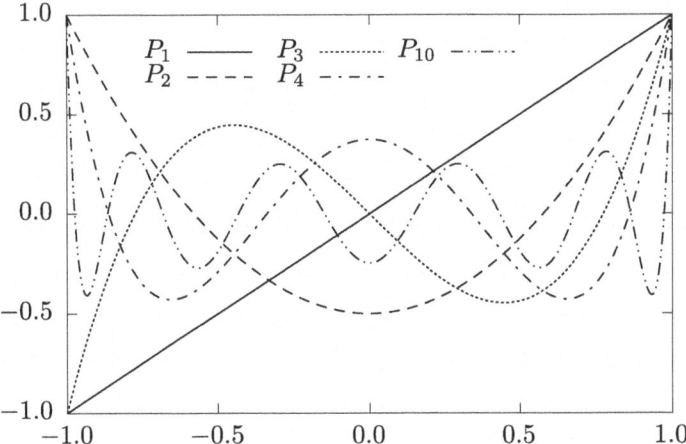

Fig. 29.10 Plots of the first Legendre polynomials P_1 to P_4, and P_{10} for comparison; $P_0 = 1$. P_l features l nodes. Using the first l polynomials one can resolve angular structures down to the order of magnitude π/l. The nth Legendre polynomial is a polynomial of degree n

the angles ϑ and φ on the surface of a sphere. This function can then be expanded in terms of spherical harmonics Y_{lm}, since the set of spherical harmonics $\{Y_{lm}\}$ forms a complete set of orthogonal functions on the surface of the sphere.

This means we have

$$\Theta(\vartheta, \varphi) = \sum_{l=0}^{\infty} \sum_{m=-l}^{l} a_{lm} Y_{lm}(\vartheta, \varphi). \tag{29.21}$$

At this point, we will skip the explicit calculation and jump to the final result. The correlation function can be represented in the form

$$C(\theta) = \frac{1}{4\pi} \sum_{l=0}^{\infty} (2l + 1) C_l P_l(\cos(\theta)). \tag{29.22}$$

It only depends on the multipole moments l, and is independent of m. The P_l are the *Legendre polynomials*. One can conceive of this relationship in a way similar to that of a Fourier expansion. When an arbitrary time-dependent signal is decomposed into its frequency spectrum, usually only the frequencies ω are of interest, which here correspond to the multipole moments l, but not any phase shifts φ_ω. These would correspond to the m-values, and then would be averaged out.

Using the expansion (29.22) one can express the correlation function in terms of the multipole moments C_l. What is important is that with the polynomial P_l, angular structures on the order of π/l can be resolved. The reason is that a Legendre polynomial of order l has l zeros on the interval $[-1, +1]$, that is in the interval $\theta \in [0, \pi]$. This is illustrated in Fig. 29.10, which shows the Legendre polynomials P_1 to P_4, and P_{10}.

29.2.3 Angular Power Spectrum

After accurately measuring the CMB, the two-point correlation function can be evaluated using the method discussed in the previous section. The data from WMAP and *Planck* cover the whole sky, therefore in principle they can be analyzed with respect to correlations up to the maximum value of $\theta = \pi$, which would correspond to $l = 1$, the dipole term. However, this case is not of particular interest, as these correlations originate from the Doppler effect. Towards small angular scales, the limit is given by the resolution of the observation instruments, which, as already mentioned, for *Planck*, is approximately $5'$.

The expansion coefficients C_l obtained are plotted as a function of the multipole moments in the so-called *angular power spectrum*. They can then be compared with theoretical, model-dependent predictions. Figure 29.11 shows the angular power spectrum based on the data of the *Planck* mission [25]. The results of the evaluation of the data are shown, together with a fit curve that corresponds to the prediction of the ΛCDM model. Figure 29.11 shows a region without pronounced structures for small l, followed by a series of several pronounced peaks beginning at about $l = 200$.

The structures with $l > 200$ are caused by effects different from those responsible for the behavior of the curve for $l < 200$. Small-angle temperature fluctuations can be explained by oscillations of the PBF, which result from local density fluctuations in the PBF. Large-angle temperature fluctuations are caused by fluctuations in the

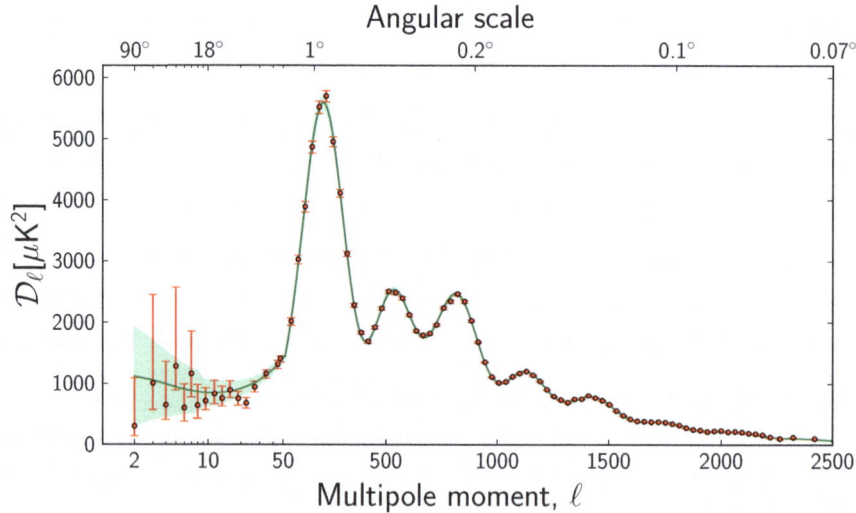

Fig. 29.11 Angular power spectrum from the data of the *Planck* mission. Seven acoustic peaks clearly stand out. Their positions can be excellently reproduced by a simple six-parameter ΛCDM model (solid curve). The variable plotted on the vertical axis is $l(l + 1)C_l/2\pi$. Credit: Planck collaboration [25], reproduced with permission ©ESO

gravitational potential of the non-baryonic dark matter. This is the *Sachs-Wolfe effect*[11,12] [31].

29.2.4 Sachs-Wolfe Effect

We will partly follow the presentation in White and Hu [36]. The time of decoupling lies in the matter-dominated phase of the universe. The total matter density of the universe consisted of a small fraction of baryonic matter and a much larger contribution from non-baryonic matter. Thus, the energy density of the non-baryonic matter was the dominant player during the phase of decoupling. We assume that the density of the non-baryonic matter possessed small fluctuations around the mean value during decoupling,

$$\varepsilon_{\rm dm}(\boldsymbol{r}) = \langle\varepsilon_{\rm dm}\rangle + \delta\varepsilon_{\rm dm}(\boldsymbol{r}). \tag{29.23}$$

These fluctuations in the energy density also induced fluctuations $\delta\phi$ in the gravitational potential of dark matter,

$$\Delta(\delta\phi) = \frac{4\pi G}{c^2}\delta\varepsilon_{\rm dm}(\boldsymbol{r}). \tag{29.24}$$

Here we are introducing a mathematical inaccuracy, as the mean value ϕ of the gravitational potential has simply "disappeared" from the Poisson equation. We will deal with this problem in another context later (see Sect. 29.2.5).

From (12.9), the metric for weak and temporally almost constant gravitational fields in the Newtonian approximation, we see that due to the relation

$$c{\rm d}\tau = \sqrt{1 + 2\frac{\phi}{c^2}}\,c{\rm d}t \approx \left(1 + \frac{\phi}{c^2}\right)c{\rm d}t \tag{29.25}$$

a gravitational potential leads to a time dilatation because of $\phi < 0$. However, a local change in time also corresponds to a local change in temperature. We can understand this when we recall that $T \sim a^{-1}$, and that $a \sim t^{2/[3(1+w)]}$ applies for the scaling factor according to (26.61) if one form of energy dominates, with the equation of state $p = w\varepsilon$.

From $Ta = {\rm const}$ we can conclude that $(a + \delta a)(T + \delta T) = aT$. It follows directly

$$\frac{a + \delta a}{a} = \frac{T}{T + \delta T} \tag{29.26}$$

---

[11] Rainer K. Sachs, 1932–2024, US-American astrophysicist of German descent.
[12] Arthur M. Wolfe, 1939–2014, US-American astrophysicist.

and because of $T/(T + \delta T) \approx 1 - \delta T/T$ also

$$\frac{\delta a}{a} = -\frac{\delta T}{T}. \tag{29.27}$$

The Taylor expansion following from (26.61)

$$a(t + \delta t) = t^{\frac{2}{3(1+w)}} + \frac{2}{3(1+w)} t^{\frac{2}{3(1+w)}-1} \delta t + \mathcal{O}(\delta t^2)$$

$$= a(t) + \delta a + \mathcal{O}(\delta t^2) \tag{29.28}$$

then implies, with $\delta t = (\phi/c^2)t$, that

$$\frac{\delta a}{a} = \frac{2}{3(1+w)} \frac{\phi}{c^2}, \tag{29.29}$$

or

$$\frac{\delta T}{T} = -\frac{2}{3(1+w)} \frac{\phi}{c^2}. \tag{29.30}$$

As the decoupling occurs in the matter-dominated phase, we have $w = 0$, and consequently

$$\left.\frac{\delta T}{T}\right|_{m_{grav}} = -\frac{2}{3} \frac{\phi}{c^2}. \tag{29.31}$$

In a radiation-dominated phase we would have $w = 1/3$, and therefore

$$\left.\frac{\delta T}{T}\right|_{r_{grav}} = -\frac{1}{2} \frac{\phi}{c^2}. \tag{29.32}$$

In a minimum of the gravitational potential, i.e., in a potential well, $\delta\phi < 0$, the small shift δa of the scaling factor towards a smaller value, which is caused by the time dilatation, has the consequence that in the well the temperature is slightly higher than in the neighborhood. A somewhat unconventional view of this situation is to interpret the time dilatation as a distortion of the horizon of the last scattering (Fig. 29.12). Photons from regions with a higher density, i.e., $\delta\phi < 0$, have escaped later (d_-), i.e., at a slightly bigger scaling factor, than those from regions with an average density (d_0), or lower density (d_+), and therefore exhibit a smaller redshift.

However, another effect is superimposed on this effect, which overcompensates the latter. When a photon starts to escape from the minimum of the potential well it has to overcome the potential barrier ϕ, and therefore experiences a time dilatation by $\delta T/T = \phi/c^2$.

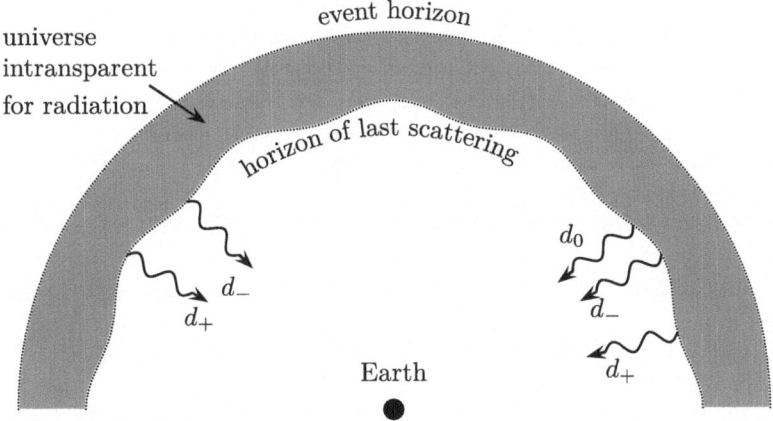

Fig. 29.12 Interpretation of the Sachs-Wolfe effect. The increase of the temperature caused by minima of the gravitational potential delays the time of decoupling, the sphere of the horizon of the last scattering is slightly distorted depending on the position on the sphere. This picture does not yet take into account the redshift caused by the escape from the gravitational potential

Therefore, when the photon arrives in the detector it exhibits a total effective temperature deviation from the average of

$$\left.\frac{\delta T}{T}\right|_{m} = \frac{1}{3}\frac{\phi}{c^2}, \tag{29.33}$$

or

$$\left.\frac{\delta T}{T}\right|_{r} = \frac{1}{2}\frac{\phi}{c^2}. \tag{29.34}$$

Its effective temperature is lower than the average temperature. Temperature fluctuations on large angular scales $\theta > 1°$ therefore contain information about the fluctuations of the gravitational potential of dark matter at the time of decoupling.

29.2.5 Acoustic Oscillations

The temporal behavior of the PBF has a strong effect on the temperature anisotropies on smaller angular scales. This temporal behavior in turn is controlled by the dominating gravitational potential of dark matter. The PBF will be compressed at places where minima of the gravitational potential of dark matter are present. However, this compression cannot grow arbitrarily high, because the radiation pressure of the photons builds up a counterpressure in the PBF, which ultimately overcompensates for the gravitational pressure and pushes the PBF out of the minima. This gives rise to a sequence of periodic phases of compression and decompression in the PBF. These phases are referred to as *acoustic oscillations*.

For an understanding of this phenomenon, we review the mathematical treatment of wave phenomena in fluids under the action of a gravitational potential. We restrict ourselves to a treatment within the framework of Newton's theory.

Wave Phenomena in Fluids

In this section we follow the arguments presented by Jeans [19]. In some way or another we already encountered them in Sect. 17.2 in the context of star formation. At the end of this section we shall discover how these seemingly completely different topics are closely related with each other. The quantitative description of wave phenomena in fluids under the action of a gravitational potential involves four ingredients. First, the mass density $\rho_m(r, t)$ (or equivalently, the energy density $\varepsilon_m = \rho_m c^2$), second, the velocity field $v(r, t)$, third, the gravitational potential $\phi(r, t)$ acting on the fluid. A fourth ingredient is the equation of state which links the pressure field to the mass density, $p = p(\rho_m)$.

Three coupled differential equations describe the fluid. First, the continuity equation

$$\nabla \cdot (\rho_m v) + \frac{\partial \rho_m}{\partial t} = 0, \qquad (29.35)$$

which in this case is the mathematical formulation for the conservation of mass. Second, we have the *Euler equation*

$$\frac{dv}{dt} = \frac{\partial v}{\partial t} + (v \cdot \nabla) v = -\frac{1}{\rho_m} \nabla p - \nabla \phi, \qquad (29.36)$$

which is essentially Newton's equation of motion for a (friction-less) fluid. Here, the total derivative of the velocity field with respect to time is given by

$$\frac{dv}{dt} = \frac{\partial v}{\partial t} + \frac{\partial v}{\partial x}\frac{dx}{dt} + \frac{\partial v}{\partial y}\frac{dy}{dt} + \frac{\partial v}{\partial z}\frac{dz}{dt}, \qquad (29.37)$$

with the components of the velocity $v_x = dx/dt$, and so on. Equation (29.36) is a compact form for writing this equation of motion. Third, for the gravitational potential ϕ we have the well-known Poisson equation of Newtonian gravitation

$$\Delta \phi = 4\pi G \rho_m. \qquad (29.38)$$

For solving the coupled differential equations (29.35), (29.36), and (29.38) we expand all variables around an as yet not explicitly specified equilibrium situation. That is we assume all functions possess an average value independent of time and position, and show a small deviation from this value which does depend on time and position:

$$\rho_m(\boldsymbol{r}, t) = \rho_{m0} + \varepsilon \rho_{m1}(\boldsymbol{r}, t) + \mathcal{O}(\varepsilon^2) \tag{29.39a}$$

$$\boldsymbol{v}(\boldsymbol{r}, t) = \boldsymbol{v}_0 + \varepsilon \boldsymbol{v}_1(\boldsymbol{r}, t) + \mathcal{O}(\varepsilon^2) \tag{29.39b}$$

$$\phi(\boldsymbol{r}, t) = \phi_0 + \varepsilon \phi_1(\boldsymbol{r}, t) + \mathcal{O}(\varepsilon^2) \tag{29.39c}$$

$$p(\boldsymbol{r}, t) = p_0 + \varepsilon p_1(\boldsymbol{r}, t) + \mathcal{O}(\varepsilon^2). \tag{29.39d}$$

The equilibrium is characterized by $\varepsilon = 0$. Now we will examine whether an equilibrium solution exists which is constant both in time and in space. That is we ask: can we find solutions with $\rho_{m0} = \text{const} \neq 0$, $\boldsymbol{v}_0 = \text{const}$, $\phi_0 = \text{const}$, and $p_0 = \text{const}$. Inserting the constant variables into (29.35) and (29.36) immediately leads to identities. Therefore, these equations are fulfilled for the constant variables. By contrast, we encounter a problem with (29.38): The left-hand side $\Delta\phi_0$ vanishes identically, and therefore the mass density must be zero, $\rho_{m0} = 0$, which we have ruled out at the beginning.

We make a bold assumption to solve this problem: Only the *deviations* from the constant density ρ_{m0} are to enter into the Poisson equation. This means we take this equation into account only beginning at the first order in ε. We have made this assumption also when deriving the Sachs-Wolfe effect. It is not easy to give a physical justification for this assumption, which is why we shall devote a separate section to its discussion below.

Without loss of generality we can set $\boldsymbol{v}_0 = \boldsymbol{0}$. This can always be achieved by a Galilei transformation. Then we obtain from (29.35), in first order,

$$\nabla \cdot [(\rho_{m0} + \varepsilon \rho_{m1}) \varepsilon \boldsymbol{v}_1] + \varepsilon \frac{\partial \rho_{m1}}{\partial t} = 0. \tag{29.40}$$

We will consider only terms linear in ε and therefore get

$$\rho_{m0} \nabla \cdot \boldsymbol{v}_1 + \frac{\partial \rho_{m1}}{\partial t} = 0. \tag{29.41}$$

We take another time derivative of this equation:

$$\rho_{m0} \nabla \cdot \frac{\partial \boldsymbol{v}_1}{\partial t} + \frac{\partial^2 \rho_{m1}}{\partial t^2} = 0. \tag{29.42}$$

Our aim will be to turn this equation into a wave equation for ρ_{m1}. To do this, we need to replace $\partial \boldsymbol{v}_1 / \partial t$ with an expression for the density. We can use the Euler equation (29.36) in first order in ε for this purpose:

$$\frac{\partial \boldsymbol{v}_1}{\partial t} = -\frac{\nabla p_1}{\rho_{m0}} - \nabla \phi_1. \tag{29.43}$$

We insert this expression in (29.42), and use $\nabla \cdot (\nabla p_1) = \Delta p_1$ and the same relation for ϕ_1. This leads to

$$- \Delta p_1 - \rho_{m0} \Delta \phi_1 + \frac{\partial^2 \rho_{m1}}{\partial t^2} = 0. \tag{29.44}$$

Using the Poisson equation, we can express $\Delta \phi_1$ by $4\pi G \rho_{m1}$. Furthermore, to replace Δp_1 we consider the linearized relation

$$p - p_0 = \left. \frac{\partial p}{\partial \rho_m} \right|_{\rho_{m0}} (\rho_m - \rho_{m0}). \tag{29.45}$$

According to the definitions in (29.39) we have $p - p_0 = \varepsilon p_1$ and $\rho_m - \rho_{m0} = \varepsilon \rho_{m1}$, and therefore

$$p_1 = \left. \frac{\partial p}{\partial \rho_m} \right|_{\rho_{m0}} \rho_{m1}. \tag{29.46}$$

We insert this expression in (29.44), multiply by $- \left(\left. \frac{\partial p}{\partial \rho_m} \right|_{\rho_{m0}} \right)^{-1}$, and shovel the $4\pi G \rho_{m0} \rho_{m1}$ to the other side:

$$\Delta \rho_{m1} - \left(\left. \frac{\partial p}{\partial \rho_m} \right|_{\rho_{m0}} \right)^{-1} \frac{\partial^2 \rho_{m1}}{\partial t^2} = -4\pi G \rho_{m0} \rho_{m1} \left(\left. \frac{\partial p}{\partial \rho_m} \right|_{\rho_{m0}} \right)^{-1}. \tag{29.47}$$

Finally we have found an inhomogeneous wave equation for the mass density ρ_{m1}. The solutions of the homogeneous equation are plane waves with velocities

$$v_{\text{sound}} = \sqrt{ \left. \frac{\partial p}{\partial \rho_m} \right|_{\rho_{m0}} }. \tag{29.48}$$

In everyday life, acoustic sound is produced by small perturbations of the otherwise constant background density and pressure of air, for example when we speak. By analogy, the perturbations of the density and pressure in fluids are referred to as sound, and the velocities (29.48) are called sound velocities. In a photon gas, $p = \varepsilon/3 = \rho_r c^2/3$ according to (26.31), and therefore $\partial p/\partial \rho_r = c^2/3$. Consequently, we obtain the value of the phase velocity

$$v_{\text{sound}} = \frac{c}{\sqrt{3}}. \tag{29.49}$$

The term sound velocity can also be justified by a very simple example:

The relations just discussed namely also apply to the ideal gas: When we write the ideal gas law in terms of the particle density n and the molecular mass of the gas particles μ, $p = nk_BT = (\rho_m/\mu)k_BT$, then we have $\partial p/\partial \rho_m = k_BT/\mu$, and arrive at the familiar form of the sound velocity $v_{sound} = \sqrt{k_BT/\mu}$ in gases. This also explains the terminology "acoustic oscillations" for the phenomena in the PBF.

We now consider the consequences of the inhomogeneity in more detail. We know that the solutions of the homogeneous wave equation are plane waves

$$\rho_{m1}(\mathbf{r}, t) = \rho_{m10}e^{i(\mathbf{k} \cdot \mathbf{r} - \omega t)}. \tag{29.50}$$

We wish to find out whether wave-like density perturbations are also possible when the inhomogeneity is taken into account. For this purpose, we insert plane waves of the form (29.50) in (29.47). We take the second derivatives and obtain

$$\left(-k^2 + \frac{1}{v_{sound}^2}\omega^2 \right) \rho_{m1}(\mathbf{r}, t) = -4\pi \frac{G\rho_{m0}}{v_{sound}^2}\rho_{m_1}(\mathbf{r}, t). \tag{29.51}$$

From this follows the *dispersion relation* for the propagation of waves in fluids with gravitational interaction:[13]

$$\omega^2(\mathbf{k}) = v_{sound}^2 k^2 - 4\pi G\rho_{m0}. \tag{29.52}$$

This result is remarkable, because we can see that for $k^2 < 4\pi G\rho_{m0}/v_{sound}^2$ the angular frequency ω turns imaginary. Physically, this corresponds to exponentially growing or decaying solutions. In this case the density perturbations can grow beyond any bounds, and the system becomes unstable. Therefore, this behavior is referred to as the *Jeans instability*.

We define the Jeans wavevector as

$$|\mathbf{k}_J| = \frac{\sqrt{4\pi G\rho_{m0}}}{v_{sound}}. \tag{29.53}$$

The corresponding Jeans wavelength is

$$\lambda_J = 2\pi/|\mathbf{k}_J|. \tag{29.54}$$

Wave-like density perturbations can occur for wavevectors $|\mathbf{k}| > |\mathbf{k}_J|$, or $\lambda < \lambda_J$. The Jeans wavelength is a characteristic length scale for density perturbations in fluids. Density perturbations with larger wavelengths cannot propagate in the fluid.

At this point we can rediscover the relation to the contraction of stars. There we had calculated in a completely analogous way the conditions under which a cloud

---

[13] A very similar dispersion relation results for density perturbations in plasmas, viz. the Bohm-Gross dispersion relation: $\omega^2(\mathbf{k}) = 3v^2k^2 + n_e e^2/(\varepsilon_0 m_e^*)$. The second term is called the plasma frequency, and m_e^* is the effective electron mass.

of gas becomes unstable, the difference being that this was exactly what we wanted
to achieve.

"Jeans Swindle"

In the foregoing derivations, we made a very dubious approximation: Since we
found a contradiction for constant quantities with non-vanishing density mass
density ρ_{m0}, we postulated that the Poisson equation only applies to the deviation
ρ_{m1} from the equilibrium density ρ_{m0}. This assumption is obviously not correct, we
simply omitted the undisturbed gravitational field.

The results of the last section therefore enjoy the questionable reputation that they
cannot be derived in a mathematically clean manner. For this reason the derivation
of the results has become known by the (in)famous sobriquet *"the Jeans swindle"*.
Nevertheless, the associated predictions have proven to be correct. In 2003, a paper
by Kiessling [20] rectified the situation by giving a mathematically clean derivation
of Jeans' dispersion relation. His starting point was the Helmholtz equation (26.33)
(instead of the Poisson equation), which Einstein had already used to justify the
introduction of the cosmological constant, see Sect. 26.6. There, the result was that
a constant gravitational potential exists for the Helmholtz equation, which results
from a constant, non-vanishing matter density everywhere. By considering the limit
$\Lambda \to 0$ in the screened potential, Kiessling could substantiate in a mathematically
clean way the derivations which we also presented in the previous section.

29.2.6 Acoustic Oscillations of the PBF

We have learned that in fluids fluctuations come into being due to the fluctuations
of the gravitational potential. Now we need to understand how these oscillations
manifest themselves as temperature fluctuations in the CMB. The fluctuations in the
density of the CDM, i.e., in the gravitational potential, cause the fluid to fall into
regions of higher density, and at the same time to rarefy in regions of lower density.
In regions of higher density the PBF is compressed, as a consequence both the
temperature and the radiation pressure increase, up to the point where the radiation
pressure reverses the motion. This generates an oscillation of the PBF the period of
which is proportional to its wavelength, which is the reason why one uses the term
"acoustic oscillations", see Fig. 29.13. Here, too, the oscillation can be decomposed
into individual modes with different frequencies.

During decoupling, the electrons recombine with the protons, and the photons
are released, the description as a PBF comes to an end. The former components of
the fluid now continue to evolve independently of each other: Matter accumulates
in regions of higher CDM density and the formation of structures of matter sets
in. The photons, on the other hand, carry the image of the temperature distribution
during decoupling to us, or to every point in the universe. At the time of decoupling,
the oscillation of the PBF is frozen. Those modes that just were at their maximum
appear strongest in the C_l diagram, the effect of other modes is weaker or even non-
existent. From this follows a sequence of peaks in the C_l diagram as sketched in

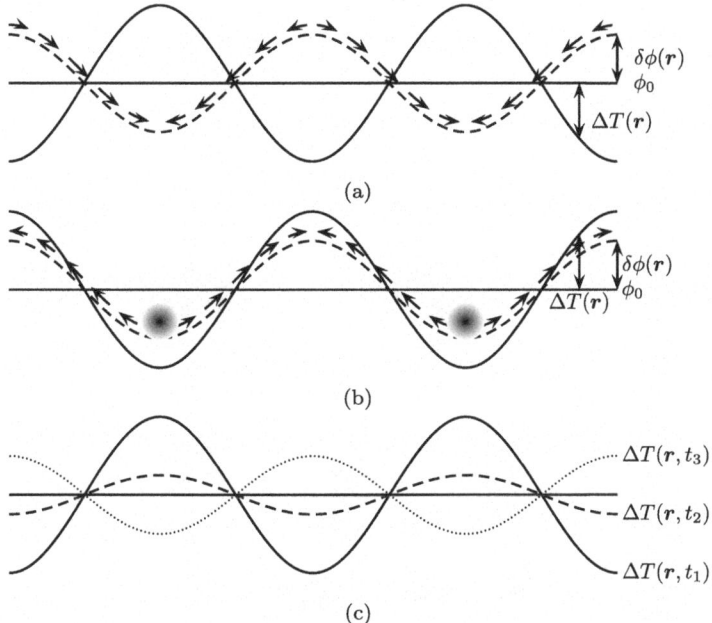

(a)

(b)

(c)

Fig. 29.13 Formation of acoustic oscillations through fluctuations $\delta\phi(\mathbf{r})$ of the mean gravitational potential ϕ_0 of the CDM distribution. (**a**) The PBF (*arrows*) accumulates in the minima of the potential $\delta\phi$ (potential well). There the temperature is increased relative to the average value. (**b**) The radiation pressure that is building up (*dark circles*) reverses the motion of the PBF, until density and temperature are lowest at the maxima of the gravitational potential (potential hill). (**c**) The temperature fluctuations $\Delta T(\mathbf{r}, t)$ correspond to a standing wave

Fig. 29.14. After the end of the inflation at ($t = t_1$), the first phase of the evolution of the universe, which we shall discuss in Chap. 30, all modes began with their oscillations. The period of the modes varies depending on the size of the mode.

The largest contribution to the fluctuations, i.e., the first peak in the C_l diagram, is provided by the mode that was at maximum compression at the time of decoupling. The second peak comes from the mode that was just at maximum decompression, and so on. The modes with frequencies in between are neither in a maximum nor in a minimum, and contribute little to the temperature distribution. The same applies to many other modes with correspondingly higher frequencies.

A further refinement of the model results when one takes into account the obvious fact that the baryons of the PBF themselves also have mass. When they accumulate in the regions of high CDM density, they also clump together due the gravitational interaction of their own masses. This effect is referred to as *baryon drag*.

The masses of the baryons produce an additional (negative) gravitational potential. Therefore, they even lower the negative deviation of the potential in the potential wells, and the temperature rises even higher. The situation is the other way round at the points of low CDM density (potential hills), when the motion

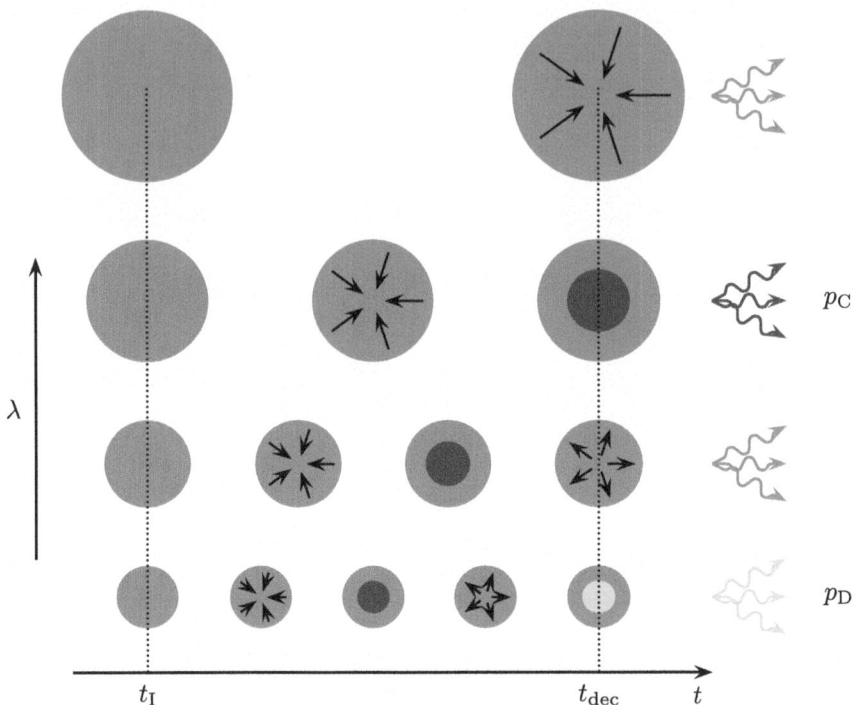

Fig. 29.14 Different modes of acoustic oscillations of the PBF. The strongest contribution to the fluctuations is generated by the mode that was maximally compressed at the time of decoupling. It produces the first peak in the C_l diagram (p_C). The second peak originates from the mode that was maximally decompressed (p_D) at the time of decoupling, and so on

has been reversed by the radiation pressure. There, the negative gravitational potential weakens the positive deviation of the gravitational potential from its mean value, the temperature deviation decreases. As a consequence, the standing wave of the temperature deviation is no longer symmetric with respect to the average temperature. It is only the modulus of the temperature deviation which enters the analysis of the CMB. Therefore, in the C_l diagram the compression peaks (recombination in potential wells) are more pronounced than decompression peaks (recombination on potential hills). In fact, the baryonic mass content can be inferred from the relative heights of the odd-number (compression) peaks and the even-number (decompression) peaks. The modifications caused by the baryonic mass are illustrated in Fig. 29.15.

The C_l diagram contains information on many cosmological parameters. For example, the position of the first peaks depends sensitively on the curvature of space. We have seen that the mode with the longest wavelength (responsible for the first peak) can be calculated using methods of hydrodynamics that are independent of cosmology. If one then compares with the size of the corresponding correlations in

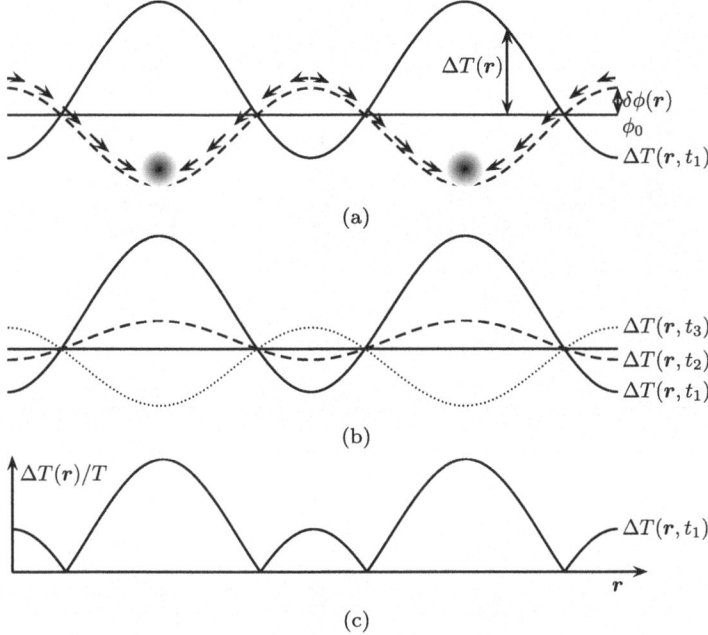

Fig. 29.15 The mass of the baryons also contributes to the gravitational potential. When this modification is taken into account the symmetry between the temperature deviations in the overdense and underdense regions is lost. (**a**) At places with higher CDM density, the mass of the baryons still deepens the potential well, and thus increases the deviation from the average gravitational potential. Conversely, at places with lower CDM density the deviation is reduced. (**b**) The standing wave of the temperature distribution $\Delta T(r, t)$ is no longer symmetric with respect to the mean temperature. (**c**) A mode that is maximally compressed shows a higher temperature deviation than a mode that is maximally decompressed

the CMB, one can extract information on the curvature using the angular diameter method (27.34) from Sect. 27.5.

29.2.7 Silk Damping

When we move to even smaller angular scales, the situation changes completely. To see this, we need to further refine our analysis of the PBF. The coupling between the photons and the baryons is not perfect: between two collisions with electrons, the photons move freely through the fluid. As a consequence, on length scales of the mean-free path of photons between two collisions with electrons no fluctuations of the temperature are possible since photons compensate these differences. If the wavelength of an acoustic oscillation mode is smaller than the mean-free-path, photons from hot, over-dense regions could fly into colder, under-dense regions, and vice versa. Therefore, such modes cannot form. Thus, the mixing of cold and

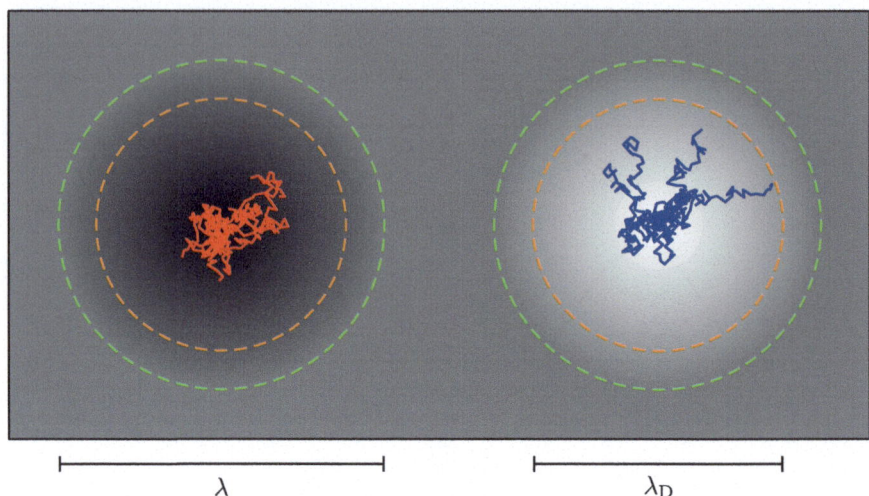

Fig. 29.16 Diffusion mixes photons from overdense and underdense regions. Therefore, anisotropy modes with wavelengths λ smaller than the diffusion mean-free path λ_D are extinguished. In this figure, the respective shade of gray symbolizes the varying density of the PBF. The red and blue curves depict the diffusion paths of a "hot" and a "cold" photon, respectively [18]

hot regions extinguishes anisotropies in the power spectrum on these length scales, as illustrated in Fig. 29.16. This effect is referred to as *Silk damping*.[14]

Another related detail concerns recombination. Up to now, we have always assumed that recombination happened at a fixed point in time. In fact, it occurred over an extended, albeit short, period in the evolutionary history of the universe, between $z \approx 1050$–950, a period of about 20,000 years. During this period, more and more hydrogen atoms formed, the degree of ionization of the PBF continuously decreased, and the mean-free path of the photons grew bigger and bigger. In this way, small-angle temperature anisotropies were exponentially damped, because hot and cold photons mixed on these length scales.

Exactly how this process happened depends on the progression of recombination, and in the literature various model assumptions have been proposed to describe the process. However, since the coupling of the photons to the baryons was still largely intact during this period, i.e., the baryons were still coupled to the photons and pulled by the electrons, also short-wave fluctuations of the baryonic matter distribution were extinguished. This sets limits on subsequent structure formation of matter in the universe.

Silk damping plays a role in the angular power spectrum from about $l \simeq 800$, all peaks at higher l are strongly attenuated compared to those at $l < 800$, as can be clearly seen in Fig. 29.11. Therefore, it is also in the range $l > 800$ that matter

---

[14] Joseph Silk, ⋆1942, British astronomer.

Table 29.1 Compilation of the results for the cosmological parameters in the ΛCDM model according to the data of the nine-year WMAP mission of 2013 [5], and the data of the *Planck* mission of 2018 [26, 27]

Parameter	Symbol	WMAP	Planck
Age of the universe [Gyr]	t_{BB}	13.772 ± 0.059	13.787 ± 0.020
Hubble constant [km s^{-1} Mpc^{-1}]	H_0	69.32 ± 0.80	67.66 ± 0.042
Baryon density	Ω_{bm}	0.04628 ± 0.00093	0.04897 ± 0.00031
Matter density	Ω_m	$0.2865^{+0.0096}_{-0.0095}$	0.3111 ± 0.0056
Dark energy density	Ω_Λ	$0.7135^{+0.0095}_{-0.0096}$	0.6889 ± 0.0056
Curvature	Ω_q	$-0.0027^{+0.0039}_{-0.0038}$	0.0007 ± 0.0019
Redshift since decoupling	z_{dec}	1091.64 ± 0.47	
Time of decoupling [yr]	t_{dec}	$374,935^{+1731}_{-1729}$	

fluctuations are deleted. The maximum possible (Silk) mass turns out to be $M < 10^{12}$ solar masses, corresponding approximately to the mass of a galaxy today.

29.2.8 Cosmological Parameters

The solid curve in Fig. 29.11, and in particular the position of the acoustic peaks, depends very sensitively on the values of the cosmological parameters. These can therefore be extracted very precisely by comparing theoretical fits with the observed data. Table 29.1 lists the values of the cosmological parameters which were obtained by analyzing the angular power spectra of the WMAP and the *Planck* data.

Both analyses yield approximately the same age of the universe and a value of the curvature compatible with 0. However, the more recent *Planck* data provide lower values for the Hubble constant, and the density of dark energy, but a larger value for the baryon density and the total matter density. In summary, it can be stated that we most likely live in an Euclidean universe that is dominated by dark energy, and is therefore expanding at an accelerated rate, and that the density of matter is largely due to dark matter. Baryonic matter itself makes up only about 4% of all energy density. However, the nature of both dark matter and dark energy remains unexplained to this day.

Large-scale cosmological simulations of the evolution of the universe from the initial conditions of the Big Bang to the present day are the subject of the Illustris project [16].

29.3 The Hubble Controversy

With the values for the Hubble constant given in Table 29.1 we look back into the early universe, namely to a redshift of $z \approx 1000$. On the other hand, recent observations of the luminosity-redshift relation of supernovae type Ia in *nearby* galaxies up to $z \approx 0.15$ have led to the value $H_0 = 74.22 \pm 1.82$ km s^{-1} Mpc^{-1} [29]. The statistical significance of the discrepancy with the value in the early cosmos amounts to 4.4 σ!

This is the Hubble controversy.

Cepheids in the Large Magellanic Cloud

The larger value could be determined because the period-luminosity (PL) relation for Cepheids stated in Eq. (23.4) could be measured with unprecedented precision by the group of A. Riess through observing the apparent magnitudes of 70 Cepheids in the Large Magellanic Cloud[15] [29]. A great help was that the team of G. Pietrzyński[16] before had determined the distance to the Large Magellanic Cloud geometrically by observing 20 eclipsing binary stars with an uncertainty of only about 1%, namely 49.59 ± 0.63 kpc (161.7×10^3 light-years) [24]. This made it possible to infer the absolute magnitudes of the observed Cepheids, and thus to derive a more precise PL relation.

Subsequently, the periodic brightness variations of Cepheids were observed in galaxies with redshifts $0.01 < z < 0.15$, in which also supernova Ia explosions had been registered. With the help of the new PL relation, the distances of the galaxies could be determined more precisely. Since the absolute luminosity of type Ia supernovae is known, the value of the Hubble constant followed from the brightness-redshift relation (27.53), which is applicable for small z.

Multiple Images of Quasars

The larger value of the Hubble constant also is found by analyzing time-delayed light curves of quasars by the H0LiCOW collaboration.[17] A quasar consists of a supermassive black hole in the center of a galaxy, surrounded by a gaseous accretion disk. Gas in the disk falling towards the black hole heats up and releases energy in the form of electromagnetic radiation. The quasars that were used are imaged multiple times by gravitational lenses, in this case formed by galaxies [17].

The radiation emitted by quasars is not constant but "flickers". If a gravitational lens lies between us and the quasar, the light experiences a time delay when passing through the gravitational lens, which depends on the path taken by the light through the (naturally non-symmetrical) structure of the gravitational lens. Therefore, the flickering reaches us with a time delay, depending on the path taken by the rays of light.

One example is the quasar HE 0435-1223, which is imaged four times through a gravitational lens [8]. The time-delayed light curves of the four images were recorded and analyzed over several years starting in 2010.

---

[15] The Large Magellanic Cloud is a satellite galaxy of our Milky Way.

[16] Grzegorz Pietrzyński, *1971, Polish astronomer.

[17] H0LiCOW collaboration: „H0 Lenses in COSMOGRAILs Wellspring". COSMOGRAIL: „COSmological MOnitoring of GRAvItational Lenses".

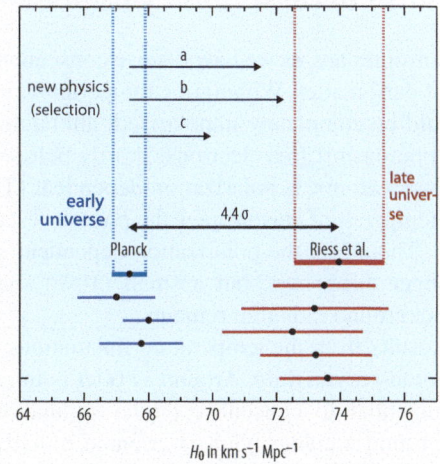

Fig. 29.17 Comparison of the values of the Hubble constant H_0 measured in the early universe (values below 70 km s^{-1} Mpc^{-1}), and in the local universe (values above 70 km s^{-1} Mpc^{-1}). The figure also shows the range of values predicted by new physics: (a) Additional relativistic particles, (b) interacting dark matter (c) time-varying dark energy and curvature of space. ©2019 Wiley-VCH GmbH [34]. Original: A. Riess et al. [27]. Editing: Physik Journal, May 2019, p. 17. Reproduced with permission

The temporal offset of the light curves depends on the composition of the gravitational lens. After constructing a mass model for the gravitational lens, the time delay can be computed and is directly proportional to the inverse of the Hubble constant. The measurements for five systems so far yield a high value of $H_0 = 72.5 \pm 2.2$ km s^{-1} Mpc^{-1} [8], compatible with the value obtained by the group of A. Riess.

The question is why the early universe expanded more slowly than today's local universe. There is very little to suggest that the discrepancy can be easily resolved [34]. One solution could be new physics in the cosmological model, for example in the form of additional energy that increases the expansion rate before recombination. This would scale down the size of the sound horizon, and thus reduce the discrepancies. However, the modifications required for this theory are substantial. But also attempts to explain the discrepancy with a further species of relativistic particles, or with a model of interacting dark matter have only a limited capacity to bridge the discrepancy of 6.6 km s^{-1} Mpc^{-1}, see Fig. 29.17.

The conference "The Hubble Controversy" dealt with this topic in November 2018. Interested readers can learn more about further proposed solutions on the conference website [38].

29.4   Polarization of the Microwave Background and Inflation

The temperature anisotropies are, as we have seen, a consequence of the fluctuations in the mass density of dark matter. Without the temperature anisotropies, the background radiation would be completely unpolarized, although the cross section for the last scattering of photons off free electrons, shortly before they recombine with protons to form hydrogen atoms, is polarization-dependent (Thompson scattering). However, due to the temperature anisotropies the photons escape at different times at different positions. Therefore, the polarization-dependent signal, averaged over all directions, no longer disappears, but a small (10%) spatially varying linear polarization of the background radiation remains.

Polarization thus results from the temperature fluctuations generated by density fluctuations of the photon-baryon fluid. Around a hotter point, a polarization pattern is formed that is tangential to concentric circles around this point, while the polarization pattern around a colder point is oriented radially, see Fig. 29.18 top. Both radial and tangential patterns are invariant under reflections about this point. In analogy with the behavior of the electric field under point reflections, these polarization patterns are designated as E modes.

The polarization of the background radiation was first detected in 2002 by the DASI instrument (*Degree Angular Scale Interferometer*) [21] stationed at the South Pole. The polarization was later confirmed by both WMAP and the *Planck* satellite. The South Pole is an ideal place for studying microwave radiation because of the prevailing low humidity.

Due to the deflection of the microwave photons in the gravitational field of mass concentrations in the universe, some of the E modes are converted into so-called B modes, whose polarization pattern, see Fig. 29.18 bottom, is antisymmetric with respect to point reflections (B, because this corresponds to the point-reflection behavior of a magnetic field). This effect is most pronounced in the angular power spectrum on angular scales of less than $1°$, as measurements at the South Pole have shown [12, 28].

An unconfirmed prediction of inflation is that it has generated a background of primordial gravitational waves [32]. Starobinskii[18] had recognized in 1979 [33] that such a phase of exponential expansion freezes quantum fluctuations of spacetime, and in particular also gravitational waves, with wavelengths greater than the Hubble distance that we introduced in Eq. (23.12). After the end of inflation, the wavelengths of these modes grow more slowly than the Hubble distance, with the consequence that they later propagate through the universe as classical gravitational waves.

Gravitational waves, which cause a quadrupole anisotropy, can generate an arbitrarily oriented linear polarization and thus an arbitrary polarization pattern, i.e., both E modes and B modes. These B modes, which are supposed to occur on angular scales larger than $1°$, would therefore be an indication of the existence

---

[18] Alexei Starobinskii, 1948–2023, Soviet and Russian astrophysicist and cosmologist.

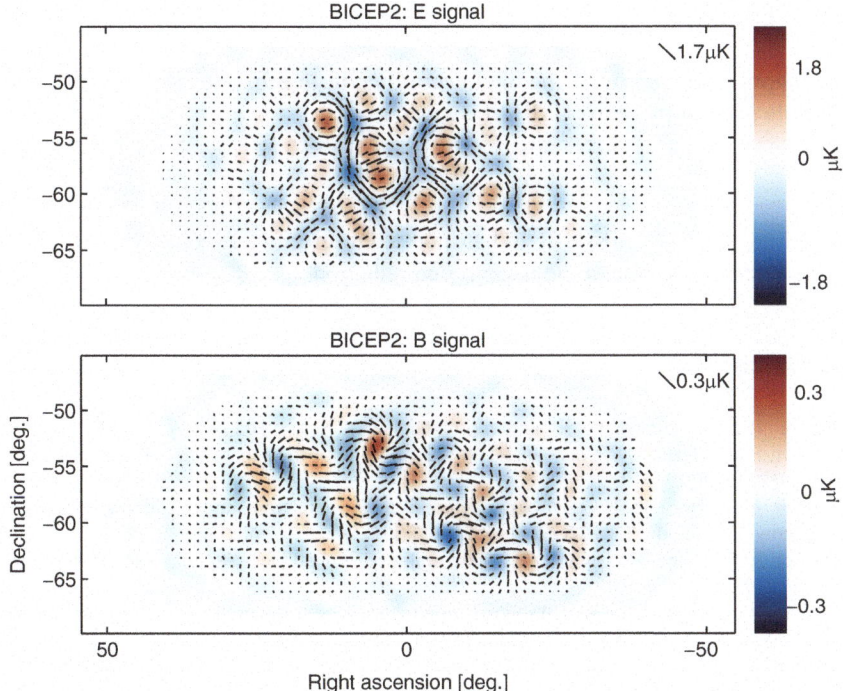

Fig. 29.18 Polarization patterns of the E and the weaker B modes of the cosmic microwave background radiation measured by the BICEP2-telescope on angular scales ranging from 1.5° to 3.6°. The black lines indicate the strength and direction of the linear polarization. From: P. A. R. Ade et al. (BICEP2 Collaboration) [6] ©APS. Reused under the terms of the Creative Commons Attribution 3.0 License

of gravitational waves that were generated during a phase of cosmological inflation in the early universe.

Discovering these primordial B modes is the goal of the BICEP2 collaboration. BICEP is an acronym for "*Background Imaging of Cosmic Extragalactic Polarization*". Beginning in 2010, BICEP2 studied for more than three years an area of 380 square degrees from the South Pole. This "southern galactic hole" was ideally suited because it is only slightly polluted with galactic dust in the foreground. In March 2014, BICEP2 reported the discovery of this kind of signal [6]. But subsequent analyses of the *Planck* satellite's observations showed that galactic foreground (especially dust) is responsible for most of the B modes observed by BICEP2 [7].

The hunt for gravitational waves caused by inflation therefore still continues.

29.5   Exercise

29.1. Dipole anisotropy of the background radiation In Fig. 29.5a we have seen the dipole anisotropy of the CMB. It is given by

$$\Delta T = \Delta T_{\max} \cos(\theta), \tag{29.55}$$

with $\Delta T_{\max} = 3.369\,\text{mK}$ [10]. What must therefore be the velocity of the Earth relative to the rest frame of the background radiation?

References

1. Alpher, R.A., Herman, R.C.: Evolution of the universe. Nature **162**, 774–775 (1948)
2. Alpher, R.A., Herman, R.C.: Remarks on the evolution of the expanding universe. Phys. Rev. **75**, 1089–1095 (1949)
3. Alpher, R.A., Herman, R.C.: Theory of the origin and relative abundance distribution of the elements. Rev. Mod. Phys. **22**(22), 153–212 (1950)
4. Alpher, R.A., Bethe, H., Gamow, G.: The origin of chemical elements. Phys. Rev. **73**(7), 803–804 (1948)
5. Bennett, C.L., et al.: Nine-year Wilkinson microwave anisotropy probe (WMAP) observations: final maps and results. Astrophys. J. Suppl. Ser. **208**(2), 20 (2013)
6. BICEP2 Collaboration: Detection of B-Mode polarization at degree scales by BICEP2. Phys. Rev. Lett. **112**, 241101 (2014)
7. BICEP2/Keck and Planck Collaborations: Joint analysis of BICEP2/*Keck array* and *Planck* data. Phys. Rev. Lett. **114**, 101301 (2015)
8. Bonvin, B., et al: H0LiCOW – V. New COSMOGRAIL time delays of HE 0435 - 1223: H_0 to 3.8 per cent precision from strong lensing in a flat ΛCDM model. Mon. Not. R. Astron. Soc. **465**, 4914 (2017)
9. Dicke, R.H., Peebles, P.J.E., Roll, P.J., Wilkinson, D.T.: Cosmic black-body radiation. Astrophys. J. Lett. **142**, 414–419 (1965)
10. Fixsen, D.J., et al.: The cosmic microwave background spectrum from the full COBE FIRAS data set. Astrophys. J. **473**, 576–587 (1996)
11. Gamow, G.: Half an hour of creation. Phys. Today **3**, 16 (1950)
12. Hansen, D., et al., (SPTpol Collaboration): Detection of B-mode polarization in the cosmic microwave background with data from the South Pole Telescope. Phys. Rev. Lett. **111**, 141301 (2013)
13. Homepage of the COBE mission: https://lambda.gsfc.nasa.gov/product/cobe/
14. Homepage of the *Planck*-Mission: https://www.cosmos.esa.int/web/planck
15. Homepage of the WMAP mission: https://science.nasa.gov/mission/wmap/
16. Homepage of the Illustris project: https://www.illustris-project.org
17. Homepage of the H0LiCOW Collaboration: https://shsuyu.github.io/H0LiCOW/site
18. Hu, W.: Wandering in the background: a cosmic microwave background explorer. PhD thesis, University of California at Berkeley (1995)
19. Jeans, J.H.: The stability of a spherical nebula. Phys. Trans. R. Soc **199**, 1–53 (1902)
20. Kiessling, M.K.H.: The "Jeans swindle" A true story – mathematically speaking. Adv. Appl. Math. **31**, 132–149 (2003)
21. Kovac, J.M., et al.: Detection of polarization in the cosmic microwave background using DASI. Nature **420**, 772 (2002)
22. Penzias, A.A.: The origin of the elements. Science **205**, 549–554 (1979)

23. Penzias, A.A., Wilson, R.W.: A measurement of excess antenna temperature at 4080 Mc/s. Astrophys. J. **142**, 419–421 (1965)
24. Pietrzyński, G., et al.: A distance to the Large Magellanic Cloud that is precise to one per cent. Nature **567**, 200 (2019)
25. Planck Collaboration, Ade, P.A.R., et al.: *Planck* 2013 results. I. Overview of products and scientific results. Astron. Astrophys. **571**, A1 (2014)
26. Planck Collaboration, Aghanim, N., et al.: *Planck* 2018 results. I. Overview and the cosmological legacy of *Planck*. Astron. Astrophys. **641**, A1 (2020)
27. Planck Collaboration, Aghanim, N., et al.: *Planck* 2018 results. VI. Cosmological parameters. Astron. Astrophys. **641**, A6 (2020)
28. POLARBEAR Collaboration: A measurement of the cosmic microwave background B-mode polarization power spectrum at sub-degree scales with POLARBEAR. Astrophys. J. **794**, 171T (2014)
29. Riess, A.G., Casertano, S., Yuan, W., Macri, L.M., Scolnic, D.: Large Magellanic Cloud cepheid standards provide a 1% foundation for the determination of the Hubble constant and stronger evidence for physics beyond ΛCDM. Astrophys. J. **876**, 85 (2019)
30. Ryden, B.: Introduction to Cosmology. Addison Wesley, San Francisco (2004)
31. Sachs, R.K., Wolfe, A.M.: Perturbations of a cosmological model and angular variations of the microwave background. Astrophys. J. **147**, 73–90 (1967)
32. Schwarz, D. : Wellen der Inflation (Waves of inflation). Phys. J. **13**(5), 16 (2014)
33. Starobinskii, A.A.: Spectrum of relict gravitational radiation and the early state of the universe. J. Exp. Theor. Phys. Lett. **30**, 682 (1979)
34. Steinmetz, M.: Die Hubble-Kontroverse (The Hubble controversy). Phys. J. **18**(5), 18 (2019). Wiley-VCH-Verlag Weinheim
35. The Galactic Magnetic Field as revealed by *Planck*: https://www.ias.u-psud.fr/soler/planckhighlights.html
36. White, M., Hu, W.: The Sachs-Wolfe effect. Astron. Astrophys. **321**, 89 (1997)
37. Wilson, R.W.: The cosmic microwave background radiation. Science **205**, 866–874 (1979)
38. https://www.we-heraeus-stiftung.de/veranstaltungen/tagungen/2018/hubble2018 (Content in English)

The First Few Moments

30

Contents

In this chapter we will look at a few aspects of the very young universe. We are making a big leap from the time of decoupling, at about $t = 380,000$ y, back to the very first fractions of a second after the Big Bang. We are thus skipping essential phases in the evolution of the universe, such as the period of *primordial nucleosynthesis*, when the atomic nuclei of light elements were formed by fusion, viz. the hydrogen isotope deuterium, various helium isotopes, and traces of lithium. The related questions are of course also of great importance to cosmology, but we will not deal with them in our context. Readers who wish to inform themselves on this topic can find an introduction and further references for example in the book by Ryden [7].

30.1 Quantum Effects

When we go back to earlier and earlier stages of the evolution of the universe, it is evident that at some point quantum effects must become important. To estimate the time and length scales for which this is the case, we need a characteristic length

associated with a certain microscopic particle. In Sect. 6.6.2 we introduced the
Compton wavelength $\lambda = \hbar/(mc)$ as a characteristic quantum length of a relativistic
particle with mass m.

Furthermore, we also have a characteristic length scale for the gravitational
interaction of a particle with mass m, namely its Schwarzschild radius $r_s = 2Gm/c^2$. It is reasonable to assume that quantum effects become important when
both length scales are approximately equal. To avoid numerical factors, we choose
the condition $\lambda = r_s/2$. We solve with respect to m, and obtain the *Planck mass* [6]:

$$m_{\mathrm{Pl}} = \sqrt{\frac{\hbar c}{G}} = 2.176\,434(24) \cdot 10^{-8} \text{ kg}. \qquad (30.1)$$

We see that this mass is neither really microscopic nor macroscopic. The equivalent
Planck energy can be defined as

$$E_{\mathrm{Pl}} = m_{\mathrm{Pl}}c^2 = 1.220\,890(14) \cdot 10^{19} \text{ GeV}. \qquad (30.2)$$

To appreciate the size of these quantities, let us compare them with the mass and the
equivalent energy of a proton: $m_p = 1.672622 \cdot 10^{-27}$ kg and $E_p = 938.272$ MeV.
More elementary quantities can be derived: The Planck length as the Compton
wavelength of a particle with the Planck mass, the Planck time as the time that
it takes light to travel the distance of the Planck length, the Planck temperature, via
the relation $k_B T_{\mathrm{Pl}} = E_{\mathrm{Pl}}$, and the Planck density $\sigma_{\mathrm{Pl}} = m_{\mathrm{Pl}}/l_{\mathrm{Pl}}^3$.

The Planck units were first introduced by Planck himself after his discovery of
the quantum of action h. In a certain sense, they represent "natural" units since,
measured in these units, the fundamental constants G, $\hbar$, and c all assume the
numerical value 1. For example, the Planck length is also found by seeking numbers
α, β, and γ, for which $l_{\mathrm{Pl}} = G^\alpha c^\beta \hbar^\gamma$ assumes the dimension of a length [11]. In
Table 30.1 all Planck units are summarized. The Planck density plays an important
role in explaining the vacuum energy density. Since the Planck units are natural
units, it is reasonable to expect an order of magnitude of the vacuum energy density
of

$$\varepsilon_{\mathrm{vac}} \sim \frac{E_{\mathrm{Pl}}}{l_{\mathrm{Pl}}^3} = c^2 \sigma_{\mathrm{Pl}}. \qquad (30.3)$$

The resulting energy density

$$\varepsilon_{\mathrm{vac}} \sim \frac{1.9560 \cdot 10^9 \text{ J}}{\left(1.6163 \cdot 10^{-35} \text{ m}\right)^3} = 4.63 \cdot 10^{113} \text{ J m}^{-3} = 2.89 \cdot 10^{132} \text{ eV m}^{-3}$$

$$(30.4)$$

Table 30.1 Summary of the Planck units [6]

Name	Term	SI units	Special units
Basic quantities			
Mass	$m_{Pl} = \sqrt{\hbar c/G}$	$2.176\,434(24) \cdot 10^{-8}$ kg	$1.31067(15) \cdot 10^{19}$ u
Length	$l_{Pl} = \sqrt{\hbar G/c^3}$	$1.616\,255(18) \cdot 10^{-35}$ m	$3.05427(34) \cdot 10^{-25}$ a$_B$
Time	$t_{Pl} = \sqrt{\hbar G/c^5}$	$5.391\,247(60) \cdot 10^{-44}$ s	
Temperature	$T_{Pl} = \sqrt{\hbar c^5/(k_B^2 G)}$	$1.416\,784(16) \cdot 10^{32}$ K	
Derived quantities			
Charge	$q_{Pl} = \sqrt{4\pi \varepsilon_0 \hbar c}$	$1.8755460378(15) \cdot 10^{-18}$ C	$1.17062376140(93) \cdot 10^1$ e
Energy	$E_{Pl} = \sqrt{\hbar c^5/G}$	$1.9560(23) \cdot 10^9$ J	$1.220\,890(14) \cdot 10^{19}$ GeV
Power	$P_{Pl} = c^5/G$	$3.62825(82) \cdot 10^{52}$ W	
Momentum	$p_{Pl} = \sqrt{\hbar c^3/G}$	$6.52478(74) \cdot 10^0$ kg m s^{-1}	
Density	$\sigma_{Pl} = c^5/(\hbar G^2)$	$5.1548(24) \cdot 10^{96}$ kg m^{-3}	

is incredibly high. For example, comparing with $\varepsilon_{c0} \approx 4.82 \cdot 10^9$ eV from (26.7), leads to the ratio

$$\frac{\varepsilon_{vac}}{\varepsilon_{c0}} \approx 5 \cdot 10^{122}. \tag{30.5}$$

The vacuum energy density actually observed is therefore tiny, from this perspective, even though it appears to be the dominant form of energy in our universe. This discrepancy between theory and observation is perhaps the worst theoretical prediction in physics today.

Quantum effects will certainly be important when the temperature reaches the order of magnitude of the Planck temperature. Analogous statements can also be made for the density, and for periods on the order of the Planck time after the Big Bang. In fact, it is believed that the Planck time is the shortest physically meaningful period of time. Very little is known about this stage in the evolution of the universe, in particular, since, so far, no theory exists that unifies gravity with the three other fundamental interactions. Readers interested in the physical reasons for the existence of shortest length and time scales will find more information in Sprenger et al. [8].

30.2 Inflation

We move a little further away from the Big Bang, but still are staying within the very first fractions of a second. The success of the description of the universe using the FLRW metric is ultimately based on the validity of the cosmological principle, the validity of which we have postulated in order to justify this description of our universe. If we want to understand in more detail how the high homogeneity of the universe comes about, we are facing a causality problem. Another problem arises if we wish to understand how the value $\Omega_0 \approx 1$ comes about, which we observe today. In this section, we will briefly discuss these and other problems of FLRW cosmology, and then introduce the theory of inflation that has been formulated to solve them. It states that our universe went through a phase of extreme expansion very shortly after the Big Bang.

30.2.1 Flatness Problem

From (26.64) we can see that $\Omega_0 \approx 1$, with the error lying in the range of a few per cent. The value $\Omega_0 = 1$, i.e., $q = 0$, is not required by or derived from any physical theory, any other value would be allowed. This finding leads to a problem when we ask what initial conditions must have prevailed in the very early evolution of the universe such that a value of $\Omega_0 \approx 1$ can be found today. To answer this question, we consider how the value of the parameter has evolved over time. Combining (26.10) with (26.11), we find

$$1 - \Omega(t) = (1 - \Omega_0)\frac{H_0^2}{H(t)^2 a(t)^2}. \tag{30.6}$$

This shows again that $\Omega_0 = 1$ is a very special initial condition, because from $\Omega_0 = 1$ follows $\Omega(t) = 1$ for all times. However, any arbitrarily small deviation from this value will lead to a strong time dependence of $\Omega(t)$, as we shall demonstrate now. We use the Friedmann equation in the form (26.58), and divide both sides by $H_0^2 a^2$,

$$\frac{H(t)^2}{H_0^2} = \sum_i \Omega_{i0} a^{-3(1+w_i)} + \frac{1 - \Omega_0}{a^2}. \tag{30.7}$$

In the early universe, the scaling factor was small, and radiation and matter provided the dominant contributions to the energy density. Therefore, in that phase

$$\frac{H(t)^2}{H_0^2} = \frac{\Omega_{r0}}{a^4} + \frac{\Omega_{m0}}{a^3}. \tag{30.8}$$

Inserting this into (30.6) leads to relation

$$1 - \Omega(a) = (1 - \Omega_0)\frac{a^2}{\Omega_{r0} + \Omega_{m0}a}. \tag{30.9}$$

Because of $a(t) \sim t^{1/2}$ in the radiation-dominated phase, and $a(t) \sim t^{2/3}$ in the matter-dominated phase, we find

$$|1 - \Omega(t)| \sim \begin{cases} t^{1/2} & \text{in the radiation-dominated phase,} \\ t^{2/3} & \text{in the matter-dominated phase.} \end{cases} \tag{30.10}$$

In any case, any deviation of Ω_0 from the value 1 that may exist today decreases steadily when we calculate back into the past. With the uncertainties of the Ω_{i0} parameters in (26.64), we can say, with high confidence, that today the inequality

$$|1 - \Omega_0| \lesssim 0.04 \tag{30.11}$$

is fulfilled. At the time of decoupling the scaling factor was $a \approx 10^{-3}$. Inserting this in (30.9) implies

$$|1 - \Omega(t_{\mathrm{dec}})| \approx 10^{-4}, \tag{30.12}$$

an already remarkably accurate agreement with the value being exactly 0. When we go even further back in time, to the period when the energy densities of radiation and matter were equal, i.e., to a scaling factor of $a_{\mathrm{rm}} \approx 2.7 \cdot 10^{-4}$ from (26.66), we already have

$$|1 - \Omega(t_{\mathrm{rm}})| \approx 2 \cdot 10^{-5}. \tag{30.13}$$

In the extreme limit of the Planck time [7] one ultimately finds

$$|1 - \Omega(t_{\mathrm{Pl}})| \approx 10^{-60}. \tag{30.14}$$

It is certainly a bit risky to extrapolate back to this extremely short time, but even for periods far above the Planck scale we still get requirements on the accuracy in the range $1{:}10^{-50}$.

This extreme specification, $\Omega(t_{\mathrm{Pl}}) = 1 \pm 10^{-50}$, is what is understood to be the *flatness problem*. This extremely high sensitivity to the initial values poses an aesthetic problem. The precise value of 1 must be inserted into the theory as a given quantity. It would be more satisfying to explain why exactly this value must occur through a physical theory.

We must realize that a deviation from this initial value would have had dramatic consequences. The result would have been either a universe that collapsed far too quickly, so that no life could have developed, or a universe that expanded so fast that no stars could have been formed, because the density of matter would have decreased far too quickly.

30.2.2 Horizon Problem

Another problem concerns the high degree of homogeneity of the universe. To us, the extreme isotropy of the CMB has been a justification of the cosmological principle, but when we really try to *explain* this isotropy, we run into problems. For example, the distance to the horizon of last scattering is only slightly smaller than

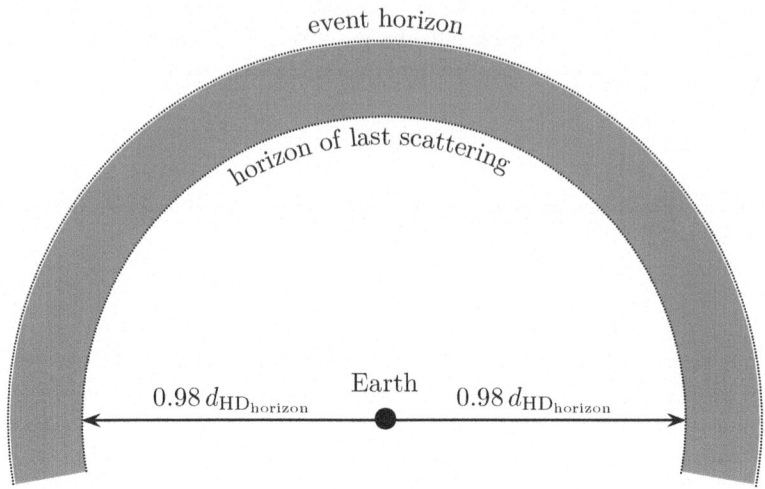

Fig. 30.1 Illustration of the horizon problem. Two points in the CMB lying opposite to each other on the celestial sphere are not causally connected, therefore they had no possibility to exchange information, and to reach thermal equilibrium with each other. Nevertheless, they possess the same temperature

the Hubble distance from (23.13), it amounts to about 98% of this distance, see Fig. 30.1.

Two points in the CMB lying exactly on opposite positions on the celestial sphere have a proper distance from each other which is significantly greater than the Hubble distance to the event horizon. Therefore, they cannot be causally connected, because the maximum distance that even photons could travel since the Big Bang is approximately the Hubble distance. It is even relatively easy to show that regions that were causally connected at the time of recombination only have an angular diameter of about 2° in the sky today, see Exercise 30.2. And yet, the microwave background radiation is very homogeneous across the entire sky. In other words: From any direction we receive a Planck spectrum with the same temperature, even though the regions concerned never had the opportunity to exchange information, and to reach thermal equilibrium.

30.2.3 Monopole Problem

This problem has its origin in considerations that lie outside the scope of this textbook.

One of the major objectives of physics today is to describe all fundamental forces in a unified theory which, unlike the standard model, also includes gravity. Theories that unite the electromagnetic, weak, and strong forces are an intermediate step on this path, the so-called *Grand Unified Theories* (GUT). All candidates for a theory of this kind predict the existence of very massive particles which represent magnetic monopoles. The energy density of these monopoles should by far surpass all other energy densities, but nothing of the sort is observed.

30.2.4 Inflationary Universe

To solve all these problems at one stroke Guth[1] proposed in 1981 [1] that at the beginning of its evolution the universe lived through a short phase of inflation: The scaling factor increased exponentially by a factor of about $\sim 10^{30}$ between 10^{-35} and 10^{-33} s after the Big Bang.

It is easy to understand why this can solve all the problems discussed. If the inflationary phase takes place after the creation of the monopoles, their density is reduced by inflation so strongly that they are completely negligible in today's universe. The seemingly causally unconnected regions in the sky today would have been much smaller prior to the inflationary phase, and would have had time enough to reach thermal equilibrium before inflation. Temperature inhomogeneities would be removed on all length scales and then be expanded to the size of the observable universe today during the inflationary phase. This phase would also explain why our universe is flat today. When the radius of a sphere is inflated by a factor of 10^{30} then an observer cannot distinguish it from an ordinary plane within the distance to his horizon. In a similar way, inflation must have increased the radius of curvature far beyond the distance to our horizon so as to make any possible curvature invisible to us.

The mechanism behind inflation is based on the assumption of the existence of an energy density ε_ϕ, which depends on the value of a scalar field ϕ, and which was dominant in the very early universe. In classical inflation theory, it is then assumed that the energy density ε_ϕ has a functional dependence on ϕ such that it is almost constant over a wide range of varying ϕ. This is referred to as a *false vacuum* (FV), see also the position denoted by FV in Fig. 30.2, as opposed to the real vacuum state, which corresponds to the minimum of ε_ϕ (region ② in Fig. 30.2).

We already know the solution of the Friedmann equation belonging to a constant energy density, it is the de Sitter metric from (26.72), with

$$a(t) = e^{H_{\mathrm{inf}}t}, \tag{30.15}$$

where in this equation we denote the Hubble constant during the inflationary phase by H_{inf}. Thus, as long as the energy density ε_ϕ dominates, the universe expands exponentially. We have already learned that in the de Sitter universe the Hubble parameter does not depend on time. But we can recognize this once again from (30.7), when we take into account just the contribution from $\Omega_{\Lambda 0}$, with $w_\Lambda = -1$. It is true that, at this point, we are not considering the vacuum energy

---

[1] Alan Guth, ⋆1947, American physicist and cosmologist.

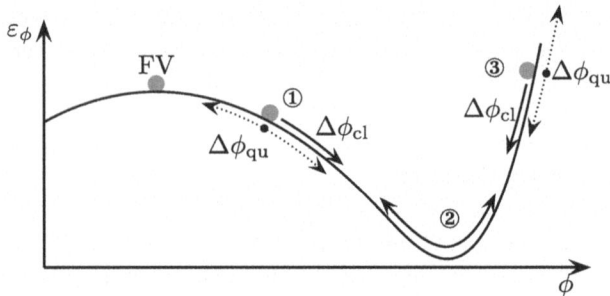

Fig. 30.2 Illustration of the theory of inflation. The essential ingredient of these models is an energy density ε_ϕ, which depends on the value of a scalar field ϕ

density ε_Λ, but ε_ϕ behaves quite analogously. When we insert $a(t) \sim \exp(H_{\mathrm{inf}}t)$ and $H(t) \sim H_{\mathrm{inf}}$ into (30.6), we obtain

$$|1 - \Omega(t)| \sim (1 - \Omega_0)\mathrm{e}^{-2H_{\mathrm{inf}}t}. \tag{30.16}$$

Unlike in the radiation-dominated and matter-dominated phase, deviations of $\Omega(t)$ from 1 are exponentially damped in the inflationary phase! Inflation models can therefore explain many of the characteristic properties of our universe. The anisotropies in the CMB are also predicted by them, and are attributed to quantum fluctuations of the initial wavefunction of the universe that have been enlarged to macroscopic or astronomical scales by inflation.

The agreement of the inflationary model with the observational results is very convincing, in particular as many other promising candidates can be ruled out on the basis of the observational data. However, one problem remains: After the inflationary phase, due to the extreme expansion, the temperature, and also the density of matter have dropped dramatically, and the universe is essentially empty. Therefore, the inflationary models predict that the particles that we see today are generated by the energy density of the scalar field released. This process is called "reheating".

In spite of its great successes, inflation theory also has problems, and raises new questions. On the one hand, of course the question of what is the nature of the postulated scalar field, for which there is no physical origin given as yet? In addition, even if one assumes the existence of this scalar field, today's view of inflation is completely different from what it was in the 1980s [2–4]. For example, models were developed in which the universe need not necessarily be in a FV (situation ③ in Fig. 30.2). The most important additional modification, however, is related to the evolution of the scalar field ϕ during the expansion. In the classical picture, ϕ moves towards the vacuum state during expansion. In Fig. 30.2 these steps are designated by $\Delta\phi_{\mathrm{cl}}$. However, quantum mechanical effects also play a role in this phase, which lead to fluctuations $\Delta\phi_{\mathrm{qu}}$, with a Gaussian distribution, and can change ϕ in any direction.

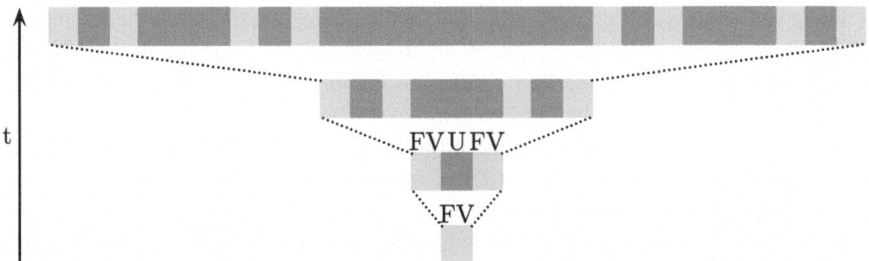

Fig. 30.3 Illustration of the theory of eternal inflation. Due to quantum fluctuations, inflation does not stop at each expansion step in parts of the universe, therefore the total region with inflation can even become larger and larger. At the same time, more and more causally separated sub-universes are created

Now, if a certain sub-volume of the universe has expanded inflationarily to such an extent that inflation would end due to the classical evolution of ϕ, then the quantum fluctuations in other sub-regions of the universe can lead to inflation continuing there. These regions become very large as a result of inflation, and the same argument can be repeated:

In some areas of these sub-regions, inflation continues, with the consequence that, once an inflationary process has begun, it will never stop. Ultimately, the region in which inflation takes place can even become larger and larger due to expansion. This phenomenon is referred to as *eternal inflation*. By this process new sub-volumes are constantly being created in which inflation has ended, and which are causally separated from the rest of spacetime. The terms "pocket universes" or "bubble universes" are used in this context. In this form, the theory of inflation therefore predicts the existence of infinitely many universes, the number of which is also constantly increasing. In Fig. 30.3 this process is demonstrated in a one-dimensional scheme. Partly because of this infinity of universes, there is also much criticism of inflation theory in the literature, see, e.g., Steinhardt [9, 10].

30.3 Final Remarks

In the previous chapters, we have obtained some insight into the theoretical methods and the important observations with which our present-day knowledge of the universe has been gained.

Apart from the question of the evolution of the universe to this day, the other question remains as to how the universe will evolve in the future. If we would like to make statements about the future of the universe, the most important question to be answered is certainly the nature of dark energy. For example, if dark energy changes over time, various types of expansion are conceivable. Ultimately, many statements about the future of the universe, like statements about the future in general, are therefore pure speculation.

However, what is certain is that the stars in the universe will use up their available nuclear fuel over time. The last stars should have burnt out in about 10^{14} years, and then the universe will become dark again, apart from the weak radiation from white dwarfs and neutron stars as well as brown dwarfs.

If, as some researchers assume, the protons are unstable, with a half-life of over 10^{33} years, after a period of this order of magnitude, the stellar remnants also all will have decayed, except for black holes.

Since black holes radiate energy through Hawking radiation, their lifetime is also finite, and lies in the range of 10^{67} years for stellar black holes, up to 10^{83} years for black holes with masses in the range of millions of solar masses, and 10^{98} years for masses in the range of a galaxy. After a period of time of this magnitude, only light leptons, photons and vacuum energy should still be present in the universe. A somewhat more detailed insight into possible future scenarios can be found in Hasinger [5].

30.4   Exercises

30.1. The meaning of the Planck charge Show that the gravitational and Coulomb forces between two particles both with a mass equal to the Planck mass and a charge equal to the Planck charge of the same sign exactly cancel.

30.2. Homogeneity of the CMB Estimate the maximum angular diameter of about $2°$, given in the text, of regions of the CMB that can be in thermal equilibrium without assuming an inflationary phase.

References

1. Guth, A.H.: Inflationary Universe: a possible solution to the horizon and flatness problem. Phys. Rev. D **23**(2), 347–356 (1981)
2. Guth, A.H.: Inflation and eternal inflation. Phys. Rep. **333**, 555–574 (2000)
3. Guth, A.H.: Eternal inflation and its implications. J. Phys. A: Math. Theor. **40**, 6811–6826 (2007)
4. Guth, A.H., Kaiser, D.I.: Inflationary cosmology: exploring the universe from the smallest to the largest scales. Science **307**, 884–890 (2005)
5. Hasinger, G.: Die Zukunft des Universums – eine Spekulation (The future of the universe – a speculation). In: Emmermann, R. (Hrsg.) An den Fronten der Forschung: Kosmos – Erde – Leben, S. 243–252. Deutscher Apotheker Verlag, Stuttgart (2003)
6. Mohr, P.J., Tiesinga, E., Newell, D.B., Taylor, B.N.: Codata Internationally Reconmmended [sic!] 2022 Values of the Fundamental Physical Constants. https://physics.nist.gov/constants
7. Ryden, B.: Introduction to Cosmology. Addison Wesley, Boston (2004)
8. Sprenger, M., Nicolini, P., Bleicher, M.: Physics on the smallest scales: an introduction to minimal length phenomenology. Eur. J. Phys. **33**, 853–862 (2012)
9. Steinhardt, P.J.: The inflation debate: is the theory at heart of modern cosmology deeply flawed? Sci. Am. **304**, 18–25 (2011)
10. Steinhardt, P.J.: The inflation debate. Sci. Am. Special Edition **23**(3s), 68 (2014)
11. Zwiebach, B.: A First Course in String Theory. Cambridge University Press, Cambridge (2009)

Numerical Values of Fundamental Physical Constants

A

See Table A.1.

Table A.1 Values of fundamental physical constants. All data taken from P. J. Mohr and E. Tiesinga and D. B. Newell and B. N. Taylor: Codata Internationally Reconmmended [sic!] 2022 Values of the Fundamental Physical Constants: https://physics.nist.gov/constants

Name	Symbol	Numerical value	Definition
Atomic mass constant	m_u	$1.660\,539\,068\,92(52) \cdot 10^{-27}$ kg	$\frac{1}{12}m\,(^{12}C)$
Bohr radius	a_B	$5.291\,772\,105\,44(82) \cdot 10^{-11}$ m	$\hbar/(\alpha m_e c)$
Boltzmann constant	k_B	$1.380\,649 \cdot 10^{-23}$ J K^{-1} (exact)	–
Compton wavelength of the electron	λ_e	$2.426\,310\,235\,38(76) \cdot 10^{-12}$ m	$h/(m_e c)$
Reduced	$\lambdabar_e$	$3.861\,592\,6744(12) \cdot 10^{-13}$ m	$\hbar/(m_e c)$
Vacuum electric permittivity (electric field constant)	ε_0	$8.854\,187\,8188(14) \cdot 10^{-12}$ F m^{-1}	–
Electron mass	m_e	$9.109\,383\,7139(28) \cdot 10^{-31}$ kg	–
Equivalent energy	$m_e c^2$	$510.998\,950\,69(16)$ keV	–
Electron volt	eV	$1.602\,176\,634 \cdot 10^{-19}$ J	–
Elementary charge	e	$1.602\,176\,634 \cdot 10^{-19}$ C (exact)	–
Fine-structure constant	α	$7.297\,352\,5643(11) \cdot 10^{-3} \approx 1/137$	$\frac{e^2}{4\pi\varepsilon_0\hbar c}$
Gravitational constant	G	$6.674\,30(15) \cdot 10^{-11}$ m^3 kg^{-1} s^{-2}	–
Speed of light	c	$299\,792\,458$ ms^{-1} (exact)	–
Vacuum magnetic permeability (magnetic field constant)	μ_0	$1.256\,637\,061\,27(20) \cdot 10^{-6}$ N A^{-2}	–
In units of $4\pi \cdot 10^{-7}$	μ_0	$0.999\,999\,999\,87(16)$	–

(continued)

S. Boblest et al., *Special and General Relativity*, https://doi.org/10.1007/978-3-662-71332-7

Table A.2 (continued)

Name	Symbol	Numerical value	Definition
Neutron mass	m_n	$1.674\,927\,500\,56(85) \cdot 10^{-27}$ kg	–
Equivalent energy	$m_n c^2$	$939.565\,421\,94(48)$ MeV	–
Neutron-proton mass ratio	-	$1.001\,378\,419\,46(40)$	m_n/m_p
Planck constant	h	$6.626\,070\,15 \cdot 10^{-34}$ J Hz^{-1} (exact)	–
Reduced	$\hbar$	$1.054\,571\,817\ldots \cdot 10^{-34}$ J s	$h/(2\pi)$
Proton mass	m_p	$1.672\,621\,925\,95(52) \cdot 10^{-27}$ kg	–
Equivalent energy	$m_p c^2$	$938.272\,089\,43(29)$ MeV	–
Proton-electron mass ratio	-	$1836.152\,673\,426(32)$	m_p/m_e
Rydberg energy	E_{Ry}	$13.605\,693\,122\,990(15)$ eV	$\alpha^2 m_e c^2/2$
Stefan-Boltzmann constant	σ	$5.670\,374\,419\ldots \cdot 10^{-8}$ W m^{-2} K^{-4}	$\frac{\pi^2 k_B^4}{60\hbar^3 c^2}$
Planck units			
Mass	m_{Pl}	$2.176\,434(24) \cdot 10^{-8}$ kg	$\sqrt{\hbar c/G}$
Length	l_{Pl}	$1.616\,255(18) \cdot 10^{-35}$ m	$\sqrt{\hbar G/c^3}$
Time	t_{Pl}	$5.391\,247(60) \cdot 10^{-44}$ s	$\sqrt{\hbar G/c^5}$
Temperature	T_{Pl}	$1.416\,784(16) \cdot 10^{32}$ K	$\sqrt{\hbar c^5/(k_B^2 G)}$
Derived Planck units			
Charge	q_{Pl}	$1.875\,546\,0378(15) \cdot 10^{-18}$ C	$\sqrt{4\pi\varepsilon_0 \hbar c}$
Energy	E_{Pl}	$1.9560(23) \cdot 10^9$ J	$\sqrt{\hbar c^5/G}$
Power	P_{Pl}	$3.628\,25(82) \cdot 10^{52}$ W	c^5/G
Momentum	p_{Pl}	$6.524\,78(74) \cdot 10^0$ kg m s^{-1}	$\sqrt{\hbar c^3/G}$
Density	σ_{Pl}	$5.1548(24) \cdot 10^{96}$ kg m^{-3}	$c^5/(\hbar G^2)$
Other derived values			
Coupling constant of GR	κ	$2.076\,66(47) \cdot 10^{-43}$ s^2 m^{-1} kg^{-1}	$8\pi G/c^4$
Compton wavelength of the neutron	λ_n	$1.319\,590\,9038(67) \cdot 10^{-15}$ m	$h/(m_n c)$
Reduced	$\lambdabar_n$	$2.100\,194\,1520(11) \cdot 10^{-16}$ m	$\hbar/(m_n c)$

Index

© The Editor(s) (if applicable) and The Author(s), under exclusive license to
Springer-Verlag GmbH, DE, part of Springer Nature 2026
S. Boblest et al., *Special and General Relativity*,
https://doi.org/10.1007/978-3-662-71332-7

The manufacturer's authorised representative in the EU is Springer
Nature Customer Service Centre GmbH, Europaplatz 3, 69115 Heidelberg,
Germany. If you have any concerns regarding our products, please
contact ProductSafety@springernature.com

Printed and bound by CPI Group (UK) Ltd, Croydon, CR0 4YY
01/06/2026
02124238-0012